Machine Shop Practice, Volume 2

K.H. MOLTRECHT
Associate Professor of Industrial
and Systems Engineering
The Ohio State University
Columbus, Ohio

INDUSTRIAL PRESS INC.
200 Madison Avenue
New York, NY 10016

Library of Congress Cataloging in Publication Data

Moltrecht, Karl Hans.
Machine shop practice.

Includes index.
1. Machine-shop practice. I. Title.
TJ1160.M66 1979 670.42'3 79-91236
ISBN 0-8311-1132-1 (v. 2)

SECOND EDITION– Second Printing

MACHINE SHOP PRACTICE—Volume 2

Preface

Many completely new and important developments in technology and machining methods have made it necessary to update *Machine Shop Practice,* resulting in this second edition of the book. The revisions in this new edition are very thorough, incorporating the latest developments in numerically controlled machine tools; the introduction of SI metric units of measurement; and many other advances in shop practices. However, great emphasis is still placed on the operation of basic machine tools and the fundamentals related to their technology. Designed for use in the classroom or for home study, this book, with its many tables and formulas, is also useful in the workshop as a reference book for the more-advanced craftsman, technician, and manufacturing engineer.

Contents

List of Tables

CHAPTER 1

Shaper Construction and Shaper Tools

The shaper is used principally to machine flat or plane surfaces with a single-point cutting tool. The cutting tool is mounted on the shaper head that is attached to the ram. The ram imparts a reciprocating motion to the tool which operates over the shaper table. The work is usually held in a shaper vise, although sometimes it is more convenient to bolt it directly onto the table. The table can be adjusted vertically. It is provided with an automatic power or hand feeding motion that is parallel to the top of the table and perpendicular to the stroke of the ram. The tool cuts on the forward stroke while the table feeds the workpiece for the next cut on the return stroke. Vertical cuts can be taken by feeding the tool with the shaper head slide. The shaper head can be set at an angle in order to take

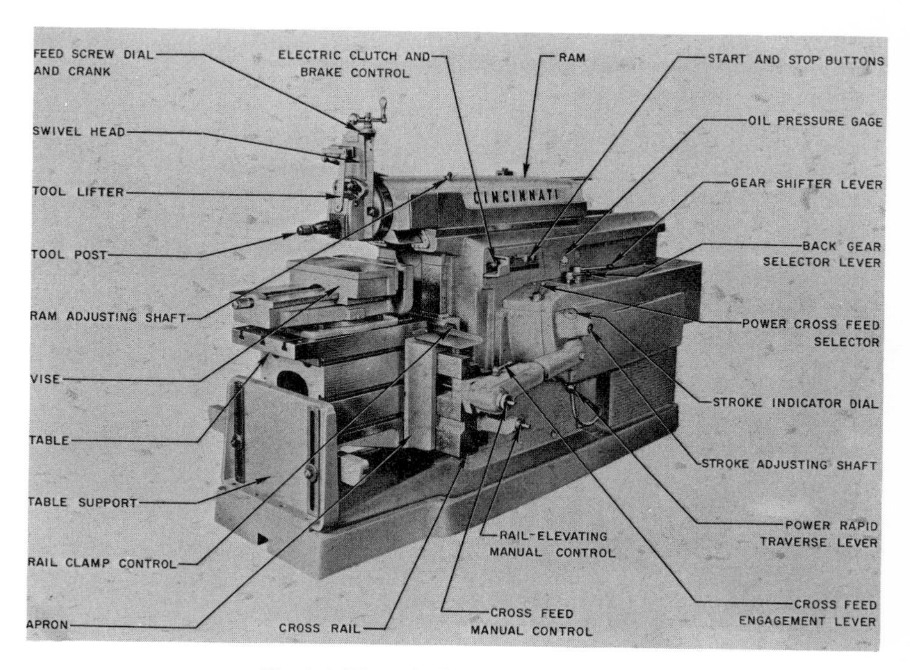

Fig. 1-1. The principal parts of a shaper.

angular cuts. Profiles can be cut by combining the feed of the shaper head tool slide and the table feed. The shaper is capable of working to close tolerances on a wide variety of work.

The Construction of the Shaper

Figure 1-1 illustrates a modern shaper and labels its principal parts. The drive gear train of a shaper is shown in Fig. 1-2, and a ram-actuating mechanism is shown in Fig. 1-3. The principal parts of the shaper and their functions will be described by referring to these illustrations. The parts of the shaper are attached to or supported by the base and a rigid housing that are made as a single casting. The power is transmitted to the shaper from an electric motor through a V-belt drive. A gear train, Fig. 1-2, is used to provide the different speeds, or strokes per minute, of the ram. These gears are shifted by the gear-shifter lever and the back gear selector lever (Fig. 1-1). The last gear in the train of gears is a large gear called the *bull gear* or *crank gear,* and some shapers (Fig. 1-2) are provided with two bull gears. Attached to the face of the crank gear in Fig. 1-3 is a crank gear screw which is rotated through bevel gears by the stroke-adjusting shaft shown in Fig. 1-1. This screw causes the crank

Fig. 1-2. The drive gear train of a modern shaper.

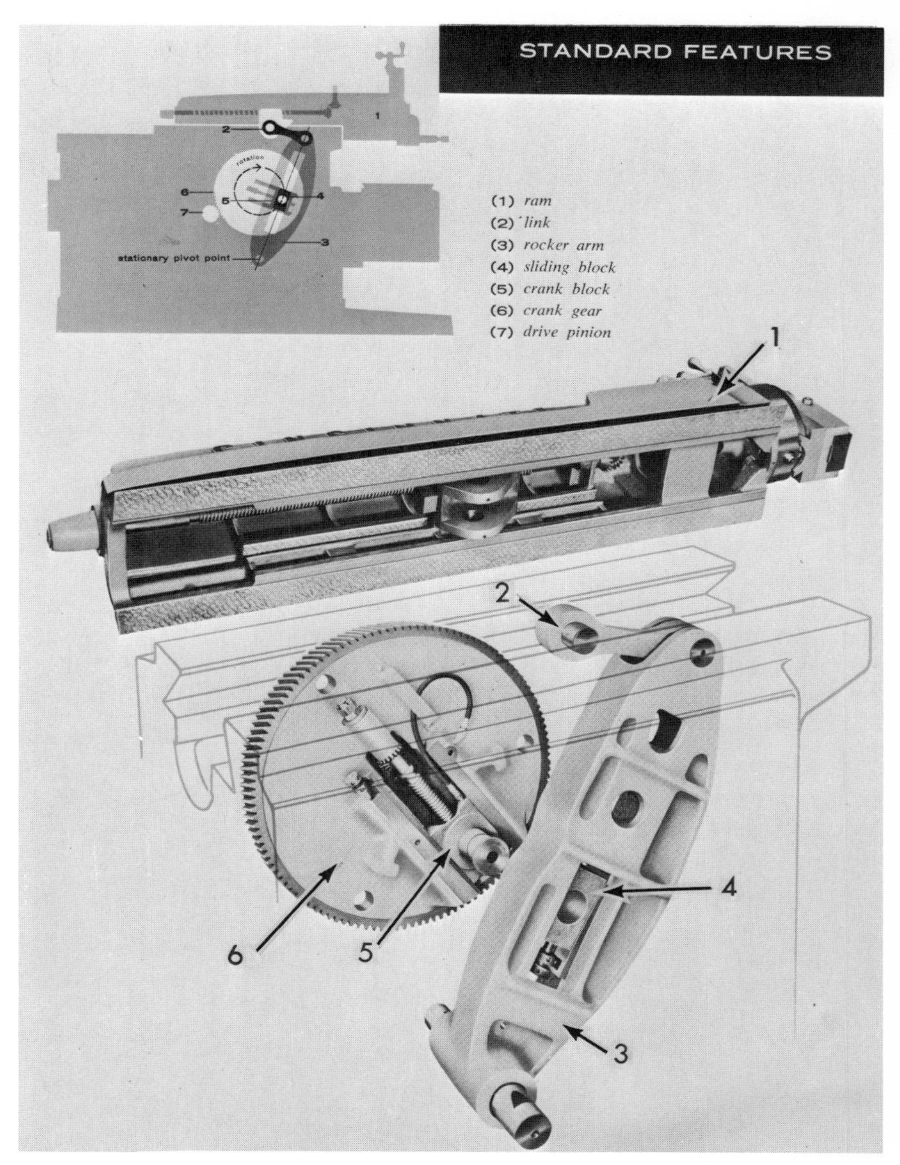

Fig. 1-3. The ram drive mechanism of a shaper.

block to move radially along the face of the crank gear. This movement of the crank block is guided by dovetail slides which are machined on the face of the crank gear. A crank pin, machined on the end of the crank

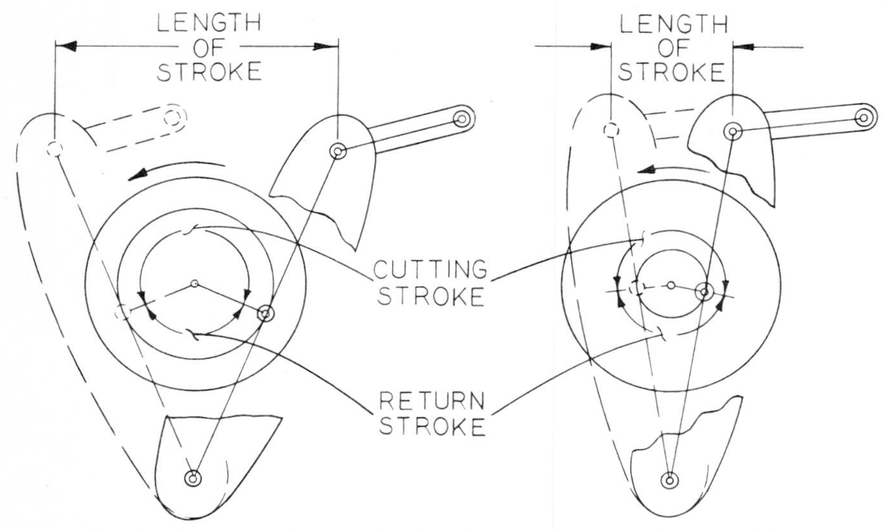

Fig. 1-4. Schematic drawing showing the effect of the crank pin position on the length of stroke.

block, fits in a sliding block with a good running fit. By sliding inside the rocker arm while the crank gear is rotating, the sliding block causes the rocker arm to rock back and forth about the pin located in the lower

Fig. 1-5. Mechanism for adjusting the position of the ram stroke.

part of the arm. This pin is attached to the shaper column. The rocking motion of the rocker arm is transmitted to the shaper ram through the link, thereby causing the ram to reciprocate back and forth.

The length of the stroke of the ram is dependent upon the radial position of the crank block as shown schematically in Fig. 1-4. An inherent feature of this mechanism is the quick return of the ram on the return or non-cutting stroke. The crank gear rotates at a constant speed. However, as is evident from Fig. 1-4, the distance that must be traveled by the crank pin on the end of the crank block is greater for the cutting stroke than for the return stroke. Thus, it will take longer for the ram to move forward when cutting than to return.

In Fig. 1-5 the link connects the rocker arm to the link block. The position of the link block inside the ram can be changed by turning the screw which is threaded through the upper portion of this block. Since the link block is held in a stationary position by the link and the rocker arm when the bull gear is not turning, it is the ram which moves forward

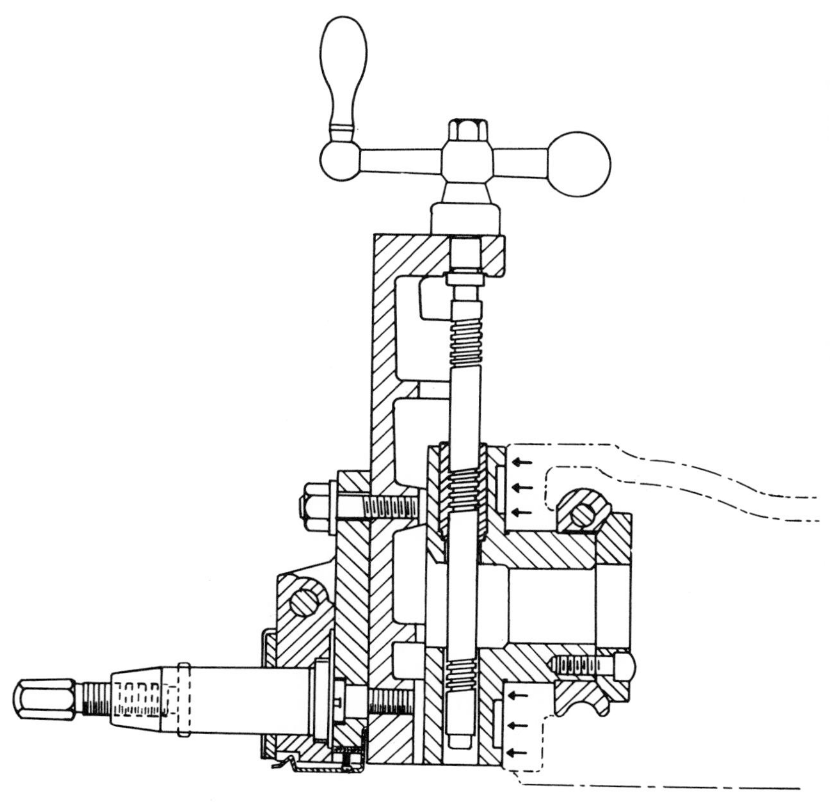

Fig. 1-6. Sectional view of a shaper head.

and back when the screw inside of the ram is turned. In this manner the position of the ram can be adjusted so that the cutting tool strokes back and forth in the right location on the workpiece. When the shaper is cutting, the link block must be clamped firmly to the ram. The ram-adjusting shaft rotates the screw through bevel gears; and the clamp shaft, located behind the ram-adjusting shaft, clamps the link block in place.

The shaper head, illustrated in Fig. 1-6, is clamped firmly to the end of the ram. Although usually clamped in a vertical position, it can be clamped at an angle to take an angular cut. The shaper head has a slide that is actuated by a feed screw. A micrometer dial located below the handle of the feed screw can be used to determine the exact amount of the movement of the shaper head slide. The clamp, provided to fix the slide to the head, should always be tightened when taking a cut using the table feed. It should be tightened very lightly, or to a snug fit, when the tool is fed by the shaper head slide. A clapper box, which can be seen in Fig. 1-7, is fastened to the front of the shaper head slide and has a channel machined on its face into which a clapper block is fitted. The fit between the clapper box and the clapper block is very close, although the clapper block is free to pivot forward and up on a taper pin which holds it in place. The purpose of this arrangement is to allow the tool to pivot away from the surface of the workpiece on its return stroke so that it will not drag. The clapper box can be tilted slightly to the right or left of the vertical position by loosening the bolts that clamp it to the face of the shaper head slide. This allows the tool to move slightly away from the shoulder formed by the cut. For example, in Fig. 1-7, the clapper box should be tilted so that its top slants away from the shoulder formed by the cut. The tool post which holds the toolholder is held in the clapper block and is clamped to it when the toolholder is clamped in place. The shaper shown in Fig. 1-1 has an automatic tool lifter that will automatically pivot the clapper block on the return stroke.

The front of the shaper column has a machined and scraped surface on which the crossrail is mounted. The crossrail should be firmly clamped to this surface when the shaper is cutting. An elevating screw located below the crossrail can move it up or down. This is done only to raise or lower the table to adjust it to hold different sizes of workpieces. It is operated by the rail-elevating manual control. The crossrail has an accurately machined and scraped surface on its forward face on which the apron (Fig. 1-8) is mounted. The apron slides horizontally along the crossrail, and this movement is perpendicular to the stroke of the ram. The shaper table, which is firmly attached to the apron, slides along with it. A table support, shown in Fig. 1-8, provides additional support to the table.

The feed of the shaper table (and the apron) is actuated through a feed screw mounted in the crossrail. The feed screw can be operated manually by turning the cross-feed manual control. All shapers have an automatic power cross-feed that moves the table a given distance during the return stroke or at the end of the stroke. Because the construction of the automatic power feed-actuating mechanism varies with different makes of

Courtesy of the Rockford Machine Tool Company

Fig. 1-7. A shaper taking a cut showing the shaper head and the clapper box (B) tilted. Clapper block is shown at A.

shapers, it will not be described. The rate of cross-feed on all shapers is given in terms of thousandths of an inch per stroke. The rate of cross-feed on the shaper shown in Fig. 1-1 can be obtained by simply turning the power cross-feed selector handle. The power cross-feed, which can be engaged by turning the cross-feed engagement lever, can move the table in either direction. Many shapers have a rapid traverse that moves the table rapidly from one position to another along the crossrail. This feature saves much labor and time which would be required to move the table manually. The rapid traverse can be started by engaging the cross-feed engagement lever and the power rapid-traverse lever simultaneously.

The shaper shown in Fig. 1-9 is called a *hydraulic shaper,* since the ram is actuated by a hydraulic piston and cylinder arrangement. The advantage of this design is that the speed of the ram can be varied infinitely

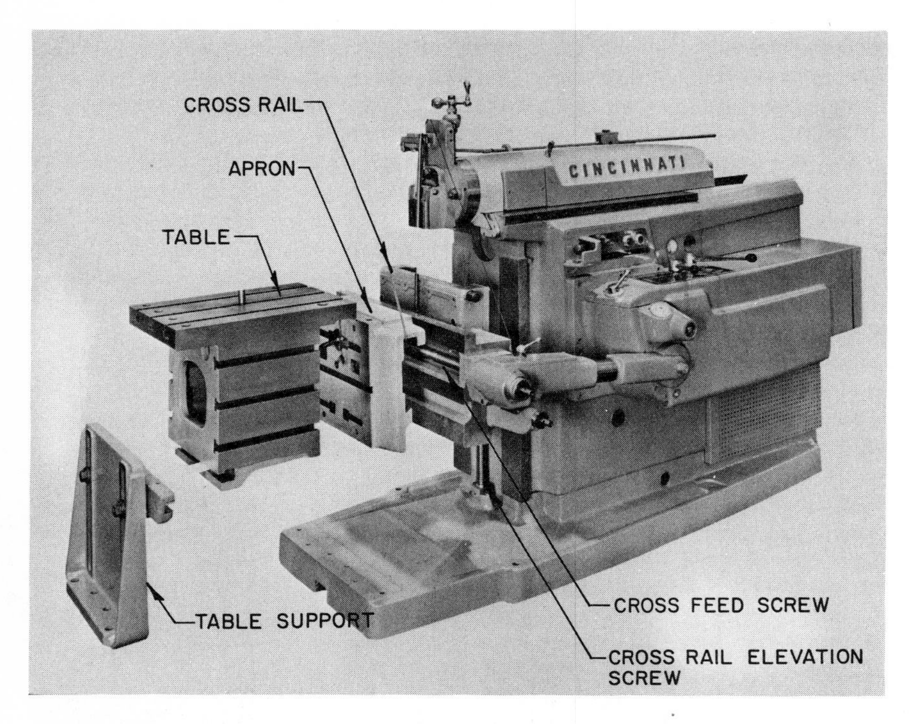

Fig. 1-8. The construction of the shaper table, apron, and cross rail.

between its fastest and slowest speed. The speed of the ram is also constant over a large portion of the stroke. The length of the stroke and the position of the stroke can be varied by simply clamping two dogs that are attached to the ram in different positions. The power cross-feed of the table is actuated by a combination of hydraulic and mechanical devices. Although the table is moved by a screw, the power for moving the screw as well as the amount that the screw turns is actuated hydraulically. The table cross-feed can also be actuated manually or by rapid traverse.

Shaper Cutting Speeds

The cutting speed is the velocity of the cutting tool as it travels through the workpiece in taking a cut. For all metal-cutting operations the cutting speed is given in terms of feet per minute. The cutting speed that should be used is dependent upon the material from which the cutting tool is made, the depth of cut, the feed rate used, the material from which the workpiece is made, and its hardness. Table 1-1 lists the recommended cutting speeds for a variety of materials that are cut on a shaper. These cutting speeds are for a depth of cut of .125 inch and a feed rate of .012 inch per stroke. A somewhat slower cutting speed should be used if deeper cuts and larger feed rates are to be used. Likewise, the cutting speed can be increased

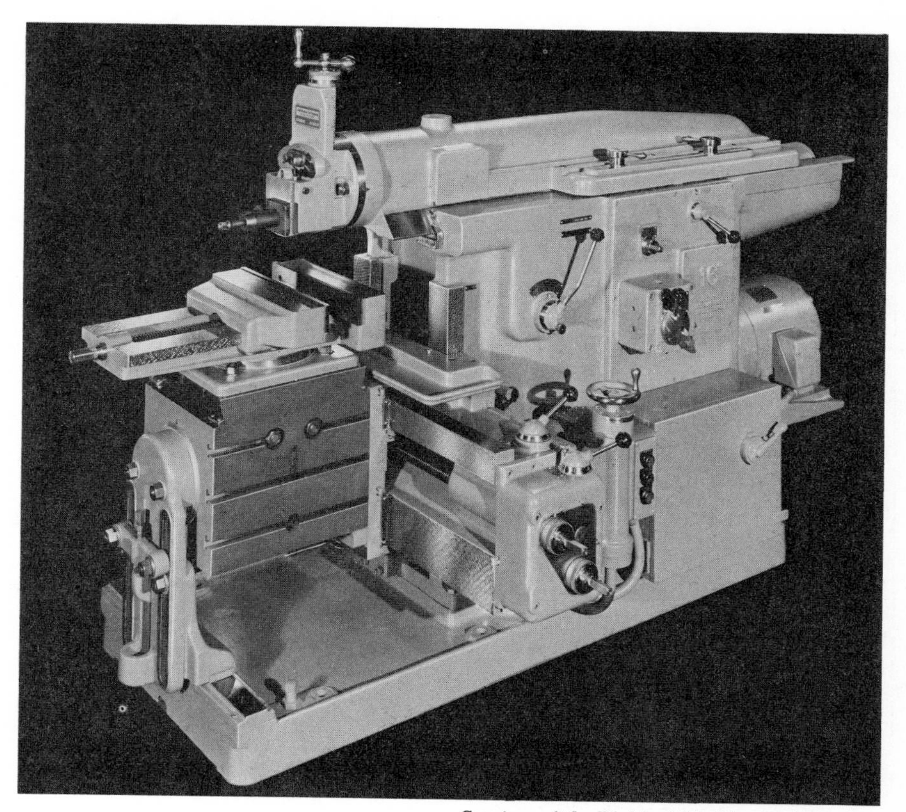

Courtesy of the Rockford Machine Tool Company

Fig. 1-9. Hydraulic shaper.

somewhat from that recommended in Table 1-1 if shallower cuts and lighter feeds are used. The recommendations are for high-speed steel cutting tools since most shaper tools are made from this material.

For shaper work the recommended cutting speed must be converted into strokes per minute of the shaper ram. Furthermore, since part of the stroke is used to return the shaper ram, the ratio of the cutting time to the return time per stroke of the ram must be known. A reasonable estimate of this ratio is 3 to 2. In effect, this means that the shaper is cutting during $3/5$ of the time per stroke and the ram is returning during $2/5$ of the time per stroke. Thus, if the shaper is operating for 1 minute, the time of the cutting stroke will be $3/5$ minute. The distance that the shaper ram travels in 1 minute during the cutting portion of the stroke is the length of the stroke multiplied by the number of strokes per minute. This can be expressed mathematically by the following equation:

$$s = L\,S_m$$

Table 1-1. Recommended Cutting Speeds for Shaping with High-Speed Steel Tools*

Material	Hardness, HB	Cutting Speed, fpm
AISI 1012, AISI 1019, and AISI 1020 Steel	100 to 125 125 to 175 175 to 220	110 110 90
AISI 1030, AISI 1040, and AISI 1050 Steel	120 to 170 170 to 200 200 to 240 240 to 300	100 85 75 65
AISI 1060, AISI 1080, AISI 1090, and AISI 1095 Steel	160 to 200 200 to 240 240 to 300 300 to 375	90 75 60 30
Alloy Steel	150 to 175 175 to 225 225 to 325 325 to 375 375 to 425	100 90 55 40 25
Maraging Steel	175 to 225 225 to 325	60 50
Gray Cast Iron ASTM Class 20, ASTM Class 30, ASTM Class 40, ASTM Class 50, and ASTM Class 60	110 to 140 140 to 190 190 to 220 220 to 260 260 to 320	120 100 75 50 30
Naval Brass, Red Brass, Yellow Brass, Nickel Silver, Manganese Bronze, Muntz Metal		140
Commercial Bronze, Phosphor Bronze		100
Water-Hardening Tool Steel W1, W2, W4, W5	150 to 200	110
Cold-Work Tool Steel A7, D1, D2, D3, D4, D5, D7	200 to 250	40
Cold-Work Tool Steel A2, A3, A4, A5, A6, A8, A10, O1, O2, O6, O7	200 to 250	60
Hot-Work Tool Steel H10, H11, H12, H13, H14, H16, H19	150 to 200 200 to 250 250 to 350	85 75 30
Hot-Work Tool Steel H20, H21, H22, H23, H24, H25, H41, H42, H43	150 to 200 200 to 250	60 50
Shock-Resisting Tool Steel S1, S2, S4, S5, S6, S7	160 to 220	60
Mold Steels P1, P2, P3, P4, P5, P6	120 to 180	75
Mold Steels P20, P21	150 to 200	60
High-Speed Steel M1, M2, M6, M10, T1, T2, T6, T7	180 to 250	60
High-Speed Steel M3 Type 1, M4, M7, M30, M33, M34, M35, M36, M41, M42, M43, M44, T4, T5, T8	220 to 280	50
High-Speed Steel M15, T9, T15, M3 Type 2	220 to 280	30

* The cutting speeds given are based on a feed of .012 inch per stroke and a depth of cut of .125 inch. HB designates Brinell hardness number.

where: s = Distance that the shaper ram travels during the cutting stroke in one minute, inches
L = Length of the shaper stroke, inches
S_m = Strokes per minute of the shaper ram

The speed of the shaper ram is the cutting speed. In order to change the speed from feet per minute, as given in Table 1-1, to inches per minute, it is multiplied by 12 or:

$$\text{Speed} = 12\ V$$

where: V = Cutting speed in feet per minute

A well-known relationship from physics is:

$$\text{Speed} = \frac{\text{Distance}}{\text{Time}}$$

Substituting the terms applicable to the shaper for speed, distance, and time results in a formula that can be used to calculate the strokes per minute of the shaper;

$$12\ V = \frac{LS_m}{3/5}$$

From which:

$$S_m = \frac{7.2\ V}{L} \qquad (1\text{-}1)$$

Example 1-1:

A piece of H-10 hot-work die steel is to be machined in a shaper. The hardness of this steel is 210 Bhn and the length of the part is 8 inches. The ram is set to overtravel the work ¾ inch at the start of the cut and ¼ inch at the end of the cut, making the length of the stroke equal to 9 inches. Calculate the strokes per minute at which the shaper should cut.

$$V = 75 \text{ (From Table 1-1)}$$

$$S_m = \frac{7.2\ V}{L} = \frac{7.2 \times 75}{9}$$

$$S_m = 60$$

Whenever possible the strokes per minute as calculated by Formula 1-1 should be used; however, it is sometimes necessary to reduce this speed because of the nature of the operation being performed or the rigidity of the setup of the workpiece. Judgment must be applied in making the final selection of the speed of the shaper ram.

Since the selection of the feed rate and the depth of cut to be used is subject to many variables, it is not practicable to give recommendations. Among the factors that must be considered are:

1. The horsepower available on the machine
2. The length of the stroke

3. The setup of the workpiece, or how rigidly it is held
4. The amount of stock to be removed from the workpiece
5. The finish required on the surface of the work.

As a general rule, rough cuts should be taken using as much feed and as deep a cut as possible so that approximately .005 to .031 inch is left on the workpiece for the finishing cut. The feed to be used for finish cutting depends upon the type of cutting tool selected for this operation. Sometimes the finish cut is taken with the same tool that was used for rough cutting; in this case a fine feed (.010 inch per stroke or less) should be used. If a broad nose tool (C in Fig. 1-14) is employed, a very coarse feed should be used.

In the metric system the cutting speed is given in terms of meters per minute and the length of stroke is in millimeters. Metric cutting speed values in terms of meters per minute can be obtained by multiplying the feet per minute values given in Table 1-1, by 0.3048. Using the same procedure as before, it will be found that the metric formula for calculating the strokes per minute of the shaper is:

$$S_m = \frac{600\,V}{L} \tag{1-2}$$

Where: S_m = Strokes per minute of the shaper ram
V = Cutting speed, m/min
L = Length of stroke, mm

Example 1-2

Calculate the strokes per minute at which the shaper ram should operate in order to cut a 25-mm-long piece of A2 cold work tool steel.

Allow an overtravel of 12 mm at the front end of the stroke and 20 mm at the back end; thus, the length of stroke is (25+12+20) 57 mm. From Table 1-1, the cutting speed for A2 cold work steel is 60 fpm or 60 x .3048= 18 m/min.

$$S_m = \frac{600\,V}{L} = \frac{600 \times 18}{57}$$
$$= 190 \text{ strokes/min}$$

Shaper Cutting Tools

Shapers utilize single-point cutting tools which are similar to those used on lathes. The basic difference in their operation is slight. The most important difference is that the lathe tools feed into the work continuously when cutting while shaper tools do not feed into the work during the cut. The shaper feed occurs during the return stroke of the ram. For this reason it is claimed that shaper tools should be made to have smaller relief angles than lathe tools; however, an increase in the rate of wear will occur if the relief angle is made too small, regardless of the fact that the cutting tool does not feed during the cut.

A shaper cutting toolholder is shown in Fig.1-10. This toolholder is used to hold high-speed steel tool bits. Fig. 1-11 shows how the tool bit can be held in this toolholder for cutting different surfaces. The long shank allows the toolholder to position the cutting tool to cut difficult-to-reach surfaces.

Some typical shaper cutting tools for cutting steel are illustrated in Fig. 1-12. The roughing tool shown at A can also be used to take finishing cuts on certain classes of work. The tool at B is a general-purpose round-nose roughing and finishing tool which is also employed to take roughing and finishing cuts on cast iron. The large-nose radius of this tool tends to eliminate feed marks, thereby producing a good surface finish on the work. Exceptionally good surface finishes can be obtained on steel surfaces with the tool shown at C. When the cutting edge is oriented 25 degrees with respect to the direction of the stroke of the ram, this tool provides a shear-like cutting action which leaves a small, rather tightly curled chip. Because of the hook ground on the face of the tool, there is a large rake angle that contributes to its ability to produce a very smooth surface finish. The cutting edge, which should be honed until it is keen, should not be blunted or rounded by the honing action. For best results the depth of cut should be approximately .001 to .003 inch per stroke, and the feed rate should be .003 to .005 inch per stroke. In some cases a heavier feed, up to .025

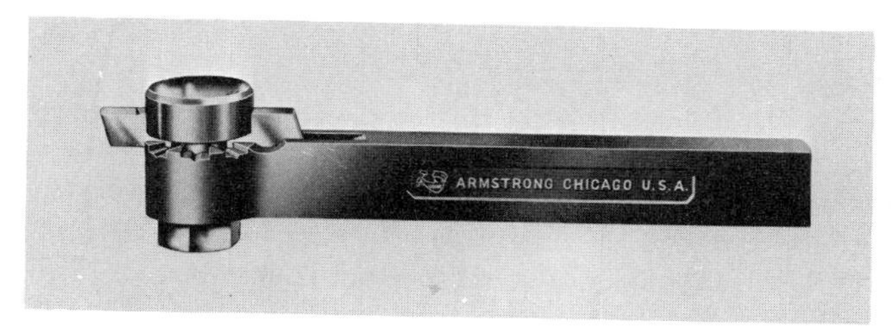

Courtesy of Armstrong Brothers Inc.

Fig. 1-10. Shaper toolholder.

Courtesy of Armstrong Brothers Inc.

Fig. 1-11. Methods of application of shaper toolholder.

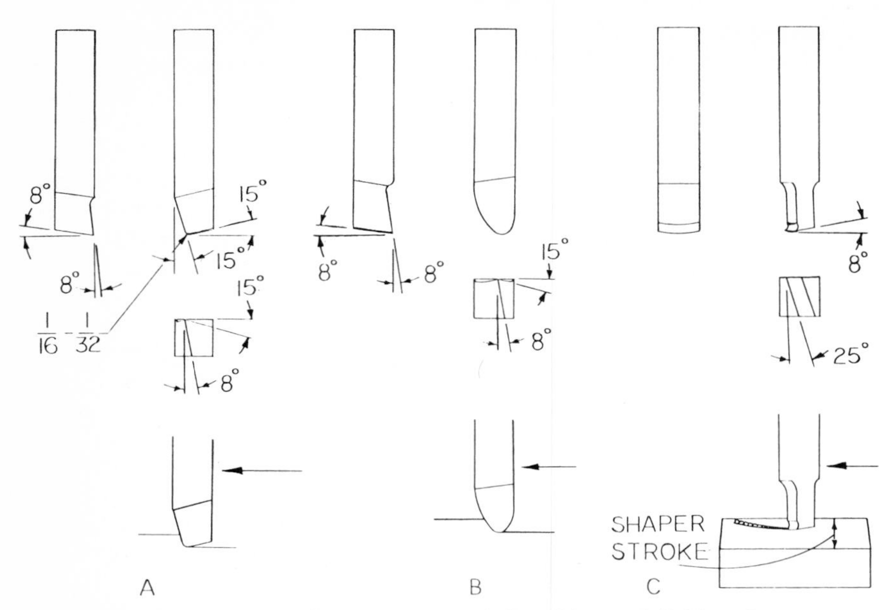

Fig. 1-12. Shaper tools for cutting steel. A. Roughing tool. B. Round-nose general-purpose tool. C. Steel finishing tool.

inch per stroke, can be used. A further improvement in the performance of this tool can be obtained by covering the surface of the work with a good grade of soluble cutting oil. Figure 1-13 shows this tool in operation.

Cutting tools intended primarily for shaping gray cast iron are shown in Fig. 1-14. A general-purpose roughing tool is shown at A. It is also used to take finishing cuts on some classes of work. The tool at B has a large side-cutting edge angle so that the tendency of the cast iron to chip at the end of the cut will be reduced. The large side-cutting edge angle allows the tool to cut using a heavier feed rate. This tool is also effective in cutting steel using a heavy feed rate. The broad-nose finishing tool at C is intended to cut cast iron only. Depending on the width of the tool's cutting edge, the feed rate should be from .250 to .750 inch per stroke. The depth of cut should be .001 to .003 inch. Since a heavy feed is used, the broad-nose tool will finish the workpiece with relatively few strokes and in a relatively short time. Thus, the tool wear is reduced, particularly when large surfaces are being shaped, and the resulting surface will be a true plane. Another advantage of this tool is that the surface it produces will be relatively free from the hardening effects of cold working, making it an easy surface to hand scrape when this is required. Hand-scraping can produce flatter surfaces than those obtained by machining. Such surfaces may be required on precision surface plates, machine tool ways, and other flat, bearing areas. Many surfaces, of course, are not hand-scraped, and the broad-nose tool will produce a very acceptable surface with characteristic feed marks parallel to each other. The two corners of the tool should

Fig. 1-13. Steel finishing tool in operation.

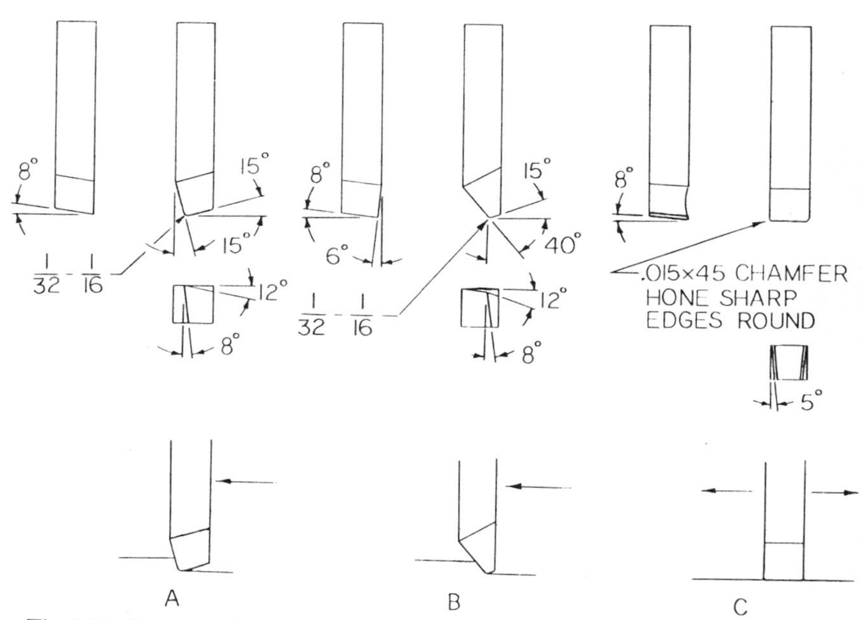

Fig. 1-14. Shaper tools for cutting cast iron. A. Roughing tool. B. High lead angle roughing tool. C. Cast-iron finishing tool.

have a .015-inch by 45-degree chamfer, and the edges of the chamfers should be lightly hand-honed to make them round. A 5-degree relief angle should be ground on each side and the end relief angle should be 8 degrees. Although a slight back rake angle is desirable, grinding a radius equal to the radius of the grinding wheel is usually sufficient. When the cutting edge is ground, the grinding wheel should be moving perpendicular and toward the cutting edge.

The tools shown in Fig. 1-15 are frequently used in shaper work. It is sometimes necessary to take a vertical cut while feeding the tool with the shaper head. This can be done with the tools shown at A. The steel tool, shown at the left, is ground with a very slight hook which provides a positive rake angle and curls the chip. The cast-iron tool, shown at the right, is ground with a 12-degree rake angle. No hook is required to curl the chip since cast iron produces a discontinuous type chip. The cast-iron side-cutting tool can be used to cut steel; however, it may be somewhat more difficult to control the chip on long cuts. Both tools should have a 10-degree end-cutting edge angle as shown.

A shoulder-finishing tool is shown at B in Fig. 1-15. If a sharp corner is required, a flat is ground on the end and on the side-cutting edge as shown. The width of this flat should be from .030 to .060 inch. If a radius is required in the corner, the flat on the side-cutting edge can be eliminated and the required radius ground on the nose of the tool. It can be used as shown

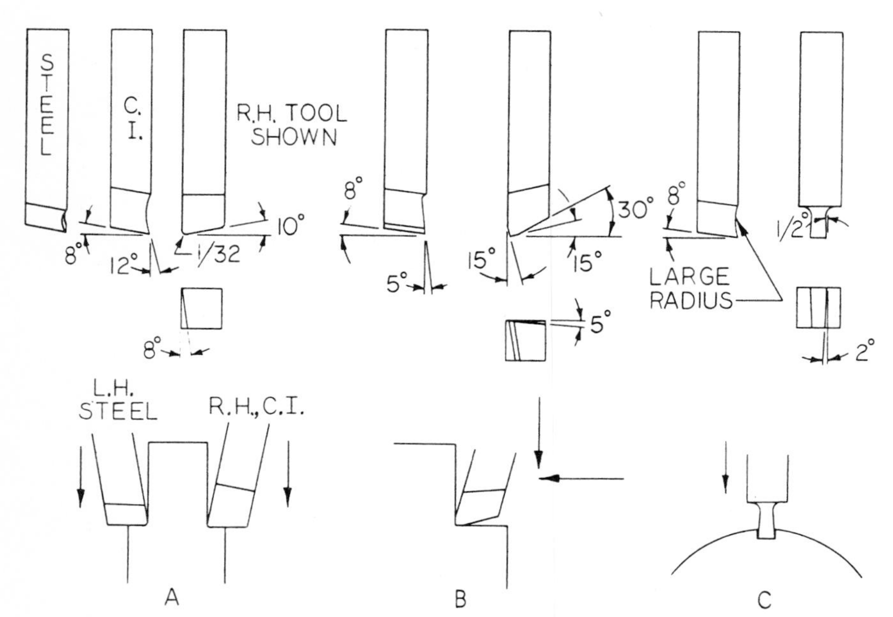

Fig. 1-15. Shaper cutting tools. A. Side cutting tool. B. Shoulder-finishing tool. C. Keyseating tool.

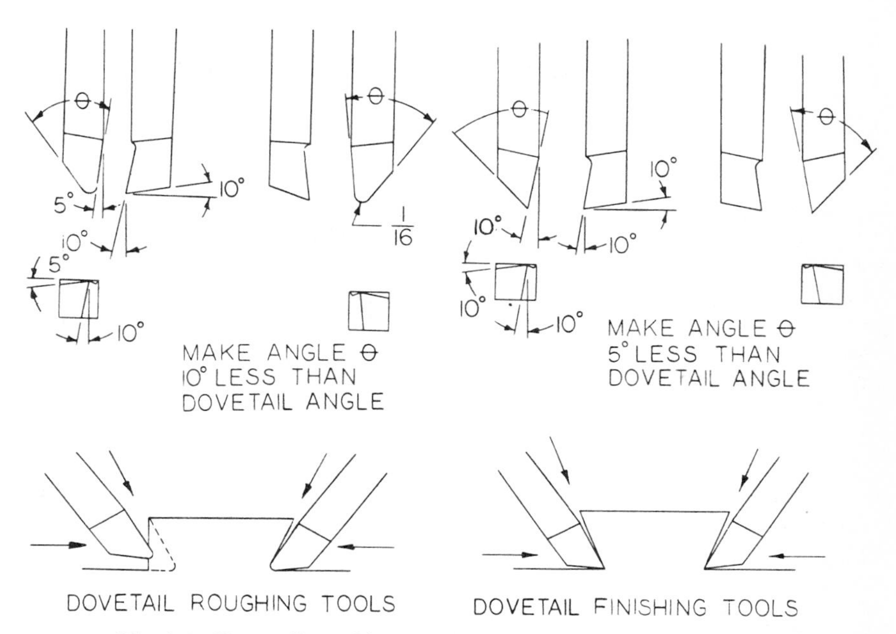

Fig. 1-16. Dovetail roughing tools (left) and finishing tools (right).

for cutting either cast iron or steel. At C the tool illustrated is called a keyseating tool, which is used to machine slots and keyseats in either cast iron or steel parts. Cutting tools for rough- and finish-shaping dovetails are shown in Fig. 1-16. These tools sometimes have a small radius at the nose instead of a sharp nose as shown in this illustration. The 5-degree angles on the sides of the roughing tools and the 10-degree angles on the sides of the finishing tools must sometimes be reduced if the angle of the dovetail being cut becomes small. These angles are very satisfactory for 60-degree dovetail angles.

The Vertical Shaper

The vertical shaper, Fig. 1-17, is also called a slotter. It is similar to a shaper except that the ram reciprocates vertically rather than horizontally. A slide, into which the ram is mounted, can be adjusted to an angle with respect to the vertical position so that the ram can reciprocate at this angle. This feature is often used when cutting clearances in metal stamping dies. The table can be moved in two perpendicular directions as well as rotated. Such a combination of table movements enables the vertical shaper to cut vertical plane surfaces, round or partially round surfaces, and surfaces with an irregular profile. Internal keyseats and slots, which are frequently necessary, can be cut, as can external surfaces.

The vertical shaper in Fig. 1-17 is a hydraulic vertical shaper. Here a hydraulic system reciprocates the ram and provides the power for actuating the power table feeding movements. There is also an indexing arrangement for obtaining angular movements of the table.

Courtesy of the Rockford Machine Tool Company

Fig. 1-17. Vertical shaper.

CHAPTER 2

Shaper Work

Shaper work embodies many of the fundamentals of machining plane surfaces that are applicable not only on the shaper but on other machine tools as well. The unique feature of machining a part that is held in a precision machine tool vise on a shaper instead of on a milling machine is that the shaper employs a relatively inexpensive single-point cutting tool instead of a multiple-point rotating cutter. Furthermore the shaper tool can easily be resharpened when required. The method of holding and aligning the part in the vise is the same in either case. Likewise, parts are sometimes clamped directly to the shaper table in the same manner as they would be if they were to be machined on a milling machine, planer, or horizontal boring mill. Shaper work is generally limited in size by the maximum length of stroke of the shaper ram. The cutting tools used on the shaper and their manner of application are similar to planer cutting tools. The shaper can machine a variety of plane surfaces, profiles, and contoured surfaces when equipped with a tracing attachment.

The Shaper Vise

Much of the work performed on shapers is done with the workpiece held in a shaper vise. It is a precision tool that must be kept in accurate adjustment. Figure 2-1 shows a part that is being machined in a shaper vise. The bottom of the shaper vise is an accurate locating surface that is parallel to the base. One jaw of the vise is a stationary jaw while the other jaw is movable. The stationary jaw forms a second locating surface. It must be kept perpendicular to the bottom of the vise. The movable jaw, on the other hand, should not be used as a locating surface at any time because the clearances that are necessary for it to slide without binding allow it to tip slightly. Both jaws have heat-treated steel inserts. The top surfaces of the jaws are also accurately machined and can be used as reference surfaces for tools, such as squares and surface gages that are used to position workpieces in the vise. The shaper vise has a swivel base which is graduated in degrees. This permits the vise jaws to be positioned parallel, perpendicular, or at an angle with respect to the stroke of the ram.

The first step in doing accurate work in a machine tool vise such as a shaper vise is to ascertain its accuracy, and before checking the accuracy of a shaper vise, the worker must be sure of the accuracy of the shaper table. To do this the table should be clamped firmly against the shaper housing

Courtesy of the Rockford Machine Tool Company

Fig. 2-1. A workpiece held in a shaper vise.

and a dial test indicator clamped to the toolholder on the shaper ram. The shaper ram is then moved slowly while the contact point of the indicator is in contact with the table. A second check should be made by placing the contact point of the indicator against a long precision parallel bar that bridges the T-slots on the shaper table. With the ram held stationary the table cross-feed is moved manually, or by rapid traverse if available.

Figure 2-2 outlines the procedure for checking the shaper vise. At A and B the procedure for checking the bottom of the vise is shown. After the dial test indicator is clamped to a toolholder on the shaper ram, the stroke of the ram is adjusted to cause the indicator to move along the bottom of the vise without touching either jaw and the ram is moved slowly back and forth as shown at A. If the bottom of the vise is parallel to the stroke of the ram, the hand of the indicator will read the same throughout the length of the stroke. The bottom of the vise should also be checked for parallelism perpendicular to the stroke of the ram as shown at B. The

precision parallel is used as shown in order to bridge the opening in the vise, and the cross-feed movement to the table is employed to indicate the vise along the parallel. If the base of the vise is found not parallel, it should be checked for dirt or nicks on its seating surfaces. The bottoms of precision machine tool vises are made parallel to their seats; however, they will occasionally not be true if they are improperly used. Similarly, the perpendicularity of the solid jaw will be affected by improper use of the vise. This can be checked with a precision machinist's square, as illustrated at C in Fig. 2-2. The beam of the square is placed on the bottom surface of the vise, and two paper feelers are used between the blade of the square and the solid jaw. When the square is held firmly against the solid jaw, either of the paper feelers should be impossible to pull out. The solid jaw can also be checked as shown at D. The precision machinist's square is clamped very lightly in the vise by placing a piece of soft wood between the beam of the square and the movable jaw. When the vise jaws are positioned perpendicular to the stroke of the ram, the indicator, which is attached to the ram, is moved along the blade of the square by slowly moving the ram. When the vise jaws are positioned parallel to the stroke of the ram, the indicator is moved from one end of the blade of the square to the other by the table cross-feed. In either case, the reading of the indicator should be the same at either end of the blade of the square.

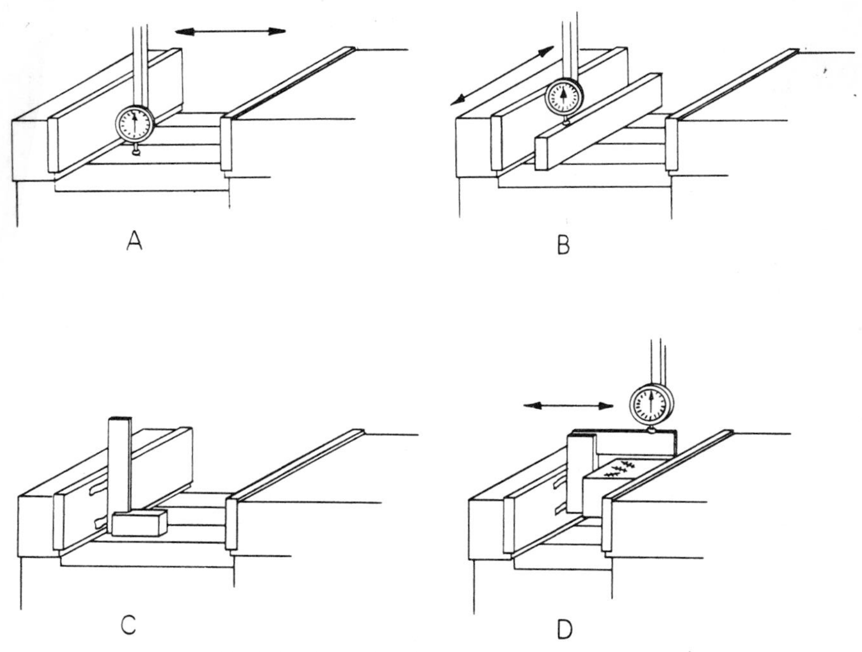

Fig. 2-2. Methods of testing the shaper vise.

Vise Work—Shaping a Plane Surface

The finished surfaces of the shaper vise and all other precision machinist's vises should always be protected from being marred by the rough unfinished surfaces of castings, forgings, and rolled parts having a rough mill scale. This is done by placing shims made of soft aluminum, copper, or brass sheet between the workpiece and the finished surfaces of the vise as shown in Fig. 2-3. The bottom of the vise should also be protected if the workpiece is large enough to seat on the bottom of the vise. If the workpiece is too small to seat on the bottom of the vise, parallels should be used between the bottom of the work and the bottom of the vise. Rough parallels made from soft metal can be employed to seat unfinished workpieces as shown in Fig. 2-3. If it is necessary to use precision parallels, a

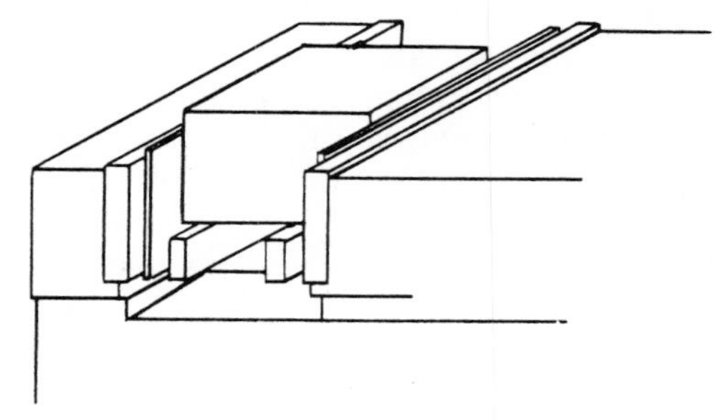

Fig. 2-3. Protecting vise jaws with soft metal shims when holding rough workpieces.

soft metal shim should be placed between the parallels and the workpiece. After the workpiece has been positioned in the approximate center of the vise, the vise is clamped tight. To seat the workpiece on the parallels give it a "dead" blow with a lead hammer. A "dead" blow is one in which the hammer is not allowed to bounce away from the workpiece. If a lead hammer is not available, a regular ball pean hammer can be used by striking a "dead" blow against a piece of soft copper or aluminum placed over the work. In either case the hammer chosen must have enough weight to seat the workpiece. Most plastic-faced hammers are too light for this purpose. Special no-bounce hammers are available for such work as seating the workpiece in the vise.

The rough workpiece is seated when at least one of the parallels is tight. Because the rough surfaces are probably not square or parallel with each other, it is generally not possible to seat the work on both parallels. The vise should not be tightened further after the workpiece has been seated as this may cause it to become unseated.

The clapper box of the shaper is tilted so that the top slants slightly away from the shoulder cut by the tool as seen in Fig. 2-4. This causes the cutting tool to swing away from the shoulder made by the tool during the cut when the ram is making its return stroke. The tool should be clamped in a vertical position so that it will swing out of the work if it slips. It should be clamped securely with the least possible overhang, as shown in Fig. 2-5. The shaper table should be adjusted to the correct height in order to prevent an excessive overhang of the tool slide of the shaper head. The length and position of the stroke of the shaper ram should be regulated to allow the cutting tool to clear the ends of the work by approximately ¾ inch at the beginning and approximately ¼ to ½ inch at the end of the cutting stroke. The shaper should be set to the

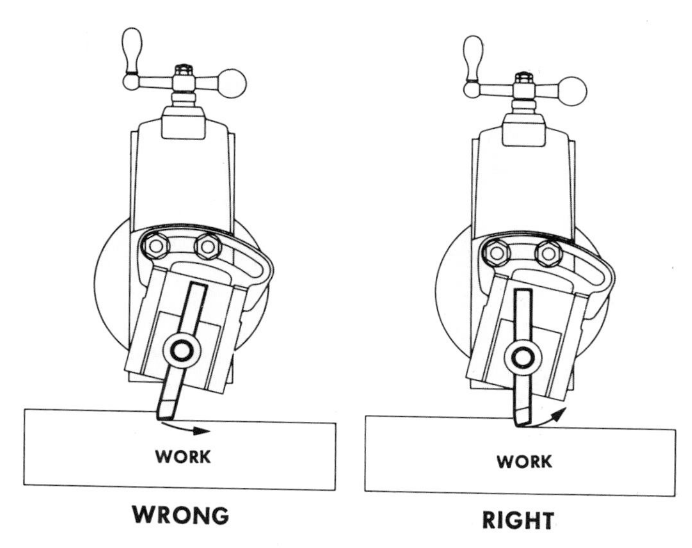

Fig. 2-4. Correct setting of the cutting tool and the clapper box.

correct number of strokes per minute or cutting speed and to the correct feed rate.

The depth of cut is obtained by the trial-cut method. The slide of the shaper head is moved down until the cutting tool just touches the top of the workpiece. The shaper table should be moved until the cutting tool is clear of the side of the work. Then the slide of the shaper head is moved down to allow the tool to take a shallow cut. If the surface of the work has a hard or abrasive scale, the first cut should always be deep enough to cut below this scale. The automatic feed is engaged and the tool is allowed to take a short cut. The ram is then stopped and the workpiece measured. By means of the micrometer dial on the shaper head slide, the tool is adjusted the required amount to cut the workpiece to the desired size. One or more roughing cuts are usually taken in this way leaving .005 to .031 inch for the finish cut. Before the final finish cut is taken, the trial-cut procedure should be used to determine that this cut will bring the workpiece to the

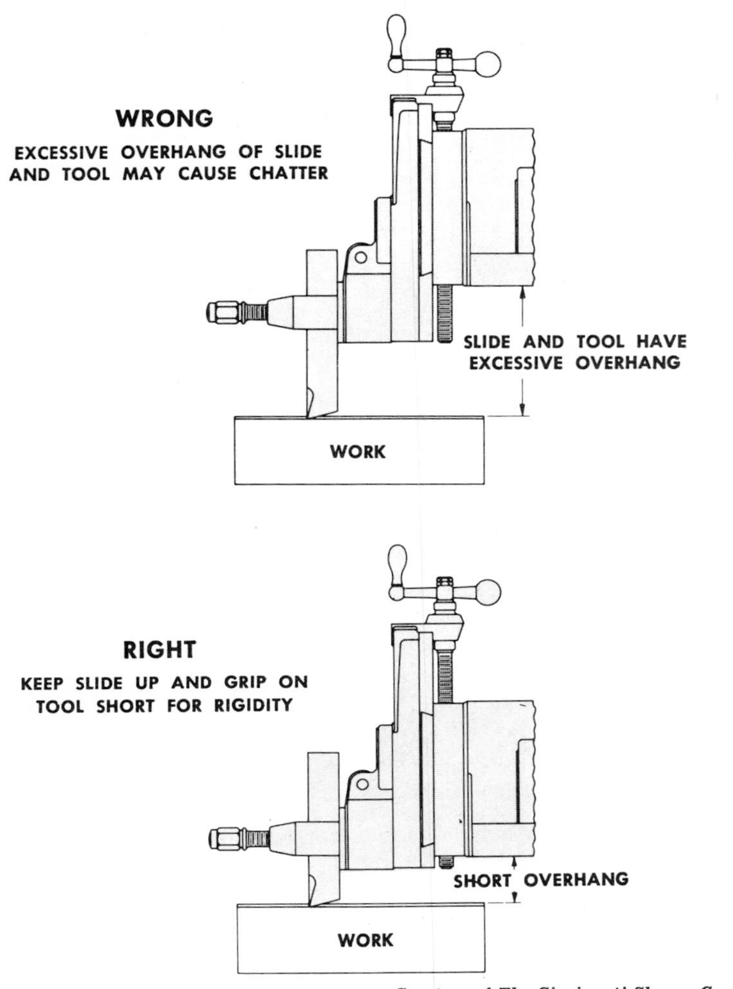

Courtesy of The Cincinnati Shaper Company

Fig. 2-5. Avoiding excessive overhang of the cutting tool and the shaper head slide.

finish size. The trial-cut procedure may be eliminated on roughing cuts if the work is first laid out to show where the finished surfaces are. In this case the depth of cut is established by using the layout lines as a guide.

Vise Work—Shaping Parallel Surfaces

When two surfaces are to be shaped parallel to each other, one of the surfaces must first be finished as described in the previous section. The finished surface must then be positioned parallel to the stroke of the shaper and to the table feed while the second surface is machined. When this is

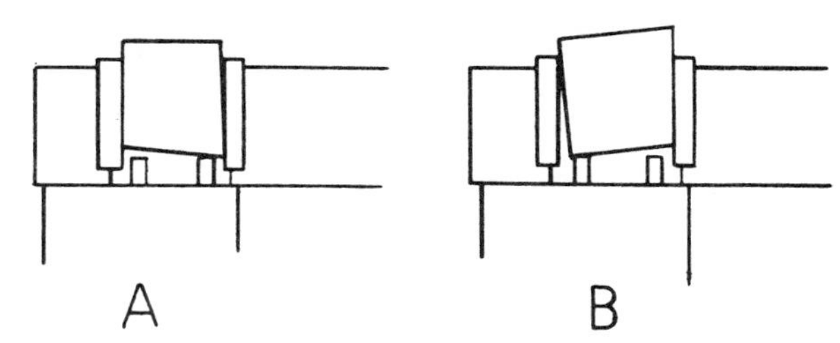

Fig. 2-6. Workpieces with sides not parallel or square cannot be accurately set at the correct angle.

done in a shaper vise, the vise must first be checked as shown at A and B in Fig. 2-2 unless it is known to be true. The previously finished surface is then made to seat on the bottom of the vise, or on precision parallels that are placed on the bottom of the vise. When the second surface is machined with the workpiece in this position, the two machined surfaces will be parallel to each other.

If the sides of the workpiece that are adjacent to the previously finished surface are not perpendicular, as shown in exaggerated form in Fig. 2-6, it will not be possible to seat the workpiece accurately in the vise when it is clamped between the two vise jaws. Other methods are therefore necessary. One method is to use hold-downs as shown in Fig. 2-7. Made from hardened and ground tool steel, hold-downs have contact edges at a slight angle, usually 5 to 8 degrees, with respect to the bottom surface. This causes the hold-downs to assume an angular position when held against the work in the vise, which directs the clamping force downward and forces the work against the parallels. Hold-downs are frequently used

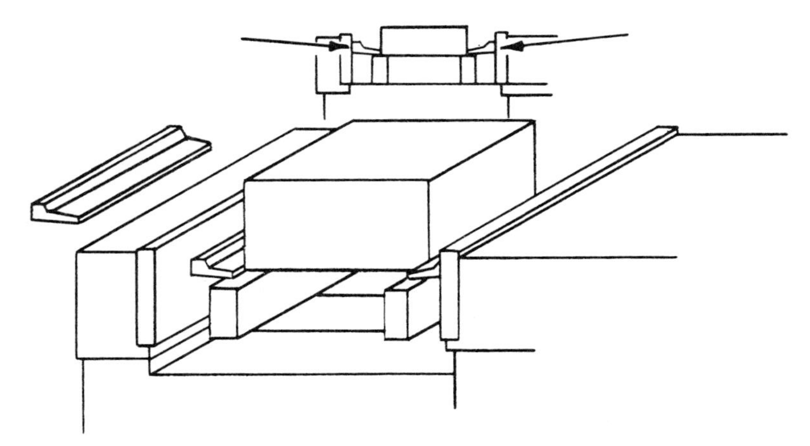

Fig. 2-7. Hold-downs used to seat work in a shaper vise.

to hold very thin workpieces which cannot otherwise be held in the shaper vise because the play in the movable jaw will prevent the work from seating on the parallel. In Fig. 2-7 the workpiece is placed on two precision parallels which are positioned so that a seat is also provided against which the hold-downs can rest. Sometimes a thin metal shim is placed between the parallels and the hold-downs. The parallels also locate the hold-downs exactly opposite each other. If the hold-downs are not opposite each other, the clamping force will cause the work to rotate, and it will be impossible to clamp the work. When the hold-downs are in position, the vise is tightened. The top surface is then shaped to size. The size of the part can be measured with depth micrometers. Performing this procedure with skill and care assures that the workpiece will be to size and that the top and bottom surfaces will be parallel to each other.

Vise Work—Shaping Perpendicular and Parallel Surfaces

Much can be learned about precision machine-tool vise work by studying the procedure used to shape the sides of a rectangular block square and parallel. Before starting to machine the block, check the vise as previously described. In the following description it will be assumed that the vise is accurate in all respects.

The first step is to machine one of the sides of the block as shown at A, Fig. 2-8. If the surfaces of the block are very rough, soft metal shims should be used to protect the vise jaws and the parallels. After the vise has

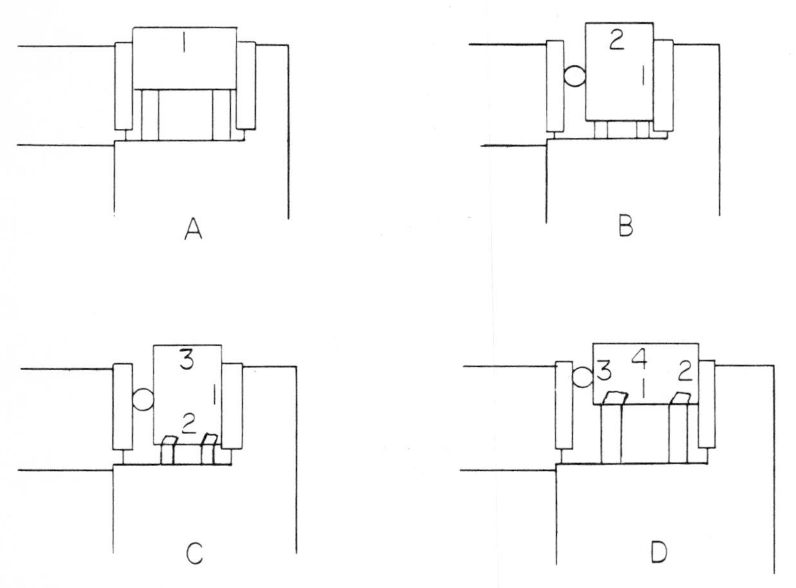

Fig. 2-8. Procedure for machining the sides of a rectangular block square and parallel with each other.

been tightened, the workpiece should be seated by striking "dead" blows with a lead hammer until at least one of the parallels is tight. The first surface is then shaped smooth. The vise is unclamped and the workpiece is removed. The next step is to clean the vise thoroughly. All chips must be removed with a brush. The final test for cleanliness should be to run the bare fingers of the hand over all of the critical locating surfaces that will affect the accuracy of the next setup. These include the solid jaw, the bottom of the vise on which the parallels rest, and the top and bottom sides of the parallels. Any small particle of dirt that is left behind by the brush will be detected by the bare fingers which have a very sensitive sense of touch. Check the edges of the machined surface on the workpiece, and remove any burrs that are present with a file. The workpiece, the parallels, and the vise are now ready for the second setup.

The objective of this setup is to have the first machined surface register against the solid jaw of the vise. To register against a locating surface means to have a surface on the workpiece, called the reference surface, contact the locating surface so that it will perform its function of locating the workpiece accurately in relation to the cutting tool for the machining operation to be performed. In this case the solid jaw of the vise is the locating surface. It is intended to locate the workpiece so that the surface machined during the second operation will be perpendicular to the first machined surface, which for the present is the reference surface on the workpiece. To accomplish this the first machined surface on the workpiece must make complete contact with the solid jaw of the vise. As shown in Fig. 2-6, this will not necessarily be accomplished if the part is simply clamped in the vise, because no other surface is square or parallel with the machined surface. In order to make certain that the machined surface will register on the solid jaw, a round rod of soft steel or brass having a diameter of ¼ to ½ inch is placed between the workpiece and the movable jaw of the vise as shown at B, Fig. 2-8. The rod should be positioned in the approximate center of the workpiece. After the vise has been tightened, the work is seated with a lead hammer until at least one of the parallels is tight. The second surface is then shaped to be perpendicular to the first machined surface.

The third setup is illustrated at C, Fig. 2-8. The vise and the parallels should be thoroughly cleaned and all burrs removed from the workpiece as previously described. Paper shims are placed between the parallels and the four corners of the workpiece as shown. In this operation the solid jaw and the two parallels act as locating surfaces. The solid jaw is used to make certain that the third machined surface is perpendicular to the first machined surface. The parallels are used to make sure that the third and second machined surfaces are parallel. A round rod is employed as before to make the workpiece register against the solid jaw. To test the setup, use paper shims to assure that all four corners on the workpiece register on the parallels. When the workpiece is properly seated on the parallels, all four paper shims will be tight. The third surface can now be machined with confidence that it will be parallel and perpendicular to the previously machined surfaces.

The fourth setup is shown in Fig. 2-8 D. It is similar to the third setup except that the positions of the reference surfaces are changed to allow the fourth side of the block to be machined. The positions of the second and third machined surfaces could, of course, be reversed. If all of the precautions described for the previous operations are carried out for this operation, the fourth surface will be cut parallel or perpendicular to the other sides.

The setup for machining the ends of the rectangular block will depend upon its length. If the block is relatively short, it can be placed on end in the vise as shown in Fig. 2-9. Any one of the previously finished machine

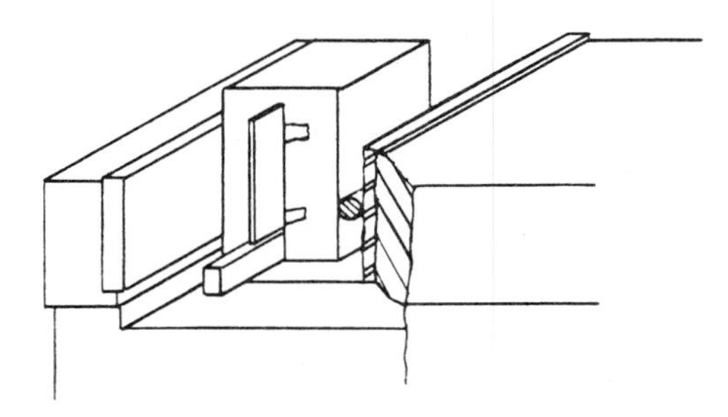

Fig. 2-9. Using a precision machinist's square to set a part perpendicular in a shaper vise.

sides can be made to register against the solid jaw using the round rod between the workpiece and the movable jaw as shown. Seated on the bottom of the vise or on thin parallels, the workpiece must be aligned in the vise with a precision machinist's square in order to position the two sides that are not held by the vise jaws. Paper feelers should be used between the blade of the square and the side of the workpiece in order to test the accuracy of this setup. The side will be square with the bottom of the vise when both of the paper feelers are tight as the precision square is held against the workpiece. The first end of the workpiece can then be shaped square with all of the sides. The setup for machining the second end is similar except that the bottom of the vise (or the parallels) will become a locating surface. Paper shims are used as feelers under the four corners of the workpiece that register on the bottom of the vise. When they are all tight, the second end can be shaped to size.

The ends of longer workpieces are shaped by using the vertical feed of the shaper head slide as shown in Fig. 2-10. Before the workpiece is set up, however, the solid jaw of the vise must be accurately aligned perpendicular with respect to the stroke of the ram. The graduations at the bottom of the vise can be used for many jobs not requiring extreme accuracy; how-

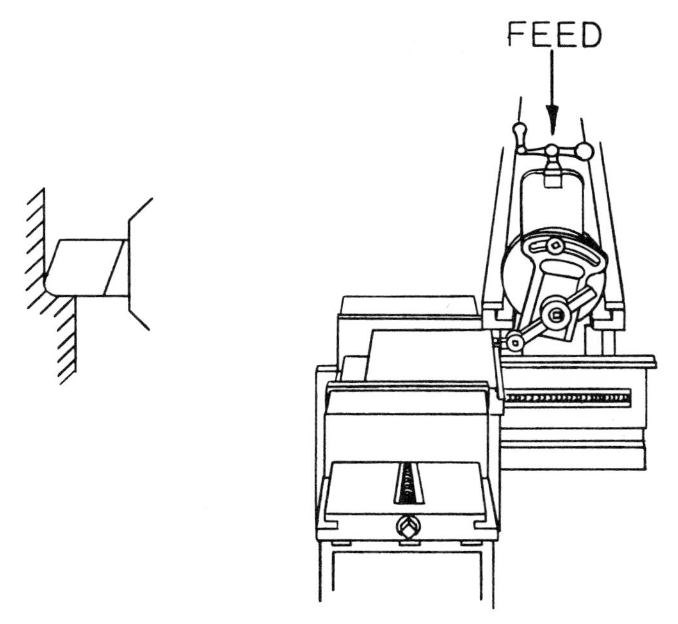

Fig. 2-10. Using the vertical feed of the shaper to cut the ends of a rectangular block.

ever, for very accurate work the solid jaw should be positioned with a dial test indicator as shown in Fig. 2-11. The dial test indicator can be clamped to the shaper head and the contact point brought into contact with the solid jaw. Since the table feed is perpendicular to the stroke of the ram, it can be used to move the vise back and forth. The vise is adjusted until the indicator reading is the same at both ends of the vise jaw.

Another method of doing this would be to lightly clamp a precision machinist's square in the vise, with a piece of soft wood placed between

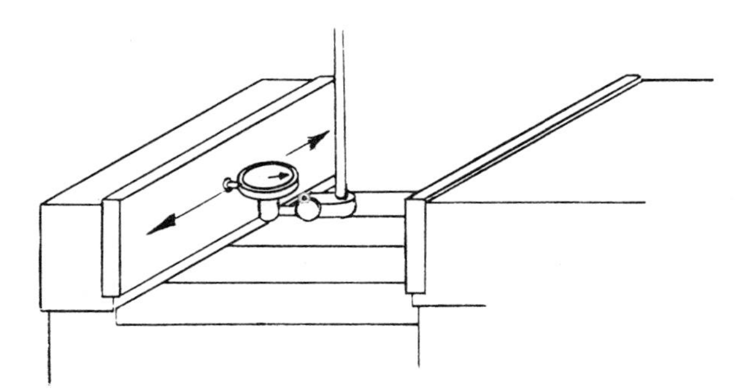

Fig. 2-11. Indicating the solid jaw of a shaper vise.

the movable jaw and the beam of the square. The square is held in a similar manner to that seen at D, Fig. 2-2, except that the blade is extended from the side of the vise instead of from the top of the vise as shown. The indicator is then moved along the blade of the square by slowly moving the shaper ram back and forth. The vise is adjusted until the indicator reading is the same at both ends of the blade of the square. The adjustment of the shaper head should also be checked to assure that the movement of the slide is vertical. The dial test indicator is attached to the slide of the shaper head and the contact point of the indicator is placed in contact with the blade of a precision machinist's square that is set in an upright position in the vise. The shaper head is then adjusted until the indicator reading is the same at both ends of the blade of the square as the shaper head slide is moved up and down.

The workpiece is clamped in the vise with one of the ends projecting beyond the end of the vise and with sides registering on the solid jaw of the vise and on the two parallels. The clapper box of the shaper head must be tilted with the top of the clapper box slanting away from the surface to be machined so that the cutting tool can swing clear of the workpiece on the return stroke. The end surface is then cut by feeding the shaper head slide manually at the end of each stroke. The second end is shaped to size in the same manner. Both ends will now be perpendicular to all of the sides and parallel to each other. The ends of some workpieces, such as in Fig. 2-12, can be shaped by clamping in one end of the shaper vise and cutting the end by using the shaper table feed. In this setup the workpiece is aligned by holding it, or sometimes by lightly clamping it, against the side of the shaper table and by the solid jaw of the vise. In order to balance the load on the movable vise jaw and to allow it to grip the workpiece uniformly, a piece of metal having the same width as the workpiece is held in the other end of the vise.

Shaper Table Work

It is sometimes not convenient to hold the workpiece in the shaper vise. The workpiece may be too large, or it may have a shape impossible to hold in the vise. Such workpieces can often be clamped directly to the shaper table, as shown in Fig. 2-13. In this operation the sides and the bottom of the channel in the steel casting are machined. When this procedure is followed, the shaper vise must be removed from the shaper table.

The work is clamped by bolts that are placed in the T-slots of the table. Although the bolts can sometimes be placed through convenient holes in the workpiece, generally they are used in conjunction with strap clamps as shown in Fig. 2-14. These strap clamps should be kept level by means of a block of metal or hard wood at one end of the clamp. Often thin metal shims are used in conjunction with the block in order to bring the strap clamp to a level position. If the block is placed too close to the bolt, most of the clamping force will be exerted against the block instead of the workpiece. The block should be placed away from the bolt so that an adequate

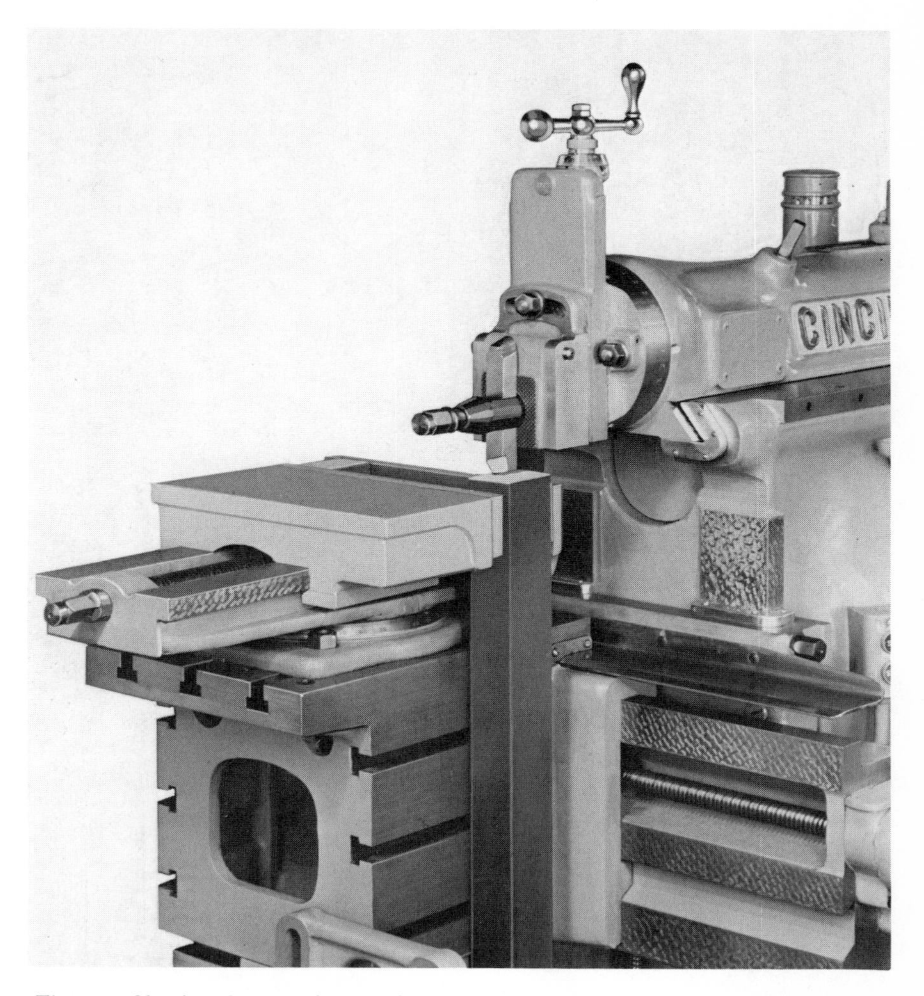

Fig. 2-12. Shaping the end of a steel bar by using the table feed. The bar is aligned by the side of the shaper table and the solid vise jaw. A steel block is clamped inside the vise, opposite the workpiece, to balance the load on a movable jaw.

amount of the clamping force is exerted against the workpiece. Thin workpieces are sometimes clamped to the shaper table with toe dogs as shown in Fig. 2-15. The toe dogs exert a downward force on the workpiece to hold it to the shaper table. Thin metal shims should be placed between the toe dogs and the table in order to allow the dogs to clamp the workpiece a short distance away from the edge. If the toe dogs contact the edge of the workpiece, they will tend to cut through the workpiece instead of clamping it. A stop should be used at the end of the workpiece, as shown,

Fig. 2-13. Shaping steel casting by clamping directly to the shaper table using strap clamps. Stops at end of casting are used to counteract thrust of shaper tool.

to prevent it from being moved lengthwise by the force of the cutting tool.

Angle plates are sometimes clamped to the shaper table against which a workpiece can be clamped in order to shape one surface perpendicular to another. The angle plate should be checked with a precision machinist's square as is shown in Fig. 2-16, to make sure that it is square before the workpiece is clamped to it. If it is not square, paper shims can be used under the corners in contact with the shaper table to make it square.

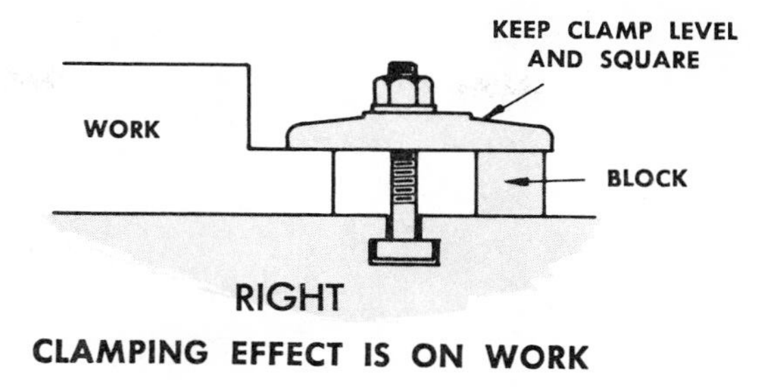

RIGHT
CLAMPING EFFECT IS ON WORK

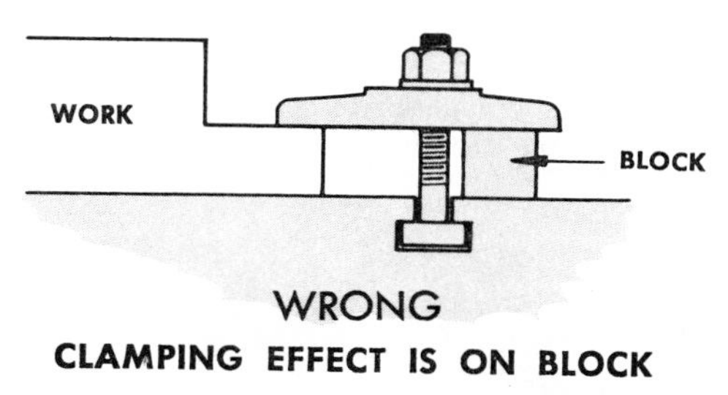

WRONG
CLAMPING EFFECT IS ON BLOCK

Courtesy of The Cincinnati Shaper Company

Fig. 2-14. Clamping work to the shaper table with bolts and strap clamps.

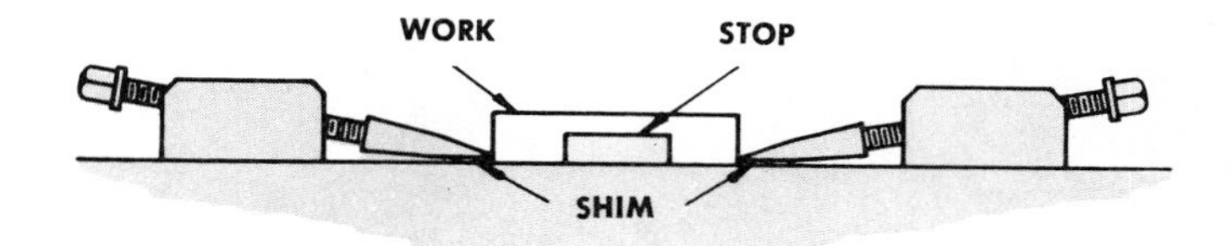

Courtesy of The Cincinnati Shaper Company

Fig. 2-15. Holding thin work to the shaper table with toe dogs.

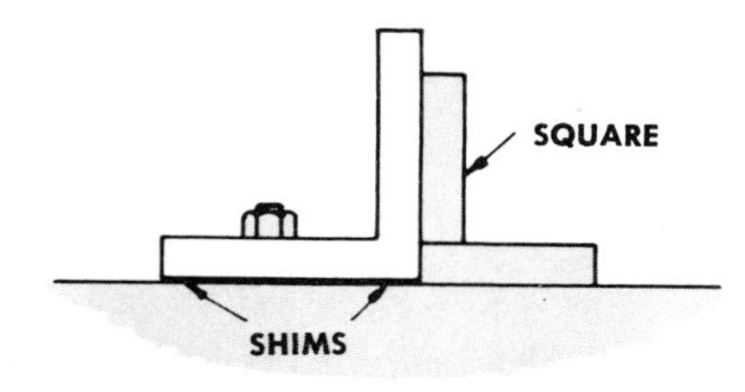

Fig. 2-16. Squaring an angle plate on a shaper table.

Fig. 2-17. Setup for finish-machining angle plate by clamping to another angle plate. Broad nose cast iron finishing tool used to finish surface of angle plate.

Fig. 2-18. A. Aligning workpiece with a surface gage in angular position in vise. B. Shaping the angular surface to the layout line.

When the time is available, however, it is best to machine the surfaces of the angle plate square with each other. One way of doing this is illustrated in Fig. 2-17. One angle plate is clamped against another, which, in turn, is clamped to the table. In this illustration, a finish cut is being taken with a broad-nose, cast iron finishing tool such as shown in view C, Fig. 1-14. The vertical surface of the angle plate provides a locating surface against which other parts can be clamped, whenever this is convenient.

Shaping Angular and Profiled Surfaces

There are several methods of producing angular surfaces on a shaper. One method is to clamp the workpiece in the vise at the required angle, as shown in Fig. 2-18. View A shows how the workpiece is aligned in the vise by using a surface gage; when it is aligned the angle is cut to the layout line with the shaper tool, as shown in view B. Some angular surfaces are cut by positioning the shaper vise at the required angle, as in Fig. 2-19. The angular surface is more frequently machined by using the shaper head slide, which is positioned at the required angle, as in Fig.

Fig. 2-19. Shaper vise positioned at an angle to shape angular surface.

2-20. The angular position can be obtained by using the graduations on the shaper head. If a more precise setting is required, a dial test indicator attached to the shaper head slide can be used to indicate against a surface known to have the correct angle, such as the blade of a vernier bevel protractor. When using this method, it is very important that the top of the clapper box is tilted away from the surface being cut so that the cutting tool will clear this surface on the return stroke. As shown in Fig. 2-21, the angular surfaces of dovetail slides can be cut by this method. Note how the clapper box is tilted in this illustration. In order to shape the

Fig. 2-20. Shaping an angular surface on a press brake forming die.

dovetail, the angular surface and the adjacent "horizontal" surface are machined in one setup. Figure 2-22 illustrates a compound angle being machined on a shaper equipped with a universal table. The workpiece is rotated about three axes to machine the compound angle. As shown, the table is rotated about two axes and the third axis is obtained by the rotation of the shaper vise. In many cases, rotation about only two axes is required to machine a compound angle.

Surfaces having an irregular profile can be machined on a shaper by

Fig. 2-21. Shaping a shaper head dovetail. Angular surface cut using feed of shaper head slide. Note correct setting of the clapper box.

Fig. 2-22. Shaping a compound angle on a shaper equipped with a universal table.

first making a layout line of the required profile on the workpiece and then cutting to this line by simultaneously manipulating the table feed and the feed of the shaper head. With skill and care, accurate results can be obtained by this method.

Shaping Keyseats and Internal Surfaces

Internal surfaces that are not cylindrical or conical in shape are often most conveniently machined on a shaper, using a boring-bar-like shaper bar, as shown in Fig. 2-23. As in boring, before an internal surface can be

Fig. 2-23. Internal shaping of a large extrusion die using a shaper bar to hold the cutting tool.

shaped, a hole large enough to insert a shaper bar of sufficient size must be present in the workpiece. The shaper bar is attached to the ram at the clapper block and a single-point cutting tool is clamped to the other end of the bar. For internal shaping a slow cutting speed is generally used. A typical example of internal shaping is illustrated in Fig. 2-23. In this illustration the opening in an extrusion die is being machined by a toolmaker who is manually manipulating the table feed and the vertical feed of the shaper head in order to cut the contour of the die opening, which is indicated by layout lines scribed on the face of the die. Many other internal shaping jobs are done in a similar manner; a job often done by this method is cutting internal keyseats.

CHAPTER **3**

Planers and Planer Work

Planers are used to machine plane surfaces that may be horizontal, vertical, or at an angle. Profiled surfaces can be planed to a layout line or to a template if the planer is equipped with a tracing attachment. Also, a planer having a tracing attachment can duplicate certain three-dimensional contours from a master. However, most of the work done on a planer involves the machining of plane surfaces on workpieces that range from medium to very large. Planer tools are single-point cutting tools that are relatively inexpensive in terms of their initial cost and are easy to resharpen.

Although some of the work formerly done on planers is now done on planer-type milling machines using large-face milling cutters, there are still many jobs that can be carried out better and more economically on the planer. For example, angular surfaces are often easier to machine on a planer. Planing is also the most effective method of machining some parts because of their shape and for machining long and narrow surfaces. Flat-bearing and reference surfaces that are to be hand-scraped after machining are preferably machined by planing. Here the reason for the desirability of planers is that surfaces produced by a face milling cutter tend to be glazed or work hardened because of the pressure of the face milling cutter on the work. This glazed or work-hardened surface is difficult to hand scrape. Planed surfaces, on the other hand, are easily scraped since work-hardening is kept to a minimum on the surfaces produced by a single-point cutting tool. Because many plane surfaces on machine tools require hand scraping, planers find wide application in machine tool plants.

The operation of a planer requires a high degree of mental effort and mechanical skill. The planer operator must often exhibit ingenuity in determining the best and safest method of setting up a complex part on the planer table. He must have a sense of responsibility for his work, since many of the parts which are machined on planers are very expensive. If such parts had to be scrapped because of poor workmanship, there would be considerable financial loss. Good judgment must be exercised in selecting and positioning the cutting tools as well as in selecting the cutting speed and the feed rate. For these reasons, planer work is both challenging and interesting.

The Construction of the Planer

The two basic types of planers—the double-housing planer, Fig. 3-1, and the open-side planer, Fig. 3-2—are made up of the following principal

Fig. 3-1. Large double-housing-type planer.

Fig. 3-2. Open side planer.

parts: the bed, the table, the housings, the top brace, the cross rail, and the toolheads. These do not include the drive motor and drive mechanism, which are of equal importance.

Planer Bed. This is a large casting which rests on the floor and upon which the planer table moves back and forth. On very long planers the bed is sometimes made from two or more castings bolted together in perfect alignment with each other (see Fig. 3-15). The bed must be somewhat more than twice as long as the planer table to support the table at all positions during the stroke. The top of the bed has two flat bearing surfaces, called the *ways* of the planer, on which the table slides. On mechanical or gear-driven planers the ways are in the form of a V-groove. On hydraulically actuated planers the ways are in the form of a dovetail slide. Since planers are usually heavy machines which must be set on a firm foundation, it is recommended that hold-down bolts passing through holes in the base of the planer bed be grouted into the foundation to hold the bed firmly in place. Leveling screws, screwed into threaded holes in the planer bed and resting on steel plates, are used to level the planer bed. The manufacturer's instructions should be closely followed for this operation, which must be done with great care so that the bed and table ways will retain their accuracy indefinitely. If the planer is not kept level, in a short time the bed and table can become badly damaged. A new planer on a new foundation should be releveled at least once a week for the first month. Thereafter it should be checked periodically for being level, the period depending upon the stability of the foundation.

Driving Motor. Usually located at the rear of the right side housing when viewed from the front, the motor on mechanical planers is a variable speed-reversing motor which causes the reversal of the planer table and provides the different table speeds. Because the speed of the driving motor, and thus the table, can be varied independently for the cutting stroke and the return stroke, a quick return stroke of the table can be effected to minimize the noncutting time. A relatively simple gear train transmits the motion from the motor to the planer table, and a large bull gear engages a rack bolted to the bottom of the table, thus moving the table back and forth. The gears that move the table are located inside the planer bed.

Hydraulic Drive. Hydraulically actuated planers have a large cylinder fastened to the bed. The piston is moved back and forth inside the cylinder by alternately causing oil to flow under pressure into one end of the cylinder and then into the other end. The piston rod, attached to the piston which moves it back and forth, extends outside of the cylinder and is attached to the planer table. A variable-speed piston pump delivers the oil to the cylinder, thus providing the different table speeds. The speed of the cutting stroke is independent of the speed of the return stroke. On hydraulic as well as mechanical planers the length of the cutting stroke is controlled by adjustable dogs located on the right side of the table.

Table Top. The planer table top is an accurate locating surface upon which workpieces or angle plates and other work-holding devices can be clamped. It is also used as a reference surface for surface gages, height gages, and precision squares when parts are aligned on the planer table in making the setup. To do accurate work the planer table must be kept free of scratches, dents, and burrs. Accurately machined T-slots, machined lengthwise on the planer table, are used to anchor clamping bolts to the table. A group of holes is drilled and reamed into the table top at regularly spaced intervals, and stop pins for clamping the workpiece are placed in these holes. To prevent chips from falling onto the ways of the bed troughs have been placed at each end of the table.

Housings. The planer housings, which are bolted to the planer bed, are used to carry the cross rail. Located inside the housings are heavy counterweights for the cross rail, and some planers have additional counterweights for the cross rail toolhead slides. As its name implies, the double-housing planer has two housings to support the cross rail at each end. The open-side planer has only a single housing, sometimes called a column, which carries the cross rail like a cantilever beam.

Double-housing planers have a large casting at the top, called the top brace, which connects the two housings together. Ways are machined on the face of the housings on which the cross rail can slide when it is adjusted up or down. This up or down movement of the cross rail is accomplished by rotating an elevating screw mounted on the outside face of the housing. Side toolheads can also be mounted on the ways of the housing. Similar in construction to shaper heads except for being provided with automatic power feeds, the side heads can be moved vertically along the ways of the housing and horizontally, or at an angle, by means of the toolhead slide. The side toolheads have a clapper box and clapper block to prevent the tool from rubbing on the workpiece on the return stroke.

Cross Rail. The construction of the toolheads that are mounted on the cross rail is essentially the same as that of the side toolheads. Both the power and the manual feeds may be operated from the operator's position on the right side of the machine and also on the left side of some planers. Thus, it is not necessary for the operator to climb over the machine to operate the cross rail toolheads. On some planers the toolheads may be moved about rapidly by means of a rapid traverse mechanism. Planers can have one or two toolheads mounted on the cross rail.

The cross rail may be raised or lowered to accommodate different sizes of workpieces; however, it must always be firmly clamped to the housings when taking a cut.

To plane a surface parallel to the top of the planer table, the toolhead must feed the cutting tool parallel to the table top. This can only be done when the cross rail ways, upon which the toolhead slides, are parallel to the table top. Some modern planers have an automatic cross rail leveling device which positions the cross rail parallel to the table top before it is clamped to the housings. Most planers found in industry today, however, do not have this feature.

The alignment of the cross rail with the table top should be periodically checked. Before this alignment is verified, however, the toolhead or toolheads should be positioned in the approximate center of the cross rail. The cross rail is then loosened and lowered until a dial test indicator attached to a toolholder can reach the table top. Next, the cross rail is clamped to the housings. (The toolheads should be centered on the cross rail whenever it is raised or lowered as this equalizes the strain on the elevating screws.) After a dial test indicator is fastened to a toolholder which is clamped in a toolhead, the toolhead is moved along the cross rail thereby moving the indicator across the table top. Finally, the indicator is read, and adjustments are made to the cross rail if required.

The method of adjusting the cross rail depends upon the construction of the planer. On some modern planers one of the elevating screws operates in a nut that can be rotated when the screw is stationary. This rotation moves one end of the cross rail up or down while the other remains stationary.

Planers not equipped with the adjustable elevating nut feature are more difficult to adjust. If the parallelism error of the cross rail on such planers exceeds approximately .012 inch, the cross rail may be adjusted by disengaging the gears that operate the elevating screw and indexing the gears one tooth. If the error is from .005 to .010 inch it is corrected by replacing the thrust washer under one of the elevating screws with a thicker washer. If the error is less than .005 inch—perhaps the error has been reduced to this amount by one of the other methods—a light truing-up cut should be taken across the planer table with a broad-nose finishing tool as at G, Fig. 3-4.

On open-side planers the cross rail is connected to a large casting called the knee, which in turn clamps to the housing. The cap screws connecting the cross rail to the knee have sufficient clearance in the cap-screw holes to permit leveling of the cross rail, which pivots on a stud located near its right end. With the cap screws that clamp the cross rail to the knee loosened slightly, two adjusting screws located at the left end of the rail are turned until the cross rail is level. On open-side planers the toolhead, or toolheads, should be positioned on the right side of the cross rail when it is being raised or lowered.

Planer Cutting Tools

Single point cutting tools similar in design to lathe and shaper tools are used on planers. Almost all planer tools are made from high-speed steel or cemented carbides, and are generally made large enough to withstand the loads imposed on them by the heavy cuts that are frequently taken. Gray cast iron can be cut with straight tungsten carbide. Plain carbon and alloy steels must be cut with crater-resisting grades of cemented carbides.

Planer tools are sometimes made from large, solid high-speed steel bars that are ground to a cutting edge at one end. This practice, although it provides a strong and rigid cutting tool, is expensive. Toolholders made from less expensive heat-treated steel can be used to hold high-speed steel inserts. Two toolholders, a straight and a gooseneck holder, and some typical high-speed steel-cutting tools are shown in Fig. 3-3. The high-speed steel tools, forged and ground to shape, are held in the toolholder by the wedge action of the serrated locking seat.

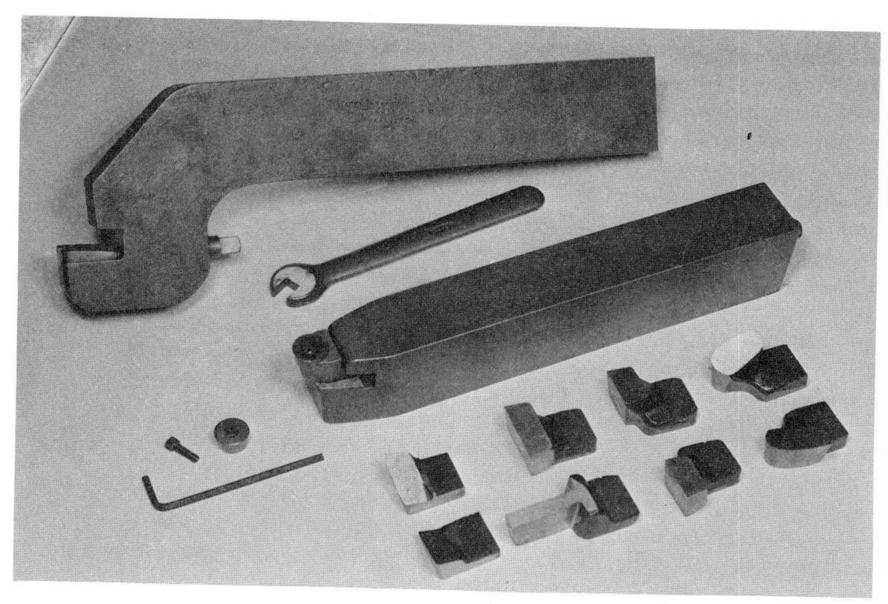

Courtesy of The Apex Tool & Cutter Co. Inc.

Fig. 3-3. Straight and gooseneck-type planer toolholders with cutting tools.

A group of planer tools that are used with the toolholder in Fig. 3-3 are illustrated in Fig. 3-4. A "straight round-nose tool," shown at A, is used to take roughing and finishing cuts on steel and cast iron. The "straight side roughing tool" at B is employed for cutting up to shoulders and for other operations where the shoulder formed by the cut must be approximately square. The "diamond-point tool" at C can take roughing cuts where a square shoulder is not required. This tool has the advantage of a large lead angle that allows cutting with a somewhat greater feed rate. The "hog-nose tool" at D is similar to the round-nose tool at A, except that its larger nose radius has a greater tendency to eliminate feed marks. Without taking a final finishing cut, the hog-nose tool can cut surfaces finished to size by taking one or two fairly deep cuts. The tool at E, called a "bottom-finishing tool," can take vertical and angular cuts when feeding with the toolhead slide. Designed to cut primarily with the end-cutting edge instead of the side-cutting edge, this tool can be used to cut large shoulders.

The "dovetail tool" at F is used to machine dovetails on the planer. Cast-iron surfaces are best finished with a "broad-nose finishing" tool like the one at G. A gooseneck type of toolholder should be used to hold this tool. All of the cutting is done by the end cutting edge. When finish planing with this tool the depth of cut should be very shallow and the feed per stroke should be coarse—almost equal to the width of the cutting edge. The broad-nose cast-iron finishing tool produces a good surface that can easily be hand-scraped if required.

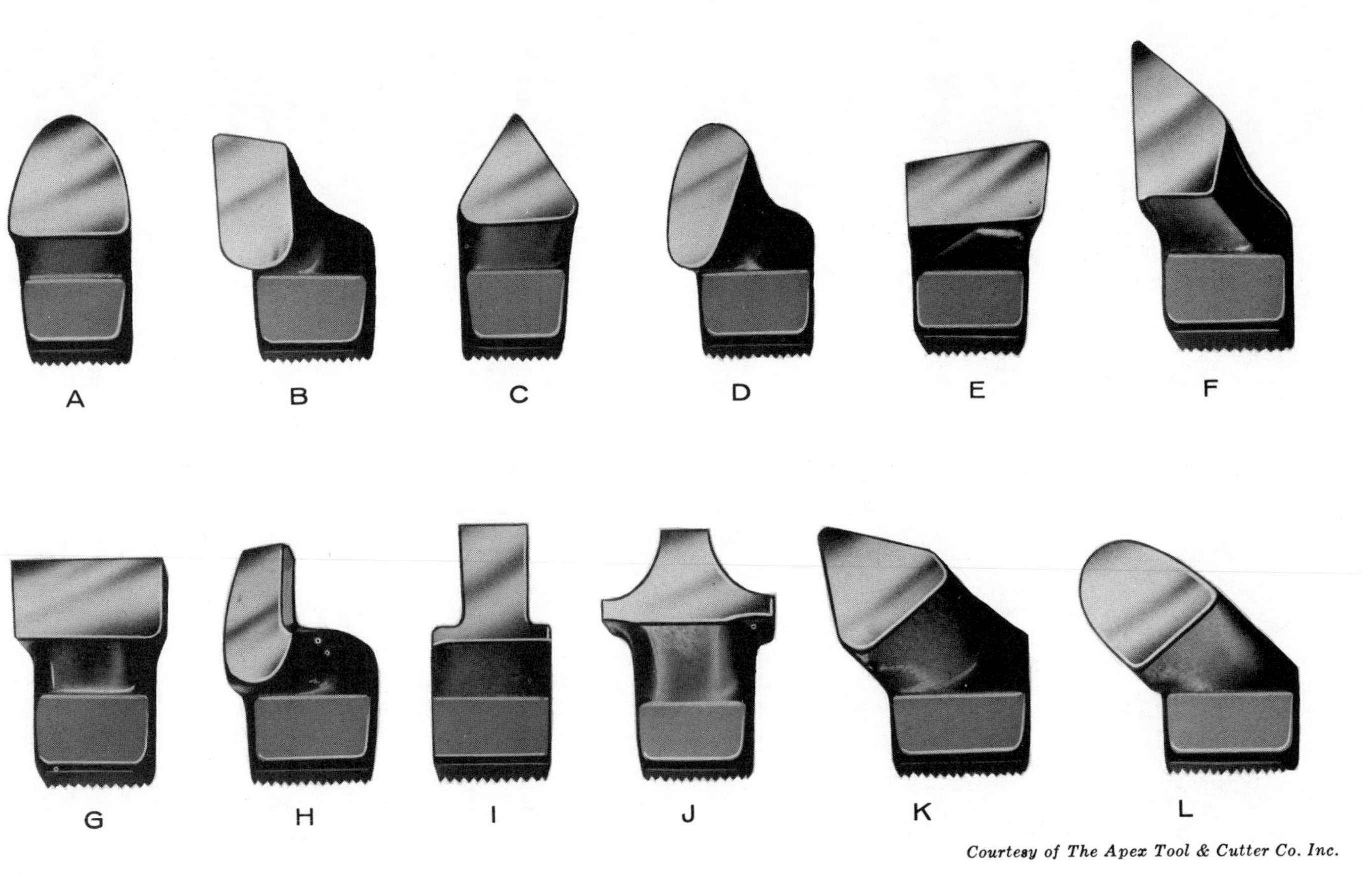

Courtesy of The Apex Tool & Cutter Co. Inc.

Fig. 3-4. Typical single point planer cutting tool shapes.

A "steel-finishing tool" is found at H. Similar to the tool shown at C (see Fig. 1-12), this tool has a cutting edge on the face on the side rather than the face shown at the front. The end-cutting edge of the tool produces a tightly curled chip with a shearlike cutting action. When properly used, the steel-finishing tool provides an excellent surface finish. The tool at I is a key-seating tool used to cut keyseats and slots. Radii are formed with the "double radius-forming tool" at J. The "offset diamond-point tool" at K and "offset round-nose tool" at L are used to take heavy vertical and angular roughing cuts.

High metal-removal rates can be achieved on modern planers using cemented-carbide cutting tools. Every effort must be made to clamp the workpiece securely to the planer table in order to eliminate vibration when planing with carbide cutting tools. If two or more workpieces are planed simultaneously in a "string" or tandem fashion, as in Fig. 3-16, the ends should be butted very closely together or separated at least 6 to 8 inches. Butting the workpieces together results in a continuous cut with less shock to the cutting tool. Tool breakage can result, however, from chips trapped between the workpieces. This problem is not likely to occur when the parts are spaced further apart.

In carbide planing it is very important for the correct grade of carbide to be used. Quite often an incorrect grade will fail rapidly while another grade will cut for a long time before requiring replacement. For heavy-duty carbide planing, toolholders with mechanically held inserts are recommended. When deep cuts on ductile metals are taken, carbide tool breakage often occurs as the last few cuts are made on the workpiece. This breakage is caused by the deflecting of the thin flange of metal remaining to be cut. The flange springs away from the tool because of the cutting forces. As the tool continues through its cut, the flange will spring back and rub against the flank of the tool, generating heat as a result of the friction. Furthermore, at a subsequent stroke, the springback of the flange will cause the tool to take a heavier cut, which is equivalent to increasing the feed. These actions, which can cause the carbide tool to fail, can be eliminated by disengaging the automatic feed for the last four or five strokes and feeding the tool manually to finish the cut. A large lead angle will reduce the tendency of the thin flange to spring back because it will have a thicker root. The lead angle, the angle between the side-cutting edge and a plane perpendicular to the surface being cut (see Chapter 7, Volume 1), should be from 30 to 45 degrees. Also, the length of the planer stroke should be adjusted to provide an even cutting speed for the length of the cut.

Cutting Speeds for Planing

Many factors must be considered when selecting the cutting speed for planing. The size and the weight of the workpiece must be given consideration because it is difficult to start and to stop a large mass. Compensation must be made when it is necessary to use a large overhang of the cutting tool to reach inaccessible surfaces. Possible variations in the hardness of castings must be taken into account. These variations are caused by heavy and thin sections in castings which cool at different rates. Weldments may have hard spots near the weld which can quickly dull the cutting edge of the tool. Other considerations peculiar to the part or the manner in which it is set up on the planer may influence the selection of the cutting speed. All of these factors must be evaluated in selecting the cutting speed for planing. The tables of recommended cutting speeds for planing provide useful information on the best cutting speed at which to operate the tool; whenever possible, this speed should be used. In many situations, however, it is necessary to use a reduced cutting speed for the reasons given.

The recommended cutting speeds for planing are given in Tables 3-1 and 3-2. These tables are based on the ability of the tool to cut at these speeds, yet have a satisfactory tool life. This is determined primarily by the type of work material, the hardness of the work material, the type of cutting tool material, the feed rate, and the depth of cut. For customary inch units, the cutting speed is given in terms of feet per minute (fpm); in metric units, it is in terms of meters per minute (m/min). Inch and metric cutting speeds can be converted from one to the other as follows:

Table 3-1. Recommended Cutting Speeds for Planing Plain Carbon and Alloy Steels*

Material	Hardness, HB	Cutting Speed, fpm	
		H.S.S.	Carbide
AISI 1012, AISI 1019, AISI 1020	100 to 125 130 to 175 180 to 225	80 70 60	300 250 200
AISI 1030, AISI 1040, AISI 1050	120 to 170 175 to 200 205 to 240 245 to 300	70 60 50 40	240 200 175 150
AISI 1060, AISI 1080, AISI 1090, AISI 1095	160 to 200 205 to 240 245 to 300 305 to 375	60 50 35 25	190 170 140 110
AISI 1320, AISI 2317, AISI 2517, AISI 3120, AISI 3316, AISI 4016, AISI 4118, AISI 4128, AISI 5020, AISI 6120, AISI 8620, AISI 9315	150 to 175 180 to 220 225 to 275 280 to 325	60 50 40 20	225 190 150 80
AISI 1330, AISI 1340, AISI 2330, AISI 2340, AISI 3130, AISI 3140, AISI 3150, AISI 4030, AISI 4060, AISI 4130, AISI 4140, AISI 4150, AISI 5140, AISI 5150, AISI 6240, AISI 6250, AISI 6440, AISI 8640, AISI 8650, AISI 8740, AISI 9840	175 to 225 230 to 275 280 to 325 330 to 375 380 to 425	50 40 30 20 —	190 160 130 80 40
Ferritic Stainless Steel	130 to 190	50	150
Austenitic Stainless Steel	130 to 190	40	110
Martensilic Stainless Steel	130 to 180 185 to 220 225 to 300	50 40 25	150 110 60

* Based on feed rate of .030 inch per stroke and depth of cut of .250 inch.

Table 3-2. Recommended Cutting Speeds for Planing Cast Metals*

Material		Hardness, HB	Cutting Speed, fpm	
			H.S.S.	Carbide
Gray Cast Iron	Class 20	110 to 140	70	240
	Class 30	140 to 190	60	170
	Class 40	190 to 220	50	150
	Class 50	220 to 260	35	110
	Class 60	260 to 320	20	50
Cast Steel — Plain Carbon AISI 1020, AISI 1025		120 to 150	70	300
Cast Steel — Plain Carbon AISI 1030, AISI 1040, AISI 1050, AISI 1070		125 to 170	70	240
		175 to 220	60	170
		225 to 280	35	140
Cast Steel — Low Alloy AISI 1320, AISI 2320, AISI 2325, AISI 3125, AISI 4020, AISI 4120, AISI 4620, AISI 5120, AISI 8420, AISI 8620, AISI 9520, AISI 9525		150 to 200	60	220
		205 to 245	45	200
		250 to 300	30	130
Cast Steel — Low Alloy AISI 1330, AISI 1340, AISI 2330, AISI 2340, AISI 3130, AISI 3140, AISI 4030, AISI 4040, AISI 4130, AISI 4140, AISI 4330, AISI 4340, AISI 5130, AISI 5140, AISI 8030, AISI 80B30, AISI 8430, AISI 8440, AISI 8630, AISI 8640		170 to 220	55	200
		225 to 275	40	170
		280 to 320	35	140
Nodular Cast Iron (Ductile Iron)		140 to 180	70	180
		185 to 220	55	160
		225 to 260	40	125
		265 to 325	10	50
		330 to 400	—	25

* Based upon a feed of .030 inch per stroke and a depth of cut of .250 inch.

Table 3-3. Feed and Depth-of-Cut Factors for Planer Cutting Speeds*

Feed Factors, F_f		Depth of Cut Factors, F_d	
Feed, Inch per Stroke	F_f	Depth of Cut, Inch	F_d
.005	1.55	.016	1.40
.008	1.52	.031	1.34
.010	1.50	.062	1.21
.012	1.47	.094	1.15
.015	1.34	.125	1.10
.018	1.24	.156	1.06
.020	1.19	.200	1.03
.025	1.08	.250	1.00
.028	1.03	.375	.94
.030	1.00	.437	.92
.036	.93	.500	.90
.040	.89	.625	.88
.045	.84	.750	.86
.050	.81	.875	.84
.062	.74	1.000	.82
.078	.67	1.250	.80
.094	.62	1.500	.78

* To be used with Tables **3-1** and **3-2**.

To obtain the cutting speed in m/min, multiply fpm by .3048; to obtain fpm, multiply m/min by 3.28, or divide m/min by .3048. In Tables 3-1 and 3-2, HB, Brinell Hardness Number, is also sometimes designated, "Bhn." The values in these tables will give a satisfactory tool life when the feed rate is .030 in. per stroke (0.76 mm/stroke) and the depth of cut is .250 in. (6.35 mm). To obtain the same approximate tool life when the feed and depth of cut are different, the feed and depth-of-cut factors in Table 3-3 should be used as follows:

$$V_f = V_i F_f F_d \tag{3-1}$$

Where V_f = The modified cutting speed, fpm or m/min
V_i = The cutting speed from Tables 3-1 and 3-2, fpm or m/min
F_f = Feed factor from Table 3-3
F_d = Depth of cut factor from Table 3-3

This formula can be used with either inch or metric units; however, only inch units or metric units can be used in any one case.

Example 3-1

A casting made from a Class 30 gray cast iron is to be machined on a planer using a high speed steel cutting tool. The feed rate will be .015 inch per stroke and the depth of cut is to be .500 inch. Calculate the modified cutting speed.

V_i = 60 fpm (From Table 3-2)
F_f = 1.34 (From Table 3-3)
F_d = .90 (From Table 3-3)
$V_f = V_i F_f F_d = 60 \times 1.34 \times .90$
V_f = 72 fpm

The cutting speeds calculated by applying the two given factors in Table **3-3** will sometimes exceed the maximum speed available on the machine. When this occurs, use the maximum table speed. The horsepower available on the machine, the rigidity of the setup, and the strength of the cutting tool impose limitations on the size of the cut that can be taken. The final selection of planer cutting speeds and feed rates must be tempered by sound judgment. In other words all aspects of the job must be considered.

The best method of finish-planing gray cast iron surfaces is to use a broad-nose cutting tool such as shown at G, Fig. **3-4**. The end-cutting edge of this tool is positioned parallel with the table or surface to be machined. When a good quality surface finish is desired, the depth of cut should be from .001 to .003 inch; when the quality of the finish is less important, a depth of cut up to .005 inch can be used. The feed rate usually varies from ¼ to 1 inch per stroke or approximately .75 by the width of the cutting tool. The recommended cutting speed for this operation can be found in Table **3-4**. An excellent surface finish can be obtained which is easily hand-scraped when required. The large feed rate causes the tool to traverse over the workpiece in a relatively few number of strokes and in a compar-

Table 3-4. Recommended Cutting Speeds for Finish Planing Gray Cast Iron with Broad-Nose Finishing Tool*

Gray Cast Iron Grade	Cutting Speed, fpm	
	H.S.S.	Carbide
Class 20	60	220
Class 25	55	200
Class 30	40	190
Class 35	35	130
Class 40	30	110
Class 45	20	90
Class 50	15	80
Class 55	10	75
Class 60	10	70

* Recommended depth of cut is .001 to .002 inch. Recommended feed rate is .75 × width of tool.

atively short time. Thus, there is less wear on the cutting edge, and the planed surface will be a true plane. Since the wear on the cutting edge is distributed over the greater length of the cutting edge, the effective wear at any one point is reduced.

Setting Up Work

Setting up a workpiece on a planer table, like selecting and positioning cutting tools, requires skill, experience, good judgment, and a high degree of ingenuity. This, however, serves to make this work most interesting and challenging. Each workpiece encountered is unique, but certain basic knowledge must be applied in determining the best setup.

The tooling used to make setups on a planer is called the planer furniture. A plentiful supply of this furniture, shown in Fig. 3-5, should always be at hand. In Fig. 3-5, CO-1 through CO-4 are strap clamps; CO-1 is a plain strap clamp, CO-2 an offset strap clamp, CO-3 a U-strap clamp with a pin end, and CO-4 a pin strap clamp. The pins on the strap can be used to set a strap clamp in a cored or drilled hole. CO-6 is a T-slot bolt with a nut and a washer. The nut in CO-8, a removable T-slot nut, can be lifted in or out of the T-slots on the planer table by a simple turn—without running the nut along the length of the slot. Stud CO-7 is screwed into this nut to make a very effective clamp screw. CO-9 is a stop pin primarily used to prevent the workpiece from sliding along the planer table. The round stop pin is placed in one of the drilled and reamed holes on the planer table. Stops can also be placed in T-slots, as for example the T-slot stop bracket CO-5 and the T-slot stop block CO-11. A screw jack is shown at CO-12. CO-10 is a chisel point.

An end view of a workpiece set up for having the top and the right sides planed is shown in Fig. 3-6. The workpiece, CN-1, is clamped to the table by strap clamps CN-6 and CN-7. The offset clamp at CN-6 allows the cutting tool to clear the clamp stud CN-8. Planer setups must be designed

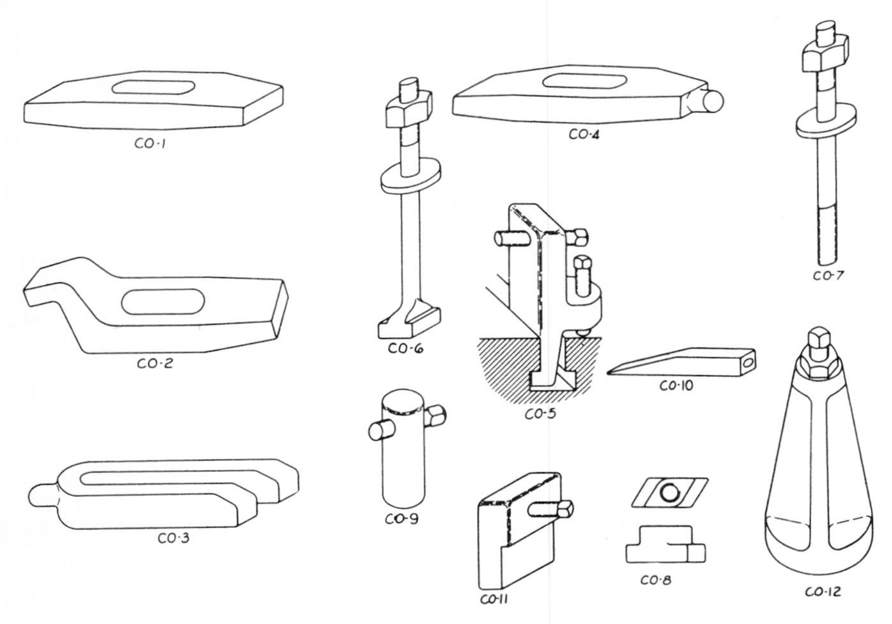

Fig. 3-5. Optional typical planer "furniture" used for setting-up workpieces.

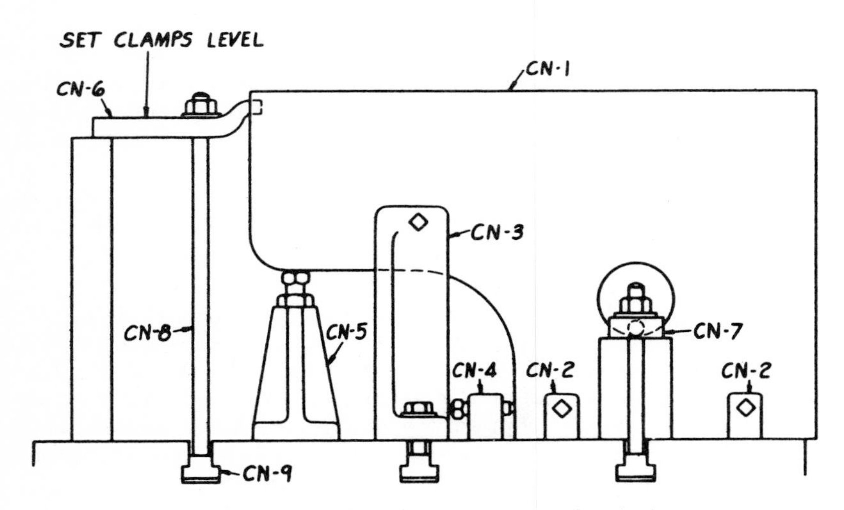

Fig. 3-6. An end view of a casting set up for planing.

so that the surfaces to be machined and the path of the cutting tool are free of clamps and obstructions. The pin clamp CN-7 allows the cored hole in the workpiece to be used as a clamping surface. The strap clamps should be set level, with metal or hardwood heel blocks used to support the heel of the strap clamps. The proportional amount of the clamping force of the bolt or stud that is transmitted to the work is directly related to the dis-

Courtesy of the Armstrong Bros. Tool Co.

Fig. 3-7. Setup of a draw die for planing. The die is placed on screw jacks because of the contoured shape of the lower part of the die. The upper surfaces must be kept free for machining.

tance between the bolt and the workpiece, and the distance between the bolt and the heel blocks. In order to transmit the maximum available clamping force onto the workpiece, the bolt or stud should be positioned as close to the work and as far from the heel block as possible. The screw jack CN-5 is used to prevent the clamp CN-6 from bending or springing the overhanging part of the workpiece. Without the screw jack the top surface of the workpiece would not be finished to a plane surface, for when the clamp CN-6 is released, the overhanging portion of the work springs back up and the finished surface on the top is no longer plane or flat. Care must always be taken to make certain that the clamping action of strap clamps does not bend the workpiece. Metal shims should be placed between the table and the workpiece underneath the clamps if the work does not seat firmly on the table and if the opening is too small for a screw jack. Stop pins CN-2 and CN-4, and the angle bracket CN-3, are used to prevent the workpiece from moving as a result of the cutting forces transmitted by the tool.

Figure 3-7 illustrates a large draw die set up for machining the upper surface on a planer. It is placed on screw jacks because of the contoured shape of the lower surface. Since the upper surface must be kept free of clamps so that it can be planed, the die is clamped down with pin-type strap clamps placed in holes drilled in each end for this purpose. An adjustable block supports the other end of the strap clamps. Bracing jacks, also called telescoping braces, are used as end stops to counteract the thrust of the cutting tool. In Fig. 3-7 the braces can be seen sloping from the table to the ends of the die. While their purpose is to counteract the cutting force, braces must be installed at both ends of the die to counteract each other when the planer is not cutting.

Thin workpieces are held by toe dogs or by chisel points. Their names are often used interchangeably, and toe dogs or chisel points can be used with equal effectiveness on most setups. In Fig. 3-8, a toe dog is shown at *A* and two styles of chisel points are shown at *B* and *C*. They exert a downward force on the workpiece to hold it firmly against the planer table. When clamping thin workpieces, two toe dogs or chisel points should always be placed opposite each other on either side of the workpiece, as at *D*, Fig. 3-8; and, as many pairs as required by the length of the workpiece are used. When the workpiece is somewhat thicker, as at E, a toe dog or chisel point can be used on one side of the workpiece while it is held by a stop the other side. The clamping force is obtained by tightening the screw on the screw bracket or screw pin placed behind each toe dog and each chisel point. Screw brackets have a rectangular shape and fit into the T-slots while screw pins are round and fit into holes provided in the planer table. All of the toe dogs or chisel points should contact the sides of the workpiece at very nearly the same height above the planer table. This is accomplished by placing the same size shim under each toe dog or chisel point while it is being clamped in position, after which the shim may be left in place or, if loose, it may be removed. The shim also prevents the toe dogs or chisel points from contacting the edge of the workpiece, which may be crushed by the clamping pressure thereby releasing the clamping force. One or more stop pins must always be placed against the front end of the workpiece to resist the thrust of the cutting tool and to hold the workpiece in place during the cut. Toe dogs and chisel points make slight indentations in the sides of the workpiece. When this is not permissible, the setup shown at F in Fig. 3-8, can be used. In this setup the workpiece is sandwiched between pieces of moist paper, such as drawing paper, and pieces of mild steel, such as key stock. A deep center punch mark is made in each piece of key stock to contain the points of the toe dogs. When the clamping force is applied, the toe dogs hold the key stock and the workpiece firmly against the table. The friction created by the moist paper holds the workpiece against the key stock, which in turn is clamped by the toe dogs.

Toe clamps, Fig. 3-9, can also be used to clamp thin workpieces, and also those that are somewhat thicker, in a manner similar to chisel points and toe dogs. The clamp jaw, or toe, is screwed down at 45 degrees to exert an equal downward and outward clamping force against the workpiece. Toe clamps are available in different sizes and as they are not confined to planers, they can be used to clamp workpieces on drill presses, shapers, milling machines, and on lathe faceplates.

In setting up the workpiece on the machine tool table, it must be aligned before it is clamped in place. Surfaces that have been previously machined or otherwise finished are helpful in making the alignment. Finished surfaces are excellent reference surfaces; they can be placed against the top

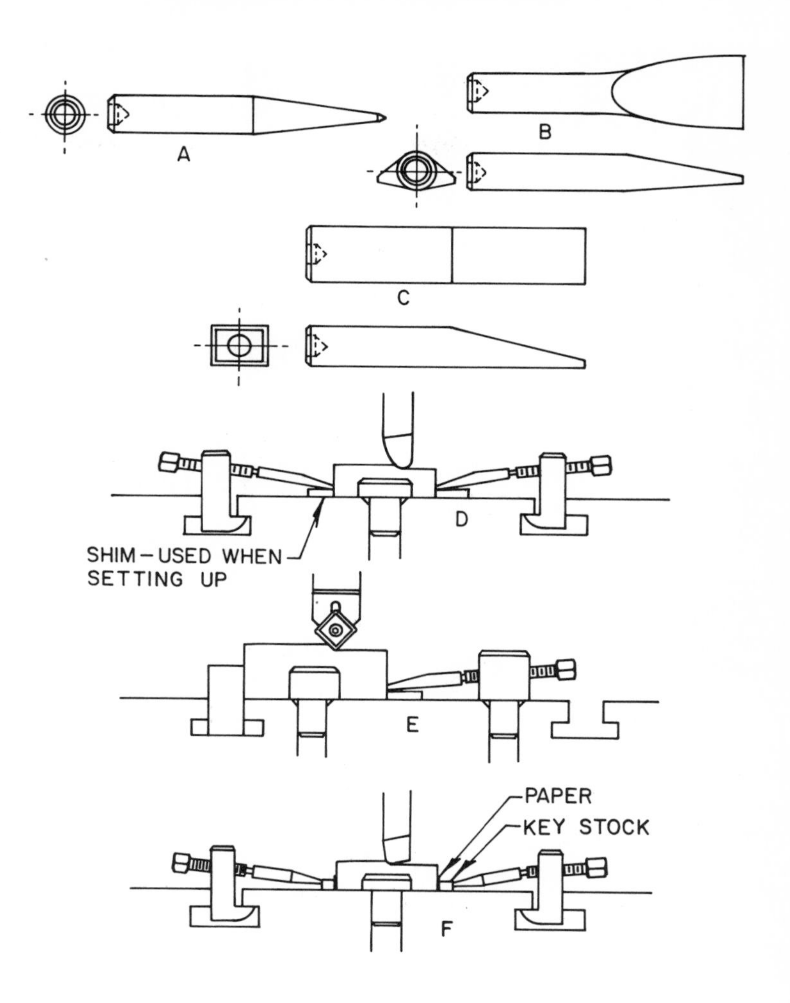

Fig. 3-8. A. Toe dog; B. Chisel point; C. Chisel point; D. Setup for planing thin workpieces; E. Setup for planing slightly thicker workpieces; F. Setup for planing thin workpieces. Key stock and moist paper used to prevent toe dogs from indenting the sides of the workpiece.

of machine tool tables or against the face of an angle plate that has been aligned beforehand. Dial test indicators can be used to indicate on finished surfaces to align the workpiece with respect to the movements of the machine tool table or the path of the cutting tool. The advantage of a previously machined or finished surface should always be utilized when aligning the workpiece on the machine.

Castings and forgings as received from the foundry or the forge shop do not have finished surfaces. With all of the surfaces being more or less

Courtesy of Northwestern Tools, Inc.

Fig. 3-9. Toe clamps used to clamp workpieces to a planer table while performing a straddle planing operation.

rough, it can be a difficult task to align the casting on the machine tool when setting up in preparation for the first machining operation. Although there are exceptions, it is usually best to make a layout on the castings before they are machined, as described in Vol. I, Chapter 3, of *Machine Shop Practice*. In many cases a layout should also be made on rough forgings. A layout will reduce the time and simplify the work involved in setting up the workpiece; it will also lessen the chance of error which might result in having to scrap the workpiece.

As an example, in Fig. 3-10 a rough, unmachined casting is shown being aligned on a planer table. A layout was made on this casting with S_V, S_H, and S_A as setup lines, while M_A and M_B are lines defining the location of surfaces to be machined; they are used as a guide when cutting these

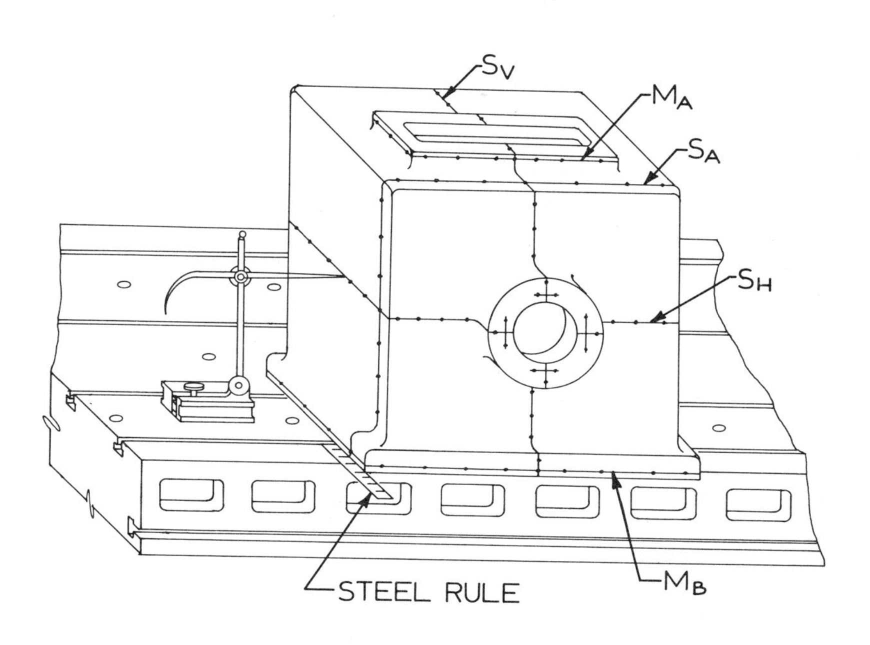

Fig. 3-10. Procedure for aligning a rough casting on a machine tool (planer) table.

surfaces. The first step is to align the casting so that S_V is perpendicular to the side of the planer table, or to its direction of travel; the second step is to align the casting so that S_H is parallel to the top of the planer table. S_V could be aligned by placing a large square against the side of the planer and with a steel rule, measuring the distance between the blade of the square and S_V, both at the front and at the rear of the casting. When both measurements are equal, the casting is aligned. This job can be greatly simplified by a third, or auxiliary, setup line, S_A, on the casting, which is scribed perpendicular to S_V. As shown in Fig. 3-10, the casting can be aligned by simply using a steel rule to measure the distance from the side of the planer table, or from the edge of a T-slot, to S_A. This measurement is made on both sides of the casting, and when these measurements are equal the casting is aligned. S_H is aligned parallel to the top of the planer table by using a surface gage as a height gage, as shown in Fig. 3-10. If necessary, place shims or wedges below the casting until the scriber point of the surface gage can touch any part of the layout line, S_H, on all four sides of the casting. In the setup shown, it will be sufficient if the surface gage scriber point can touch S_H at any point on the two sides and at the back of the casting. Both alignments should be checked before clamping the casting in place and again after it has been clamped.

M_A and M_B (see Fig. 3-10) should not be used to align the casting; they are used as a guide when machining the surfaces that they encompass. In some cases, however, a layout line is intended to serve the dual purpose of a setup line and as a guide for machining (see Fig. 3-14 in *Machine Shop Practice,* Vol. I). Sometimes a setup line is used as a reference line to which a surface is machined; however, the surface is not machined directly to this line. For example, in Fig. 3-10 the face of the boss must be machined. A layout line to which the boss can be machined is not made on the boss because it is difficult to reach the boss with the surface gage scriber point when laying out the casting. In this case the layout line, S_A, is accurately scribed with respect to a predetermined feature of the casting, which also determines the location of the face of the boss. The boss is then machined until the distance between its face and S_A is the required amount, measured with a steel rule and a straightedge placed across the face of the boss.

A rough surface must sometimes be aligned parallel to the top of a machine tool or layout table. A dial test indicator cannot be used for this purpose because the rough surface would cause too great variations in the readings to obtain a representative result. These normal irregularities can be averaged out by the feel of the bent point of the scriber attached to the surface gage, when the surface gage is used as a height gage, as in Fig. 3-11. Set up as shown, the surface gage is adjusted until the bent point of the scriber just touches, but does not scratch, the surface to be aligned. Then the entire surface is scanned, using the feel transmitted by the scriber to determine the parallelism of the surface. The surface will be as parallel to the table as it is possible for a rough surface to be when the surface irregularities are felt to be generally uniform, and when no constantly increasing drag is felt in any direction as the surface is being scanned, with the exception of local high spots.

Stresses of sufficient magnitude to bend large or small metal parts can exist inside metals. These stresses, called residual stresses, are usually balanced by other residual stresses in the workpiece. When the material containing these stresses is removed by a metal-cutting operation, the opposite stresses become unbalanced and distort the part until a balance of stresses is again achieved. The machining operation is one of the causes of additional residual stresses, which further complicates matters. The residual stresses due to machining can, in some cases, be large enough to bend

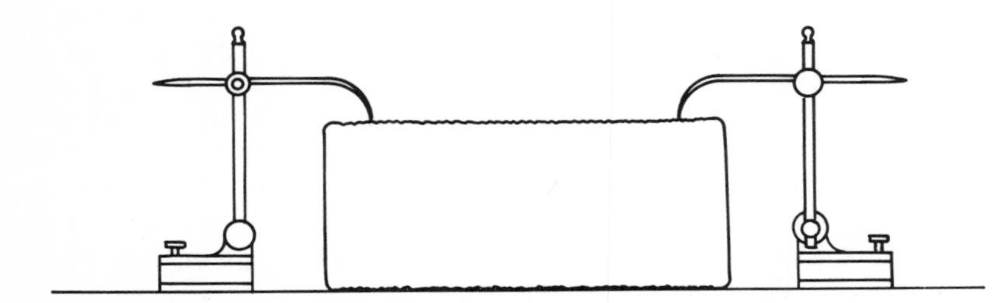

Fig. 3-11. Using a surface gage to test a rough surface for parallelism with the table of a machine tool.

the part. Other causes of residual stresses may be unequal cooling rates during solidification in heavy and thin sections of castings, working the metal beyond its yield strength such as cold rolling or cold drawing, and unequal cooling rates in heat treatment. The residual stresses can be removed from all metals by a stress-relieving heat treatment, although this procedure is not always practical or economical. Furthermore, the heat treatment may cause significant losses in mechanical properties of the metal, which in some cases cannot be tolerated. Therefore, many parts must be machined which have "locked in" residual stresses.

It is not possible to determine how much or, sometimes, in what direction a part will bend because of residual stresses. In some cases, depending on its shape, the part will not distort any measurable amount at all—as for example, a cube-shaped piece of metal. In other instances the distortions caused by residual stresses can result in very large dimensional inaccuracies. When present, their effect can be overcome by using the following procedure. The workpiece should first be rough-machined all over. It is then unclamped completely from the machine and allowed to distort. It is then reclamped, with great care taken not to allow the force exerted by the clamps to bend or spring the part. Metal shims should be placed between the table and the workpiece below the clamps and in other areas that do not seat on the table to prevent springing the part. Finishing cuts are then taken to remove the effects of the distortion and produce a plane or flat surface. Thin workpieces, such as shown in Fig. 3-8, will often distort an excessive amount if the "locked-in" residual stresses are great. Such workpieces must be machined by removing a small layer of metal from one side and then by turning the part over and removing a small layer from the other side. Turn the part several times, planing one side and the other in order to minimize and, if possible, neutralize the distortion caused by the imbalance of the residual stresses when a layer of metal is removed.

Planer Work

The variety of work and the variety of setups that can be done on planers are without limit. The setups and jobs illustrated and discussed in this section suggest ways and means of doing other planer work that might be encountered. In addition to reading the subject matter, it is suggested that a careful study of the illustrations be made, with attention given to such details as the positioning of cutting tools and the setup of the workpiece on the planer.

A die block is being planed on the open-side planer shown in Fig. 3-12. Heavy simultaneous cuts are being taken by the crossrail and side toolheads on the top and side surfaces of the workpiece. The planer in Fig. 3-13 is planing a vertical surface using the vertical feed of the crossrail toolhead. The clapper box is tilted to allow the cutting tool to clear the side of the workpiece as the clapper block is lifted on the return stroke. It is interesting to note that the workpiece in this illustration is a planer table. The cutting tool is cemented carbide, which must take a difficult interrupted cut as it cuts across the cored openings in the side.

Courtesy of the Rockford Machine Tool Company

Fig. 3-12. Planing a die block. The horizontal and vertical surfaces are cut simultaneously.

In Fig. 3-14 the base of a cylindrical grinding machine is being planed with cemented-carbide cutting tools. The ways on the grinding machine table are being planed by the crossrail toolheads and the pads on the side of the table are planed with the side toolhead on the housing. Figure 3-15 illustrates a large casting being machined on an open-side hydraulic planer. The planer bed is made of three sections that are bolted together. Strap clamps are used to hold the workpiece to the table, and stop pins are used

Fig. 3-13. Taking a vertical cut when planing the side of a planer table on a planer.

to prevent it from shifting. Large telescoping braces are positioned against the end of the workpiece. These braces support the work high above the table in order to counteract the thrust of the cutting tool when the top of the workpiece is planed. Two low braces used to provide additional support serve practically the same purpose as stop pins. Telescoping braces are made from heavy pipe into which a telescoping screw is placed. An adjusting nut is used to tighten the screw against the end of the workpiece. The end of the telescoping screw that is placed against the workpiece is fork-shaped. The V-shaped groove, which forms the fork, fits firmly against

Fig. 3-14. Planing a cylindrical grinding machine base.

the corners on the workpiece. Telescoping braces are essential when planing tall workpieces like the one shown. Parts are sometimes "strung up" in tandem fashion, illustrated as in Fig. 3-16. In this way two or more parts can be planed simultaneously with a considerable reduction in time and cost of operation. The parts that are being planed in this illustration are planer clapper boxes or tool boxes as they are also called. The two toolheads are taking simultaneous vertical "straddle" cuts with the left-hand head planing the boss and the right-hand head planing the side of the clapper block seat. When the other side of the clapper box is planed, the left-hand head will cut the inside face while the outside head planes a boss located on the outside face. The clapper boxes on the planer that are doing the cutting are tilted to allow the cutting tools to clear the workpieces on the return stroke. The workpieces are clamped to the table by means of chisel points, and stop pins are placed at the ends of each casting to prevent any movement caused by the thrust of the tool.

A very interesting job is shown in Fig. 3-17. Two steel bars are clamped to the planer table by chisel points and stop pins. Two cemented-carbide cutting tools, a roughing and a finishing tool, are set to cut simultaneously on the special double cutting head. The roughing tool is set behind the finishing tool, even though it leads the finishing tool in taking the first cut

Courtesy of the Rockford Machine Tool Company

Fig. 3-15. Planing a large casting on an open end planer.

as the toolhead is fed from right to left. The finishing tool, a steel-finishing tool similar to that at H in Fig. 3-4 and C in Fig. 1-11, produces the long curled chips which can be seen on the planer table.

In planer work much of the total time required to do a job is spent in setting up the workpiece on the planer table. Naturally, during this setting-up process the planer is necessarily idle and does not cut. To increase the amount of time a planer spends cutting, the planer can have duplex tables, as shown in Fig. 3-18. Work is set up on one table while the part held on the other table is planed. A quick-acting precision latch arrangement moves each table from the cutting to the loading position. The rear table is, of course, loaded from the rear of the planer.

Angular cuts may be taken on a planer by tilting the planer head at the required angle and feeding the tool with the head. An angular cut is being taken with each cross rail head in Fig. 3-19 to cut a dovetail slide on three castings which are "strung up" in tandem fashion on the planer table. The castings are clamped by bolts placed through openings inside the castings. Angular cuts may also be taken by the side heads of the planer.

Modern planers are sometimes equipped with tracer attachments which enable them to machine contours. These attachments are essentially copying attachments that reproduce the form of a master or a template. As an

Fig. 3-16. Planing small planer parts (clapper boxes) in a string or tandem fashion.

example, in Fig. 3-20, contoured surfaces are being planed on two workpieces that are set up end-to-end on the planer table. A template mounted on the planer cross rail controls the vertical movement of the cutting tool as it is fed across the workpiece by the planer feed mechanism (or manually). In this manner the template contour is reproduced on the workpieces.

Estimating the Power Required for Planing and Shaping

Since planing and shaping are very similar operations, the power required to perform these operations can be estimated using the same formulas, which are given below. Tables 5-10, 5-11, 5-12, 5-13, and 5-15, provided in Chapter 5, must be used with these formulas. The tool wear factor, W, is always equal to 1.00 when sharp cutting tools are used; since the cutting tools are normally used after some wear has taken place on the cutting edge, a tool wear factor of 1.30 is normally used.

For inch units only:

$$Q = 12\, V\, f\, d \tag{3-2}$$

Fig. 3-17. Rough and finish planing steel bars simultaneously with a double cutting head.

For SI metric units only:

$$Q = \frac{V}{60} f d \tag{3-3}$$

Fig. 3-18. Duplex planer tables.

Fig. 3-19. Two angular cuts taken simultaneously by the cross rail heads on three castings "strung up" in tandem fashion.

Courtesy of Mimik Tracers Inc.

Fig. 3-20. Contour planing using a tracer attachment to guide the path of the cutting tool.

For either inch or SI metric units:

$$P_c = K_p\, C\, Q\, W \tag{3-4}$$

$$P_m = \frac{P_c}{E} = \frac{K_p\, C\, Q\, W}{E} \tag{3-5}$$

Where:

P_c = Power at the cutting tool; hp or kW
P_m = Power at the motor; hp or kW
K_p = Power constant (See Tables 5-10, 5-11, and 5-12)
Q = Metal removal rate; in.^{3}/min or cm^3/s
W = Tool wear factor
C = Feed factor for power constant (See Table 5-13)
E = Machine tool efficiency factor (See Table 5-15)
V = Cutting speed; fpm or m/min
f = Feed rate; in./stroke, or mm/stroke
d = Depth of cut; in. or mm

Example 3-2:

A 175-200 HB steel casting is to be planed, removing ⅜ inch of stock in one cut. The cutting speed is to be 60 fpm and the feed rate is to be .030 in./stroke. Estimate the power required to take this cut.

$K_p = .78$ (From Table 5-11); $C = .83$ (From Table 5-13); $E = .80$ (From Table 5-15)

$$Q = 12\,V\,f\,d = 12 \times 60 \times .030 \times .375 = 8.1 \text{ in.}^3/\text{min}$$

$$P_c = K_p\,C\,Q\,W = .78 \times .83 \times 8.1 \times 1.3 = 6.8 \text{ hp}$$

$$P_m = \frac{P_c}{E} = \frac{6.8}{.8} = 8.5 \text{ hp}$$

CHAPTER 4

Milling Machine Construction

The milling process is used to produce a variety of surfaces by using a circular-type cutter with multiple teeth or cutting edges which successively produce chips as the cutter rotates. These cutting edges are located on the periphery and also often on the face of the cutter. The shape of the milling cutter and the path that it takes determine the shape of the surface produced. The function of the milling machine is to provide the means of holding and rotating the milling cutter, of holding and feeding the workpiece into the cutter, and of transmitting the necessary power to cut the metal at the desired rate. The great variety in the size and shape of parts, as well as advances in technology, has led to the development of many different types and styles of milling machines. They are extensively used at the present time for machining both large and small workpieces.

Classification of Milling Machines

Milling machines, as a class of machine tools, are very versatile. They are capable of machining economically one or two piece lots as well as parts on a large-volume production basis. The inherent advantage of the milling process is the circular cutter, which is economical in first cost and which has a high metal removal rate since it can bring a large number of cutting edges into the cut in a relatively short space of time. This advantage has led to the design of a large variety of machine tools primarily for the purpose of milling. These can be classified as follows:

1. Knee-and-column- (or column-and-knee-) type milling machines
2. Fixed-bed-type milling machines
3. Planer-type milling machines
4. Special milling machines.

Further classifications of the milling machines are made on the basis of the type of control used, such as numerically controlled milling machines, and on the basis of the position of the spindle—i.e., horizontal or vertical.

Knee-and-Column-Type Milling Machines

The knee-and-column-type milling machine is also frequently called a column-and-knee-type milling machine. As a very versatile machine capable of performing a wide variety of operations, it is extensively used in machine shops and tool and die shops. The distinguishing characteristic of knee-and-column-type milling machines is that the table can be moved three directions in space, as can be seen in Fig. 4-2. Thus, a workpiece

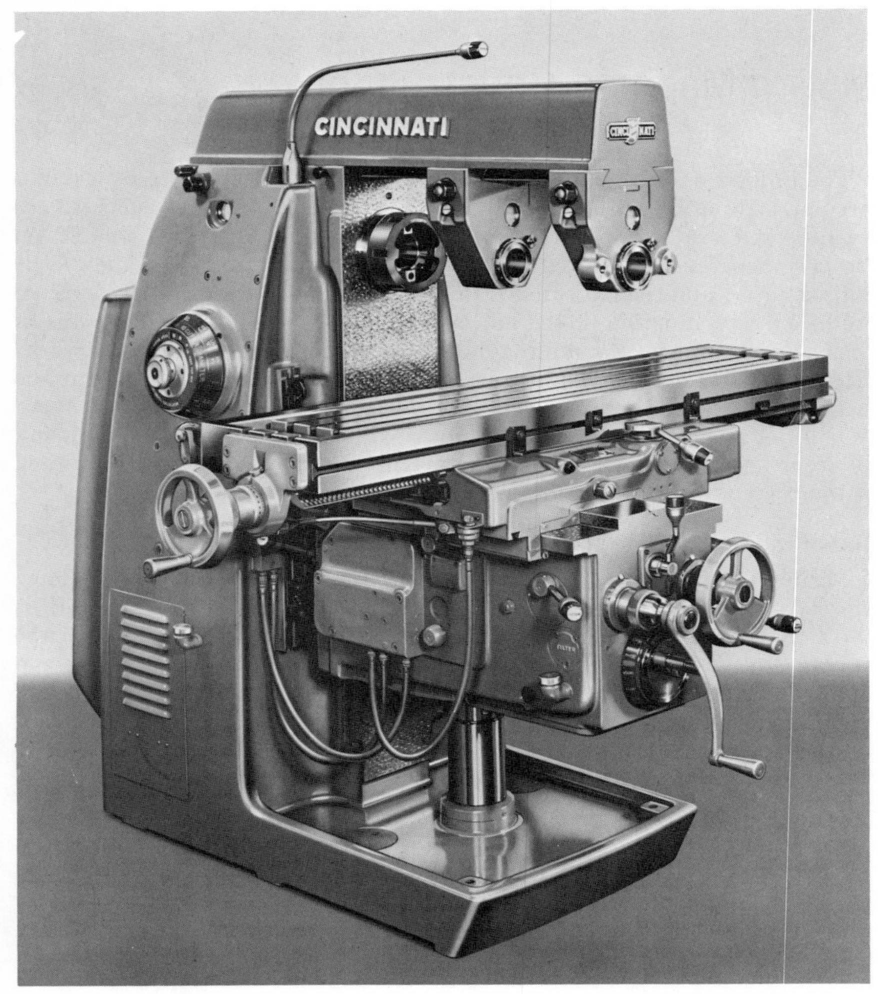

Courtesy of Cincinnati Milacron

Fig. 4-1. Plain knee-and-column-type milling machine.

mounted on the table of the milling machine can be easily and accurately positioned relative to the spindle which contains the cutting tools, and many different kinds of operations can be performed. There are several different types of knee and column milling machines which are described here.

Plain Knee and Column Milling Machine. A plain knee and column milling machine is illustrated in Figs. 4-1 and 4-2. The main supporting frame of knee and column milling machines is the column, including the base. The front face of the column is a machined and precision hand-scraped surface which supports and guides the knee. Being a precision bearing, the column face should be kept free of scratches, nicks, and other damage which would impair its accuracy.

The knee slides up and down on the column face. This movement is actuated by the elevating screw seated on the base of the machine. The

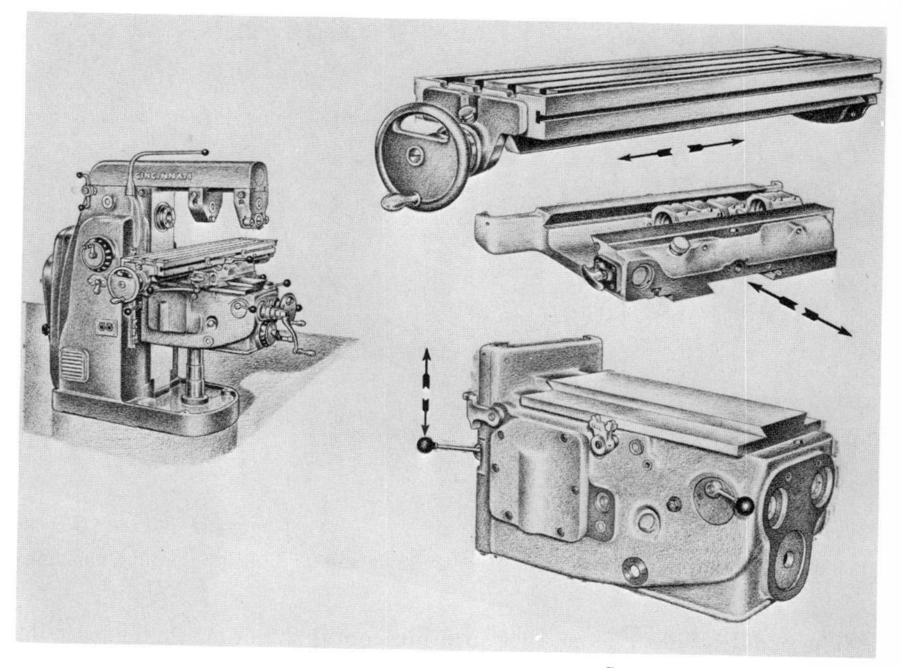

Courtesy of Cincinnati Milacron

Fig. 4-2. Principal components of a plain knee-and-column-type milling machine.

elevating screw is a telescoping screw—i.e., one screw working inside another screw—which can be seen, in part, toward the front of the machine in Fig. 4-3. The elevating screw can be turned manually to raise or to lower the knee by turning the large hand crank on the front of the knee (Fig. 4-1). A micrometer dial behind this crank permits accurate vertical adjustments to be made. The vertical feed can be power-actuated on all but light-duty knee and column milling machines. The top surfaces of the knee are precision-machined and hand-scraped to form a bearing surface on which the saddle can slide. This surface should be protected from damage; i.e., wrenches and other tools should not be placed on it.

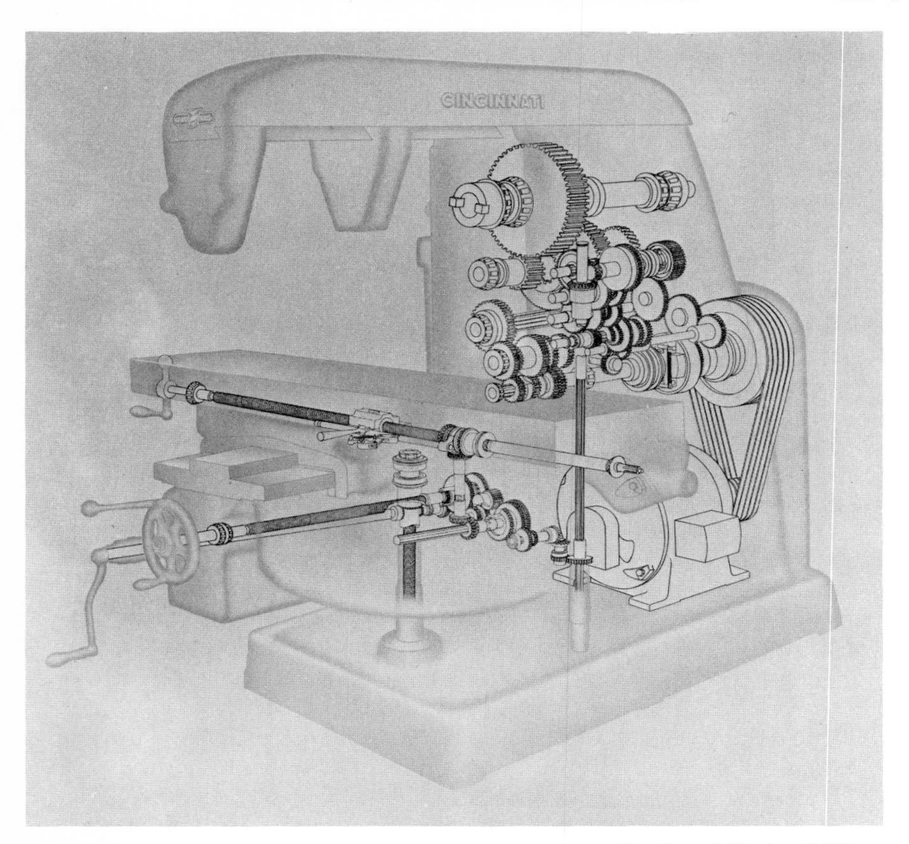

Courtesy of Cincinnati Milacron

Fig. 4-3. Illustration of power train of a plain knee-and-column-type milling machine.

The saddle slides on the knee in a horizontal direction that is parallel to the spindle. This feed direction is called the *transverse feed.* The handwheel in front of the knee (Fig. 4-1) is used to actuate the transverse feed by turning the transverse feed screw, which can be seen in Fig. 4-3. A micrometer dial behind the handwheel permits accurate adjustments to be made in this direction. On all except light-duty machines, the transverse feed can be actuated by power. The surfaces on the top of the saddle are precisely machined and hand-scraped to form a slide that is perpendicular to the lower slide, which works on the knee.

The milling-machine table moves over the upper slide of the saddle in a direction perpendicular to the axis of the spindle. This feeding movement, called the *longitudinal feed,* is actuated by a handwheel or crank located at the end of the table. Thus the longitudinal feed screw or lead screw (Fig. 4-3) is caused to rotate in a stationary nut attached to the saddle. A micrometer dial behind the longitudinal feed handwheel or crank permits accurate adjustments of the table to be made in this direction. A power longitudinal feed is also available on most milling machines. The tabletop is a precision reference and locating surface used to locate workpieces, vises, angle plates, and other work-holding fixtures in a plane that is parallel to the axis of the spindle. Since the accuracy of the operation being performed often depends upon the condition of the tabletop, it should not be abused by carelessly clamping rough work surfaces to it or by placing heavy tools on it. A nick could result in a scraped workpiece. T-slots milled into the top of the table are used to retain or to anchor bolts for clamping workpieces and attachments to the table. The T-slots, which are machined parallel to the lengthwise direction of the table, can be used as reference surfaces when workpieces and accessories are set up on the table.

The power train of a knee and column milling machine is shown in Fig. 4-3. The electric drive motor is located in the column of the machine, in a position where the heat that it generates has the least effect on the precision-sliding movements. The power is transmitted from the motor through a large multiple V-belt to a shaft containing a clutch that starts and stops the spindle. Then the power goes directly to the spindle through a selective speed gear transmission built into the column of the machine. This transmission provides the different spindle speeds that are available on the machine. The power for the three table feeds is taken off to the vertical shaft by means of a pair of bevel gears. The vertical shaft has a long spline upon which a gear slides. This gear actuates a gear train that provides all of the power table feeds. A selective change gear box located either in the column or in the knee provides for the different feed rates. In modern milling machines the feed rate is given in terms of *inches per minute* of table travel. In addition to the three directions of power feeds, a rapid traverse in three directions is usually available. The rapid traverse, used to rapidly position the table, spares much manual effort.

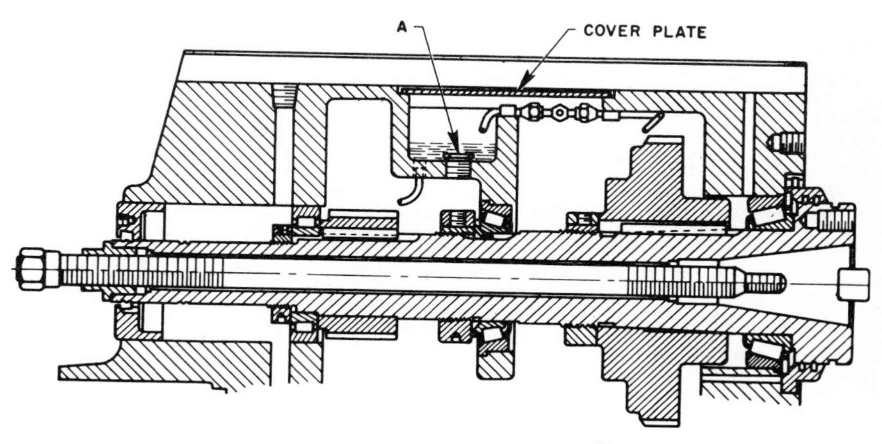

Courtesy of Cincinnati Milacron

Fig. 4-4. Section through a milling-machine spindle.

The spindle is mounted on precision machine-tool grade antifriction bearings which are housed in the column, as seen in Fig. 4-4. The spindle is hollow. A draw-in bar, shown being placed through the spindle, is used to hold milling cutters, arbors, and adaptors in the spindle. The inside of the spindle nose has an American Standard Milling Machine Spindle Nose Taper which is 3.500 inches per foot for all sizes. This is a self-releasing taper, used to locate the cutters, arbors, or adaptors. Two lugs or keys on the face of the nose act as drivers for driving the cutters. There are four standard sizes (No. 30, 40, 50, and 60) of milling-machine spindle nose tapers, the size used depending upon the size of the machine. The outside diameter of the spindle nose is ground to a very close tolerance. Large face milling cutters, which are mounted on the outside of the spindle nose, are held to the spindle by four socket-head cap screws which fit into threaded holes in the face of the spindle.

The overarm is mounted on the top of the column (Fig. 4-2) or inside two bored holes located on the top of the column (Fig. 4-6). The purpose of the overarm is to support and align the arbor supports and various other attachments. The overarm in Fig. 4-2 is a heavy rectangular casting with a built-in vibration dampener. Some milling machines can be equipped with a rectangular overarm with a built-in overhead spindle or vertical head as shown in Fig. 4-30. This permits the machine to be used as either a vertical or a horizontal milling machine. If desired, the horizontal and

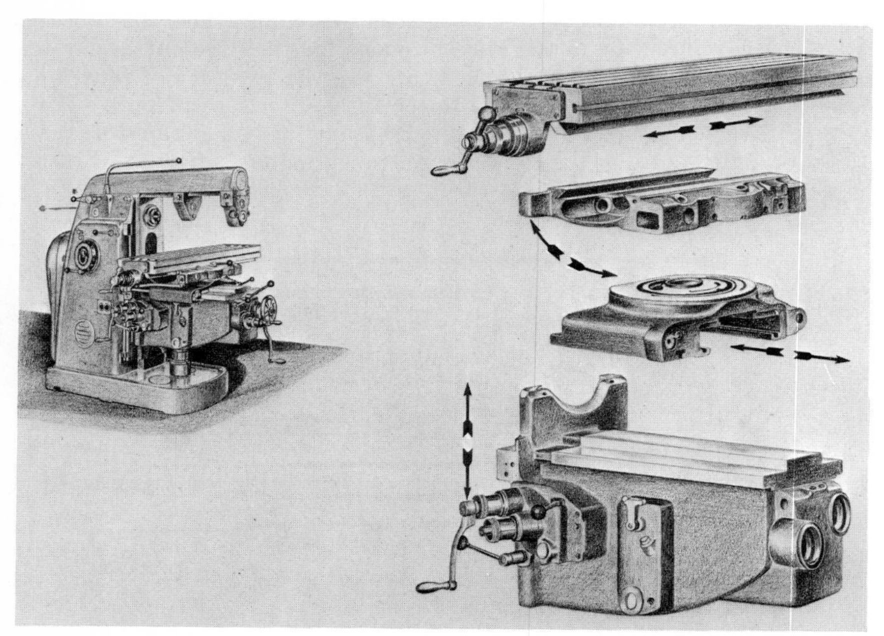

Courtesy of Cincinnati Milacron

Fig. 4-5. Principal components of a universal knee-and-column-type milling machine.

Courtesy of the Brown & Sharpe Manufacturing Company

Fig. 4-6. Universal column and knee milling machine milling helical flutes in a milling cutter blank.

overhead spindles could be used simultaneously. Another overarm design used on modern milling machines is the double overarm shown in Figs. 4-6 and 4-7. The two solid cylindrical bars that form the overarm are contained in two holes bored in the upper part of the column.

Arbor supports mounted on the lower side of the overarm are used to hold the outer ends of long arbors on which milling cutters are mounted. They can be seen mounted in position in Figs. 4-1 through 4-7. Two styles of arbor supports can be seen in Fig. 4-7. The one mounted on the inside has a larger bearing and must be used when a Style B arbor (Fig. 4-18) is used. This arbor support is also used on the large bearing collar of a Style A arbor. The support shown on the outside in Fig. 4-7 can only support the outer end of a Style A arbor. The inside arbor support in Fig. 4-7, having

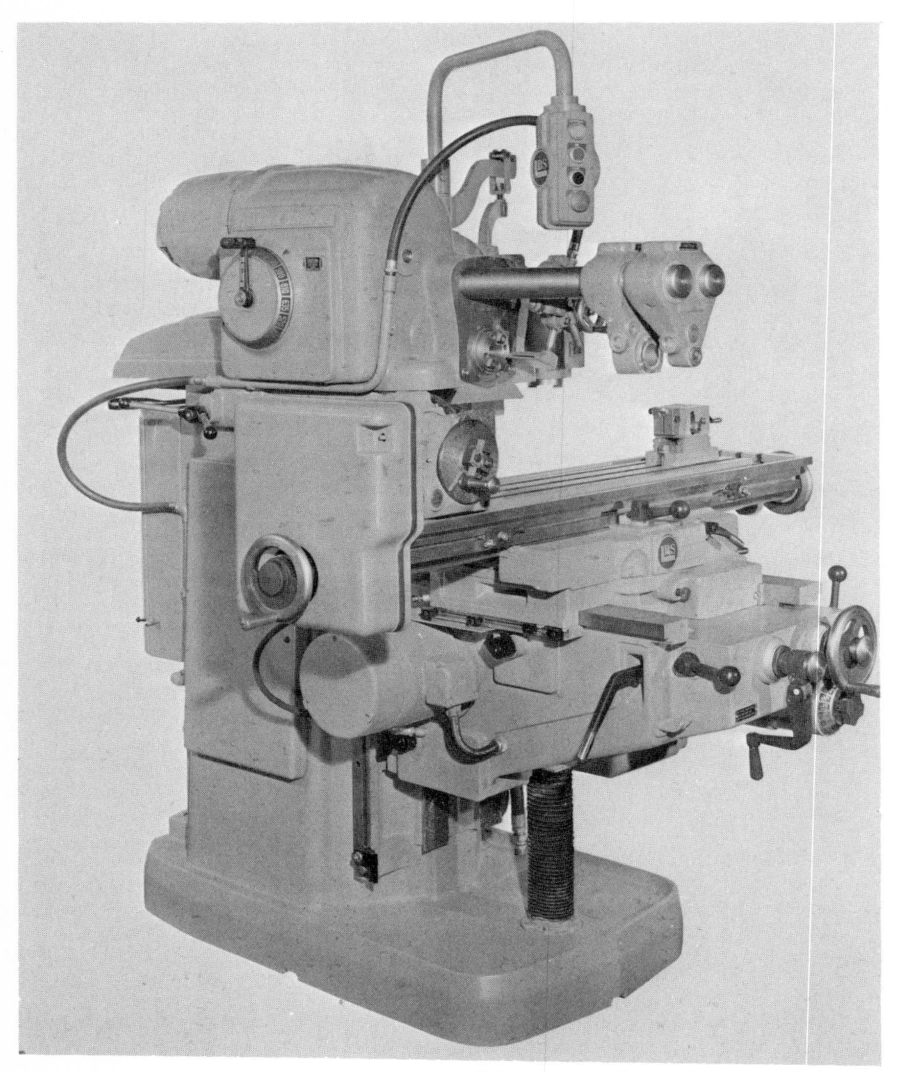

Courtesy of the Brown & Sharpe Manufacturing Company

Fig. 4-7. Sliding-head-type universal column-and-knee-type milling machine.

a larger bearing, can support a heavier cutting load and can be positioned anywhere along the length of the arbor. Thus, this arbor support can be placed as close to the cutter as possible to provide maximum support. The arbor support on the outside in this illustration can only support a Style A arbor at its extreme outer end, where the arbor is reduced in diameter. This type of arbor support, however, has the advantage of providing more clearance below the arbor so that workpieces, vises, clamps, etc., can be brought closer to the arbor. This is often an advantage when parts are milled with a cutter having a small diameter.

Universal Knee-and-Column-Type Milling Machine. A universal knee-and-column-type milling machine, shown in Figs. 4-5 to 4-7, is very similar in construction to a plain knee-and-column-type milling machine. The principal difference is in the construction of the saddle and in the addition of swiveling table housing (Fig. 4-6), which allows the table to be

Courtesy of Cincinnati Milacron

Fig. 4-8. Vertical knee-and-column-type milling machine.

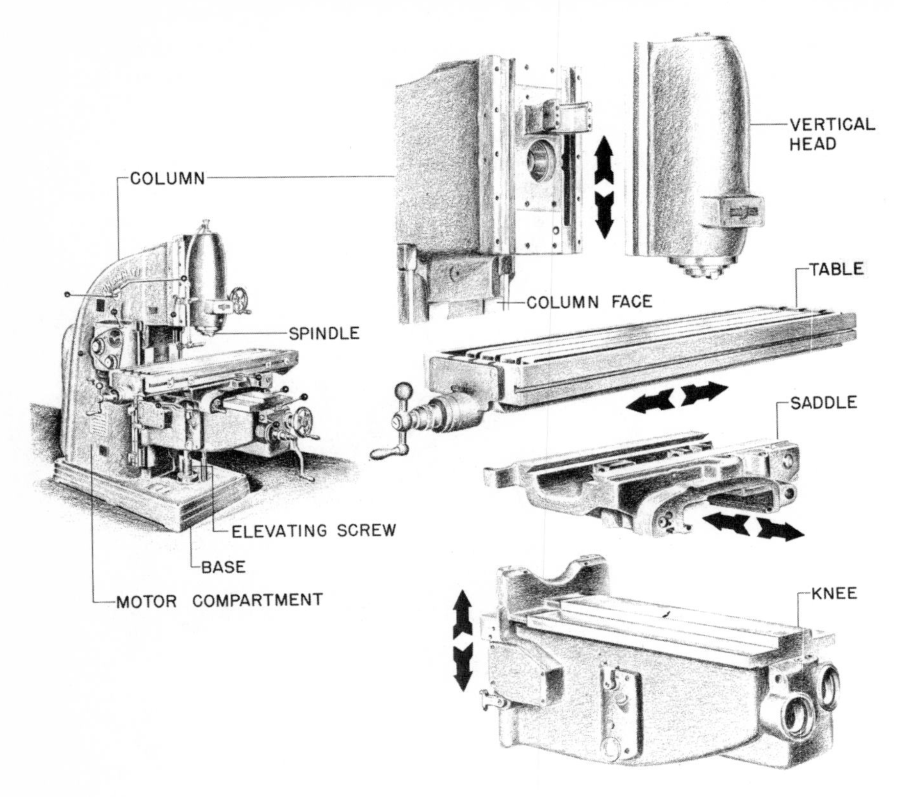

Courtesy of Cincinnati Milacron

Fig. 4-9. Principal components of a vertical knee-and-column-type milling machine.

swiveled at an angle to the axis of the spindle. Generally the tables of universal milling machines will swivel approximately 45 degrees in both directions from the normal position perpendicular to the axis of the spindle. This feature extends the operating range of the milling machine by making it possible to cut helical grooves with cutters mounted on the milling-machine arbor. In Fig. 4-6, a universal column-and-knee-type milling machine is shown cutting a helical flute in a milling cutter blank. A universal spiral index head is used to rotate the milling cutter blank as it is fed into the cutter by the longitudinal table feed. The index head is also used to space the flutes around the circumference. Note that the table is swiveled to the helix angle of the flute. A sliding-head-type universal milling machine is shown in Fig. 4-7, where the spindle head is mounted on a dovetail slide on top of the column. Although the spindle is rigidly held in a fixed position inside the spindle head, it can be moved toward or away from the face of the column by moving the entire spindle head on the dovetail slide. In this way the nose of the spindle can be brought as close to the workpiece as possible so that arbor-mounted cutters can be placed close to the nose. Thus the rigidity of the setup is improved. The sliding head can also reach over the table to allow spindle-mounted cutters such as end mills and face mills to reach surfaces on the workpiece that cannot be positioned close to the column. The electric drive motor and the entire speed change gear train are mounted on the sliding head. A

Courtesy of Bridgeport Machines, Inc.

Fig. 4-10. Toolroom-type vertical milling machine equipped with a milling head and a vertical shaping head.

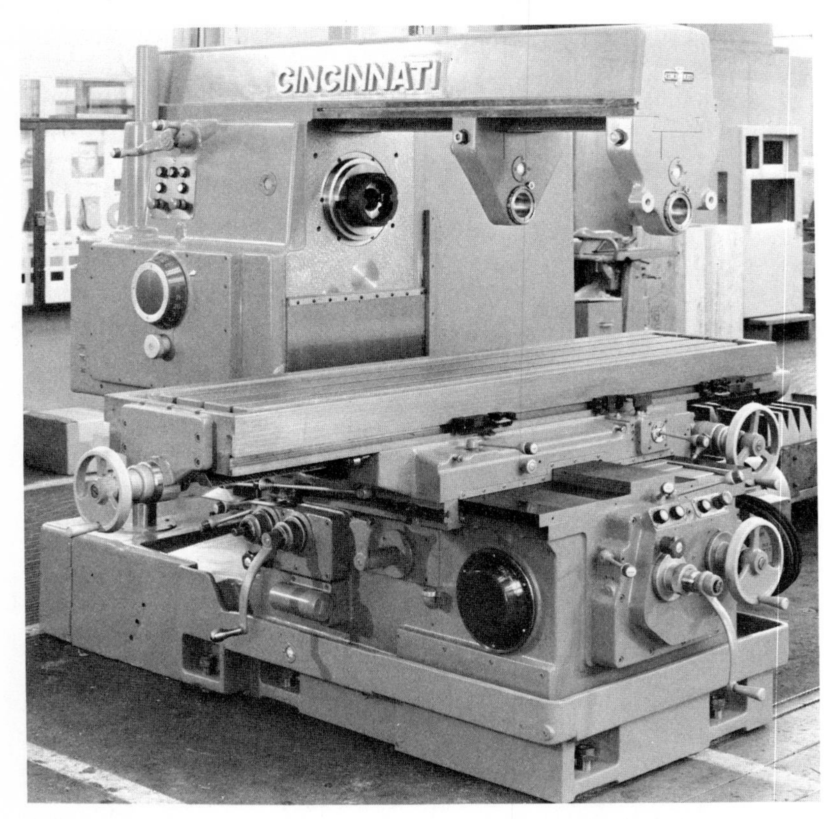

Courtesy of Cincinnati Milacron

Fig. 4-11. Fixed-bed-type milling machine for general-purpose work.

separate motor, which can be seen extending out from the left side of the knee, provides the drive for the power feeds.

Vertical Knee-and-Column-Type Milling Machine. A vertical-type knee and column milling machine is shown in Figs. 4-8 and 4-9. The spindle of the vertical milling machine is located vertically, parallel to the face of the column, and perpendicular to the top of the table. The vertical head can be moved up and down by hand or by power feed. This machine is especially suitable for performing operations which require the use of end mills and face milling cutters such as milling dies, cutting profiles, milling molds and for locating and boring holes in jigs and fixtures. The vertical milling machine shown in Fig. 4-10 is a light-duty milling machine for which there is a wide range of application in toolroom work as well as for performing other light-duty milling operations. Two heads are mounted on a ram which can be swiveled to bring either head into the operating position over the table. Several different heads are available. The machine in Fig. 4-10 has a milling head and a vertical shaping head shown at the rear of the machine. Both heads can be set in angular positions. The spindle of the milling head is of the quill-type construction which can be moved up and down like a drill press spindle. The table is fed by hand, although a longitudinal power feed can be obtained as an attachment. The sensitivity and ease of handling of this machine make it especially adaptable for doing fine intricate work.

Fixed-Bed-Type Milling Machines

The distinguishing characteristic of all fixed-bed-type milling machines is the absence of the knee construction found in the knee-and-column-type milling machines. Some fixed-bed-type milling machines are designed for general purpose work while others are intended for high-production work. A general-purpose fixed-bed-type milling machine is shown in Fig. 4-11.

Courtesy of Cincinnati Milacron

Fig. 4-12. Spindle carrier of fixed-bed milling machine.

The table can move in a longitudinal and in a transverse direction. The vertical position of the spindle with respect to the table is obtained by moving the spindle carrier up and down along the side of a column which is called the headstock. This arrangement is more clearly evident in Fig. 4-12. The rigid fixed bed beneath the table and the saddle minimizes deflections and permits very heavy cuts to be taken. Also available is a vertical-spindle milling machine with the two-way movement of the bed.

Courtesy of Cincinnati Milacron

Fig. 4-13. Cincinnati Hydro-Tel milling machine.

A Cincinnati Hydro-Tel milling machine is shown in Fig. 4-13. This is a vertical fixed-bed-type milling machine; however, the table on this machine rests directly on the bed and moves in the longitudinal direction only. The transverse relationship of the spindle to the table is obtained by a massive cross slide onto which the spindle carrier is mounted. The spindle carrier moves up and down on the cross slide in order to position the spindle vertically with respect to the top of the table.

The unique feature of the Hydro-Tel milling machine is its sensitive controls in relation to its size. Such machines can perform conventional vertical milling operations; however, they are very frequently equipped with a tracer attachment for contour and profile milling. An example of this work is found in Fig. 4-14, where a section of a drawing die is being reproduced from a plaster of Paris master. The Hydro-Tel milling machine can also be obtained equipped with a numerical control system.

A fixed-bed production-type milling machine is shown in Fig. 4-15. The table of this machine rests directly on the bed and moves in the longitudinal direction only. Designed to perform repetitive operations, the machine

Courtesy of Cincinnati Milacron

Fig. 4-14. Reproducing the contour of a plaster of Paris master onto a section of a drawing die.

is set up to follow a certain cycle of table movements repeatedly. The spindle can be made to rotate continuously or to stop when the table feed is stopped. A spindle carrier provides the movement required to position the spindle in a vertical direction. The spindle is carried in a quill which can be moved in and out of the spindle carrier so that the milling cutter can be positioned in a transverse direction.

Planer-Type Milling Machines

A planer-type milling machine is shown in Fig. 4-17. These are very large milling machines which do a class of work similar to that done on a

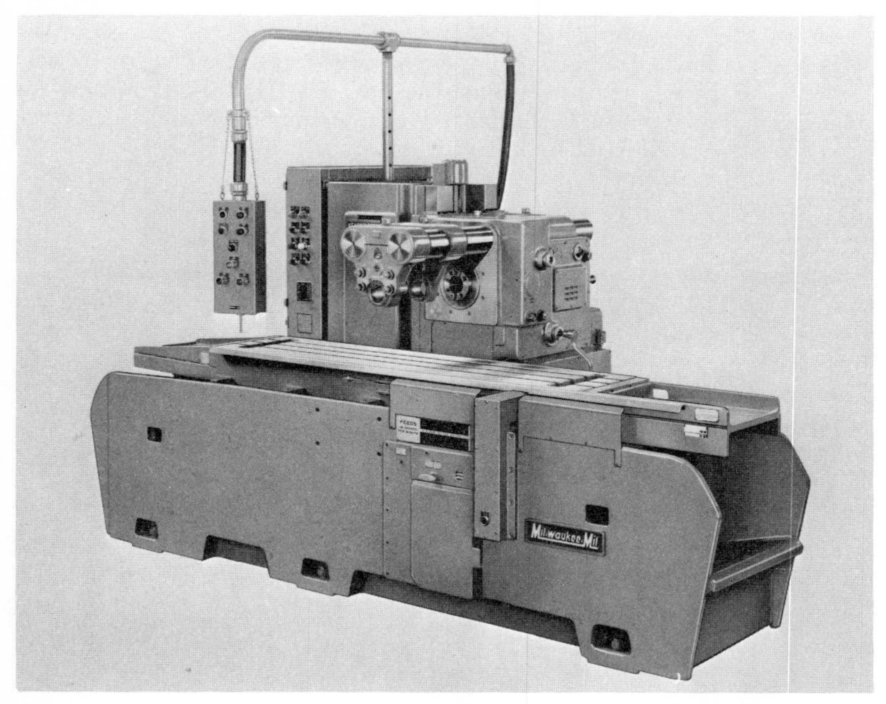

Courtesy of the Kearney & Trecker Corporation

Fig. 4-15. Fixed-bed production-type milling machine.

planer. The construction of the planer-type milling machine resembles that of a planer. The table is mounted on the bed. Astride the bed and the table are two vertical housings which, like the planer, are connected together at the top by a top brace. The crossrail is mounted on the housing, and one or two cutter heads are mounted on the crossrail. Additional side cutter heads are mounted on one or both of the housings. Although slab milling cutters are sometimes used on these machines, the majority of the cutting is done with large-face milling cutters. The planer-type milling machine in the illustration has two tables. On one table a cut can be taken on the workpiece, while simultaneously a part is set up on the other table. The noncutting time of the machine can thereby be reduced. When a very large workpiece is to be machined, the two tables can be attached to each other and used as one.

Special Milling Machines

There is a great variety of special milling machines. Usually these machines have been designed to machine one particular part or a group of very similar parts. For example, the special milling machine illustrated in Fig. 4-17 is used to mill a family of large screws. It will handle any screw from 4 to 24 inches in diameter and up to 24 feet long. The helix is rough cut with the large cutter above the part and finished with an end mill, which is held in a horizontal spindle located behind the part. A control system developed by the builder of this machine operates the work-driving fixture rotation in exact relation to the table travel so that a number of different and highly accurate right- and left-hand leads can be

Courtesy of The Ingersoll Milling Machine Company

Fig. 4-16. Planer-type milling machine.

Courtesy of Cincinnati Milacron

Fig. 4-17. Special helix milling machine.

cut. Special milling machines are extensively used in the automotive and aircraft industries.

Milling-Machine Accessories

Milling-machine accessories not only extend the usefulness of the machine; some are essential to the performance of certain milling operations. Some of the most important accessories will be described in this section.

Milling-Machine Arbors. Milling-machine arbors are used to hold milling cutters in the milling machine. The cutters are driven by a key on the arbor and are held in place by collars which fit over the arbor. The collars are clamped together by a nut on the end of the arbor. There are two types of arbors, Style A and Style B, which are shown in Fig. 4-19. The Style A arbor has a small-diameter cylindrical surface at its end. This cylindrical surface fits in the arbor support bushing; and its small size permits the end of the arbor support to be made smaller. Thus the surface of the work, or of the vise which holds the work, can be brought closer to the arbor than would be possible with the Style B arbor. This has an advantage when a small-diameter milling cutter is used which necessitates that the workpiece be brought up close to the arbor. The Style B arbor has one or more collars of a comparatively large diameter which are used in the arbor support. The large arbor support bushings in which these collars are placed provide a better support for the arbor and allow higher cutting

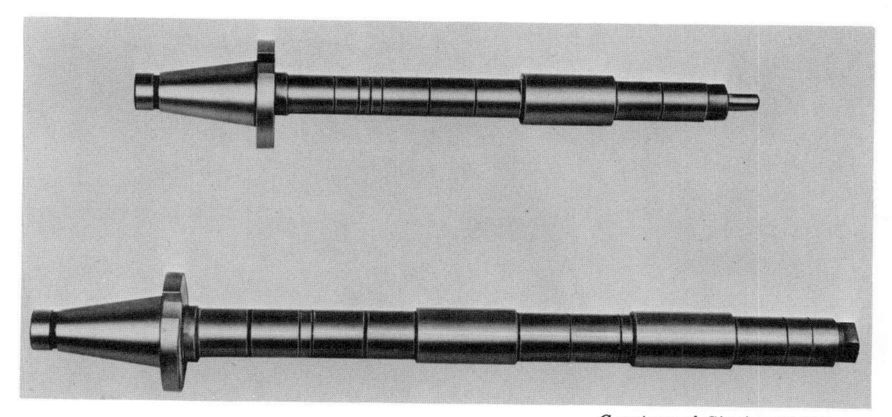

Courtesy of Cincinnati Milacron

Fig. 4-18. Standard milling-machine arbors. Upper view, Style A arbor; lower view, Style B arbor.

loads to be supported. Milling-machine arbors are made in various lengths and in standard diameters of $\frac{7}{8}$, 1, $1\frac{1}{4}$, and $1\frac{1}{2}$ inches. The shank of the arbor has an American Standard Milling Machine Taper of a size corresponding to the taper in the spindle of the milling machine on which it is

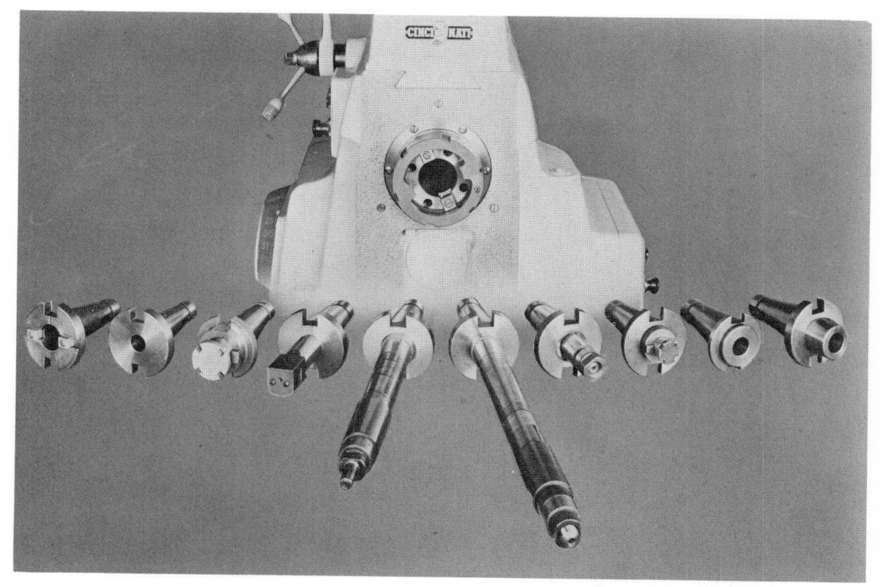

Courtesy of Cincinnati Milacron

Fig. 4-19. A variety of arbors, stub arbors, and adaptors that are held in the milling-machine spindle.

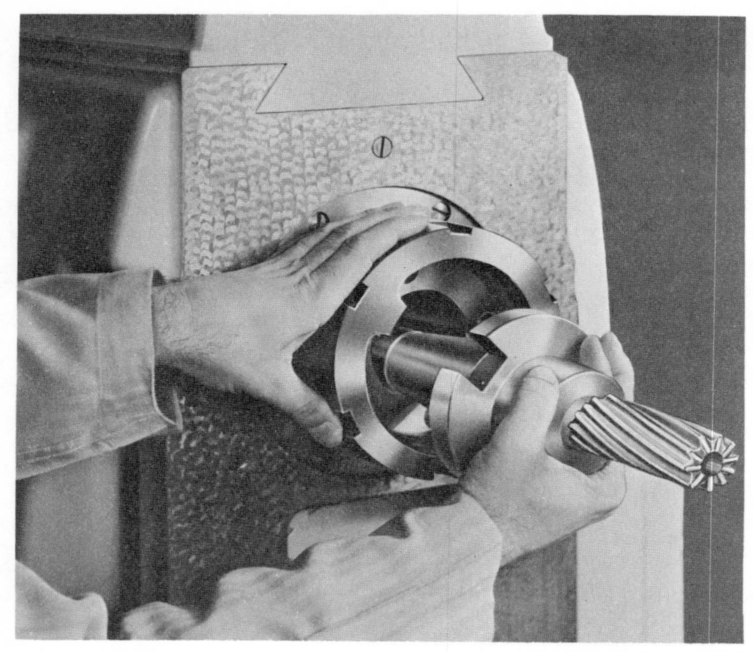

Courtesy of Cincinnati Milacron

Fig. 4-20. The Cincinnati Arbor-Loc® spindle-nose arrangement for providing a rapid change of tooling.

to be used. The arbor is located by the taper and is driven by the keys or lugs on the spindle nose which fit the slots on the arbor flange. The arbor is held in the spindle by the draw-in bolt.

Spindle-Nose Tooling. Examples of the variety of spindle-nose tooling available are shown in Fig. 4-20. Such tooling is required to hold milling

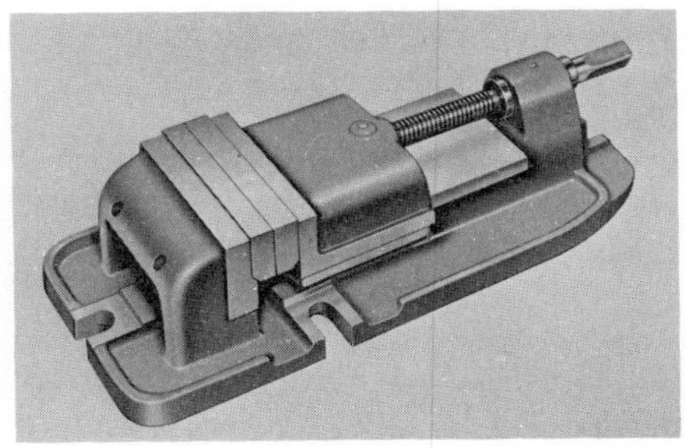

Courtesy of Cincinnati Milacron

Fig. 4-21. Plain milling-machine vise.

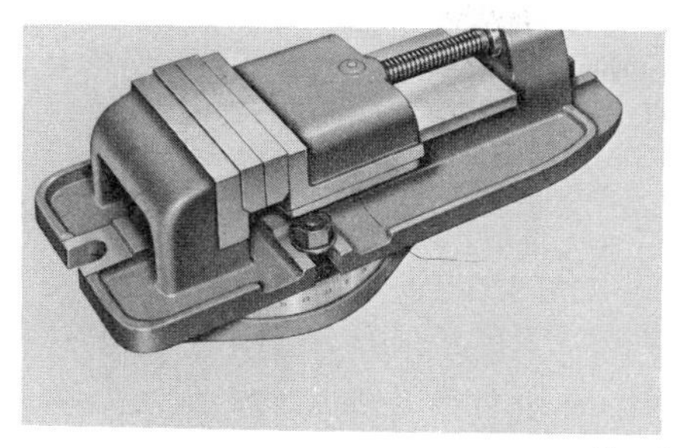

Courtesy of Cincinnati Milacron

Fig. 4-22. Swivel-type milling-machine vise.

cutters, drills, reamers, drill chucks, collet chucks, and other cutting tools. The amount of tooling required depends upon the type of work to be done on the milling machine. Fig. 4-21 shows an end milling cutter which is held in an adaptor being inserted into the spindle of a milling machine. This type of spindle nose provides a rapid method of changing such tooling, which is very helpful when a job requires a variety of operations such as drilling, boring, and milling without the setup's being changed. To change

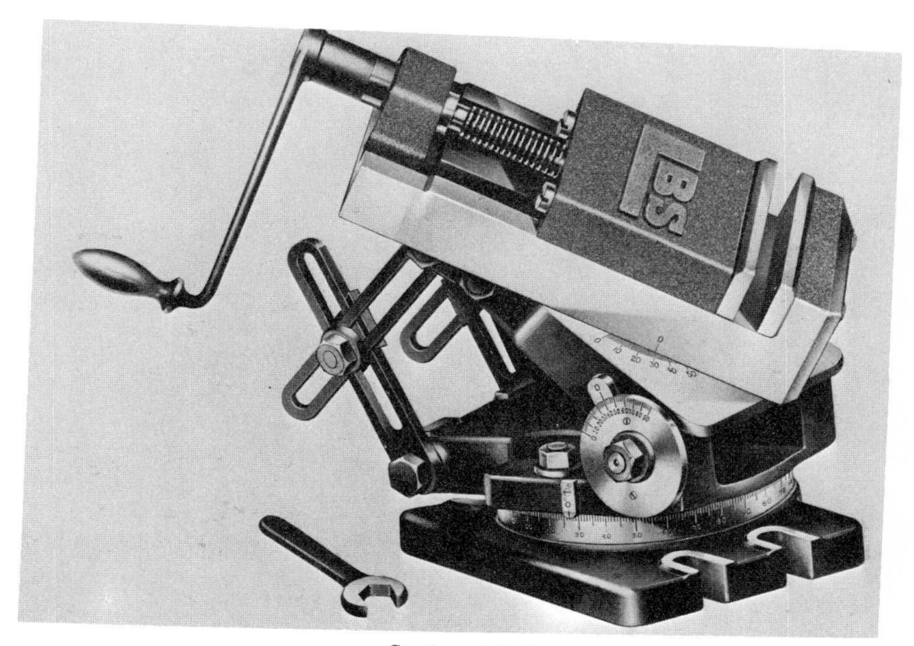

Courtesy of the Brown & Sharpe Manufacturing Company

Fig. 4-23. Toolmakers' universal milling-machine vise.

a tool the adaptor nut is loosened with a spanner wrench and the cutting tool removed. The next tool is inserted; and after the adaptor nut is snugged up by hand, the tool is then tightened with the spanner wrench.

Milling-Machine Vises. Milling-machine vises are an important accessory to the milling machine. These precision tools, which should be handled with care, are used to hold and to locate the workpiece in the correct position for a wide variety of milling-machine operations. A plain

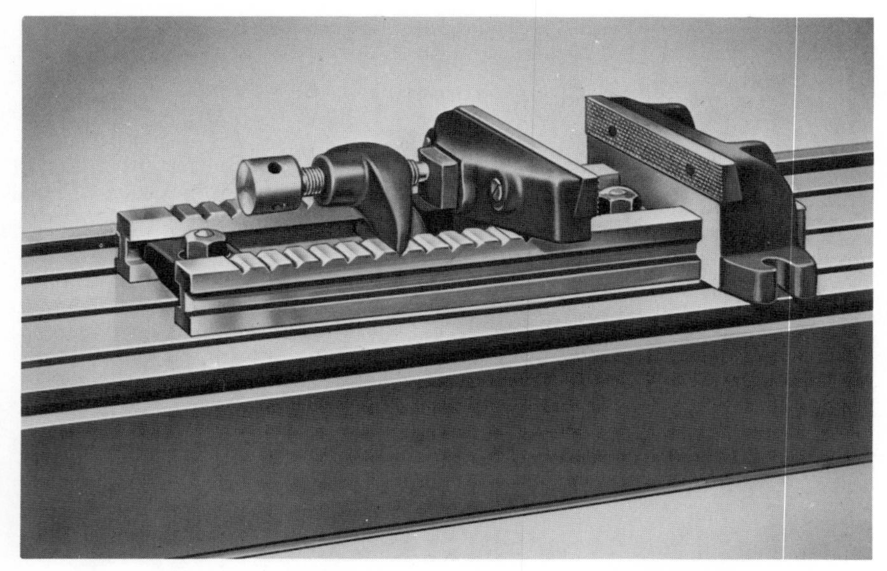

Courtesy of Cincinnati Milacron

Fig. 4-24. Cincinnati All-Steel vise for holding rough workpieces.

milling-machine vise is shown in Fig. 4-21. Two keys located on the bottom surface of the vise fit into the t-slots on the milling-machine table so that the vise jaws are positioned with respect to the milling-machine spindle. The keys may be attached to position the jaws parallel to the spindle or they may be removed and reattached to position the jaws perpendicular to the spindle. Swivel vises, Fig. 4-22, have the same construction as plain vises with the exception of the swivel plate under the body. The swivel vise may be easily set to any angle with respect to the spindle. A toolmakers' universal vise, Fig. 4-23, allows the body of the vise to be rotated in three planes so that the work can be held for compound angles to be machined. This vise can be used as a swivel vise when it is not holding a part at a compound angle. The construction of all of the aforementioned vises is similar to that of the shaper vise. The solid jaw is accurately made perpendicular to the base and the work-seating surfaces inside the vise. It can be used as an accurate locating surface when a workpiece is set up in the vise. The movable jaw should not be used as a locating surface.

Figure 4-24 illustrates a Cincinnati All-Steel Vise which is used to hold rough workpieces. Rough castings and forgings cannot be held firmly enough in the conventional smooth-jaw precision vises, especially when

Courtesy of the Brown & Sharpe Manufacturing Company

Fig. 4-25. A universal vertical milling attachment used to mill an angular surface.

heavy roughing cuts are to be taken. The serrated jaws of the all-steel vise are hardened and can hold the workpiece firmly in place. The movable jaw can swivel, thereby adapting itself to the irregularities in rough castings or forgings and to clamping surfaces that are not parallel.

Vertical Milling Attachments. Vertical milling attachments (Fig. 4-25) greatly extend the range of work that can be done by horizontal-spindle milling machines. They are generally used to hold end milling cutters, face milling cutters, and sometimes arbor-mounted cutters which are held in

Courtesy of the Volstro Manufacturing Co., Inc.

Fig. 4-26. Rotary milling head.

place with stub arbors. Made in several different styles and sizes, some vertical milling attachments are designed to take heavy cuts with face milling cutters; others are designed to have a high spindle speed for use with smaller end milling cutters. The universal vertical milling attachment in Fig. 4-25 is milling an angular surface on a part held in a swivel vise. The two graduated swivels of this attachment allow the spindle to be

Courtesy of the Volstro Manufacturing Co., Inc.

Fig. 4-27. Multi-angle milling head milling a sharp 90-degree internal corner.

set accurately by one-half degree increments to any desired angle in any plane. When a vertical milling attachment is frequently needed, a milling machine equipped with an independent overhead spindle permanently mounted on the overarm can be obtained. Figure 4-29 shows such a machine. This overhead spindle, which eliminates the need for a vertical

Courtesy of the Kearney & Trecker Corporation

Fig. 4-28. Slotting attachment.

attachment, has the advantage of being readily available for use. Some smaller high-speed vertical milling attachments have a quill-type spindle.

Rotary Milling Head. The rotary milling head shown in Fig. 4-26 is attached to the spindle of a vertical milling machine. It is used to round corners and to make special tools, dies, and molds requiring intricate combinations of angles and radii, angles tangent to radii, and to make one radius blend into another. The head will rotate 360 degrees; it has a vernier type scale graduated in 5-minute intervals. A micrometer dial attached to the cross-feed screw allows precise settings of the cross slide to be made.

Multi-Angle Milling Head. A multi-angle milling head is shown in Fig.

4-27. This head can be used to machine internal corners to a sharp 90-degree configuration with respect to three intersecting surfaces; i.e., two sides and a bottom surface. It is an especially useful attachment for making molds and certain types of dies. Other types of angular milling heads for milling internal surfaces, such as keyways, are also available.

Slotting Attachment. A slotting attachment, shown in Fig. 4-28, imparts a reciprocating motion to a single-point cutting tool. It is primarily used to cut internal surfaces such as keyways, splines, slots, and other internal geometric configurations.

Circular Milling Attachment. A circular milling attachment is also shown in Fig. 4-28. Used to hold and rotate the workpieces which are being slotted or milled, circular milling attachments are also called circular tables and rotary tables. They impart a rotary motion to the workpiece. Some circular milling attachments can be provided with an index plate, as seen in Fig. 4-28, which is used to divide a part into predetermined segments or to turn the table an exact angular distance. Power table feeds can also be provided on some circular milling attachments.

Dividing Head. A universal dividing head is shown mounted on the

Courtesy of Cincinnati Milacron

Fig. 4-29. A universal milling machine equipped with a universal dividing head, a long and short lead attachment, and an overarm mounted independent overhead spindle.

table of the milling machine in Fig. 4-29. It may be employed to index a part at a given angle or to index a part into a specified number of divisions such as gear teeth. The tailstock, which can be used to hold the workpiece between centers, is seen mounted on the opposite end of the table from the dividing head. A support for slender workpieces is shown placed between the tailstock and the dividing head. The dividing head is also called an index head.

Helical Milling Driving Mechanism. A helical milling driving mechanism is shown mounted at the rear of the dividing head in Fig. 4-29. This mechanism causes the spindle of the dividing head to rotate while the table advances, thereby producing a helix on the workpiece. Enclosed inside of the housing are change gears that make an exact relationship possible

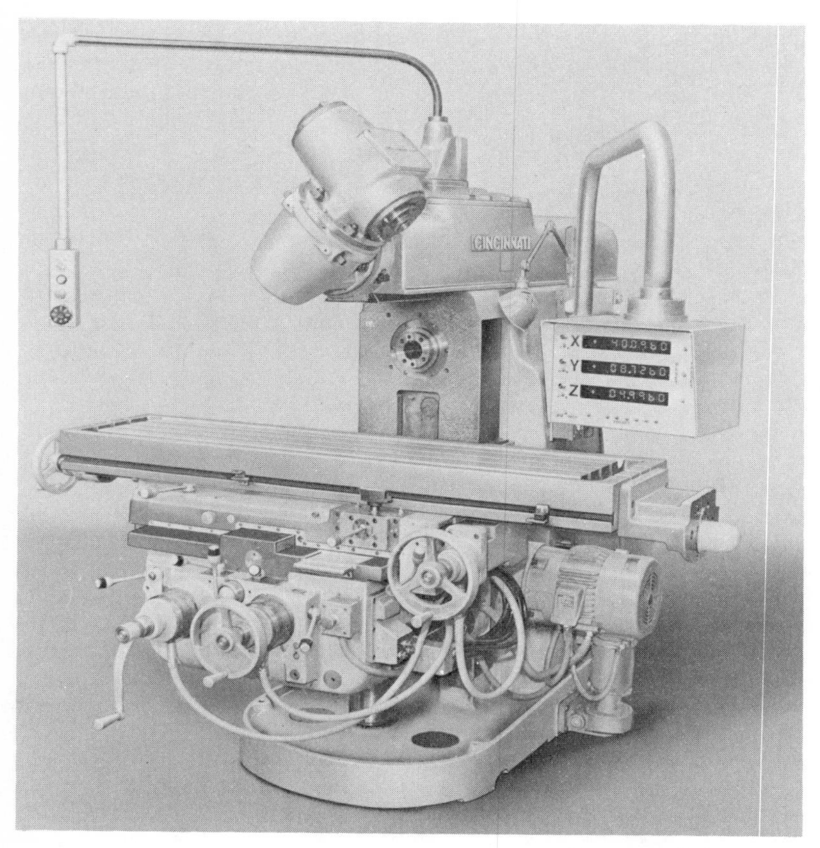

Courtesy of Cincinnati Milacron

Fig. 4-30. Digital position readout with inch and metric capability, attached to a milling machine with independent overhead spindle.

between the rotation of the dividing head and the table feed. The driving mechanism illustrated in Fig. 4-29 is a short- and long-lead mechanism with a range of leads from .010 to 1,000 inches.

Digital Position Readout. Shown in Fig. 4-30 is a digital position readout which will display numerically the position of the milling machine table in two horizontal (X and Y) directions and in the vertical (Z) direction, relative to a reference position selected by the operator. The readout may be switched to display either inches or millimeters to a resolution of .0005 in. or 0.01 mm. Thus, when equipped with this attachment, parts dimensioned in either customary inch or metric units can be machined. The numerical display shows the exact position of the table at a glance and the time needed to make precision measurements can be reduced when machining parts on a milling machine.

Backlash Eliminator. The backlash eliminator, shown in Fig. 4-31, is used to enable the milling machine to take climb milling cuts. This attachment, which must be built into the machine, eliminates the backlash due to the clearance between the longitudinal table feed screw and the nut in which it works. The backlash eliminator consists of two racks and a spur gear. The rack at the left is actuated by turning a thumb screw which extends out through the saddle casting. This causes the nut engaged to the rack and the gear to rotate. The gear actuates the other rack, which in turn rotates the second nut. Threads on the outside of the nuts cause them to move closer together, bringing the internal threads in the nuts into contact with the sides of the feed screw threads so that there is no backlash. The thread in one nut rests on one side of the feed screw thread and the thread in the other nut rests against the other side of the feed screw thread. The force exerted by the thumb screw is limited by a

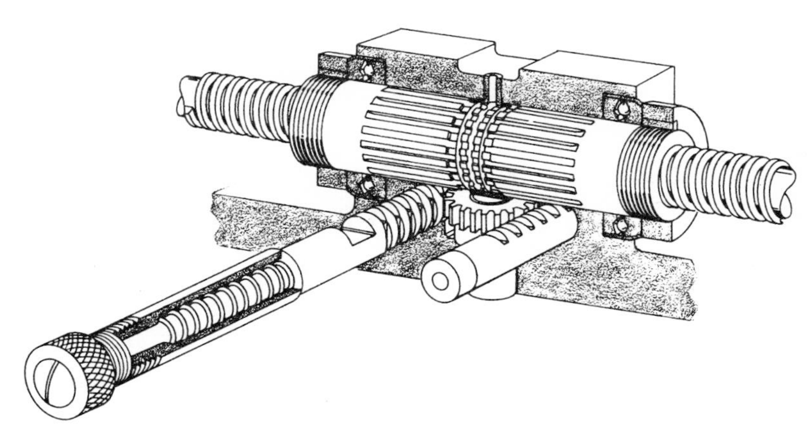

Courtesy of Cincinnati Milacron

Fig. 4-31. Backlash eliminator for eliminating the backlash between the feed screw and the nut.

spring, preventing the nuts from excessive binding against the feed screw. Turning the thumb screw to release the force will cause the two nuts to be released from the side of the feed screw thread as the feed screw is rotated.

CHAPTER **5**

Milling Cutters

Milling cutters can roughly be classified as arbor mounted cutters, end milling cutters, and face milling cutters. There are various standard types, styles, and sizes within each classification from which a selection can be made to suit most applications. Special milling cutters are made to meet some high-production requirements or to fulfill a particular need. It should be pointed out to avoid confusion, that milling cutters are also called *mills* and an end milling cutter is often called an *end mill.* The word "mill" is also used, of course, to designate the actual milling operation, e.g., "to mill a surface."

Equally important in becoming familiar with the types of milling cutters that are available is an understanding of the operating conditions at which each should be used. To use these cutters successfully, it is necessary to know the proper speed and feed rate at which they should be operated. Where maximum productive output is required, it is necessary to be able to estimate the power required to take the cut. These and other such topics are treated in this chapter.

Milling Cutter Materials

The cutting edges of milling cutters are primarily made of high-speed steel or from cemented carbides. Many milling cutters are made from solid high-speed steel, including most general-purpose end mills and arbor-mounted cutters. End milling cutters and arbor mounted cutters made for high production often have cemented carbide cutting edges. Face milling cutters also may have high-speed steel or cemented carbide cutter blades, or cemented carbide indexable inserts.

High-Speed Steel. High-speed steels are a group of highly alloyed tool steels characterized by their ability to retain a high level of hardness and wear-resistance at temperatures up to approximately 1100 F (590 C), where other tool steels will soften and fail. When annealed, high-speed steel can be machined into the shape of the cutting tool and then hardened again. High-speed steels are very deep-hardening; thus, they can be sharpened many times without a significant loss in hardness.

There are many different types of high-speed steels which are roughly classified into two types: those having only tungsten as a principal alloying element (designated by the letter "T" preceding the steel number), and those having tungsten and molybdenum as principal alloying elements (designated by the letter "M"). The alloying elements in high-speed steel are carbon (C), tungsten (W), molybdenum (Mo), chromium (Cr), and vanadium (V). Some high-speed steels also contain cobalt (Co) which raises the hardness and wear-resistance at high temperatures, but also increases its brittleness and heightens the tendency of the cutting edge to chip. Most milling cutters are made from the type M2 high-speed steel. For milling high-temperature alloys and other metals having a hardness of approximately 370 HB, and harder, milling cutters are made from M42, M43, or T15 high-speed steel, which has been developed for this purpose. Some milling cutters are made of high-speed steel that is a product of the particle metallurgy process, whereby atomized molten steel is recombined under extreme pressure.

Cemented Carbides. Cemented carbides are harder than high-speed steel and can retain their hardness at a higher temperature. As a result, much faster cutting speeds can be used when milling with cemented carbides. Cemented carbides, however, are more brittle and less shock-resistant. For this reason, greater care must be used in designing and specifying cemented carbide milling cutters on the job.

A most important factor in using cemented carbides is the selection of the correct grade for the application. There are four general classes of carbides; straight tungsten carbides, crater-resistant (steel cutting) carbides, titanium carbides, and coated carbides. Within each class there are many grades. Since the grades made by each carbide producer will differ from those made by another, the actual carbide producer should be consulted when selecting a grade. In general, straight tungsten carbides are recommended for milling gray cast iron, ferritic malleable iron, austenitic stainless steels, high-temperature alloys, copper, brass, bronze, aluminum, zinc die-cast alloys, and plastics. Crater-resistant grades of carbides and coated carbides are recommended for milling plain carbon steels, alloy steels, alloy cast iron, pearlitic malleable iron, nodular iron, Monel metal, martensitic and ferritic grades of stainless steel, and tool steels. In addition to these materials, coated carbides are used to mill gray cast iron. Coated carbides have a thin coating of titanium carbide, titanium nitride, or aluminum oxide bonded to a cemented carbide insert, called the *substrate.* Since grinding will remove the coating, they can be obtained only in the form of indexable inserts. Where applicable, coated carbides can operate at faster cutting speeds, usually 20 to 40 per cent, but sometimes as much as 50 per cent faster. Solid titanium carbide is very brittle and has a low resistance to mechanical shock; therefore, its use in milling is restricted to finishing inserts mounted on the face of some indexable insert face milling cutters. Carbide grades for milling should have good thermal shock-resistance since the cutting edge on the cutter heats and cools rapidly on entering and leaving the cut. For this reason the use of a coolant is not recommended when milling with carbides, except under certain special conditions.

Cermets. Cermets are a mixture of approximately 70 per cent aluminum oxide and 30 per cent titanium carbide. Face milling cutters with cermet indexable inserts are used on some very-high-speed milling operations.

Arbor-Mounted Milling Cutters

Arbor-mounted milling cutters are characterized by a hole in the center into which the milling-machine arbor is inserted. The arbor hole is very accurately ground to the required size. A keyseat in the hole engages a key on the arbor which is used to drive the cutter. All arbor-mounted cutters have teeth on their periphery, and some of them have additional teeth on their side faces.

Plain Milling Cutters. Plain milling cutters have teeth only on their periphery and are used to mill plane, or flat, surfaces. Milling a flat surface with a plain milling cutter is called *slab milling,* and for this reason, plain milling cutters are also called "slab milling cutters." Two light-duty plain milling cutters are shown in Fig. 5-1, views A and B. Light-duty plain milling cutters less than ¾ inch (19 mm) wide, as in view A, have straight teeth; they are used to mill narrow plane surfaces and shallow slots. When the light-duty plain milling cutters are ¾ inch wide, and over (view B, Fig. 5-1), they have helical teeth and are used exclusively to perform slab milling operations. The helix angle of the teeth is 18 to 20 degrees. Heavy-duty slab milling cutters, such as shown in Fig. 5-1, view C, are the most useful plain milling cutters, since good results can readily be obtained when taking either heavy- or light-duty cuts. The helix angle of the teeth on these cutters is 45 degrees, causing more teeth to be cutting at a given time, and providing a smooth entry and exit into the cut. The helical teeth also tend to produce a smooth finish on the milled surface. A helical, or high-spiral, plain milling cutter is shown in Fig. 5-1, view D. The helix angle of the teeth of this cutter is 52 degrees. This large helix angle causes the cutter to absorb much of the cutting load as end thrust; for this reason there will be less tendency of the cutting forces to impel the cutter and the workpiece away from each other, and there will be less tendency to spring back and dig into the work on entering and leaving a cut. Helical plain milling cutters are recommended particularly for taking light cuts on frail, thin-walled parts and intermittent cuts on fragile pieces.

Side Milling Cutters. Side milling cutters are arbor-mounted cutters having teeth on the periphery and on one or both sides. The side teeth are ground slightly concave to prevent them from dragging against the workpiece. While the peripheral teeth do most of the metal cutting, the side teeth take a light scraping cut to prevent the cutter from binding. Side milling cutters are used to cut slots such as keyseats, to perform side milling operations on the sides of parts, and to cut right-angle corners or shoulders. Frequently a pair of side milling cutters separated by a spacing collar is mounted on an arbor to cut two sides simultaneously to a predetermined width. This operation is called *straddle milling* and is shown at B, C, and E, in Fig. 5-2.

Straight-tooth side milling cutters are shown in views A and B, Fig. 5-2.

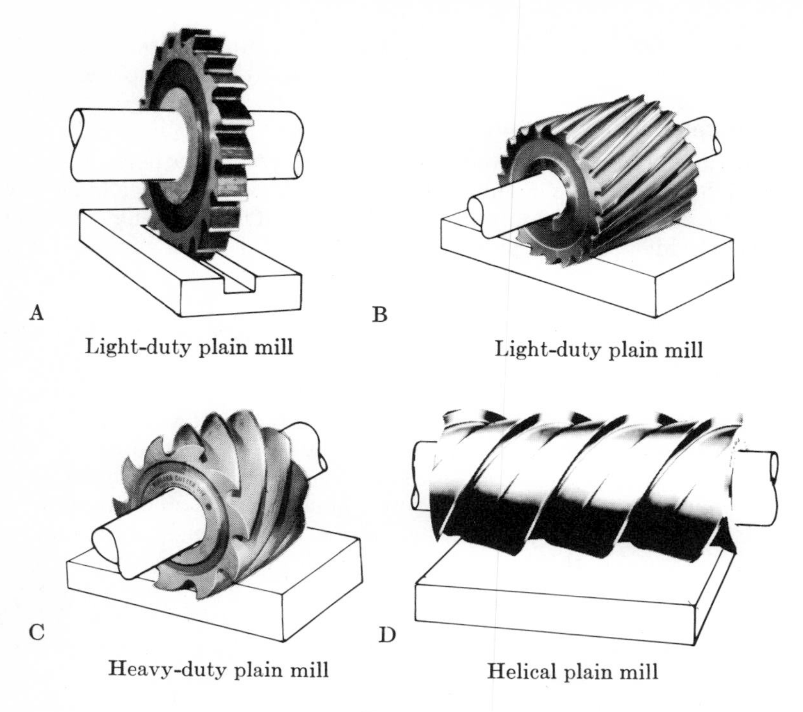

A Light-duty plain mill

B Light-duty plain mill

C Heavy-duty plain mill

D Helical plain mill

Fig. 5-1.

The peripheral teeth are straight while the side teeth are at an angle with respect to the axis of the cutter. They are at an angle so that the rake surface will form a positive rake angle on the peripheral teeth. Although used for all side milling cutter operations, they are especially recommended for milling thin walled sections, where the maximum number of teeth should be engaged. Straight-tooth side milling cutters are sometimes converted into form milling cutters by grinding the teeth to the required outline, such as a corner radius. Half side milling cutters, shown in Fig. 5-2, view C, have teeth on one side only. The 15-degree helix angle on the peripheral teeth is an advantage, providing a rake angle on the side teeth and a smoother cutting action than straight teeth. They cannot be used to cut slots, but can perform side milling and straddle milling operations. Staggered-tooth side milling cutters, shown in Fig. 5-2, views D, E, and F, have a 10-degree positive axial rake or helix angle, right- and left-hand on alternate teeth. Alternate side teeth are removed so that only those side teeth remain that have an effective positive axial rake angle. These are smooth and free-cutting cutters that are used to perform all side milling cutter operations. They are especially recommended for milling deep slots and for deep face milling operations. A most important feature of the design of these cutters is that they can be interlocked; i.e., they can be mounted on the arbor, side by side, with the peripheral teeth over-

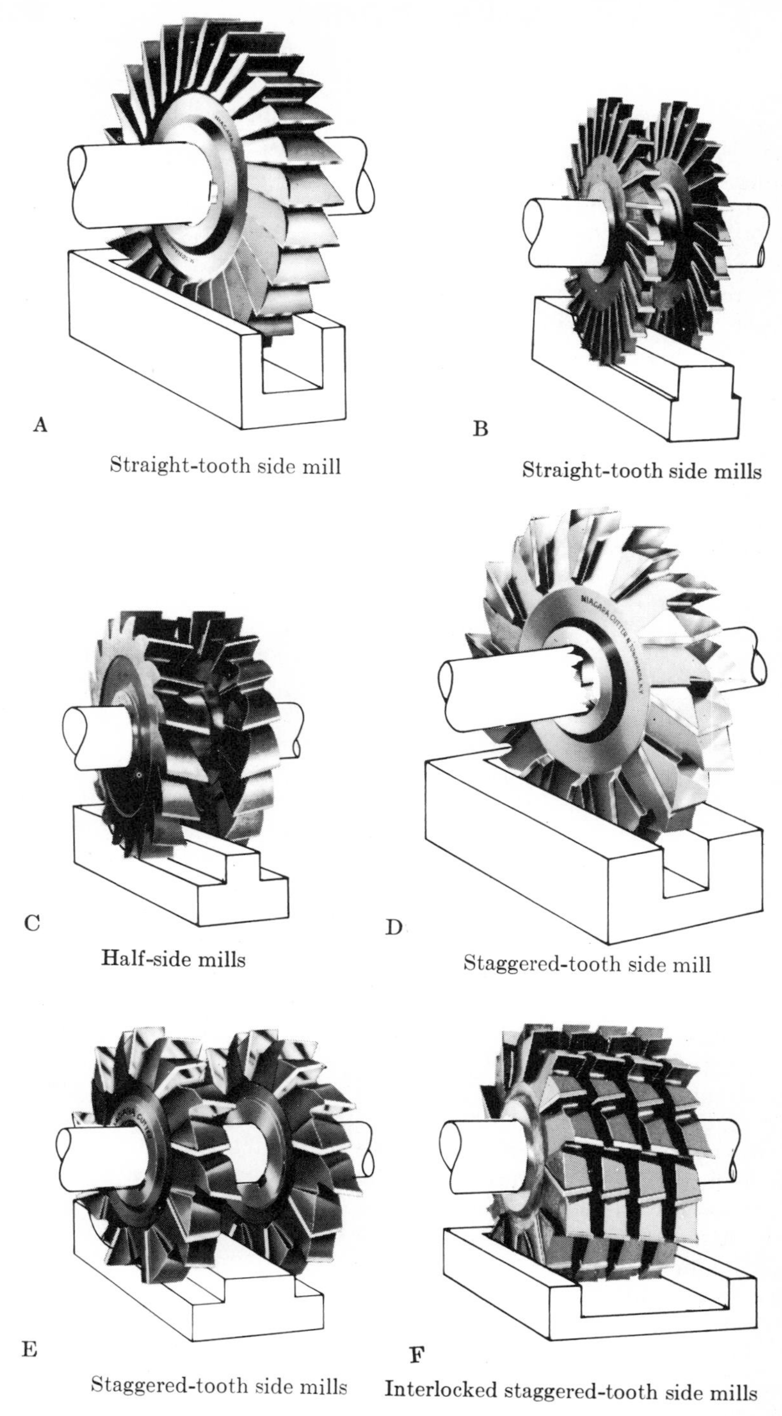

A Straight-tooth side mill

B Straight-tooth side mills

C Half-side mills

D Staggered-tooth side mill

E Staggered-tooth side mills

F Interlocked staggered-tooth side mills

Fig. 5-2.

lapping so that a plane surface will be cut by the peripheral teeth. Interlocking staggered-tooth side milling cutters are shown milling a wide slot in view F. By selecting the right combination of cutters and by the use of shims spaced between the cutters, or by slightly modifying the width of a cutter, the slot width can be closely controlled.

Angle Milling Cutters. A single-angle milling cutter is shown at A in Fig. 5-3. The side teeth are ground slightly concave and all of the teeth have zero-degree axial and radial rake angles. They are standard in 45- and 60-degree included angles. Single-angle side milling cutters are used to mill notches, angles, and chamfers. A double-angle milling cutter is shown at B in Fig. 5-3. They are made with 45-, 60-, and 90-degree included angles and have zero-degree axial and radial rake angles. Double-angle cutters are used for milling threads, notches, serrations, angles, and chamfers. Single-angle milling cutters with shanks are also available. These cutters are held in collet chucks and are not a type of arbor-mounted cutters.

Metal Slitting Saws. Plain metal slitting saws, view C, Fig. 5-3, have teeth on the periphery only and are ground concave on the sides to avoid binding in the cut. They range in diameter from 2½ to 8 inches (63.5 to 203 mm) and from 1⁄32 to 1⁄8 inch (0.8 to 3.2 mm) in width. These saws are designed for slitting and cutting-off operations of all kinds. Straight-tooth and staggered-tooth metal slitting saws having side teeth are available in widths ranging from 1⁄16 to ¼ inch (1.6 to 6.4 mm). A staggered-tooth metal slitting saw is shown in view D, Fig. 5-3. These saws are designed for deep slotting operations and for sinking-in type cuts.

Form Milling Cutters. Form milling cutters have teeth with a profile that reflects the profile of the surface to be milled. There are two kinds of form milling cutters, "profile ground," and "form relieved." Profile ground form milling cutters have teeth which are similar to the teeth on plain and side milling cutters, on which the relief angle is ground on the land behind the cutting edge. These cutters are sharpened by grinding this relief angle while maintaining the required profile; an often difficult job. The teeth on form relieved cutters (views E and F, Fig. 5-3) are cut on an engine lathe equipped with a back-off attachment. This attachment replaces the compound rest and is geared to the headstock spindle. First the cutter blank is prepared by boring the hole, rough turning the profile, and milling the flutes. A single point, form-cutting tool which will produce the required profile on the form milling cutter is then made. With the lathe set up and the cutter turning slowly, a cam on the back-off attachment causes the single-point form tool to move in and out an equally measured distance on each tooth while cutting the profile on the teeth of the milling cutter, thereby also cutting the relief behind the cutting edges on the cutter. Cut in this manner, the profile of the form relieved milling cutter will not change as the diameter is decreased when the cutter is sharpened. The cutters are sharpened by grinding the radial rake angle on the face of the teeth. They can be sharpened in this manner many times—until the teeth are too weak to withstand the cutting load—without changing the profile of the cutter. Care must be taken, however, to maintain the original radial rake angle, otherwise the profile on the cutter will be altered.

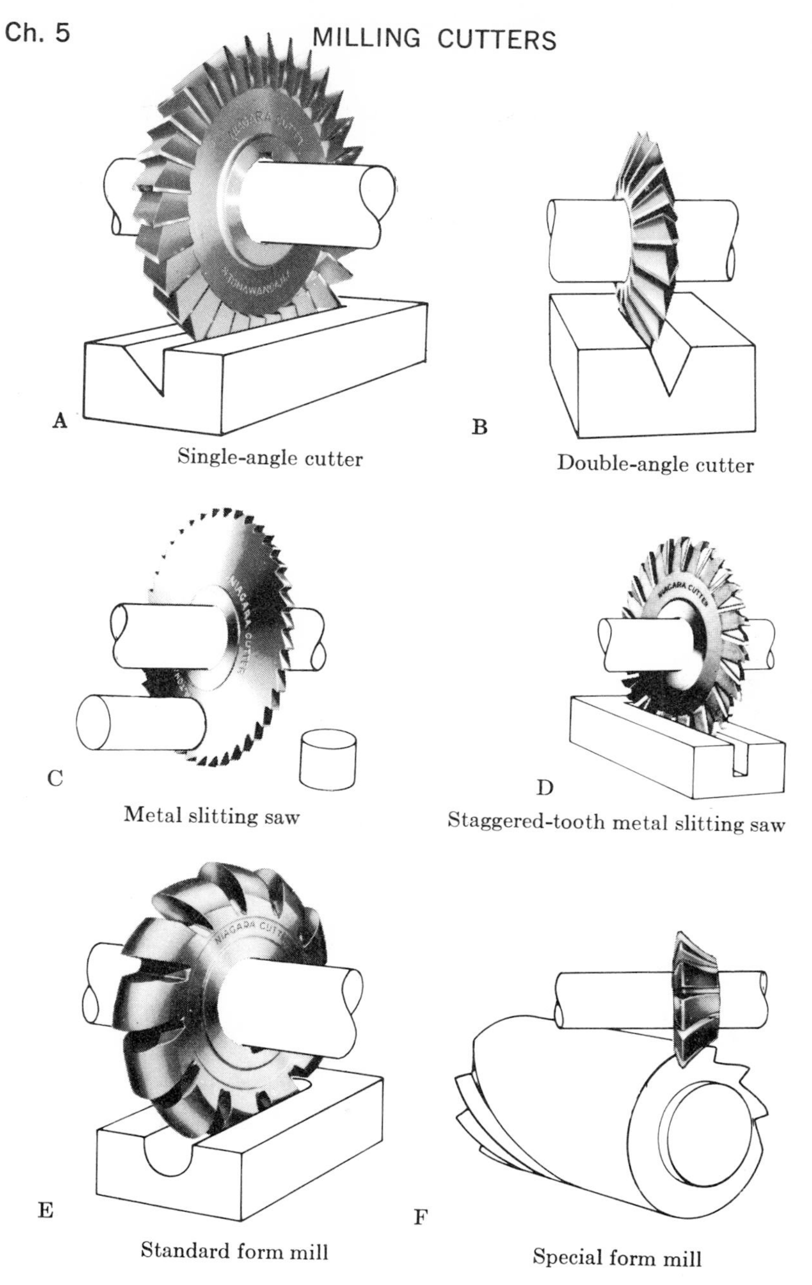

Single-angle cutter

Double-angle cutter

Metal slitting saw

Staggered-tooth metal slitting saw

Standard form mill

Special form mill

Fig. 5-3.

A few form relieved milling cutters can be obtained as standard cutters. Standard form milling cutters are available for milling some convex and concave surfaces, for corner rounding, for milling involute gear teeth and sprocket teeth. A standard form relieved cutter is shown at E in Fig. 5-3.

Many form relieved cutters are specially made to suit a particular job. A special form relieved cutter designed to cut the flutes on a plain milling cutter is shown in view F, Fig. 5-3. The illustration in Fig. 5-4 shows a form milling operation; the cutter here is a special interlocking type, form-relieved milling cutter.

Elements of Arbor-Mounted Cutters. The *nomenclature* of the elements of arbor-mounted cutters is given in Fig. 5-5, and the terms are self-explanatory. Most of the elements are built into the cutter and usually are not changed, except for the relief angles which are ground when the cutter is sharpened and thus, are subject to possible change. Relief angles provide relief behind the cutting edges to allow them to penetrate the work and form a chip. If the relief angles are worn to the extent that there is no relief behind the cutting edges, the cutter would rub and fail to penetrate the work. Excessive wear and heat indicate that the relief angles may be too small, while chatter in the cut indicates that the relief angles may be too large.

The peripheral relief angle provides relief behind the peripheral teeth on the cutter, that do the bulk of the cutting. For average materials such as mild steel and gray cast iron, the recommended peripheral relief angle is 4 to 7 degrees; for harder and tougher materials, such as tool steel, it is 3 to 6 degrees; and for softer materials such as soft brass, aluminum, magnesium, and plastics, it is 7 to 12 degrees. On side milling cutters the side teeth also require relief, although less relief is necessary since these teeth are not intended to take a heavy cut. Furthermore, care must be exercised to maintain the required cutter width within allowable limits. For this reason two practices are followed:

1. A very light "hairline" land, or margin, is left on the teeth without relief behind which a relief angle of 2 to 4 degrees is ground, except on thin saws where the relief angle is increased to 3 to 5 degrees.
2. The relief angle is ground to a sharp edge but is reduced to approximately 1 to 2 degrees. This method is not recommended for saws.

The side teeth should also be dished, the amount of concavity being .001 to .0015 inch per inch (0.02 to 0.03 mm per 20 mm) for side milling cutters and .005 to .0075 inch per inch (0.10 to 0.15 mm per 20 mm) for saws. The surface formed by the relief angle on the periphery and on the side is called the "land," which on many cutters is raised. The width of the land is limited by grinding a secondary clearance angle behind the land. On arbor mounted cutters the width of the land may vary from about $3/64$ inch (1.2 mm) on smaller cutters, to $1/8$ inch (3.2 mm) on very-large-diameter cutters.

The inclination of the tooth face affects the ease with which the chip is formed. Since all of the cutting on plain milling cutters and most of the cutting on side milling cutters is done with the peripheral teeth, it is the radial rake angle that has the greatest effect on the formation of the chip. Most high-speed steel arbor-mounted milling cutters have a radial rake angle of about 10 degrees positive, except metal cutting saws, which may have a radial rake angle varying from 0 to 10 degrees positive. The recommended radial and axial rake angle for cemented-carbide side milling cutters is the same as for the carbide face milling cutters given in Table 5-1. On high-speed steel side milling cutters the axial rake angle is usually 10 to 15 degrees, except for straight-tooth side milling cutters. The axial rake angle on plain or slab milling cutters may be from 0 to 52 degrees.

The plain milling cutter in Fig. 5-6 has straight teeth. The chip formed by this cutter can be seen in the illustration. The thickness of the chip increases as the tooth of the milling cutter progresses into the work. Figure 5-7 illustrates the theoretical profile of a chip that is undeformed. It can

Courtesy of the Brown & Sharpe Manufacturing Company

Fig. 5-4. Form-milling operation using a form relieved milling cutter.

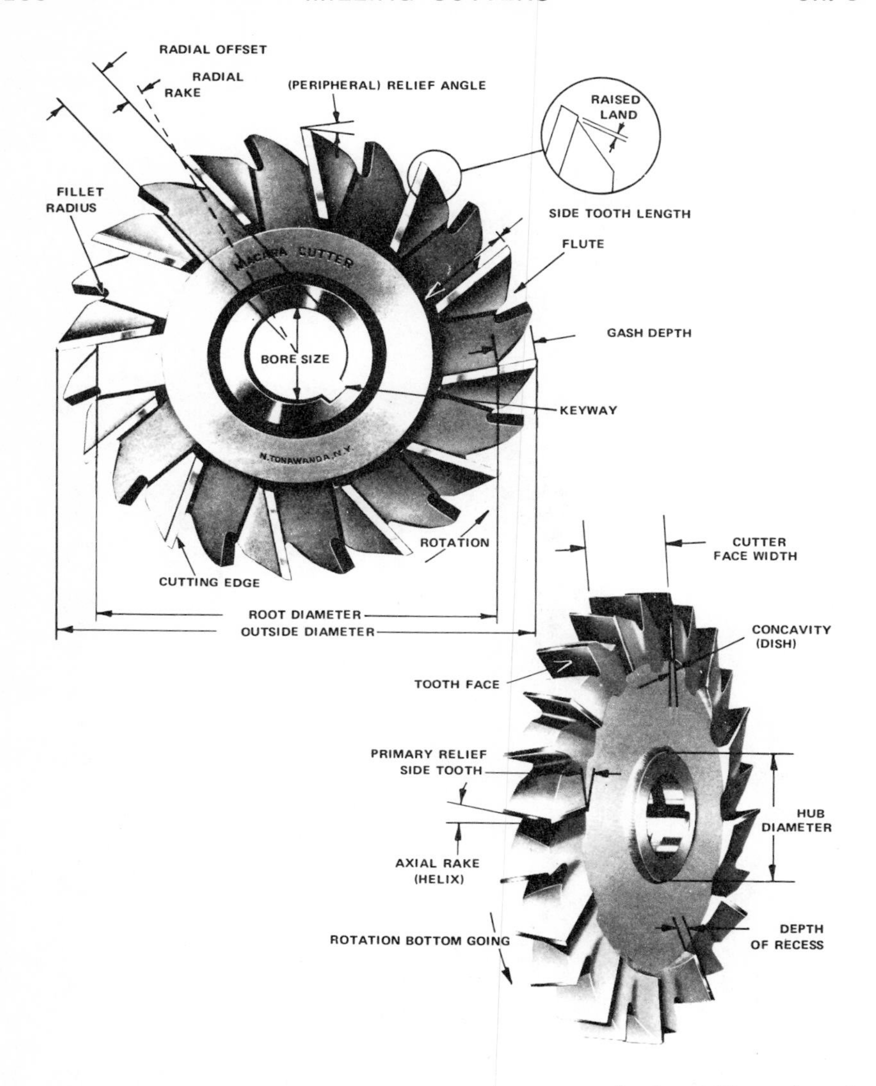

Courtesy of Niagara Cutter Inc.

Fig. 5-5. Milling cutter nomenclature.

be seen from this figure that the chip thickness, t, increases as the tooth penetrates into the work, or from position 1 to position 3. The advance, f, of the cutter, however, is the same for all positions and is equal to the feed per tooth of the milling cutter. Figure 5-7, therefore, also shows that when the table feeds the work into the rotating milling cutter at a uniform rate, the feed per tooth is constant regardless of the depth of the cut.

Plain milling cutters with a width of less than ¾ inch have straight

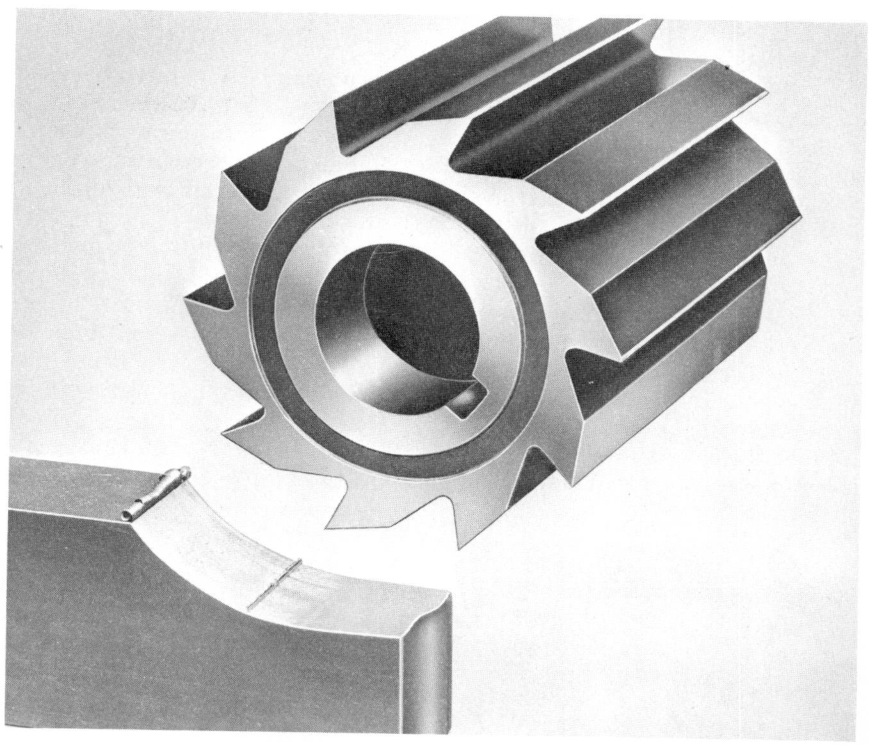

Courtesy of Cincinnati Milacron

Fig. 5-6. Plain milling cutter showing chip formation.

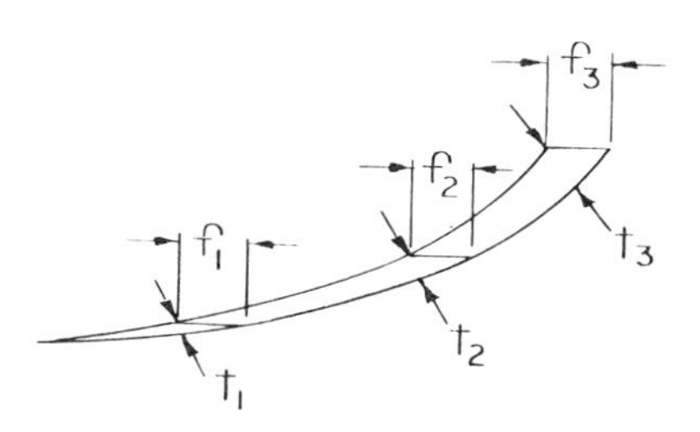

Fig. 5-7. Profile view of an undeformed milling chip.

teeth. Straight tooth cutters cause the chip to form uniformly along the entire width of cut as shown in Fig. 5-6. The cutting load on the tooth builds up uniformly but rapidly. When the tooth leaves the work, the cutting load drops suddenly. This sudden drop in the load is accompanied by a shock which causes a reaction in the entire milling-machine structure but most particularly in the spindle driving mechanism. The repetition of this shock as the loads on the teeth of a milling cutter rotating at a uniform speed suddenly drop in rapid succession can set up the vibratory condition called chatter.

The tendency for chatter to occur can be substantially reduced by forming the cutting edge into a helix, as shown in Fig. 5-8. At the beginning of the cut, the length of the cutting edge penetrating into the workpiece is small. As the cutter continues to rotate, the length of the cutting edge in contact with the workpiece increases, gradually forming a widening chip until the maximum is reached. The depth of the cut and the helix angle of the cutter will determine the length of tooth travel during which the maximum tooth contact and chip width occur. The tooth contact and chip width then gradually decrease until zero is reached as the tooth leaves the work. Thus, this gradual building up and release of the cutting load will avoid the shock that accompanies a sudden release of the cutting load. In addition, in most instances two teeth are simultaneously cutting, as in Fig. 5-8, to further stabilize the cutting load.

Since the chip is formed at an angle when the teeth are on a helix angle, it is often assumed that a shearing action is taking place between the cut-

Courtesy of Cincinnati Milacron

Fig. 5-8. Heavy duty plain milling cutter with a 45-degree helix angle showing chip formation.

ting edge and the chip. It must be emphasized that this does not occur, for there is no axial sliding motion between the cutting edge and the chip. The width of the chip is at all times equal to the length of the cutting edge that is in contact with the work, and the cutting action is the same for helical teeth as for straight teeth.

End Milling Cutters

End milling cutters constitute a large group of milling cutters made to a variety of shapes and sizes. The group of typical end milling cutters displayed in Fig. 5-9 are characterized by having cutting edges on the end face as well as on the periphery. Also, they are always held in the milling-machine spindle by a collet chuck or some kind of adaptor. Among the most versatile of cutting tools, end mills are used to mill plane surfaces, slots, profiles, and three-dimensional contours.

The elements of an end milling cutter are shown in Fig. 5-10. The radial rake angle is generally small, as it is limited by the necessity of keeping the end cutting edges approximately radial. The helical rake angle and

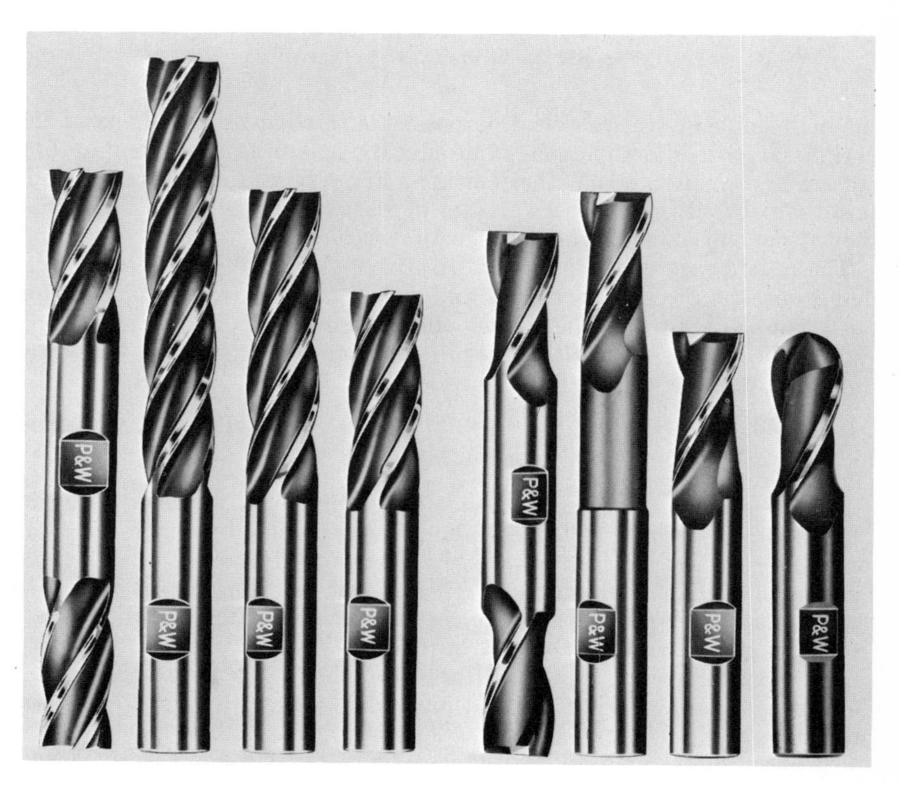

Fig. 5-9. A group of different end milling cutters.

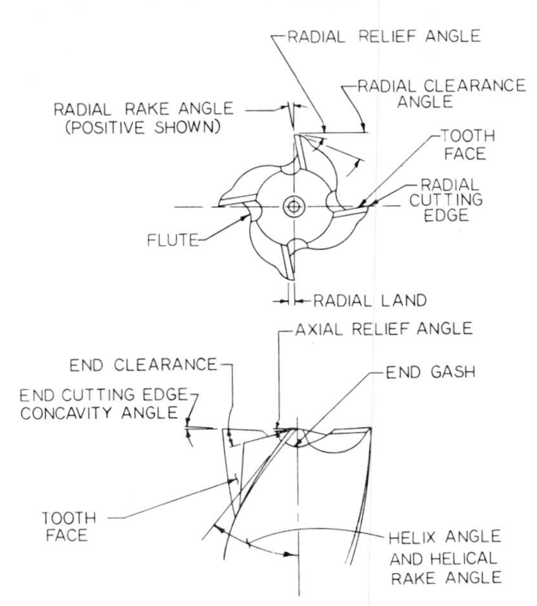

Fig. 5-10. Elements of end milling cutters.

the helix angle are for practical purposes the same angle. The helix of the peripheral cutting teeth has the same effect on end mills as on plain milling cutters in providing a smoother cutting action. There is, however, a limitation on the size of the helix angle. It must not be made so large that the cutting edge of the end cutting teeth is weakened.

The relief angle of the peripheral teeth, or the "radial relief angle," is determined by the size of the end mill and by the material to be cut. For most applications, including most steels and cast irons, the following radial relief angles (given in terms of the cutter diameter first followed by the radial relief angle) are recommended:

1/16″—22°, 1/8″—17°, 3/16″—14°, 1/4″—12°, 3/8″—12°, 1/2″—11°, 3/4″—10°, 1″—9°, 1 1/2″—8°, 2″—7°

More detailed recommendations for the radial relief angle are provided in Chapter 14, Table 14-1. Depending on the method by which the radial relief angle is ground, three types of relief are used; namely, concave, flat, and eccentric. The three types of relief are shown in Fig. 14-5, Chapter 14.

On most end mills the relief angle on the end cutting teeth—the axial relief angle—should be about 4 degrees. Two fluted center cutting end mills should have an axial relief angle of about 7 degrees because they are often fed endwise, or plunged into solid stock, to be used as a twist drill.

The flutes in an end mill may have a right- or left-hand cut as shown in Fig. 5-11. The cut refers to the side of the flute on which the face of the teeth are located. Furthermore, end mills with a right-hand cut are designed to cut while rotating counterclockwise when viewed from the end having the teeth. Left-hand-cut end mills are designed to cut while rotating clockwise when viewed from this end. Although the helix may be right- or left-handed for either cut, usually a right-hand-cut flute has a right-hand helix and a left-hand-cut flute has a left-hand helix. In this way the end-cutting edges have a positive rake angle.

Most standard end milling cutters are made to have two or four flutes in sizes up to approximately 1 inch. Larger end mills, up to 2 inches in diameter, are made with six or eight flutes. Three-fluted, center-cut-type end mills are made with diameters up to 3 inches. Increasing the number

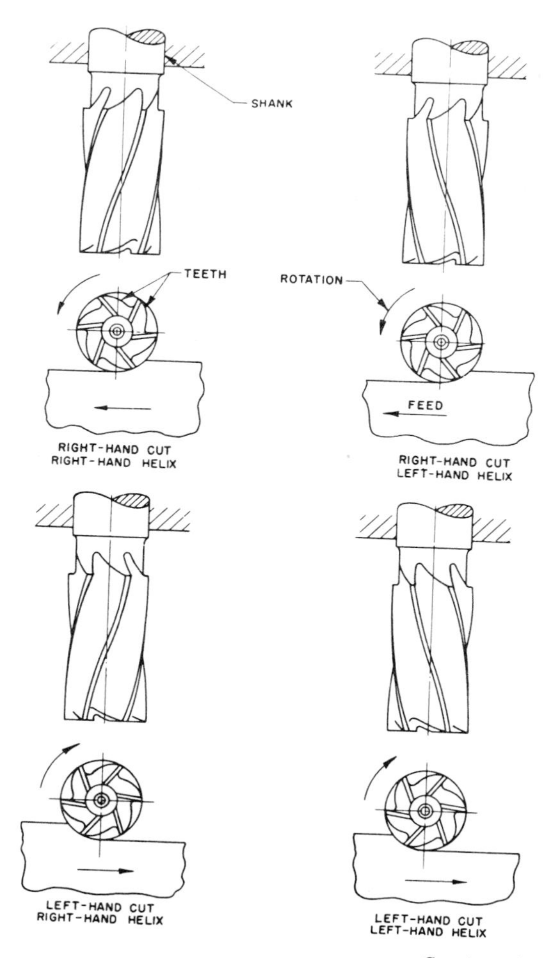

Courtesy of Cincinnati Milacron

Fig. 5-11. The four combinations of hand of helix and hand of cut for end mills.

of flutes on the end mill helps to stabilize the cutter when milling slots and allows a faster feed (inches per minute) to be used. The flutes must, however, be large enough to provide adequate space for the chips.

Figure 5-12 illustrates the three different types of construction which can be used for the end teeth of end milling cutters with four or more teeth. The end mill at the bottom has a cupped-type end on which the end teeth extend only to approximately the bottom of the flute. The conventional square end mill in the center has a notch in the end which permits the end teeth to extend to the center hole. The added length of the end teeth allows radii or forms to be ground on the end when required. The end mill at the top has two end teeth which join in the center. This type, called a *center-cutting end mill,* can plunge directly into a workpiece to create an opening or a hole in much the same manner as a twist drill. After it has been plunged to the required depth, it can be made to cut a slot such as a keyseat. The four peripheral teeth produce a good finish on the workpiece and allow a reasonable feed rate to be used.

The most common center-cutting-type end mill is the *two-fluted end mill,* which in some shops is called a *two-lipped end mill.* Four typical two-fluted end mills are shown at the right in Fig. 5-9. The conventional square end two-fluted end mill is the least expensive and the most easily sharpened of the center-cutting-type end mill. As it can easily be sunk directly into the workpiece like a drill, it is frequently used to cut keyseats

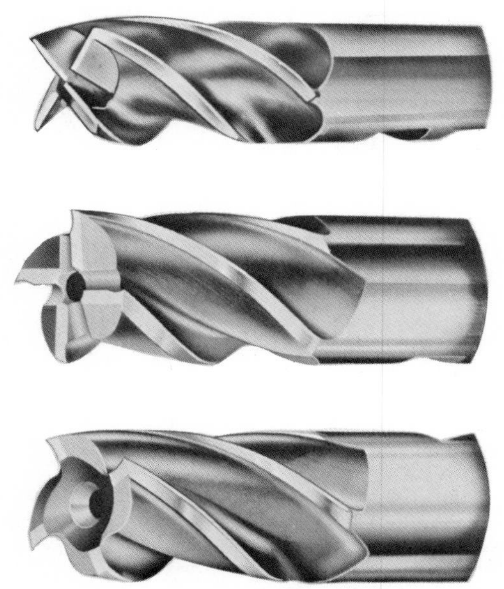

Courtesy of The Metal Cutting Tool Institute

Fig. 5-12. General-purpose four-fluted end mills showing three types of construction for the end cutting edges.

and other slots that do not extend to an open end or shoulder. Two-fluted-ball end mills, such as shown at the right in Fig. 5-9, are used to cut complex three-dimensional contours such as are encountered in dies and molds. The operation for which they are used is called *die-sinking.*

Although taper-shank end mills are still being made, the majority of the end mills have a straight shank. Some straight-shank end mills have cutting teeth on both ends, with the shank in the center of the cutter. The majority of the taper-shank end mills have a Brown & Sharpe taper.

Rough-Cutting End Mills. These end mills have been developed primarily to remove a large amount of stock, as shown in Fig. 5-13. Their teeth, shaped in the form of a radius which produces a small chip that does not load the flutes, are positioned around the cutter in the form of a left-hand helix like a left-hand thread. This makes the teeth overlap each other so that the surface produced is flat. The teeth are form-relieved and are sharpened by grinding only the face of the flutes.

Indexable Insert End Mills. Cemented-carbide, indexable-insert end milling cutters, Fig. 5-14, operate at a faster cutting speed and a faster table feed than conventional high-speed-steel end mills. Indexable insert end mills are available with one or two cemented-carbide inserts. Each

Fig. 5-13. Rough cutting end mill taking a typical cut in a workpiece.

insert has four cutting edges; when a cutting edge is worn the insert can rapidly be indexed to provide a sharp cutting edge until all the available cutting edges are used up. Both single and double insert end mills are designed to perform the operations shown in Fig. 5-14, which includes plunge cutting (partially or completely through the workpiece), slot milling, pocket milling, peripheral milling, and ramping. In addition, they can also be used to counterbore and to spot face.

Shell End Milling Cutters

Standard shell end milling cutters are available in sizes from $1\frac{1}{4}$-inch diameter to 6-inch diameter. They are intended for taking surfacing cuts and corner cuts and are not generally used to cut slots. As shown in Fig. 5-15, shell end mills are mounted on shell end mill arbors, which are fitted directly into the spindle of the milling machine. One obvious advantage of the shell end mill is that when the body of the cutter is worn, the shank can be reused on another shell end milling cutter. Shell end milling cutters

Courtesy of the Valenite Div. of the Valeron Corp.

Fig. 5-14. Indexable insert end mill cutter showing the type of cuts that can be taken with this cutter.

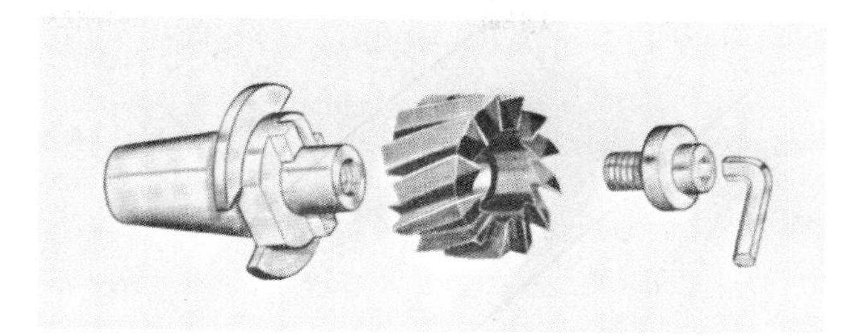

Courtesy of The Metal Cutting Tool Institute

Fig. 5-15. Shell end mill and shell end mill adaptor.

tend to bridge the gap between face milling cutters and end milling cutters.

The corner of the shell end milling cutter is the point where the cutting edges on the periphery meet the cutting edges on the end. The corner may be square, rounded, or chamfered. The most frequently used is the square corner, which is easiest to grind. However, this corner does tend to be a focal point of edge wear. The rounded corner, while working very well, is difficult to grind. The chamfered corner is best because it is easy to grind and wears well. The chamfer (or radius in the case of a rounded corner) should not exceed $\frac{1}{16}$ inch in length, since the small chamfer will not interfere with the main flow of the chip which is approximately perpendicular to the peripheral cutting edges. The chip flow from a larger chamfer will tend to interfere with the peripheral cutting-edge chip, resulting in an increase in the cutting temperature, accelerated tool wear, shorter cutter life, and quite often a deterioration in the resulting surface finish.

T-Slot Cutter. The milling cutter shown in Fig. 5-16 is used to cut T-slots in machine tool tables and accessories. A slot must first be cut in the work-

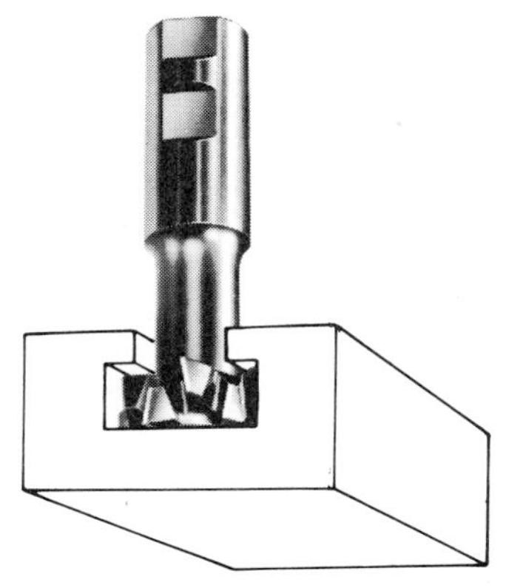

Courtesy of Niagara Cutter Inc.

Fig. 5-16. T-slot cutter.

piece so that the neck of the T-slot cutter may be clear to cut the T-slot. These cutters have teeth on the periphery and on two sides, with alternate side teeth removed in the manner of staggered tooth side milling cutters. The side teeth have a slight concavity to avoid dragging in the cut.

Woodruff Keyseat Cutters. Woodruff keys are small half-round shaped keys that are fitted into a half-round keyseat on shafts to transmit torque. The keyseats are cut by sinking the cutter directly into the workpiece at the desired location—as shown in Fig. 5-17. Smaller Woodruff keyseat cutters (1/4 to 1 1/2 inch diameter) are shank type cutters such as shown in the illustration; the cutters for larger diameter Woodruff keys are arbor mounted, being similar in appearance to staggered tooth side milling cutters.

Face Milling Cutters

Face milling cutters are primarily used to take surfacing cuts in producing flat surfaces. Generally 6 inches and larger in diameter, they are mounted directly on the nose of the milling-machine spindle. All face milling cutters have inserted cutting teeth. A cemented-carbide face milling cutter is shown in Fig. 5-18. The very fast metal cutting rates obtainable with cemented-carbide face milling cutters make high production rates possible.

The nomenclature of a face milling cutter is given in Figs. 5-19 and 5-20. Cutting edges are ground on the end face and on the periphery of the cutter. The point where these two cutting edges meet is called the corner of the cutting edge. A sharp corner should be avoided as it would wear rapidly and reduce the life of the cutter. In order to improve the performance of the face milling cutter the corner should be chamfered. A single chamfer should be used, and the chamfer should not exceed 1/16 inch in width unless it is made large enough so that the entire chip is formed by the chamfer as at A, Fig. 5-21. If the entire chip is formed by the chamfer, it will flow freely and approximately perpendicular to the cutting edge. The chip at B, Fig. 5-21, is formed partially by the cutting edge on the chamfer and partially by the cutting edge on the periphery. Since the direction of flow of the chip is approximately perpendicular to the two cutting edges, the

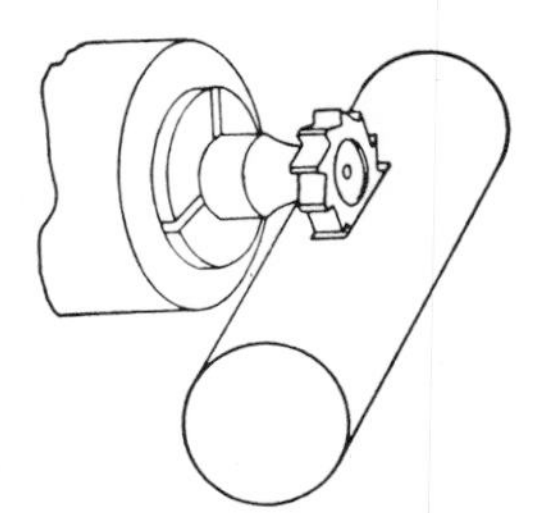

Courtesy of the National Twist Drill & Tool Div., Lear Siegler Corp.

Fig. 5-17. Woodruff keyseat cutter.

Courtesy of The Ingersoll Milling Machine Co., Cutting Tool Div.

Fig. 5-18. Indexable insert cemented carbide face milling cutter.

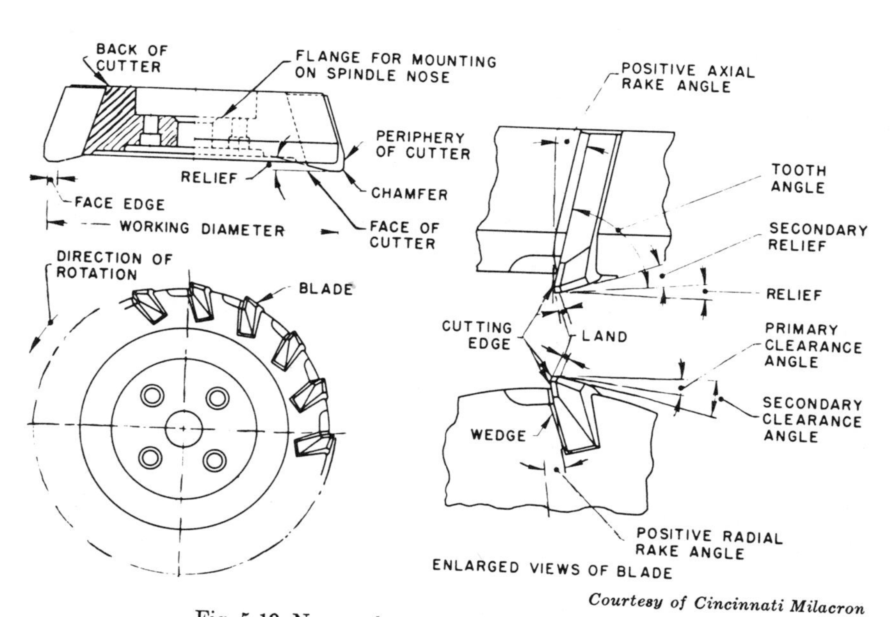

Courtesy of Cincinnati Milacron

Fig. 5-19. Nomenclature of a face milling cutter.

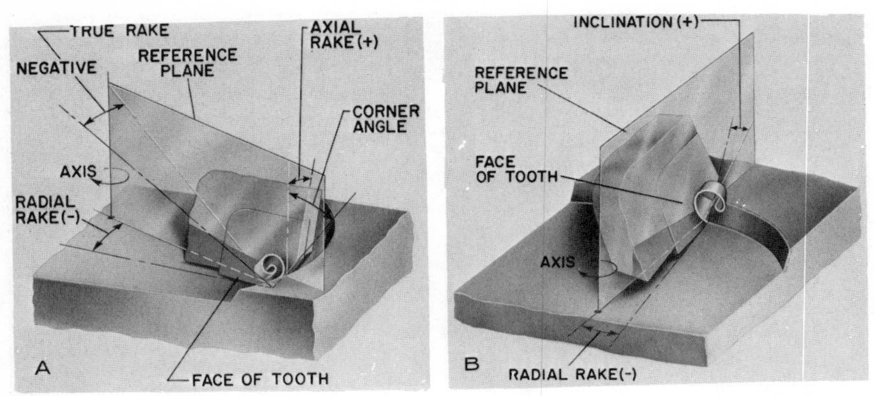

Courtesy of Cincinnati Milacron

Fig. 5-20. A. True rake angle. B. Angle of inclination of milling cutter tooth.

chips formed will obviously interfere with each other. Such interference will result in a less efficient cutting action, an increase in the tool wear, and a decrease in the life of the cutter. A double chamfer will result in similar undesirable effects. If, however, the width of the chamfer is less than 1/16 inch, the small chip formed by the chamfer will not seriously interfere with the flow of the chip formed by the peripheral cutting edge.

The angle of the chamfer in the corner is known as the *corner angle* (see A, Fig. 5-20), or sometimes the *lead angle* of the cutter. The corner angle has a pronounced effect on the performance of the face milling cutter. For example, a large corner angle will permit the use of a faster feed rate, thereby increasing the production rate of the milling operation. Figure 5-22 shows two face milling cutters, one with a 30-degree corner angle and

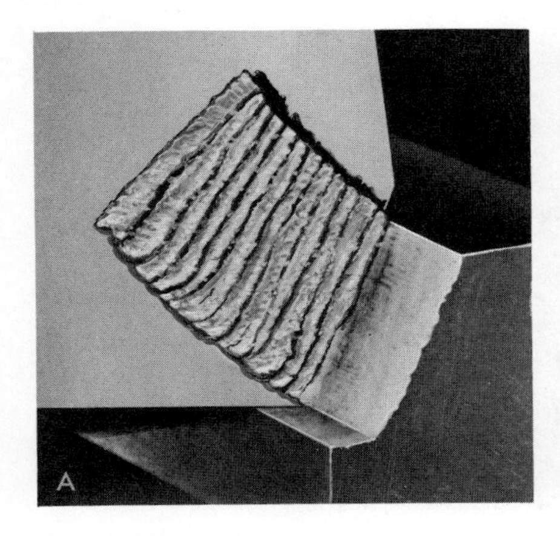

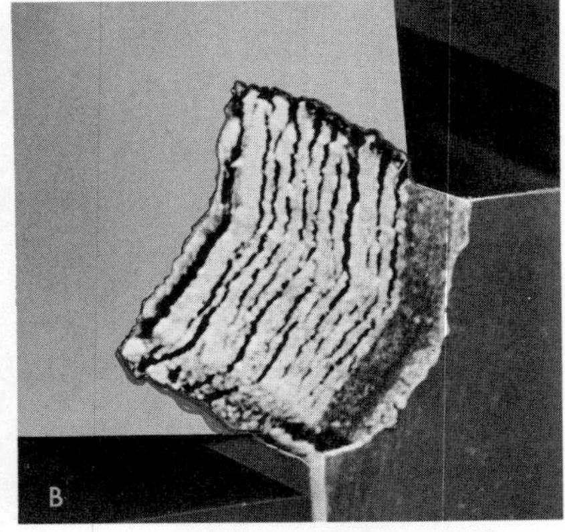

Courtesy of Cincinnati Milacron

Fig. 5-21. A. Flow of chip formed by chamfer on milling cutter tooth. B. Chip formed by chamfer and peripheral cutting edge showing interference to flow.

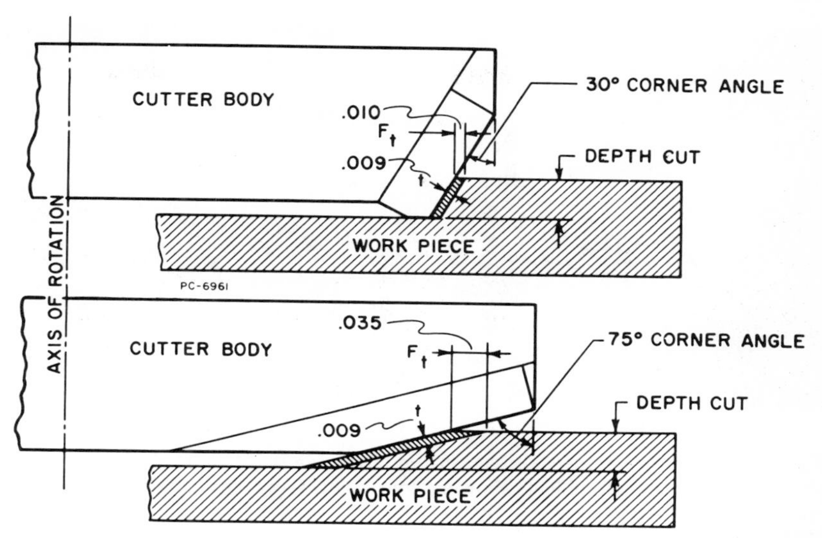

Courtesy of Cincinnati Milacron

Fig. 5-22. Greater feed per tooth can be obtained without increasing the chip thickness by increasing the corner angle.

the other with a 75-degree corner angle. In each case the feed of the machine is adjusted so that the chip thickness will be .009 inch per tooth. The feed per tooth which will result in this chip thickness is .010 inch per tooth for the cutter with the 30-degree corner angle and .035 inch per tooth for the 75-degree corner-angle cutter. Thus, the feed rate of the 75-degree corner-angle cutter can be up to 3.5 times faster than the feed rate of the 30-degree corner-angle cutter. A further advantage is that the chip will have a greater tendency to flow axially along the length of the cutter and to flow clear of the cutter. On the other hand, the 75-degree cutter will exhibit a greater tendency to chatter when cutting.

Other important face milling cutter elements are the radial rake angle, the axial rake angle, the true rake angle, and the inclination of the cutting edge (see Fig. 5-20). The true rake angle is perpendicular to the cutting edge that is forming the chip. Since the flow of the chip will be approximately perpendicular to the cutting edge, the true rake angle is the angle along which the chip will flow. The true rake angle has a great influence on the tool life of the cutter, the surface finish obtained, the power consumption, and the deflections resulting from the cutting forces. The inclination of the cutting edge influences the direction of the chip flow. It should always be positive so that the chips will be directed away from the cutter body and the finish machined surface on the workpiece, as in Fig. 5-23.

Cemented-carbide face milling cutters should have positive axial rake angles, negative radial rake angles, and large corner angles which in combination provide a negative true rake angle and a positive inclination. The recommended cutter angles for face milling different materials are

Courtesy of The Ingersoll Milling Machine Co., Cutting Tool Div.

Fig. 5-23. Shear-Clear face milling cutter having positive axial and negative radial rake angles milling a die block, showing flow of chips away from workpiece and cutter.

shown in Table 5-1. The recommended clearance and relief angles (Fig. 5-19) are given in Table 5-2. High-speed-steel face milling cutters, Fig. 5-24, are usually provided with a larger axial rake angle than is recommended in Table 5-1. When it is necessary to mill a square shoulder, as in Fig. 5-25, the face milling cutter must have a zero-degree corner angle. This type of indexable insert cutter has negative axial and radial rake angles; it is provided with ample chip space to enable it to handle the chips.

Table 5-1. Recommended Rake and Inclination Angles for Cemented Carbide Face Milling Cutters

Material	Hardness, HB	Axial Rake Angle	Radial Rake Angle	Corner Angle	True Rake Angle	Inclination Angle
Plain Carbon Steels	120 to 260	+5°	−5°	30°	−2°	+6°
Alloy Steels	160 to 400	+5°	−10°	45°	−3°	+10°
Nitriding Steels	260 to 320	−5°	−10°	45°	−10°	+3°
Hot-Work Die Steels	180 to 300	−5°	−10°	45°	−10°	+3°
Tool Steels	160 to 220	+5°	−10°	45°	−3°	+10°
High-Speed Steels	210 to 280	+5°	−10°	45°	−3°	+10°
Stainless Steels	160 to 280	+7°	−7°	45°	0°	+9°
High-Temperature Alloys	260 to 320	+5°	−10°	50°	−2°	+11°
Nickel Alloy	160 to 250	+5°	−5°	45°	0°	+7°
Gray Cast Iron	140 to 220	+5°	−10°	45°	−3°	+10°
Nodular Cast Iron	140 to 220	+5°	−10°	45°	−3°	+10°
Ferritic-Malleable Cast Iron	120 to 160	+10°	−5°	30°	0°	+11°
Pearlitic-Malleable Cast Iron	180 to 260	+5°	−10°	45°	−3°	+10°
Magnesium Alloys	40 to 80	+15°	+15°	30°	0°	+5°
Aluminum Alloys	60 to 110	+15°	+15°	30°	0°	+5°
Copper Alloys	120 to 210	+5°	+5°	30°	+6°	+2°
Titanium	180 to 300	0°	−10°	30°	−7°	+5°

Table 5-2. Recommended Clearance and Relief Angles for Cemented Carbide Face Milling Cutters

Material	Primary Clearance Angle Peripheral Cutting Edges	Primary Clearance Angle Corner Cutting Edges	Primary Relief Angle Face Cutting Edges
Steel	4 to 5°	4 to 5°	3 to 4°
Gray Cast Iron	7°	7°	5°
Aluminum	10°	10°	10°
Titanium	12°	12°	12°

When a face milling cut (or an end milling cut) is taken, the bulk of the metal is removed by means of the large chips that are formed by the teeth on the periphery of the face milling cutter. This part of the cut can be called the *primary cut*. The face edges on the face of the cutter teeth (see Fig. 5-19) perform a very important function by taking a very light *secondary cut*. The heavier chip load of the peripheral teeth in taking the primary cut causes a slight but nevertheless positive deflection of the cutter and the workpiece away from each other. When the primary cut comes to an end, the chip load is released and the cutter and the work spring back toward each other. This combined spring-back provides the "depth of cut" which is taken by the face edges of the cutter on the "back" part of the cutter revolution. The secondary cut is a light, more or less scraping cut, taken primarily by the face edges of the cutter. The feed

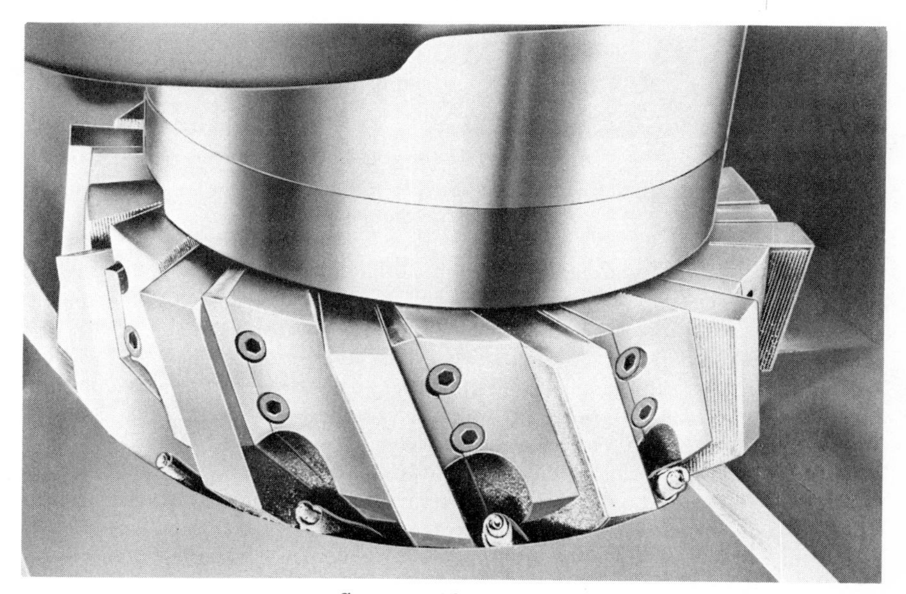

Courtesy of The Ingersoll Milling Machine Co., Cutting Tool Div.

Fig. 5-24. High-speed steel, blade type, face milling cutter.

Courtesy of The Ingersoll Milling Machine Co., Cutting Tool Div.

Fig. 5-25. Cemented carbide indexable-insert face milling cutter with 0° corner angle for milling square shoulders.

marks left by face milling and end milling cutters provide evidence of the secondary cut.

Whenever possible the automatic table feed should be kept in engagement until the surface that has been cut is completely clear of the face milling (or end milling) cutter. This procedure will allow the secondary cut made by the face edges to be completed over the entire surface, thereby producing a more uniformly flat or plane surface.

In many instances the finish on the milled surface must be very smooth and the feed marks such as described in previous paragraphs would not be tolerated. On those face milling cutters which have removable blades (see Figs. 5-23 and 5-24) instead of indexable inserts (Fig. 5-18), one blade can be set about .003 to .005 inch (0.08 to 0.13 mm) above the other blades. This blade, called a *wiper* blade, is ground to have a "flat" on the face, which is actually a broad ellipse with a slight chamfer on each side. The flat should be parallel to the surface to be cut and it should be at least $\frac{1}{16}$ inch (1.6 mm) wider than the feed. When cutting, the wiper blade will produce the finish on the milled surface, scraping off the feed marks. This method can be used to mill steel and cast iron. When possible, tilting the milling machine spindle very slightly into the cut to allow the cutter to contact the workpiece only on the cutting portion of its rotation, will help to improve the finished surface.

Indexable insert face milling cutters cannot utilize this method because the inserts must remain in a fixed position in their pocket. Such cutters can, however, be provided with pockets on the face in which special wiper blades, or finishing inserts, can be clamped, as shown in Fig. 5-26. For balance at high speeds, two finishing inserts are clamped to the face, which are positioned .003 to .005 inch above the roughing blades on the periphery. The cutter can be used to rough and finish cut in one pass. Better results are obtained, however, by first taking a roughing cut, with the finishing inserts replaced by dummy inserts that do not cut. The finishing inserts are then attached and a second cut across the workpiece is taken without changing the depth of cut. Indexable insert face milling cutters with finishing blades can be used on steel, cast iron, and some other metals. Face milling cutters having finishing inserts only, and no roughing teeth, are available.

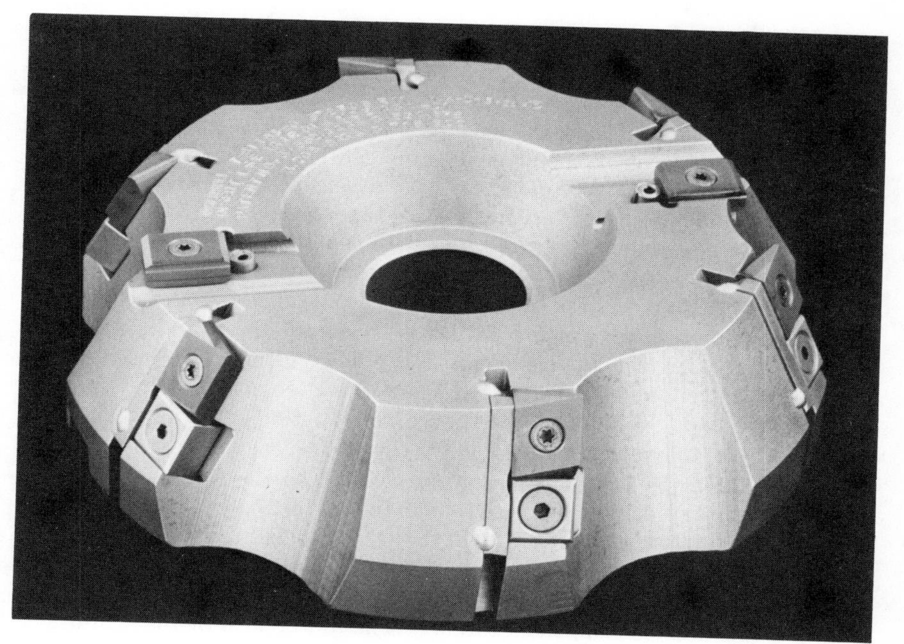

Courtesy of The Ingersoll Milling Machine Co., Cutting Tool Div.

Fig. 5-26. Cemented carbide, indexable insert, face milling cutter with finishing blades held in pockets on the face of the cutter.

Mounting Milling Cutters

In order to obtain a high degree of accuracy in performing milling operations, the milling cutters must be correctly mounted on the milling machine. To facilitate the accurate mounting of milling cutters, the nose of the milling machine is provided with the following:

1. An accurately ground tapered hole for locating arbors, adaptors, and shank-type cutters
2. An accurately ground outside diameter for locating face milling cutters with respect to the axis of rotation of the spindle
3. An accurately ground face which is perpendicular to the axis of the spindle
4. Four threaded holes for holding the clamping screws of face milling cutters; and
5. Two driving keys for driving face milling cutters, adaptors, and arbors.

The taper in the spindle nose is a self-releasing taper, and its only function is to locate objects in the spindle. The power is transmitted through two radial keys attached to the face of the spindle.

Arbor-Mounted Cutters. Milling-machine arbors are finish ground on all surfaces in order to provide an accurate method of mounting the cutters. An arbor holding two straddle-milling cutters correctly mounted in the milling machine is shown in Fig. 5-27. The correct procedure for mounting the arbor and the cutter is outlined as follows:

1. Clean the tapers in the spindle and on the arbor. If the taper in the spindle or on the arbor is damaged it must be repaired. Nicks and burrs should be removed with a small fine abrasive stone. All dirt must be removed from both tapers. Make a final check by feeling the surface of the taper with the bare hand. The bare hand will detect small particles of dirt or lint that would be undetected with a rag.
2. Place the arbor in the spindle and insert the draw-in bolt into the threads in the tapered end of the arbor. Turn the draw-in bolt nut until 1 or 2 inches of thread are engaged. When placing the arbor in the spindle, make certain that the radial keys on the spindle engage the slots on the arbor.
3. Tighten the arbor in the spindle by turning the threaded collar on the draw-in bolt with a wrench. In Fig. 5-27, the threaded collar can be seen in position against the end of the spindle.
4. Clean the face of the arbor flange opposite the taper. This face must be absolutely clean, otherwise the arbor will be bent slightly when the arbor nut is tightened. The final check of the cleanliness of this surface should also be done with the bare hand.
5. Place an end collar and spacing collars over the arbor up to the position where the milling cutter, or cutters, are to be mounted.

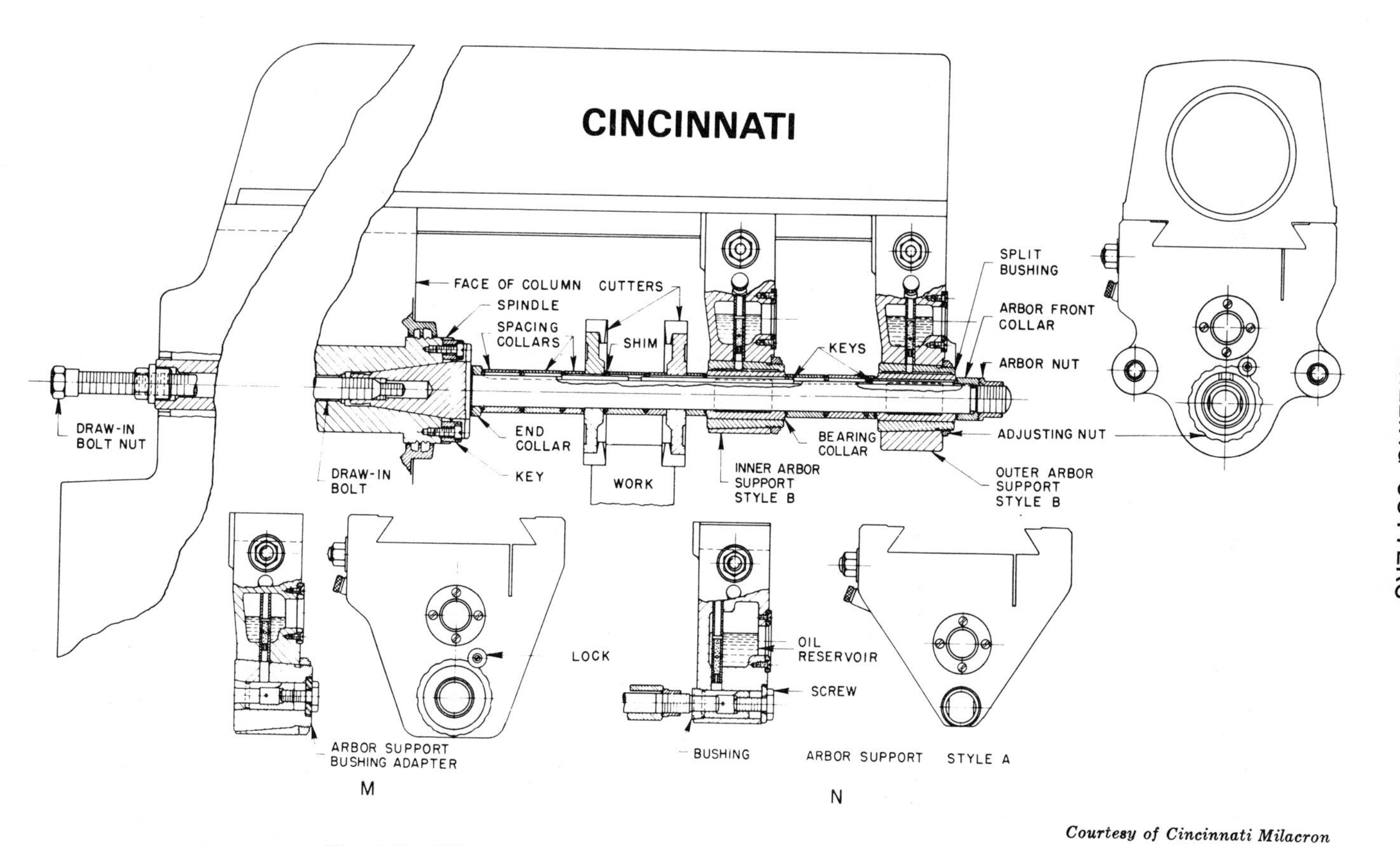

Courtesy of Cincinnati Milacron

Fig. 5-27. Milling-machine arbor correctly mounted in a milling machine with two straddle milling cutters in place.

Various lengths of spacing collars are available. The faces on each end of the collar are ground to extremely close tolerances of parallelism and perpendicularity with respect to the axis of the collar. An inaccuracy in these surfaces or any dirt or foreign matter between the faces of adjacent collars on the arbor will result in the deflection of the arbor when the arbor nut is tightened. The milling cutter will not run true if the arbor is deflected, and a few teeth on the cutter will do most of the cutting. This procedure will result in inaccurate work and rapid wear of the cutting edges of the milling cutter. For this reason it is necessary to clean the faces of each collar with the bare hand before placing them on the arbor.

6. Place the milling cutter on the arbor. The faces of the milling cutter should be cleaned with the bare hand in the same manner as the spacing collars were. The cutter should be placed over a key which is inserted in the keyseat on the arbor. The milling cutter should always be driven by a key, and the friction of the spacing collars against the face of the cutters should not be depended upon to provide the drive. If not driven by a key, the cutter could slip on the arbor during the cut with the result that the teeth of the cutter would be broken.
7. Place additional spacing and bearing collars behind the cutters. Bearing collars, which are slightly larger in diameter than the spacing collars, are made in one length corresponding to the length of the arbor-support bearing. Bearing collars provide the primary support of the Style B arbors. They may be used to provide additional support for a Style A arbor (see Fig. 6-26), although there will be a resulting loss of clearance below the arbor. Make certain that the faces of the collars are clean as previously described.
8. Place the nut on the end of the arbor, but DO NOT tighten. The face of the nut should be cleaned in the same manner as the collars. If a wrench is used to tighten the nut before an arbor support is in place, the arbor would very likely be bent and ruined.
9. Place the arbor support over the bearing collars. Make certain that the oil reservoir of the arbor support is filled with the proper grade oil. The inner arbor support allows a closer approach to the cutter than an outer arbor support. One or both arbor supports may be used in a given setup. Style A arbor supports must be used when a Style A arbor is used. They are placed over the small end, or pilot end, of the Style A arbor. When in position, the arbor supports should be clamped firmly to the overarm.
10. Tighten the arbor nut. The arbor is now rigidly supported by the arbor supports so that the torque applied by the wrench will not bend and damage the arbor.

The cutter and the arbor are removed by performing the procedure just described in the reverse order. When removing the arbor, position the two keys on the spindle in a horizontal position. In this way they will tend to support the arbor when it is loosened and keep it from falling out of the spindle. Sometimes two or more cutters must be spaced along the arbor to a dimension that cannot be obtained with standard collars. In this event thin metal shims can be used. Shims made from a good grade of paper, with a uniform thickness, can be substituted if metal shims are unavailable. The spacers and shims should be carefully stored and kept free of nicks and burrs.

Straight shank end mills can be held by a collet chuck as shown in Fig. 5-28. The collet chuck is held in the spindle of the milling machine. Straight shank end milling cutters are also held in adaptors which have an accurately ground cylindrical hole to allow the straight end mill shank to be inserted in it with very little clearance. The end milling cutter is then clamped in the hole with a set screw which contacts a flat spot that has been machined on the shank. Cam-lock and bayonet-lock shanks and end mill holders have also been developed to hold the end mill in place. The shank of the end mill must correspond with the type of holder used. Shell end mills are mounted on shell end-mill arbors as shown in Fig. 5-15. The shell end-mill adaptor is held directly in the milling-machine spindle by the draw-in bar.

Face Milling Cutters. There are three methods of mounting face milling cutters shown in Figs. 5-29 and 5-30. Shell end mills bridge the gap between end mills and face milling cutters; however, their method of mounting is similar to that for face milling cutters. Shell end milling cutters and some small face milling cutters are mounted on a flange type centering shank (or shell end mill adaptor) such as is illustrated in Fig. 5-15 and in view A, Fig. 5-29. The adaptor, or centering shank, is held in the spindle by the draw-in bar and is driven by the keys on the nose of the milling machine. Two keys on the flange of the adaptor drive the cutter, which is held on the adaptor by a bolt and washer. Medium and large face milling cutters, however, are mounted on centering shanks which do not have a flange, as seen in view B, Fig. 5-29. The cutter is held on the centering shank by

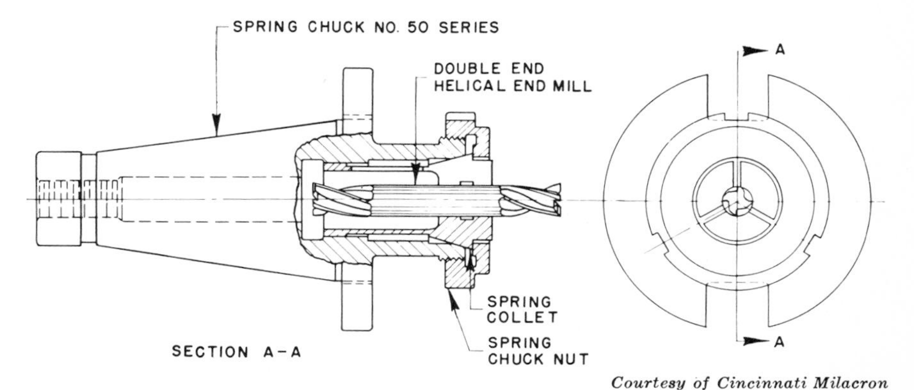

Courtesy of Cincinnati Milacron

Fig. 5-28. Spring collet chuck for holding straight shank end mills.

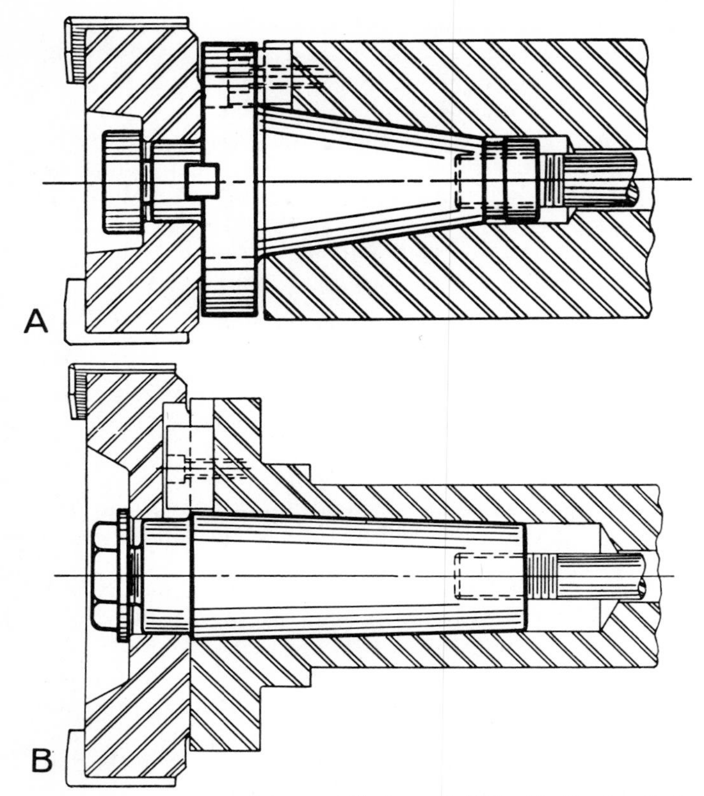

Courtesy of The Ingersoll Milling Machine Co., Cutting Tool Div.

Fig. 5-29. A. Flange type centering shank for mounting face milling cutters and shell end mills. B. Centering shank without flange for mounting face milling cutters.

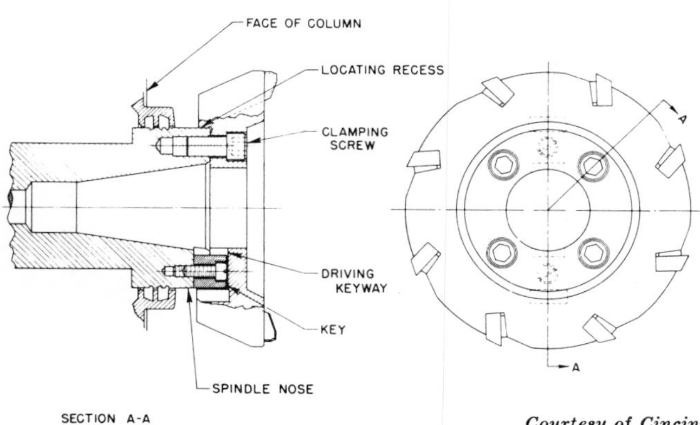

Courtesy of Cincinnati Milacron

Fig. 5-30. Face milling cutter mounted on the milling-machine spindle.

a bolt and washer, and is driven directly by the keys on the spindle nose of the milling machine. The National Standard Drive is shown in Fig. 5-30. Face milling cutters mounted in this manner must have a counterbore on the back side that fits very closely over the outside diameter of the milling machine spindle. The cutter is mounted directly on the spindle nose, bolted to it by four clamping screws and the drive is provided by the two keys on the spindle. Some very-small-diameter carbide face milling cutters have a straight shank which is held in an adaptor or a collet.

Fly Cutters and Fly-Cutter Holders. Fly cutters are single-point cutting tools used in lieu of milling cutters to mill various kinds of surfaces. They are held in a fly-cutter holder as shown in Fig. 5-31. Some fly-cutter holders are made longer than the one in the illustration. These longer fly-cutter holders are designed to be supported by an arbor support mounted on the overarm.

Fly cutters have the disadvantage of having only one cutting edge, which restricts the feed rate with which they can be used. They are, however, very useful in jobbing shops and in tool and die shops. Different shapes can be "form-milled" by simply grinding the cutting edge of the single-point cutting tool to the profile required. When only a few parts are to be made, this method is more economical than having an expensive special form milling cutter made.

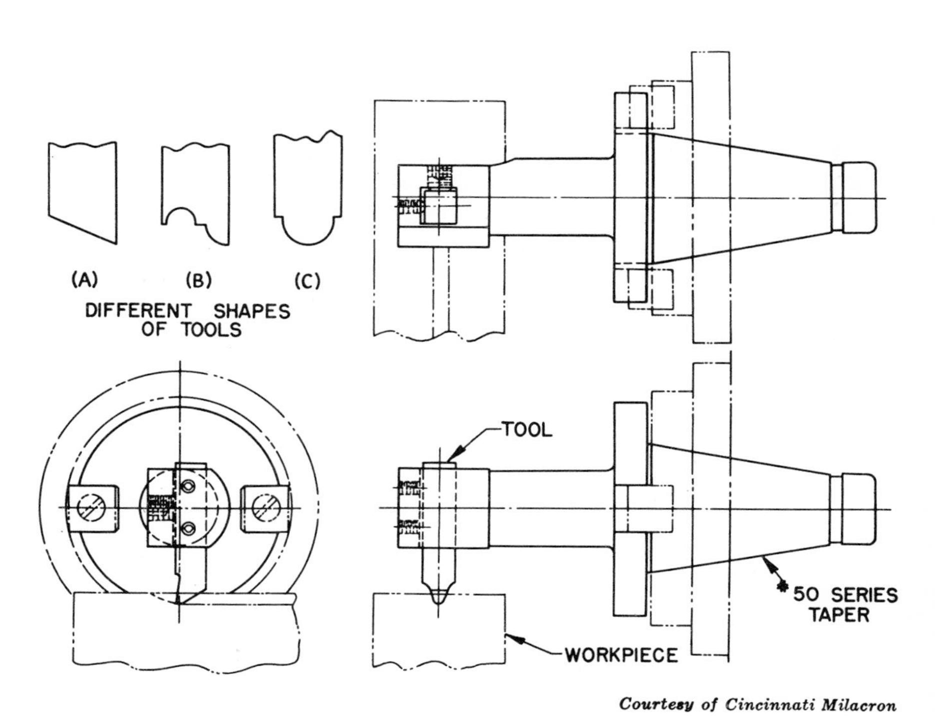

Courtesy of Cincinnati Milacron

Fig. 5-31. Fly cutters and fly cutter holder.

Another type of fly cutter is shown in Fig. 5-32. It offers many advantages in performing face milling operations for which it is intended. The cutting tool is a standard single-point, high-speed-steel tool bit which can be sharpened by hand. When used with a fine table feed rate, this cutter will produce an excellent surface finish on most materials. Since only a single cutting edge engages the workpiece, the cutting force is light, enabling frail parts to be milled and setups to be used which are somewhat less secure than required when milling with a face milling cutter. On light, low powered milling machines, relatively large surfaces can be milled in a single pass that would otherwise require a series of passes with an end milling cutter.

Cutting Speed for Milling

The cutting speed for milling is the speed at the periphery of the cutter as it is rotating. In the customary inch system of units the cutting speed is given in terms of feet per minute, or fpm, which is sometimes called surface feet per minute, or sfpm. In the metric system the cutting speed

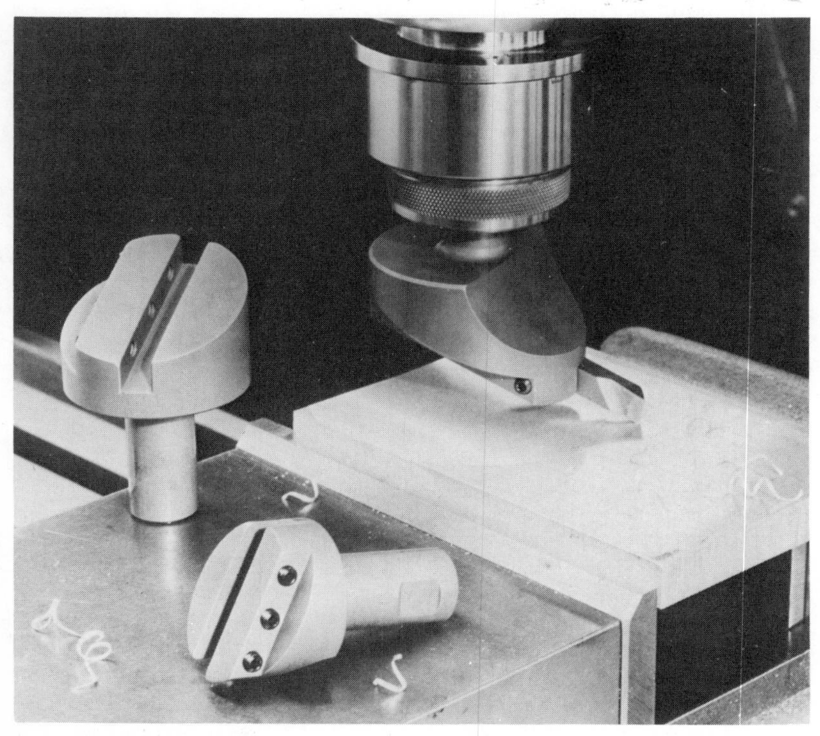

Courtesy of the City Tool Die & Mfg. Co., Inc.

Fig. 5-32. Fly cutter for performing face milling operations with a single point tool.

is in meters per minute, or m/min. The recommended cutting speeds feet per minute are given in Tables 5-3 through 5-7 for milling various materials with high-speed steel and cemented carbide cutters. To obtain meters per minute, multiply feet per minute by .3048. For each material a range of values is given to account for the shop variables encountered, which will be discussed in the following paragraph.

In addition to the cutting tool material, the cutting speed depends primarily on the work material and its hardness. The hardness range for which the listed cutting speed is valid in the case of each material is given in the tables. In general, an increase in the hardness of a material reduces the speed at which it can be cut. Since the hardness of a material is not always known in the shop, the material condition that is associated with a corresponding hardness in the table is given. The cutting speed is also influenced by the feed rate, and to a lesser extent by the depth of cut. Heavier cuts using a heavy feed require a slower cutting speed than do lighter cuts. Since the cost of replacing and sharpening a milling cutter is more than the cost of a single-point cutting tool, a longer tool life is more desirable for milling than for turning; therefore, the cutting speed for milling should be somewhat slower than for turning under the same tool and work material conditions. When using cemented carbide milling cutters the grade of carbide used has an influence on the cutting speed that can be used. The correct grade, as recommended by the carbide producer or cutter manufacturer, must be used. Where they can be used, coated carbides can often cut successfully using a cutting speed that is 20 to 40 per cent and sometimes up to 50 per cent higher than the values given in the cutting speed tables. In general, cemented carbide cutters having indexable inserts are operated at a somewhat faster cutting speed than those having brazed-on carbide tips, or blades to which the carbide is brazed. Other factors to consider in selecting the cutting speed are the design of the milling cutter and the rigidity of the workpiece, the setup, and the machine. When starting out to mill a new material, it is usually advisable to start at the lower end of the range of values given in the table; then, as experience is gained, the cutting speed may be increased.

Calculating the Cutting Speed

The formulas for calculating the speed of the milling machine spindle and the cutter are given below for inch and for metric units. Since the calculated speed may not be available on the machine, the closest available speed should be used. On some machines the range between speeds is large and it may be advisable to use the closest lower speed available.

$$N = \frac{12\,V}{\pi\,D} \qquad \text{(Inch units only)} \qquad (5\text{-}1)$$

$$N = \frac{1000\,V}{\pi\,D} \qquad \text{(Metric units only)} \qquad (5\text{-}2)$$

Table 5-3. Cutting Speed in Feet per Minute for Milling Plain Carbon and Alloy Steels

Material AISI and SAE Steels	Hardness, HB*	Material Condition*	Cutting Speed, fpm	
			HSS	Carbide
Free Machining Plain Carbon Steels (Resulphurized), 1212, 1213, 1215	100-150	HR, A	140	600
	150-200	CD	130	550
1108, 1109, 1115, 1117, 1118, 11120, 1126, 1211	100-150	HR, A	130	550
	150-200	CD	115	500
1132, 1137, 1139, 1140, 1144, 1146 1151	175-225	HR, A, N, CD	115	450
	275-325	Q and T	70	290
	325-375	Q and T	45	200
	375-425	Q and T	35	170
Free Machining Plain Carbon Steels (Leaded), 11L17, 11L18, 12L13, 12L14	100-150	HR, A, N, CD	140	600
	150-200	HR, A, N, CD	130	625
	200-250	N, CD	110	400
Plain Carbon Steels, 1006, 1008, 1009, 1010, 1012, 1015, 1016, 1017, 1018, 1019, 1020, 1021, 1022, 1023, 1024, 1025, 1026, 1513, 1514	100-125	HR, A, N, CD	110	425
	125-175	HR, A, N, CD	110	400
	175-225	HR, N, CD	90	350
	225-275	CD	65	250
1027, 1030, 1033, 1035, 1036, 1037, 1038, 1039, 1040, 1041, 1042, 1043, 1045, 1046, 1048, 1049, 1050, 1052, 1524, 1526, 1527, 1541	125-175	HR, A, N, CD	100	375
	175-225	HR, A, N, CD	85	325
	225-275	N, CD, Q and T	70	225
	275-325	Q and T	55	200
	325-375	Q and T	35	160
	375-425	Q and T	25	140
1055, 1060, 1064, 1065, 1070, 1074, 1078, 1080, 1084, 1086, 1090, 1095, 1548, 1551, 1552, 1561, 1566	125-175	HR, A, N, CD	90	350
	175-225	HR, A, N, CD	75	300
	225-275	N, CD, Q and T	60	200
	275-325	Q and T	45	160
	325-375	Q and T	30	145
	375-425	Q and T	15	125
Free Machining Alloy Steels (Resulphurized), 4140, 4150	175-200	HR, A, N, CD	100	400
	200-250	HR, N, CD	90	350
	250-300	Q and T	60	280
	300-375	Q and T	45	220
	375-425	Q and T	35	160
Free Machining Alloy Steels (Leaded), 41L30, 41L40, 41L47, 41L50, 43L47, 51L32, 52L100, 86L20, 86L40	150-200	HR, A, N, CD	115	425
	200-250	HR, N, CD	95	375
	250-300	Q and T	70	260
	300-375	Q and T	50	210
	375-425	Q and T	40	180
Alloy Steels, 4012, 4023, 4024, 4028, 4118, 4320, 4419, 4422, 4427, 4615, 4620, 4621, 4626, 4718, 4720, 4815, 4817, 4820, 5015, 5117, 5120, 6118, 8115, 8615, 8617, 8620, 8622, 8625, 8627, 8720, 8822, 94B17	125-175	HR, A, N, CD	100	400
	175-225	HR, N, CD	90	350
	225-275	CD, N, Q and T	60	250
	275-325	Q and T	50	200
	325-375	Q and T	40	175
	375-425	Q and T	25	150

* Abbreviations designate: HR, hot rolled; CD, cold drawn; A, annealed; N, normalized; Q and T, quenched and tempered; HB, Brinell hardness number; and HRC, Rockwell C scale hardness number.

Table 5-3 (*Cont.*) Cutting Speed in Feet per Minute for Milling Plain Carbon and Alloy Steels

Material AISI and SAE Steels	Hardness, HB*	Material Condition*	Cutting Speed, fpm	
			HSS	Carbide
Alloy Steels, 1330, 1335, 1340, 1345, 4032, 4037, 4042, 4047, 4130, 4135, 4137, 4140, 4142, 4145, 4147, 4150, 4161, 4337, 4340, 50B44, 50B46, 50B50, 50B60, 5130, 5132, 5140, 5145, 5147, 5150, 5160, 51B60, 6150, 81B45, 8630, 8635, 8637, 8640, 8642, 8645, 8650, 8655, 8660, 8740, 9254, 9255, 9260, 9262, 94B30	175-225	HR, A, N, CD	75	310
	225-275	N, CD, Q and T	60	260
	275-325	N, Q and T	50	210
	325-375	N, Q and T	35	180
	375-425	Q and T	20	140
Alloy Steels, E51100, E52100	175-225	HR, A, CD	65	300
	225-275	N, CD, Q and T	60	250
	275-325	N, Q and T	40	130
	325-375	N, Q and T	30	100
	375-425	Q and T	20	60
Ultra High Strength Steels (Not AISI) AMS 6421 (98B37 Mod.), AMS 6422 (98BV40), AMS 6424, AMS 6427, AMS 6428, AMS 6430, AMS 6432, AMS 6433, AMS 6434, AMS 6436, AMS 6442, 300M, D6 ac	220-300	A	60	250
	300-350	N	45	180
	350-400	N	20	130
	43-48 HRC	Q and T	..	100
	48-52 HRC	Q and T	..	60
Maraging Steels (Not AISI) 18% Ni Grade 200, 18% Ni Grade 250, 18% Ni Grade 300, 18% Ni Grade 350	250-325	A	50	250
	50-52 HRC	Maraged	..	60
Nitriding Steels (Not AISI) Nitralloy 125, Nitralloy 135, Nitralloy 135 (Mod.), Nitralloy 225, Nitralloy 230, Nitralloy N, Nitralloy EZ, Nitrex 1	200-250	A	60	280
	300-350	N, Q and T	25	200

* Abbreviations designate: HR, hot rolled; CD, cold drawn; A, annealed; N, normalized; Q and T, quenched and tempered; HB, Brinell hardness number; and HRC, Rockwell C scale hardness number.

Where: N = Spindle and milling cutter speed; rpm
V = Cutting speed; fpm, or m/min
D = Diameter of milling cutter; in. or mm
$\pi = 3.14$ (pi)

Example 5-2

A ½-inch (12.7 mm) diameter high-speed-steel end mill has four teeth and is to cut an O2 oil-hardening tool steel having a hardness of 200-220 HB. The cutting speed for this steel is 50 fpm ($50 \times .3048 = 15.2$ m/min). Calculate the spindle speed, using both inch and metric units.

Table 5-4. Cutting Speed in Feet per Minute for Milling Tool Steels

Material Tool Steels (AISI Types)	Hardness, HB*	Material Condition*	Cutting Speed, fpm	
			HSS	Carbide
Water Hardening W1, W2, W5	150-200	A	85	250
Shock Resisting S1, S2, S5, S6, S7	175-225	A	55	215
Cold Work, Oil Hardening O1, O2, O6, O7	175-225	A	50	200
Cold Work, High Carbon High Chromium D2, D3, D4, D5, D7	200-250	A	40	150
Cold Work, Air Hardening				
A2, A3, A8, A9, A10	200-250	A	50	200
A4, A6	200-250	A	45	160
A7	225-275	A	40	140
Hot Work, Chromium Type H10, H11, H12, H13, H14, H19	150-200	A	60	250
	200-250	A	50	200
	325-375	Q and T	30	150
	48-50 HRC	Q and T	—	80
	50-52 HRC	Q and T	—	60
	52-54 HRC	Q and T	—	40
	54-56 HRC	Q and T	—	20
Hot Work, Tungsten Type H21, H22, H23, H24, H25, H26	150-200	A	55	200
	200-250	A	45	170
Hot Work, Molybdenum Type H41, H42, H43	150-200	A	55	180
	200-250	A	45	140
Special Purpose, Low Alloy L2, L3, L6	150-200	A	65	300
Mold				
P2, P3, P4, P5, P6	100-150	A	75	350
P20, P21	150-200	A	60	300
High-Speed Steel				
M1, M2, M6, M10, T1, T2, T6	200-250	A	50	175
M3-1, M4, M7, M30, M33, M34, M36, M41, M42, M43, M44, M46, M47, T5, T8	225-275	A	40	150
T15, M3-2	225-275	A	30	130

* Abbreviations designate: A, annealed; Q and T, quenched and tempered; and HB, Brinell hardness number.

$$N=\frac{12\,V}{\pi\,D}=\frac{12\times 50}{\pi\times .5}$$
$$=382 \text{ rpm}$$

$$N=\frac{1000\,V}{\pi\,D}=\frac{1000\times 15.2}{\pi\times 12.7}$$
$$=381 \text{ rpm}$$

Milling Machine Table Feed Rate

The milling machine table feed rate should always be calculated in order to obtain the maximum production rate from the cutter and the machine. Using a machine table feed rate which has been determined by

Table 5-5. Cutting Speed in Feet per Minute for Milling Stainless Steels

Material Stainless Steels	Hardness, HB*	Material Condition*	Cutting Speed, fpm HSS	Cutting Speed, fpm Carbide
Free Machining Stainless Steels				
(Ferritic), 430F, 430F Se	135-185	A	95	375
(Austenitic), 203EZ, 303, 303 Se, 303MA, 303Pb, 303Cu, 303 Plus X	135-185	A	90	325
	225-275	CD	75	300
(Martensitic), 416, 416 Se, 416 Plus X, 420F, 420F Se, 440F, 440F Se	135-185	A	95	375
	185-240	CD	80	325
	275-325	Q and T	50	225
	375-425	Q and T	20	100
Stainless Steels				
(Ferritic), 405, 409, 429, 430, 434, 436, 442, 446, 502	135-185	A	75	275
(Austenitic), 201, 202, 301, 302, 304, 304L, 305, 308, 321, 347, 348	135-185	A	60	200
	225-275	CD	50	180
(Austenitic), 302B, 309, 309S, 310, 310S, 314, 316, 316L, 317, 330	135-185	A	50	200
(Martensitic), 403, 410, 420, 501	135-175	A	75	325
	175-225	A	65	275
	275-325	Q and T	40	175
	375-425	Q and T	25	100
(Martensitic), 414, 431, Greek Ascoloy	225-275	A	55	225
	275-325	Q and T	45	180
	375-425	Q and T	25	100
(Martensitic), 440A, 440B, 440C	225-275	A	50	180
	275-325	Q and T	40	140
	375-425	Q and T	20	100
(Precipitation Hardening) 15-5PH, 17-4PH, 17-7PH, AF-71, 17-14Cu Mo, AFC-77, AM-350, AM-355, AM-362, Custom 455, HNM, PH13-8, PH14-8Mo, PH15-7Mo, Stainless W	150-200	A	60	200
	275-325	H	50	180
	325-375	H	40	110
	375-450	H	25	75

* Abbreviations designate: A, annealed; CD, cold drawn; Q and T, quenched and tempered; H, precipitation hardened; and HB, Brinell hardness number.

a guess can cause the milling cutter teeth to be overloaded or drastically underloaded, each of which will have an adverse effect on the cutter. The milling machine table feed rate is expressed in terms of inches per minute (in./min), or millimeters per minute (mm/min) on metric machines. To convert from one to the other: multiply in./min by 25.4 to obtain mm/min; divide mm/min by 25.4 to obtain in./min.

Chip Load per Tooth. The basic measure of the feed rate of milling cutters is the chip load per tooth, which for customary inch units is expressed in terms of inch per tooth (in./tooth); in SI metric units it is expressed in terms of millimeters per tooth (mm/tooth). They can be converted from one into the other as follows: multiply in./tooth by 25.4 to obtain mm/tooth; divide mm/tooth by 25.4 to obtain in./tooth. Recommended values of the basic feed rate are given in Tables 5-8 and 5-9 for different types of milling cutters and for different materials. Feed rates

Table 5-6. Cutting Speed in Feet per Minute for Milling Ferrous Cast Metals

Material Ferrous Cast Metals	Hardness, HB*	Material Condition*	Cutting Speed, fpm	
			HSS	Carbide
Gray Cast Iron				
ASTM Class 20	120-150	A	100	425
ASTM Class 25	160-200	AC	80	325
ASTM Class 30, 35, and 40	190-220	AC	70	250
ASTM Class 45 and 50	220-260	AC	50	190
ASTM Class 55 and 60	250-260	AC, HT	30	110
ASTM Type 1, 1b, 5 (Ni-Resist)	100-215	AC	50	200
ASTM Type 2, 3, 6 (Ni-Resist)	120-175	AC	40	190
ASTM Type 2b, 4 (Ni-Resist)	150-250	AC	30	180
Malleable Iron				
(Ferritic), 32510, 35018	110-160	MHT	110	475
(Pearlitic), 40010, 43010, 45006, 45008, 48005, 50005	160-200	MHT	80	375
	200-240	MHT	65	250
(Martensitic), 53004, 60003, 60004	200-255	MHT	55	225
(Martensitic), 70002, 70003	220-260	MHT	50	200
(Martensitic), 80002	240-280	MHT	45	130
(Martensitic), 90001	250-320	MHT	25	110
Nodular (Ductile) Iron				
(Ferritic), 60-40-18, 65-45-12	140-190	A	75	425
(Ferritic-Pearlitic), 80-55-06	190-225	AC	60	325
	225-260	AC	50	200
(Pearlitic-Martensitic), 100-70-03	240-300	HT	40	160
(Martensitic), 120-90-02	270-330	HT	25	90
	330-400	HT	—	30
Cast Steels				
(Low Carbon), 1010, 1020	100-150	AC, A, N	100	375
(Medium Carbon), 1030, 1040, 1050	125-175	AC, A, N	95	375
	175-225	AC, A, N	80	325
	225-300	AC, HT	60	250
(Low Carbon Alloy), 1320, 2315, 2320, 4110, 4120, 4320, 8020, 8620	150-200	AC, A, N	85	325
	200-250	AC, A, N	75	300
	250-300	AC, HT	50	225
(Medium Carbon Alloy), 1330, 1340, 2325, 2330, 4125, 4130, 4140, 4330, 4340, 8030, 80B30, 8040, 8430, 8440, 8630, 8640, 9525, 9530, 9535	175-225	AC, A, N	70	300
	225-250	AC, A, N	65	250
	250-300	AC, HT	50	200
	300-350	AC, HT	30	180
	350-400	HT	..	125

* Abbreviations designate: A, annealed; AC, as cast; N, normalized; HT, heat treated; MHT, malleablizing heat treatment; and HB, Brinell hardness number.

Table 5-7. Cutting Speed in Feet per Minute for Milling Light Metals and Copper Alloys.

Materials	Material Condition *	Cutting Speed, fpm	
		HSS	Carbide
Light Metals			
All Wrought Aluminum Alloys	CD	600	1200
	ST and A	500	1100
All Aluminum Sand and Permanent Mold Casting Alloys	AC	750	1400
	ST and A	600	1200
All Aluminum Die Casting Alloys †	AC	125	550
	ST and A	100	450
† except Alloys 390.0 and 392.0	AC	80	500
	ST and A	60	425
All Wrought Magnesium Alloys	A, CD, ST and A	800	2000
All Cast Magnesium Alloys	A, AC, ST and A	800	2000
Copper Alloys			
314 Leaded Commercial Bronze; 332 High Leaded Brass; 340 Medium Leaded Brass; 342 High Leaded Brass; 353 High Leaded Brass; 356 Extra-High Leaded Brass; 360 Free-Cutting Brass; 370 Free-Cutting Muntz Metal; 377 Forging Brass; 385 Architectural Bronze; 485 Leaded Naval Brass; 544 Free-Cutting Phosphor Bronze	A	300	650
	CD	350	600
226 Jewelry Bronze; 230 Red Brass; 240 Low Brass; 260 Cartridge Brass, 70%; 268 Yellow Brass; 280 Muntz Metal; 235 Low Leaded Brass; 365 Leaded Muntz Metal; 368 Leaded Muntz Metal; 443 Admiralty Brass (inhibited); 445 Admiralty Brass (inhibited); 651 Low Silicon Bronze; 655 High Silicon Bronze; 675 Manganese Bronze; 687 Aluminum Brass; 770 Nickel Silver; 796 Leaded Nickel Silver	A	200	500
	CD	250	550
102 Oxygen-Free Copper; 110 Electrolytic Tough Pitch Copper; 122 Phosphorus Deoxidized Copper; 170 Beryllium Copper; 172 Beryllium Copper; 175 Beryllium Copper; 210 Gilding, 95%; 220 Commercial Bronze; 502 Phosphor Bronze, 1.25%; 510 Phosphor Bronze, 5%; 521 Phosphor Bronze, 8%; 524 Phosphor Bronze, 10%; 614 Aluminum Bronze; 706 Copper Nickel, 10%; 715 Copper Nickel, 30%; 745 Nickel Silver; 752 Nickel Silver; 754 Nickel Silver; 757 Nickel Silver	A	100	200
	CD	110	225

* Abbreviations designate: A, annealed; AC, as cast; CD, cold drawn; and ST and A, solution treated and aged.

Table 5-8. Feed in Inches per Tooth (f_t) for Milling with High-Speed Steel Cutters

Material	Hard-ness, HB	End Mills							Plain or Slab Mills	Form Reliev-ed Cutters	Face Mills and Shell End Mills	Slotting and Side Mills
		Depth of Cut, .250 in.			Depth of Cut, .050 in.							
		Cutter Diam., in.			Cutter Diam., in.							
		½	¾	1 and up	¼	½	¾	1 and up				
		Feed per Tooth, inch										
Cast Steel	100-180 180-240 240-300	.001 .001 .001	.003 .002 .002	.003 .003 .002	.001 .001 .0005	.002 .002 .002	.003 .003 .002	.004 .003 .002	.003-.008 .003-.008 .002-.006	.004 .004 .003	.003-.012 .003-.010 .003-.008	.002-.008 .002-.006 .002-.005
Zinc Alloys (Die Castings)	...	.002	.003	.004	.001	.003	.004	.006	.003-.010	.005	.004-.015	.002-.012
Copper Alloys (Brasses & Bronzes)	100-150 150-250	.002 .002	.004 .003	.005 .004	.002 .001	.003 .003	.005 .004	.006 .005	.003-.015 .003-.015	.004 .004	.004-.020 .003-.012	.002-.010 .002-.008
Free Cutting Brasses & Bronzes	80-100	.002	.004	.005	.002	.003	.005	.006	.003-.015	.004	.004-.015	.002-.010
Cast Aluminum Alloys—As Cast	...	.003	.004	.005	.002	.004	.005	.006	.005-.016	.006	.005-.020	.004-.012
Cast Aluminum Alloys—Hardened	...	.003	.004	.005	.002	.003	.004	.005	.004-.012	.005	.005-.020	.004-.012
Wrought Aluminum Alloys—Cold Drawn	...	.003	.004	.005	.002	.003	.004	.005	.004-.014	.005	.005-.020	.004-.012
Wrought Aluminum Alloys—Hardened	...	.002	.003	.004	.001	.002	.003	.004	.003-.012	.004	.005-.020	.004-.012
Magnesium Alloys	...	.003	.004	.005	.003	.004	.005	.007	.005-.016	.006	.008-.020	.005-.012
Ferritic Stainless Steel	135-185	.001	.002	.003	.001	.002	.003	.003	.002-.006	.004	.004-.008	.002-.007
Austenitic Stainless Steel	135-185 185-275	.001 .001	.002 .002	.003 .003	.001 .001	.002 .002	.003 .002	.003 .002	.003-.007 .003-.006	.004 .003	.005-.008 .004-.006	.002-.007 .002-.007
Martensitic Stainless Steel	135-185 185-225 225-300	.001 .001 .0005	.002 .002 .002	.002 .002 .002	.001 .001 .0005	.002 .002 .001	.003 .002 .002	.003 .003 .002	.003-.006 .003-.006 .002-.005	.004 .004 .003	.004-.010 .003-.008 .002-.006	.002-.007 .002-.007 .002-.005
Monel	100-160	.001	.003	.004	.001	.002	.003	.004	.002-.006	.004	.002-.008	.002-.006

Table 5-8 (*Cont.*) Feed in Inches per Tooth (f_t) for Milling with High-Speed Steel Cutters

Material	Hardness, HB	End Mills: Depth of Cut, .250 in., Cutter Diam., in. ½	¾	1 and up	End Mills: Depth of Cut, .050 in., Cutter Diam., in. ¼	½	¾	1 and up	Plain or Slab Mills	Form Relieved Cutters	Face Mills and Shell End Mills	Slotting and Side Mills
		Feed per Tooth, inch										
Free Machining Plain Carbon Steels	100-185	.001	.003	.004	.001	.002	.003	.004	.003-.008	.005	.004-.012	.002-.008
Plain Carbon Steels, AISI 1006 to 1030; 1513 to 1522	100-150	.001	.003	.003	.001	.002	.003	.004	.003-.008	.004	.004-.012	.002-.008
	150-200	.001	.002	.003	.001	.002	.002	.003	.003-.008	.004	.003-.012	.002-.008
AISI 1033 to 1095; 1524 to 1566	120-180	.001	.003	.003	.001	.002	.003	.004	.003-.008	.004	.004-.012	.002-.008
	180-220	.001	.002	.003	.001	.002	.002	.003	.003-.008	.004	.003-.012	.002-.008
	220-300	.001	.002	.002	.001	.001	.002	.003	.002-.006	.003	.002-.008	.002-.006
Alloy Steels having less than 3% Carbon. Typical examples: AISI 4012, 4023, 4027, 4118, 4320, 4422, 4427, 4615, 4620, 4626, 4720, 4820, 5015, 5120, 6118, 8115, 8620, 8627, 8720, 8822, 9310, 93B17	125-175	.001	.003	.003	.001	.002	.003	.004	.003-.008	.004	.004-.012	.002-.008
	175-225	.001	.002	.003	.001	.002	.003	.003	.003-.008	.004	.003-.012	.002-.008
	225-275	.001	.002	.003	.001	.001	.002	.003	.002-.006	.003	.003-.008	.002-.006
	275-325	.001	.002	.002	.001	.001	.002	.002	.002-.005	.003	.002-.008	.002-.005
Alloy Steels have 3% Carbon or more. Typical examples: AISI 1330, 1340, 4032, 4037, 4130, 4140, 4150, 4340, 50B40, 50B60, 5130, 51B60, 6150, 81B45, 8630, 8640, 86B45, 8660, 8740, 94B30	175-225	.001	.002	.003	.001	.002	.003	.004	.003-.008	.004	.003-.012	.002-.008
	225-275	.001	.002	.003	.001	.001	.002	.003	.002-.006	.003	.003-.010	.002-.006
	275-325	.001	.002	.002	.001	.001	.002	.003	.002-.005	.003	.002-.008	.002-.005
	325-375	.001	.002	.002	.001	.001	.002	.002	.002-.004	.002	.002-.008	.002-.005
Tool Steel	150-200	.001	.002	.002	.001	.002	.003	.003	.003-.008	.004	.003-.010	.002-.006
	200-250	.001	.002	.002	.001	.002	.002	.003	.002-.006	.003	.003-.008	.002-.005
Gray Cast Iron	120-180	.001	.003	.004	.002	.003	.004	.004	.004-.012	.005	.005-.016	.002-.010
	180-225	.001	.002	.003	.001	.002	.003	.003	.003-.010	.004	.004-.012	.002-.008
	225-300	.001	.002	.002	.001	.001	.002	.002	.002-.006	.003	.002-.008	.002-.005
Ferritic Malleable Iron	110-160	.001	.003	.004	.002	.003	.004	.004	.003-.010	.005	.005-.016	.002-.010
Pearlitic-Martensitic Malleable Iron	160-200	.001	.003	.004	.001	.002	.003	.004	.003-.010	.004	.004-.012	.002-.008
	200-240	.001	.002	.003	.001	.002	.003	.003	.003-.007	.004	.003-.010	.002-.006
	240-300	.001	.002	.002	.001	.001	.002	.002	.002-.006	.003	.002-.008	.002-.005

Table 5-9. Feed in Inches per Tooth for Milling with Cemented Carbide Cutters

Material	Hardness, HB	Face Mills	Slotting and Side Mills
		Feed per Tooth, inch	
Free Machining Plain Carbon Steels	100-185	.008-.020	.003-.010
Plain Carbon Steels, AISI 1006 to 1030, 1513 to 1522	100-150	.008-.020	.003-.010
	150-200	.008-.020	.003-.010
Plain Carbon Steels, AISI 1033 to 1095, 1524 to 1566	120-180	.005-.020	.003-.010
	180-220	.005-.020	.003-.010
	220-300	.003-.012	.003-.008
Alloy Steels having less than .3% Carbon content. Typical examples: AISI 4012, 4023, 4027, 4118, 4320, 4422, 4427, 4615, 4620, 4626, 4720, 4820, 5015, 5120, 6118, 8115, 8620, 8627, 8720, 8822, 9310, 93B17	125-175	.006-.020	.003-.010
	175-225	.006-.020	.003-.010
	225-275	.006-.016	.003-.010
	275-325	.004-.012	.003-.008
	325-375	.003-.008	.003-.007
Alloy Steels having .3% Carbon content, or more. Typical examples: AISI 1330, 1340, 4032, 4037, 4130, 4140, 4150, 4340, 50B40, 50B60, 5130, 51B60, 6150, 81B45, 8630, 8640, 86B45, 8660, 8740, 94B30	175-225	.005-.020	.003-.010
	225-275	.004-.012	.003-.008
	275-325	.003-.010	.003-.008
	325-375	.003-.008	.003-.007
Tool Steels	200-275	.004-.012	.003-.007
	275-325	.003-.010	.003-.006
	36-45 HRC	.003-.006	.002-.005
	45-55 HRC	.003-.005	.002-.003
Ferritic Stainless Steels	110-160	.005-.015	.003-.010
Austenitic Stainless Steels	135-185	.005-.012	.003-.010
	185-275	.005-.010	.003-.008
Martensitic Stainless Steel	135-185	.005-.015	.003-.010
	185-225	.005-.010	.003-.008
	225-300	.004-.008	.003-.007
Precipitation Hardening Stainless Steels	Annealed	.004-.012	.003-.010
	275-350	.003-.008	.002-.005
	350-450	.002-.005	.002-.004
Cast Steel	100-180	.008-.020	.003-.010
	180-240	.005-.016	.003-.010
	240-300	.004-.012	.003-.008
Gray Cast Iron	140-185	.008-.020	.005-.012
	185-225	.008-.016	.005-.010
	225-300	.005-.012	.004-.008
Ferritic Malleable Iron	110-160	.005-.020	.004-.012
Pearlitic-Martensitic Malleable Iron	160-200	.005-.020	.003-.010
	200-240	.005-.016	.003-.010
	240-300	.004-.010	.003-.008
Nodular (Ductile) Iron	140-200	.008-.020	.003-.010
	200-275	.006-.014	.003-.008
	275-325	.005-.012	.003-.007
	325-400	.003-.008	.002-.004
Copper Alloys (Brasses and Bronzes)	100-150	.005-.020	.003-.012
	150-250	.004-.014	.003-.010

Table 5-9 (*Cont.*) Feed in Inches per Tooth for Milling with Cemented Carbide Cutters

Material	Hardness, HB	Face Mills	Slotting and Side Mills
		Feed per Tooth, inch	
Wrought and Cast Aluminum Alloys	....	.005-.020	.005-.020
Wrought and Cast Magnesium Alloys	....	.005-.020	.005-.020
Superalloys	....	.003-.010	.002-.006
Titanium Alloys	....	.003-.010	.002-.006
Nickel Alloys	....	.003-.010	.002-.006
Monel	....	.003-.010	.002-.006
Plastics, Hard Rubber, etc.	....	.003-.015	.003-.012

less than approximately .001 in./tooth (0.03 mm/tooth) should not be used except with small end mills and when milling certain very hard materials. At such low feed rates the teeth will tend to rub against the workpiece instead of penetrating to form a chip, resulting in excessive tool wear. With a dull cutter this may occur at even higher feed rates.

Table Feed Rate Formula. The formula for the table feed rate, given below, can be used either with customary inch or SI metric units, but not both at the same time.

$$f_m = f_t\, n_t\, N \qquad (5\text{-}3)$$

Where: f_m = Feed rate of table; in./min, or mm/min
f_t = Feed rate of cutter; in./tooth, or mm/tooth
n_t = Number of teeth on cutter
N = Spindle speed, or cutter speed; rpm

Example 5-2

A ½-inch (12.7 mm) diameter high-speed steel end mill having four teeth is to be used to mill a .250-inch (6.35-mm) deep slop in D2 high carbon, high chrome, tool steel which has been annealed to a hardness of 200-210 HB. The spindle speed to be used is 300 rpm. Calculate the table feed rate by using the inch units and then by using the metric units.

The feed rate selected from Table 5-8 for a ½-inch end mill is .001 in./tooth, or 0.025 mm per/tooth.

Inch units: $f_m = f_t\, n_t\, N = .001 \times 4 \times 300 = 1.2$ in./min
Metric units: $f_m = f_t\, n_t\, N = 0.025 \times 4 \times 300 = 30$ mm/min (1.2 in./min)

Estimating Milling Power

The available power on any machine tool places a limit on the size of the cut it can take. When a large amount of metal must be removed from the workpiece it is necessary to estimate the maximum size of the cut that can be taken without overloading the machine. But since many machining operations require that only light cuts be taken for which the machine obviously has ample power, in such cases, estimating the power required to take the cut would be a wasteful effort. The formulas which follow can be used to estimate the power required for milling. Since conditions in different shops may vary and machine tools are not all designed alike, the calculated results may not correspond precisely with the results actually obtained on the job; however, the calculations do provide a reasonable estimate which will suffice in most practical situations.

The measure of power in customary inch units is the horsepower; in SI metric units the kilowatt is the measure of both mechanical and electrical power. The power required to cut a material is dependent on the rate at which it is being cut and upon the power constant of the material, K_p. Each material has a power constant which will vary with the hardness of the material. The power constant is determined experimentally and is equal to the horsepower required to cut a material at a rate of one cubic inch per minute; in SI metric units the power constant is equal to the power in kilowatts required to cut a material at a rate of one cubic centimeter per second, or 1000 cubic millimeters per second ($1 cm^3 = 1000$ mm^3). Different values of the power constant are required for use with customary inch and with SI metric units, which are related as follows: to obtain the SI metric power constant, *multiply* the inch power constant by 2.73; to obtain the inch power constant, *divide* the SI metric power constant by 2.73. The power constant is also called the unit horsepower, unit power, and the specific power consumption. Values of the power constant are provided in Tables 5-10, 5-11, and 5-12.

The values of the power constant are essentially unaffected by the cutting speed, the depth of cut, and the cutting tool material when operating at normal cutting conditions. There are, however, factors that do affect the value of the power constant and thereby the power required to cut a material. They include the hardness and microstructure of the work material, the feed rate, the rake angle, and the condition, either sharp or dull, of the cutting edge. Power constant factors for different feed rates are given in Table 5-13. The values in the power constant tables are for sharp cutting tools; however, all metal cutting tools wear as they are used and require more power to cut the material as they become dull or worn. Factors to provide for tool wear are given in Table 5-14. In this table, extra-heavy-duty face milling occurs only in the case of certain high production operations, such as are found in the automotive industry. Most other face milling operations are in the light- and medium-duty category. The effect of the rake angle can usually be disregarded for average milling applications. The basic rake angle for most values found in the power constant tables is positive 14 degrees. Only when the deviation from this angle is great, is it necessary to consider the effect of the rake angle. If

Table 5-10. Power Constants, K_p, for Wrought Steels, Using Sharp Cutting Tools

Material	Brinell Hardness Number	K_p Inch Units	K_p SI Metric Units
Plain Carbon Steels			
All Plain Carbon Steels	80-100	.63	1.72
	100-120	.66	1.80
	120-140	.69	1.88
	140-160	.74	2.02
	160-180	.78	2.13
	180-200	.82	2.24
	200-220	.85	2.32
	220-240	.89	2.43
	240-260	.92	2.51
	260-280	.95	2.59
	280-300	1.00	2.73
	300-320	1.03	2.81
	320-340	1.06	2.89
	340-360	1.14	3.11
Free Machining Steels			
AISI 1108, 1109, 1110, 1115, 1116, 1117, 1118, 1119, 1120, 1125, 1126, 1132	100-120	.41	1.12
	120-140	.42	1.15
	140-160	.44	1.20
	160-180	.48	1.31
	180-200	.50	1.36
AISI 1137, 1138, 1139, 1140, 1141, 1144, 1145, 1146, 1148, 1151	180-200	.51	1.39
	200-220	.55	1.50
	220-240	.57	1.56
	240-260	.62	1.69
Alloy Steels			
AISI 4023, 4024, 4027, 4028, 4032, 4037, 4042, 4047, 4137, 4140, 4142, 4145, 4147, 4150, 4340, 4640, 4815, 4817, 4820, 5130, 5132, 5135, 5140, 5145, 5150, 6118, 6150, 8637, 8640, 8642, 8645, 8650, 8740	140-160	.62	1.69
	160-180	.65	1.77
	180-200	.69	1.88
	200-220	.72	1.97
	220-240	.76	2.07
	240-260	.80	2.18
	260-280	.84	2.29
	280-300	.87	2.38
	300-320	.91	2.48
	320-340	.96	2.62
	340-360	1.00	2.73
AISI 4130, 4320, 4615, 4620, 4626, 5120, 8615, 8617, 8620, 8622, 8625, 8630, 8720	140-160	.56	1.53
	160-180	.59	1.61
	180-200	.62	1.69
	200-220	.65	1.77
	220-240	.70	1.91
	240-260	.74	2.02
	260-280	.77	2.10
	280-300	.80	2.18
	300-320	.83	2.27
	320-340	.89	2.43
AISI 1330, 1335, 1340, E52100	160-180	.79	2.16
	180-200	.83	2.27
	200-220	.87	2.38
	220-240	.91	2.48
	240-260	.95	2.59
	260-280	1.00	2.73

Table 5-11. Power Constant, K_p, for Ferrous Cast Metals, Using Sharp Cutting Tools

Material	Brinell Hardness Number	K_p Inch Units	K_p SI Metric Units	Material	Brinell Hardness Number	K_p Inch Units	K_p SI Metric Units
Gray Cast Iron	100-120	.28	0.76	Malleable Iron			
	120-140	.35	0.96	Ferritic	150-175	.42	1.15
	140-160	.38	1.04	Pearlitic	175-200	.57	1.56
	160-180	.52	1.42		200-250	.82	2.24
	180-200	.60	1.64		250-300	1.18	3.22
	200-220	.71	1.94				
	220-240	.91	2.48	Cast Steel	150-175	.62	1.69
Alloy Cast Iron	150-175	.30	0.82		175-200	.78	2.13
	175-200	.63	1.72		200-250	.86	2.35
	200-250	.92	2.51	...	...	...	...
				...	...	...	...

Table 5-12. Power Constant, K_p, for High-Temperature Alloys, Tool Steel, Stainless Steel, and Nonferrous Metals, Using Sharp Cutting Tools

Material	Brinell Hardness Number	K_p Inch Units	K_p SI Metric Units	Material	Brinell Hardness Number	K_p Inch Units	K_p SI Metric Units
High Temperature Alloys							
A286	165	.82	2.24	Stainless Steel	150-175	.60	1.64
A286	285	.93	2.54		175-200	.72	1.97
Chromoloy	200	.78	3.22		200-250	.88	2.40
Chromoloy	310	1.18	3.00	Zinc Die Cast Alloys	...	.25	0.68
Inco 700	330	1.12	3.06	Copper (pure)	...	.91	2.48
Inco 702	230	1.10	3.00	Brass			
Hastelloy-B	230	1.10	3.00	Hard	...	.83	2.27
M-252	230	1.10	3.00	Medium	...	.50	1.36
M-252	310	1.20	3.28	Soft	...	.25	0.68
Ti-150A	340	.65	1.77	Leaded	...	.30	0.82
U-500	375	1.10	3.00	Bronze			
Monel Metal	...	1.00	2.73	Hard	...	.91	2.48
				Medium	...	.50	1.36
				Soft	...	.33	0.90
Tool Steel	175-200	.75	2.05	Aluminum			
	200-250	.88	2.40	Cast	...	.25	0.68
	250-300	.98	2.68	Rolled (hard)	...	.33	0.90
	300-350	1.20	3.28				
	350-400	1.30	3.55	Magnesium Alloys	...	.10	0.27

the rake angle used is more positive, the power required will decrease approximately one per cent per degree; if the rake angle used is more negative the power required will increase, again approximately one per cent per degree.

The machine tool serves to transmit the power from the driving motor to the cutter, where it is used to cut the workpiece. A measure of the efficiency by which this is done is the machine tool efficiency factor, E. Average values of this factor are given in Table 5-15. Cutting fluids will usually decrease the power required to cut the material when operating in the lower range of cutting speeds. It is not, however, possible to provide specific recommendations for the effect of cutting fluids because each cutting fluid exhibits its own characteristics.

Formulas for estimating the power at the cutter and at the motor are

Table 5-13. Feed Factor, C, for Power Constants

Inch Units				SI Metric Units			
Feed in.*	C	Feed in.*	C	Feed mm†	C	Feed mm†	C
.001	1.60	.014	.97	0.02	1.70	0.35	.97
.002	1.40	.015	.96	0.05	1.40	0.38	.95
.003	1.30	.016	.94	0.07	1.30	0.40	.94
.004	1.25	.018	.92	0.10	1.25	0.45	.92
.005	1.19	.020	.90	0.12	1.20	0.50	.90
.006	1.15	.022	.88	0.15	1.15	0.55	.88
.007	1.11	.025	.86	0.18	1.11	0.60	.87
.008	1.08	.028	.84	0.20	1.08	0.70	.84
.009	1.06	.030	.83	0.22	1.06	0.75	.83
.010	1.04	.032	.82	0.25	1.04	0.80	.82
.011	1.02	.035	.80	0.28	1.01	0.90	.80
.012	1.00	.040	.78	0.30	1.00	1.00	.78
.013	.98	.060	.72	0.33	.98	1.50	.72

* Turning–in./rev; Milling–in./tooth; Planing and Shaping–in./stroke; Broaching–in./tooth.
†Turning–mm/rev; Milling–mm/tooth; Planing and Shaping–mm/stroke; Broaching–mm/tooth.

Table 5-14. Tool Wear Factors, W

Type of Operation		W
For all operations with sharp cutting tools		1.00
Turning:	Finish turning (lightcuts)	1.10
	Normal rough and semi-finish turning	1.30
	Extra-heavy duty rough turning	1.60-2.00
Milling:	Slab milling	1.10
	End milling	1.10
	Light and medium face milling	1.10-1.25
	Extra-heavy duty face milling	1.30-1.60
Drilling:	Normal drilling	1.30
	Drilling hard-to-machine materials and drilling with a very dull drill	1.50
Broaching:	Normal broaching	1.05-1.10
	Heavy duty surface broaching	1.20-1.30

For planing and shaping, use values given for turning.

given below. In using these formulas it will also be necessary to use Formulas 5-1, 5-2, and 5-3, which have been provided previously in this chapter.

For inch units only:

$$Q = f_m \, w \, d \tag{5-4}$$

For SI metric units only:

$$Q = \frac{f_m \, w \, d}{60{,}000} \tag{5-5}$$

For either inch or SI metric units:

$$P_c = K_p \, C \, Q \, W \tag{5-6}$$

$$P_m = \frac{P_c}{E} = \frac{K_p \, C \, Q \, W}{E} \tag{5-7}$$

Table 5-15. Machine Tool Efficiency Factors, E

Type of Drive	E	Type of Drive	E
Direct Belt Drive	.90	Geared Head Drive	.70-.80
Back Gear Drive	.75	Oil-Hydraulic Drive	.60-.90

Where: P_c = Power at the cutter; hp, or kW
P_m = Power at the motor; hp, or kW
K_p = Power constant (See Tables 5-10, 5-11, and 5-12)
Q = Metal removal rate; in.^{3}/min, or cm^3/s
C = Feed factor for power constant (See Table 5-13)
W = Tool wear factor (See Table 5-14)
E = Machine tool efficiency factor (See Table 5-15)
f_m = Table feed rate; in./min, or mm/min
w = Width of cut; in., or mm
d = Depth of cut; in., or mm

Example 5-3

A .125-in. (3.18-mm)-deep by 3-inch (76.2-mm)-wide cut is to be taken in 180-200 HB gray cast iron with a 14-tooth, 4-inch (101.6 mm) diameter shell end milling cutter, using a cutting speed of 70 fpm (21.3 m/min) and a feed of .006 in./tooth (0.15 mm/tooth), on a milling machine having an efficiency of .80. Using both inch and metric formulas, estimate the power at the motor required to take this cut.

Inch units: $K_p = .60$ (From Table 5-11); $C = 1.15$ (From Table 5-13); $W = 1.10$ (From Table 5-14)

$$N = \frac{12\,V}{\pi D} = \frac{12 \times 70}{\pi \times 4} = 70 \text{ rpm (rounded)}$$

$$f_m = f_t\, n_t\, N = .006 \times 14 \times 70 = 6 \text{ in./min (rounded)}$$

$$Q = f_m\, w\, d = 6 \times 3 \times .125 = 2.25 \text{ in.}^3\text{/min}$$

$$P_m = \frac{K_p\, C\, Q\, W}{E} = \frac{.60 \times 1.15 \times 2.25 \times 1.10}{.80}$$

$$= 2.1 \text{ hp}$$

SI metric units: $K_p = 1.64$ (From Table 5-11); $C = 1.15$ (From Table 5-13); $W = 1.10$ (From Table 5-14)

$$N = \frac{1000\,V}{\pi D} = \frac{1000 \times 21.3}{\pi \times 101.6} = 70 \text{ rpm (rounded)}$$

$$f_m = f_t\, n_t\, N = 0.15 \times 14 \times 70 = 150 \text{ mm/min (rounded)}$$

$$Q = \frac{f_m\, w\, d}{60{,}000} = \frac{150 \times 76.2 \times 3.18}{60{,}000} = 0.606 \text{ cm}^3\text{/s}$$

$$P_m = \frac{K_p\, C\, Q\, W}{E} = \frac{1.64 \times 1.15 \times 0.606 \times 1.10}{.80}$$

$$= 1.57 \text{ kW} \quad (1.57 \times 1.341 = 2.1 \text{ hp})$$

Whenever the maximum power available on a machine tool is to be utilized the cutting conditions should be selected in the following order: 1. select the maximum depth of cut; 2. select the maximum feed rate that can be used; and 3. estimate the maximum cutting speed that will utilize the maximum power available on the machine tool. This order is based on obtaining the longest tool life, while at the same time achieving the greatest amount of production possible from the machine tool. *The life of a cutting tool is affected most by the cutting speed, then by the feed rate, and least by the depth of cut.* The maximum metal removal rate of which the machine tool is capable, is used as the basis for estimating the cutting speed that will utilize all of the power available on the machine for the material being cut.

Example 5-4

A 5-inch-wide block of 210-220 HB H10 tool steel is to be milled with an 8-inch diameter, 10-tooth indexable insert, cemented carbide face milling cutter on a 20 hp milling machine. The power constant for the H10 tool steel is .88. The amount of stock to be removed is .250 inch, which is to be removed in one cut. A feed rate of .012 in./tooth is selected as being the maximum that can be used with the cutter, for which C=1.00. The tool wear factor, W, is 1.20 for this face milling operation and the machine tool efficiency factor is estimated to be .80. Estimate the cutting speed that will utilize the maximum power available on the machine.

$$Q_{max}=\frac{P_m\,E}{K_p\,C\,W}=\frac{20\times .80}{.88\times 1.00\times 1.20} \qquad \left(P_m=\frac{K_p\,C\,Q\,W}{E}\right)$$
$$=15.15\text{ in.}^3/\text{min}$$

$$f_m=\frac{Q_{max}}{w\,d}=\frac{15.15}{5\times .250} \qquad (Q=f_m\,w\,d)$$
$$=12\text{ in./min (rounded)}$$

$$N=\frac{f_m}{f_t\,n_t}=\frac{12}{.012\times 10} \qquad (f_m=f_t\,n_t\,N)$$
$$=100\text{ rpm}$$

$$V=\frac{\pi\,D\,N}{12}=\frac{\pi\times 8\times 100}{12} \qquad \left(N=\frac{12\,V}{\pi\,D}\right)$$
$$=209\text{ fpm}$$

Example 5-5

Suggest the cutting conditions for milling the tool-steel block in the previous example if the milling machine has a 5-hp motor.

Since there is less power available in this case, the metal removal rate must be reduced to enable the cut to be taken on this machine. The width and depth of the cut remain unchanged. The metal removal rate will be reduced by reducing the feed rate to .005 in./tooth, for which C=1.19.

$$Q_{max}=\frac{P_m\,E}{K_p\,C\,W}=\frac{5\times .80}{.88\times 1.19\times 1.20}$$
$$=3.18\text{ in.}^3/\text{min}$$

$$f_m = \frac{Q_{max}}{w\,d} = \frac{3.18}{5 \times .250}$$
$$= 2.5 \text{ in./min} \quad \text{(rounded)}$$

$$N = \frac{f_m}{f_t\, n_t} = \frac{2.5}{.005 \times 10}$$
$$= 50 \text{ rpm}$$

$$V = \frac{\pi\, D\, N}{12} = \frac{\pi \times 8 \times 50}{12}$$
$$= 105 \text{ fpm}$$

This cutting speed is below that recommended for cemented carbide and very close to the cutting speed recommended for high-speed steel. On the 5 hp milling machine, it is evident that the maximum production rate can be obtained with a high-speed-steel cutter. The tool steel block may be machined by shell end milling or by slab milling; in this case slab milling is selected. A 4-inch diameter, 6-inch wide, heavy-duty plain milling cutter having 10 teeth will be used. The feed rate selected for this cutter is .005 in./tooth. From the tables, $W = 1.10$ and $C = 1.19$.

$$Q_{max} = \frac{P_m\, E}{K_p\, C\, W} = \frac{5 \times .80}{.88 \times 1.19 \times 1.10}$$
$$= 3.47 \text{ in.}^3\text{/min}$$

$$f_m = \frac{Q_{max}}{w\,d} = \frac{3.47}{5 \times .250}$$
$$= 2.75 \text{ in./min} \quad \text{(rounded)}$$

$$N = \frac{f_m}{f_t\, n_t} = \frac{2.75}{.005 \times 10}$$
$$= 55 \text{ rpm}$$

$$V = \frac{\pi\, D\, N}{12} = \frac{\pi \times 4 \times 55}{12}$$
$$= 58 \text{ fpm}$$

While this recommendation is perfectly valid for the conditions in the example that were examined, it is not a general recommendation. Each job is unique and must be examined on its own merits. Moreover, it is usually necessary to make small adjustments to suit the spindle speeds and the feed rates that are available on the machine. Generally, a set of cutting conditions can be found that will utilize the capacity of the machine when making such adjustments.

CHAPTER **6**

Milling Machine Operations

Milling machines are used to perform a large variety of machining operations. In addition to those that can be classified as strictly milling operations using milling cutters, other operations, such as slotting, drilling, boring, reaming, etc., which do not utilize milling cutters and are performed on other machine tools, are often also performed on the milling machine. An example of the variety of operations that can be performed can be obtained by studying Fig. 6-1. The different kinds of cutting tools used to machine the part that is clamped to the table can be seen in the illustration. Although much of the work done on a milling machine involves the production of plane or contoured surfaces, large- and small-diameter holes are also frequently produced. Operations involving the use of the dividing head will be treated later in chapters exclusively devoted to this topic.

Courtesy of Cincinnati Milacron

Fig. 6-1. Straddle milling the inside face of a casting. Tools for other milling operations are placed on a board to protect the milling machine and the cutting edges from damage.

Many of the principles pertaining to the operation of other machine tools discussed in previous chapters are used in conjunction with work commonly done on the milling machine. These include the principles of drilling, reaming, boring, and precision hole location. The principles of clamping in making a setup on planers apply equally well to clamping workpieces directly to the milling-machine table. The principles involved in doing accurate work in a shaper vise should be reviewed, for they also apply to milling-machine work when the part is held in a vise. Although it would be repetitious to cover all of this material again in this chapter, these principles must be kept in mind when the workpiece is set up and cuts taken on the milling machine.

Conventional and Climb Milling

Conventional milling is also called *up milling*. As illustrated in the upper view of Fig. 6-2, the direction of motion of the milling cutter tooth as it engages the work is opposite from the direction of the movement of

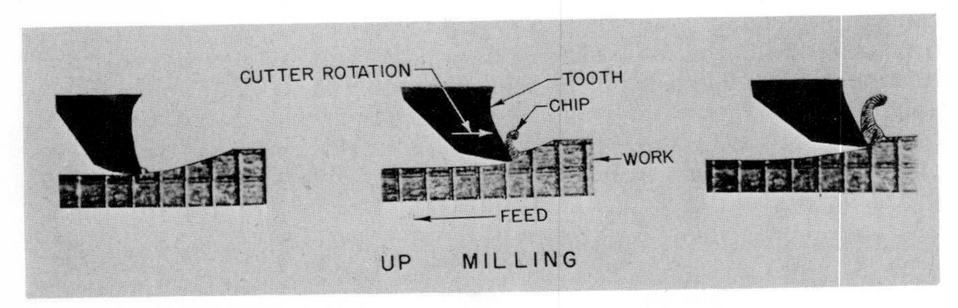

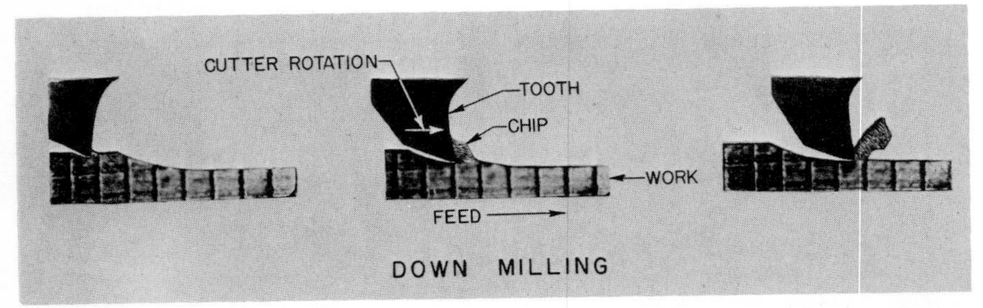

Courtesy of Cincinnati Milacron

Fig. 6-2. Upper view—conventional up milling. Lower view—climb or down milling.

the work caused by the table feed. The cutting forces resulting from this method of milling will keep the feed screw nut against the same side of the feed screw thread as when feeding the table toward the cutter without taking a cut. Thus, the table and the workpiece will never have a tendency to pull toward the cutter because of lost motion between the nut and the table feed screw.

In conventional milling a very thin chip is formed at the beginning of the cut. The thickness of the chip increases as the tooth proceeds along its path until it reaches a maximum in the position where the tooth leaves the workpiece.

Climb milling has an advantage when certain materials, such as aluminum, are milled, because it produces a much better surface finish on the workpiece than can be obtained by conventional milling. Climb milling is also called *down milling*. As the lower view in Fig. 6-2 shows, the milling cutter tooth and the workpiece move in the same direction. The velocity of the milling cutter tooth is faster than the velocity of the table feed, which moves the work into the cutter and thereby forms the chip. The cutting force resulting from climb milling is in the same direction as the feed. This will cause the feed screw, which is attached to the table, to pull away from the side of the feed screw nut against which it was bearing as the work was approaching the cutter. In effect, since the workpiece will be pulled into the cutter by the action of the cutting forces, the workpiece, the cutter, and the milling-machine arbor can all be seriously damaged. Climb milling, therefore, must not be used in most instances, unless the milling machine is equipped with a backlash eliminator. Light profiling-type cuts can often be taken with end milling cutters using the climb milling method. The magnitude of the cutting forces is usually low, and the weight of the table is sufficient to prevent the workpiece from being pulled into the cutter. Sometimes clamping the table lightly will add an additional drag to the table so that the work will not be pulled into the cutter.

Figure 6-2 shows that in climb or down milling the maximum chip thickness occurs at a point close to the position where the tooth makes the initial contact with the workpiece. As the cut continues the chip thickness decreases, reaching a minimum where the tooth leaves the workpiece.

Setting Up the Workpiece

For most jobs done on a milling machine, setting up the workpiece is the most difficult and critical part of the work. The workpiece must not only be securely clamped, but also be held on the machine in such a position that each surface to be machined will, when finished, be accurately aligned with other surfaces on the part. Accuracy in making a setup is essential on most jobs; without it, close tolerance work cannot be done, unsatisfactory workpieces that have to be scrapped will result. Each setup must be planned in advance and then carried out with care and patience.

The first step, then, is to plan the setup. While each setup is unique, inasmuch as it is made to suit the part to be milled, there are several basic types of setups, which will now be described. Large workpieces are usually placed on the top of the milling machine table, and are clamped there by means of strap clamps and T-slot bolts, as shown in Figs. 6-4, 6-5, and 6-12. The principles of applying strap clamps have been treated in detail in previous chapters and will not be repeated here, except to say that the

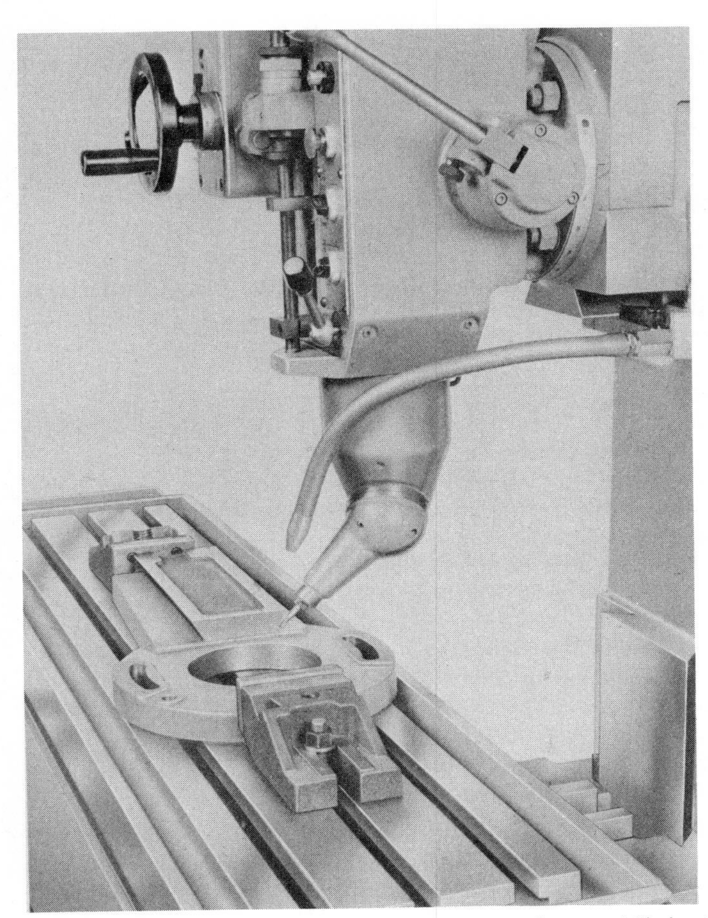

Courtesy of Cincinnati Milacron

Fig. 6-3. Workpiece clamped with table jaws while end milling casting using a universal multi-angle milling attachment on a toolroom type vertical milling machine.

heel blocks should hold the clamps in a level position and the T-slot bolts should be placed as close to the workpiece as possible. Table jaws, Fig. 6-3, are convenient for holding the workpiece against the table; they can be placed in pairs over a T-slot anywhere along the table. The jaws are serrated and when tightened exert a downward clamping force. Although it is sometimes necessary to place precision parallels between the work-piece and the table, this should be avoided whenever possible. Parallels are used in this manner when the workpiece must be raised above the table top to prevent the milling cutter or other cutting tool from cutting into the table, as shown in Fig. 6-4. The setup is less rigid when the work-

Courtesy of The Ingersoll Milling Machine Co., Cutting Tool Div.

Fig. 6-4. Workpiece mounted on parallels to provide clearance for cemented carbide end mill used to mill slots.

piece is clamped on parallels and thus is more likely to deflect or slide when subjected to the cutting forces. In Fig. 6-4 two stops are placed in the T-slot to prevent the workpiece from sliding, otherwise lighter cuts would have to be taken. The ends of several pieces of round bar stock ground parallel and to the same height are used as parallels in the setup shown in Fig. 6-5. These parallels have the advantage of requiring less space and can be positioned where they are most needed without interfering with each other or with the T-slot bolts. The strap clamps are placed directly above the parallels, which illustrates an important principle in the application of these clamps.

Frequently the workpiece can be set up by clamping it in a milling machine vise, as shown in Fig. 6-8. This vise is a precision tool which must not be abused, but is to be used correctly. The principles of using precision vises were treated in detail in Chapter 2; they should be reviewed since they apply to both milling machine and shaper vises. Whenever possible, the forces generated by the milling cutter should be directed against the solid jaw of the milling machine vise rather than against the movable jaw. On many occasions, however, the work must be held in the vise as shown in Fig. 6-8, where the largest component of the cutting force is parallel to the vise jaws. Long workpieces are sometimes held in two matched vises, as shown in Fig. 6-10.

Courtesy of Cincinnati Milacron

Fig. 6-5. Parallels made from round bar stock used to raise workpiece above table to provide clearance for cutter.

Angle plates, Fig. 6-6, provide a surface against which the workpiece can be clamped, one that is perpendicular to the top of the machine table. When a surface must be milled perpendicular to a previously machined surface, the part can be clamped with the machined surface against the vertical face of the angle plate. While there are other methods of milling surfaces perpendicular to each other, it is frequently most convenient to use an angle plate. As an example, the part in Fig. 6-6 could be machined on a horizontal spindle milling machine with the finished surface of the casting clamped directly to the table; however, when set up on a vertical milling machine as shown, the surfaces to be milled are more accessible to the operator for observing the cut.

Before actual work in making the setup is started, the top of the milling

Courtesy of Bridgeport Machines Inc.

Fig. 6-6. Casting clamped to angle plate for milling top surface perpendicular to seating surface on angle plate, and for milling inside surfaces. Parallels used to align casting.

machine table and all of the surfaces that are to seat on it must be clean and free of small chips, nicks, or burrs; the surfaces seating on the table may be on the workpiece, on the bottom of the vise, or on a leg of an angle plate. Small chips, nicks, and burrs will prevent these surfaces from being accurately seated on the table. This condition will surely be reflected on the surfaces that are machined during the setup. As a final check before placing a part on the table, run the bare fingers of the hand over the table top and the seating surface; this procedure will feel out the presence of any small chips, nicks, and burrs. A similar procedure is used before placing a part in a vise. After a part, vise, or angle plate is on the table, it must be aligned.

Although the setup for each part is unique, usually the process becomes a matter of aligning some surface or axis on the workpiece in some specified relationship with the axis of the milling-machine spindle. Fortunately, the construction of the knee and column milling machine is very helpful, if the operator remembers that the longitudinal feed is perpendicular to the spindle axis and that the transverse feed is parallel to the spindle axis. For example, a vise may be set with its jaws parallel to the axis of the spindle as shown in Fig. 6-7. The dial test indicator is held on a mag-

Courtesy of the Brown & Sharpe Manufacturing Company

Fig. 6-7. Using a dial test indicator and a magnetic indicator base to align a vise jaw parallel to the milling-machine spindle axis.

netic base which is attached to the face of the column. A precision parallel bar or a previously finish-machined surface on the workpiece is clamped against the solid jaw of the vise. With the indicator in contact with the parallel bar or finish-machined surface, the transverse feed is moved back and forth and the vise adjusted until the indicator reading is the same all along the bar. The jaws of the vise could be positioned perpendicular to the spindle axis, with the longitudinal instead of the transverse feed used when the parallel bar is being indicated.

The vise or the workpiece can be set with the solid jaw or a finished surface parallel to the spindle axis by the blade of a square being placed against the finished surface of the work while the beam of the square is held firmly against the face of the milling-machine column. This method may be used either as a fast way of setting the part up or as a rough setup before the indicator is used as described. To obtain the most accurate setup, the part should be indicated. On some milling machines the face of the column is deliberately scraped slightly low in the center in order to insure that the knee will bear against the outer surface of the face. This, of course, will limit the accuracy of aligning the workpiece with the square. The side of the milling-machine table can be used as a reference surface for aligning a vise or a workpiece perpendicular to the spindle axis. Again, however, the dial test indicator, when used as described, is so convenient that it is the preferred method of aligning a part perpendicular to the spindle axis.

Settings at an angle to the spindle axis can be made by holding a protractor against a finished surface while indicating along the blade using either the longitudinal or transverse feed. The side of the table or the table T-slots can serve as a reference surface for the protractor. The angle plate shown in Fig. 6-27 was located by this method.

The vertical table feed can also be used on occasion in aligning a workpiece with a dial test indicator. For example, the face of an angle plate could be checked for perpendicularity by placing an indicator on it and using the vertical table feed for reference. A precision square held on the table of the milling machine could also be useu to make this alignment.

From this discussion it is evident that a workpiece can be located on the milling-machine table by indicating directly against a reference surface on the workpiece, or by placing the reference surface against a surface which has a known location with respect to the spindle axis. The most difficult situation arises when a rough casting is to be machined on which there are no previously finished surfaces. In such cases a layout should be made and the workpiece set up on the machine according to the layout lines. The procedure is described in Chapters 3 and 10 for work on planers and horizontal boring machines.

After the workpiece, vise, or angle plate has been aligned, it is clamped firmly to the milling machine table. It is good practice to check the alignment again after clamping.

Slab Milling

Slab milling is the operation of producing a flat or plane surface with a plain milling cutter. For this reason plain milling cutters are sometimes called slab milling cutters. The workpiece can be held in a vise, as in Fig. 6-8, or it can be clamped directly to the table, as in Fig. 6-9. It is sometimes convenient to hold long workpieces in two vises, as shown in Fig. 6-10.

To perform the slab milling operation the workpiece must be correctly set up on the machine and the arbor and milling cutters mounted in the

Courtesy of Cincinnati Milacron

Fig. 6-8. Slab milling on a Cincinnati Toolmaster milling machine equipped with a combination horizontal-vertical spindle attached to the end of the overarm.

spindle as described in Chapter 5. After the spindle speed and the feed rate have been calculated, the machine is set up accordingly.

There are two methods of setting the depth of cut. One method is to set the table so that the milling cutter will cut below any scale on the surface of the workpiece without cutting the workpiece undersize. This setting is usually made visually. Next, a trial cut is taken and a measurement made of the surface that has been cut. The table is then moved

Courtesy of Cincinnati Milacron

Fig. 6-9. Slab milling with workpiece clamped directly to the milling-machine table.

vertically the distance required to cut the part to size as determined by the measurement. Finally, the entire surface is cut to size. When very accurate work is done, the best procedure is to take a rough cut over the entire surface leaving about $\frac{1}{16}$ inch on the part for taking a finish cut.

Another method of setting the depth of cut is shown in Fig. 6-11. The depth of the cut required to mill the part to size is first determined by taking measurements on the part. Next, a long sliver of paper is placed between the surface of the workpiece and the cutter. After the milling-machine spindle is engaged, the work is positioned below the rotating cutter. The paper feeler is held in one hand, and the table is raised with the other hand by slowly turning the elevating screw crank. The paper feeler must be long enough to allow the fingers holding it to keep a safe distance from the rotating cutter. Carefully continue to raise the table until the cutter just grazes the paper feeler without cutting into it. When this occurs each tooth will exert a light pull on the paper which can readily be felt. Then move the workpiece clear of the cutter. The table is raised a distance equal to the depth of cut required plus the thickness of the paper feeler. The cutter and the workpiece are now in such a position that the slab milling cut can be taken over the entire surface without an interruption.

Generally a good surface finish is obtained on either steel or cast-iron surfaces that are machined by slab milling. Cast-iron surfaces finished

Courtesy of the Brown & Sharpe Manufacturing Company

Fig. 6-10. Slab milling a long workpiece that is clamped in two vises.

by slab milling are usually similar in nature to shaped or planed surfaces in that they can be easily hand-scraped when required.

Side Milling

In slab milling the plane surface that is machined is parallel to the axis of the milling-machine arbor. Side milling is the operation of machining a plane surface perpendicular to the milling-machine arbor with an arbor-mounted cutter called a *side milling cutter*. Figure 6-12 illustrates the side-milling operation. The cutter shown in this operation is an inserted-tooth, cemented-carbide, side milling cutter. Since it is frequently necessary to bring the workpiece close to the arbor during the side-milling operation, extra precautions should be taken to make certain that the arbor is clear of all obstructions that might pass beneath it as the table

Fig. 6-11. Locating the milling cutter with respect to the surface of the workpiece.

Courtesy of Cincinnati Milacron

Fig. 6-12. Cutting the side of the workpiece with a side milling cutter.

feeds the workpiece past the cutter. In order to mill a flat surface with the side milling cutter, the longitudinal table feed must be continued until the cutter is completely clear of the workpiece.

In side milling, most of the metal is removed by the cutting action of the peripheral teeth of the side milling cutter. This cutting action causes a very small deflection of the cutter and the workpiece away from each other. When the primary cut is finished, the work and the teeth of the cutter spring back toward each other. Because of this spring-back, the side milling cutter takes a very light cut or a secondary cut on the back part of the turn. The side teeth assist in taking the secondary cut.

Straddle Milling

Straddle milling is the milling of two parallel plane surfaces, or faces, that are perpendicular to the axis of the milling-machine arbor. Two side milling cutters are used—spaced apart so that the distance between the milled surfaces is obtained in one cut. In the straddle-milling operation of Fig. 6-1 the inside bosses of the casting are finished in one cut. Note that the casting being machined is clamped onto parallels instead of directly to the table. This permits the sides to be milled without any danger of having the arbor support hitting the table. Another straddle-milling operation is shown in Fig. 6-13. This illustration is unique in that it shows a double straddle-milling operation. Here the outside and inside surfaces of the casting are being milled.

The accuracy of the surfaces produced by straddle milling is determined in part by the spacing of the face milling cutters on the milling-machine arbor. Since the face milling cutters must be sharpened by grinding from time to time, the spacing collars between the cutters must be adjusted to maintain the correct distance between the cutters. Sometimes special collars made from precision shim stock with a given thickness must be used in order to obtain an accurate spacing between the side-milling cutters.

Gang Milling

Gang milling is a production milling operation where two or more milling cutters are mounted together on a milling-machine arbor in order to mill two or more surfaces simultaneously. A gang-milling operation is shown in Fig. 6-14. Almost every type of arbor-mounted milling cutter can be used in a gang-milling operation. Because of the number of surfaces that can be cut simultaneously, both production cost and time can be saved by gang milling.

On a gang-milling operation the spindle speed is determined by the diameter of the largest cutter. The production rate is determined by the table feed rate. This, in turn, is determined by the spindle speed of the cutter, and by the feed per tooth and the number of teeth of the cutters. It can be seen, therefore, that the diameters of the cutters should be as nearly alike as possible and the teeth of the cutters proportioned so that they can all use the same maximum table feed rate. If, on occasion, a large-diameter cutter in relation to the other cutters must be used, the larger cutter may be designed to have cemented-carbide teeth while the other cutters are made from high-speed steel. This arrangement permits the utilization of the optimum speed of the smaller-diameter high-speed

Courtesy of the Brown & Sharpe Manufacturing Company

Fig. 6-13. Straddle milling two pairs of surfaces on a casting simultaneuosly.

Courtesy of Cincinnati Milacron

Fig. 6-14. Gang milling operation.

steel cutters. The larger cutter, having carbide teeth, will therefore not limit the spindle speed that can be used.

Face Milling

Face milling is an operation for producing plane or flat surfaces using a face milling cutter. Very fast metal-removal rates are possible with face milling, especially when cemented carbide face milling cutters are used as these can be operated at high cutting speeds. A typical cemented carbide face milling operation is shown in Fig. 6-15. The setup is designed so that the cutter will cut on the downward path of the teeth. There are two important reasons for doing this: 1. The cutting forces push the casting

Courtesy of the Brown & Sharpe Manufacturing Company

Fig. 6-15. Face milling with a carbide face milling cutter on a horizontal-spindle milling machine.

down against the table instead of against the strap clamps; and, 2. The flow of the chips removed is directed downward and away from the operator. When face milling, the cutting forces are often very large and possible movement of the workpiece must be prevented by directing these forces toward a solid object, such as the milling machine table, to solid stops firmly clamped to the table, or to the solid jaw of a vise. The flow of chips resulting from a high-speed cemented carbide face milling operation can be hazardous; they must be directed away from the operator and protection must also be provided for other persons in the vicinity. Figure 6-16 illustrates a heavy face milling operation on a vertical milling machine. Two step blocks are clamped to the table at each end of the casting to prevent it from moving. The cut is taken by feeding from the operator's

Courtesy of Cincinnati Milacron

Fig. 6-16. Face milling with a carbide face milling cutter on a vertical milling machine.

left to his right so that the chip flow is directed toward the back of the machine and away from the operator.

In face milling the automatic power feed should be kept engaged until the cutter is completely clear of the workpiece; otherwise the surface produced may not be perfectly flat. Although most of the metal is cut off by the primary cut taken by the peripheral teeth of the face milling cutter, the secondary cut taken at the "back" part of the turn as the cutter revolves will remove a small amount of metal. If the secondary cut is not taken completely across the workpiece, it is obvious that somewhat more metal will be removed from that portion of the workpiece on which the secondary cut has been taken. The secondary cut will usually leave telltale feed marks on the workpiece. On some milling machines (see Fig. 4-29) these feed marks can be eliminated by slightly tilting the spindle into the cut. When a very smooth surface finish is required, a face milling cutter having a wiper blade or finishing inserts mounted on the face of the cutter should be used. The cemented carbide face milling cutter in Fig. 6-17 has two finishing inserts mounted on its face that produce a very smooth finish on the milled surface.

Face milling cuts may be taken by straddling the cut as shown in view A, Fig. 6-18. However, it is usually better to cut on either one side or the

Courtesy of The Ingersoll Milling Machine Co., Cutting Tool Div.

Fig. 6-17. Taking a finishing cut on a surface with a face milling cutter having finishing teeth on face to produce a smooth surface.

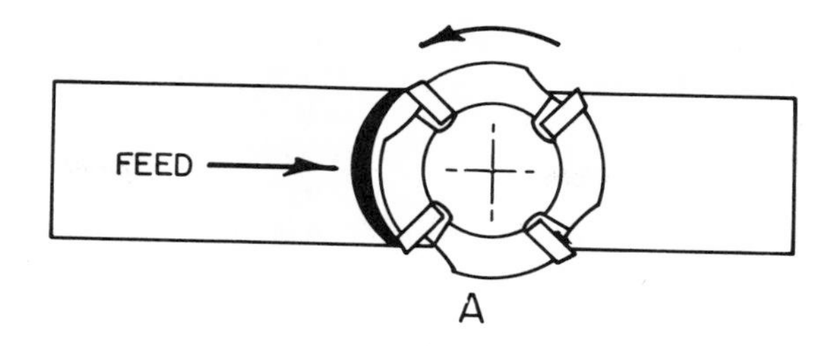

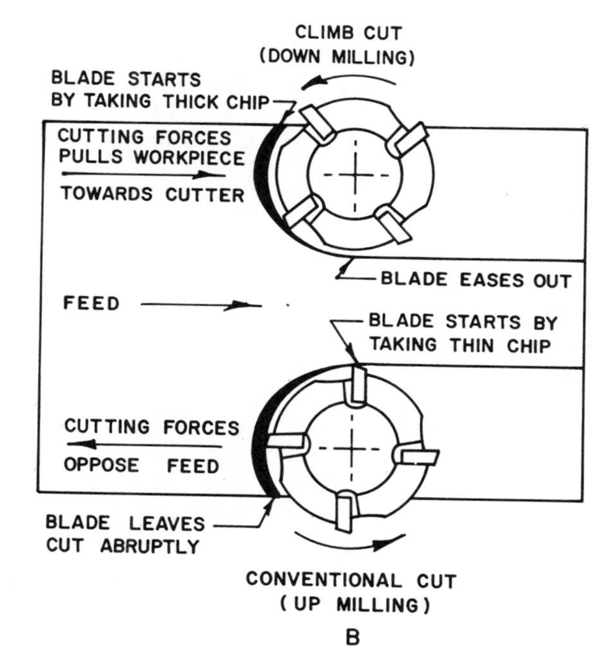

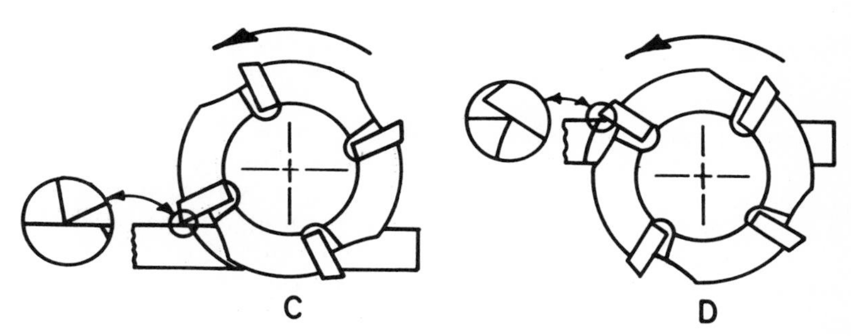

Fig. 6-18. The fundamental methods of face milling. A. Taking a straddle cut. B. Taking a climb cut and a conventional cut. C. Face milling with body outside of workpiece; blade makes initial contact at tip of edge. D. Face mill with body inside workpiece; blade makes contact away from tip of edge.

other of the face of the cutter as in view B. Whenever possible, the workpiece should be cut in one pass; the cutter diameter should be larger than the width of the cut by a ratio of 4 to 3, or 3 to 2. Very wide surfaces must be milled by a series of overlapping cuts. Overlapping face milling cuts should always be taken with the feed used to take each cut going in the same direction. The cutter is returned to the starting point for each new cut by moving the table so that the cutter passes around the outside of the surface being cut.

Face milling cuts can be taken by conventional cut (up milling) or by a climb cut (down milling), as shown in view B, Fig. 6-18. On a conventional cut the chip is initially very thin; it increases and then decreases slightly until it leaves the cut somewhat abruptly. When the face milling cutter has sharp, positive rake blades good results are obtained by this method. The cutting forces generated by the blades oppose the feed thus preventing the backlash between the feedscrew and nut from causing the workpiece to be pulled into the cutter. It is recommended that a conventional cut be used for face milling on older machines and on machines that do not have a blacklash eliminator (see Fig. 4-31). The disadvantage of the conventional cut is that dull cutters and negative-rake face milling cutters have difficulty in starting the cut in the very thin chip region. In this area dull, honed, and negative-rake cutting edges tend to rub against rather than penetrate the surface of the workpiece, thereby generating heat and causing the cutting edge to wear. Also, the rubbing action work-hardens the work surface which may, in some cases, be severe enough to make penetration by the next blade more difficult.

On a climb cut, the blade, or insert, of the cutter starts by taking a heavy cut and eases out of the cut by producing a thin chip. Since most cemented carbide face milling cutters have a negative radial-rake angle, the face milling cut taken with these cutters should be a climb cut whenever possible. The climb cut, however, tends to pull the work into the cutter; therefore, this method should not be used unless the machine is either equipped with a backlash eliminator or the workpiece is heavy enough to prevent it from being pulled by the cutter. Also, climb cuts are not recommended when the edge of the workpiece into which the initial penetration is made by the cutting edges of the cutter has a heavy abrasive scale.

A very important consideration when using cemented carbide face milling cutters is the entry angle of the cutter teeth. This is illustrated in views C and D, Fig. 6-18. Cemented carbides, although very hard, are also brittle. As a result, they do not withstand shock loads well, especially when they occur in a weak area, such as at the tip of the cutting edge. Shock loads at the tip of the cutting edge very often cause the cutting edge to fail by breaking. When taking a climb cut, the initial contact with the edge of the workpiece causes a shock load, and the position on the cutting edge where this will occur is determined by the entry angle. In view C, Fig. 6-18, the center of the face milling cutter body is outside the edge of the workpiece against which the cutting edges make their initial contact. When the cutter body is in this position, the initial load on the teeth is taken at the very tip of the cutting edge, where, as mentioned previously, it is weak and liable to break. When the center of the cutter body is inside the edge of the workpiece against which initial contact is made by the teeth, as in view D, the initial contact on the teeth is behind the tip of the cutting edge where the teeth have greater strength and are better able to withstand shock loads. Therefore, whenever possible, when using a cemented carbide face milling cutter, the center of the cutter body should be inside the edge of the workpiece against which the teeth make their initial contact, as in views B and D.

Another serious problem that sometimes occurs when face milling with cemented carbides is *chip sticking*. A chip may stick or be welded to the face of a carbide blade, or insert, as it leaves the cut and be carried around until the blade or insert starts into the next cut. When this blade or insert starts to penetrate the workpiece, the welded-on chip sets up a very high load which frequently causes the cutting edge to break. A prime cause of chip welding is using too light a feed per tooth; and here increasing the feed will often help. Other causes of chip sticking may be due to the grade of carbide used, choosing a speed that is too slow when cutting a soft material, too high when cutting a hard material, or taking a cut that is too wide. These causes suggest the steps to be taken to overcome this problem. Occasionally, changing the cut from a conventional to a climb cut will be helpful.

Face milling cutters, depending on their size, are capable of very high metal removal rates which require a large amount of power. The available power on the machine often places a limit on the size of cut and the type of face milling cutter that can be used. When planning to take a heavy face milling cut it is always advisable to determine in advance whether the machine has the power to take the cut, using the method described in Chapter 5 of this volume.

End Milling

End milling cutters are very versatile cutters that can perform a wide variety of operations. Their usefulness is extended by the many types of end milling cutters that are available. An end milling cutter is shown taking a facing cut on an inside surface of a casting in Fig. 6-19. The casting is clamped in an All-Steel vise, and the cut is taken by engaging the transverse automatic power feed. The facing cut shown is taken by using only the peripheral teeth or cutting edges of the end milling cutter, making the cut similar to slab milling. In Fig. 6-19 a pocket that has previously been milled out with the end mill can be seen just behind the cutter. When cutting the pocket the peripheral teeth of the end mill remove most of the metal, but the end teeth do take a light scraping cut which is similar to the action of the end teeth on a face milling cutter.

End mills are often used to cut slots, as shown in Fig. 6-20. One problem that can occur when slotting with an end milling cutter is the "par-

Courtesy of Cincinnati Milacron

Fig. 6-19. Facing a surface inside a casting with an end mill.

allelogram" or "wobble" slot, which is a slot with sides that are parallel to each other but not perpendicular to the bottom of the slot. This condition occurs most frequently when a two-fluted end mill is used with a large helix angle. It also occurs when an excessive flute length is projecting from the spindle in which the end mill is held. The principal cause of the parallelogram slot is the deflection of the end milling cutter brought about when one flute is cutting into the material while the other flute is not cutting and is unsupported by a side of the slot. This condition is prevented by increasing the spindle speed and decreasing the feed rate so that the chip load on each tooth is reduced. Decreasing the length of the end that is projecting from the end of the spindle will improve the rigidity of the setup and thereby reduce the tendency of the end mill to deflect and to produce a parallelogram slot. The parallelogram slot can also be prevented by using a four-fluted center-cutting type of end mill which receives better support from the sides of the slot. When the correct spindle speed and table feed rates are used, and when the setups of the work and the cutting tool are rigid, a straight slot can be cut with a two-fluted end mill.

Courtesy of Cincinnati Milacron

Fig. 6-20. Milling a slot with an end milling cutter.

Shell end milling cutters are usually used like face milling cutters to mill plane surfaces, and the principles of applying the shell end milling cutters are the same as previously described for face milling cutters. They are also frequently used to cut square corners and to mill wide slots. To mill plane surfaces they are sometimes used as a plain milling cutter by cutting with their peripheral teeth only. A typical shell end milling operation is shown in Fig. 6-21.

Milling Keyseats

Keyseats are slots that are cut lengthwise in shafts in which keys are held. The keys are used to transmit the driving torque that is conveyed to or from the shaft by pulleys, gears, or sprockets which are attached to the shaft. When required, keyseats are also used to align various machine elements which are sometimes attached to shafts. In addition, keys are

Courtesy of the Brown & Sharpe Manufacturing Company

Fig. 6-21. Face milling a flat surface with a shell end mill.

occasionally used on machine elements other than shafts and parts that are attached to shafts.

Milling machines are usually used to machine keyseats in shafts. The keyseat may be cut with a double side milling cutter (Figs. 6-22 and 6-23) used on a horizontal milling machine. Keyseats are frequently cut with end milling cutters, in which case either a horizontal- or a vertical-spindle milling machine can be used. A vertical milling machine or a vertical spindle attachment is preferred for cutting keyseats, because it is easier to align the cutter with respect to the workpiece, and the operation is easier for the milling-machine operator to observe. Keyseats in shafts that are too large to be handled on milling machines are usually cut on horizontal boring machines. In this case milling cutters are used in the horizontal boring machine, and the procedures followed are essentially the same as those used on the milling machine.

In general, staggered-tooth side milling cutters are preferred to plain side milling cutters for milling keyseats, because the staggered-tooth cutter provides a smoother cutting action. Double side milling cutters can be used to cut keyseats which start and end in the central portion of the

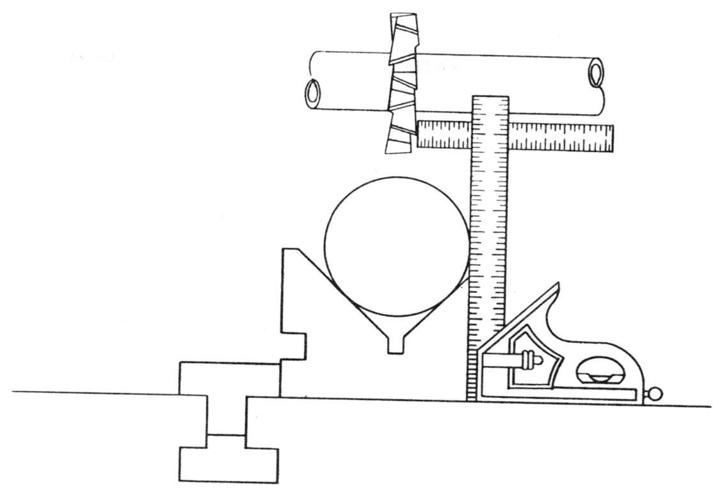

Fig. 6-22. Using a rule and a square to position a double side milling cutter in the center of a shaft in preparation for milling a keyseat.

shaft without extending to the end of the shaft or to the end of the shoulder. This is possible because double side milling cutters can be plunged or sunk directly into the shaft at any point.

Keyseats that are cut with ordinary four-fluted end milling cutters must extend to the end of the shaft or to the end of a shoulder in order to allow the cutter to enter the workpiece. These cutters cannot be sunk into the workpiece in the manner of a twist drill because the end teeth do not extend to the center of the cutter. Center-cutting-type end milling cutters on which the end cutting teeth extend to the center of the cutter are available. These cutters can have two, three, and four flutes, although the two-fluted center-cutting-type end mills are the most common. Since these cutters can plunge directly into solid metal, they can be used to mill keyseats in the central portions of shafts. A type of key with a half-moon shape, called a Woodruff Key, is sometimes used. The keyseats for these keys are also half-moon shaped slots which must be cut with a special Woodruff Keyseat cutter.

The shaft in which the keyseat is to be milled can be aligned and held in a milling machine in several different ways. It can be held in a milling-machine vise, or, if it is very long, in two vises. Larger shafts are sometimes clamped directly onto the table over a T-slot, which helps to align the shaft on the milling machine. The shaft can also be clamped in a V-block, or in a matched pair of V-blocks, as shown in Figs. 6-22 and 6-23. The clamps holding the shaft in place are not seen in these illustrations. V-blocks can be aligned on the milling-machine table by placing them against a slot block, as shown in Fig. 6-22. In shops where round parts such as shafts are frequently machined on a milling machine, V-

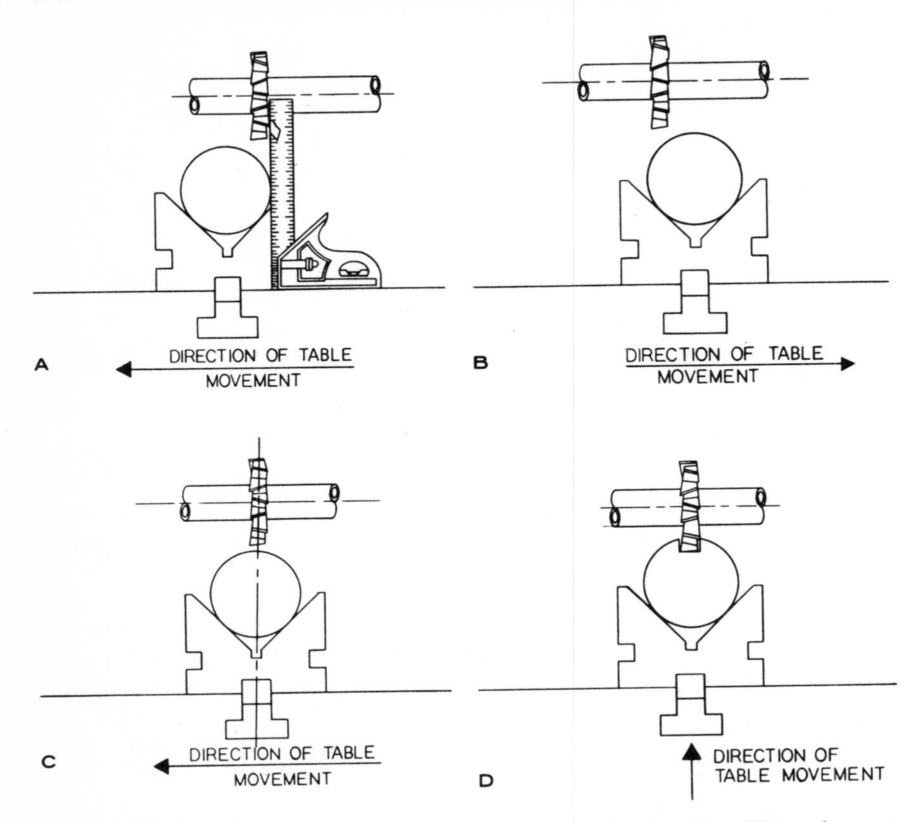

Fig. 6-23. Procedure for centering a milling cutter in preparation for milling a keyseat. A. Touching up against blade of square using a feeler. B. Compensating for lost motion in feed screw by moving beyond center. C. Positioning the table to center the cutter. D. Cutting the keyseat.

blocks with a key in their base which fits into the T-slots of the table (see Fig. 6-24) are a useful accessory. A most convenient method of holding shafts for milling keyseats is between the centers on the dividing head and the dividing head tailstock. This method accurately aligns the workpiece and provides a free access for the cutter to the surfaces to be milled.

Before the keyseat can be cut, the milling cutter must be aligned with the center of the shaft. This can be done by first making a layout of the keyseat on the shaft and adjusting the table in the transverse direction until the cutter is within the layout lines. Frequently, however, the keyseat is cut without a layout in order to save time. In this case one of three procedures described in the following paragraphs can be used to center the cutter with respect to the shaft.

A square and a steel rule, as shown in Fig. 6-22, can be used to align the cutter with the center of the shaft. The square is held against the

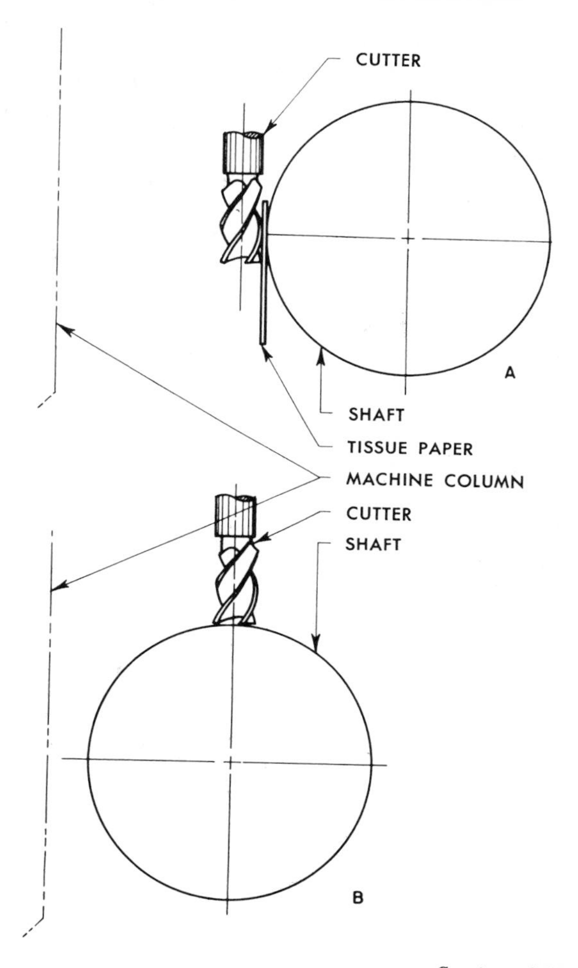

Courtesy of Cincinnati Milacron

Fig. 6-24. Procedure for centering a milling cutter in preparation for milling a keyseat by touching up against the side of the workpiece.

side of the shaft, and the distance from the blade of the square to the side of the milling cutter is measured with a rule. The milling cutter should, of course, not be rotating. The table of the milling machine is adjusted until the measurement made by the rule is equal to one-half of the difference obtained when the width of the side milling cutter or the diameter of the end milling cutter, if used, is subtracted from the diameter of the shaft. For example, if the width of the side milling cutter in Fig. 6-22 is .375 inch and the diameter of the shaft is 1.500 inches, the cutter will be centered when the reading on the rule is equal to .5625 or $\frac{9}{16}$ inch.

The other procedures depend on making accurate movements of the table which are obtained by reading the micrometer dial of the transverse feed screw. One method is to hold the blade of a square against the side of the workpiece and to move the table until the milling cutter, which must not be rotating, touches the blade as shown at A in Fig. 6-23. The arrows in Fig. 6-23 indicate the direction that the table must be moved in order to reach the positions shown. It is helpful to place a thin paper feeler between the cutter and the square to gage the contact between the blade of the square and the milling cutter. The paper feeler should slip; however, a drag should be perceptible when it is pulled. The cutter and the blade of the square are then the thickness of the paper feeler apart. The distance that the table should be moved in order to center the cutter over the shaft should then be equal to one-half of the difference between the diameter of the shaft and the thickness of the side-milling cutter (or the diameter of an end milling cutter) minus the thickness of the paper feeler. For example, if the width of the cutter is .500 inch, the diameter of the shaft is 2.000 inches, and the thickness of the paper feeler is .003 inch, the distance that the table must be moved is $(2.000 - .500) \div 2 - .003 = .747$ inch. Before the table is moved to the center position, as determined by the reading of the micrometer dial, it should be moved beyond this position as shown at B in Fig. 6-23. The table is then moved to the center position illustrated at C in Fig. 6-23. This procedure must be used to eliminate the error that can result from the lost motion between the feed screw and the feed screw nut. It should be noted that the micrometer dial must be read when the table has reached positions A and C in order for it to move the exact distance required to align the cutter and the shaft. Also, observe that the table is moved in the same direction to reach both position A and position C. The table is then moved vertically to obtain the required depth of cut, and the keyseat is cut to the required length as shown at D, Fig. 6-23.

The third procedure can be used when it is possible to touch the side of the shaft with the milling cutter, as shown in Fig. 6-24. In this illustration a four-fluted end milling cutter held in a vertical spindle is used. The milling cutter should be rotating while a long strip of paper is held between the cutter and the shaft with one hand. Simultaneously the other hand is used to turn the transverse feed handwheel in order to move the shaft toward the cutter slowly and carefully. When the teeth of the cutter just graze the paper without cutting into it, the table movement is stopped. The table is then centered by moving it a distance equal to the sum of the radius of the shaft, the radius of the end milling cutter, and the thickness of the paper feeler. For example, if the diameter of the shaft is 1.500 inches, the diameter of the end milling cutter is .500 inch, and the thickness of the paper feeler is .003 inch, the table movement required to center the cutter would be $\frac{1.500}{2} + \frac{.500}{2} + .003 =$ 1.003 inches.

Sometimes a vertical spindle is not available on a milling machine when a keyseat is to be cut. In this case the keyseat must be cut with the end milling cutter held in the horizontal spindle. The procedure used to align the shaft and the cutter is similar to that described in the previous paragraph except that the table movements are vertical instead of transverse. The table is first raised until the rotating cutter just grazes a long paper feeler which is held between the top of the shaft and the cutter. Next, center the table by raising it an amount that is equal to the radius of the shaft, the radius of the end milling cutter, and the thickness of the paper feeler. If the shaft is too large to use this method, a layout must be made on the shaft and the keyseat cut to the layout lines.

Milling Angular Surfaces

There are several methods of milling two surfaces at an angle to each other. A very common method is illustrated in Fig. 6-25, where a slot is being milled at an angle with respect to the sides while held in a swivel-base vise. The solid jaw of the vise, which is the locating surface, can be positioned at any desired angle with respect to the direction of the table feed and the cutter. Graduations at the base of the vise facilitate setting the vise at the desired angle. Workpieces that cannot be conveniently held in a vise may be clamped at an angular position on the table in order to mill an angular surface. Complex or compound angles can be

Courtesy of Cincinnati Milacron

Fig. 6-25. Using a swivel vise for angular straddle milling.

milled by holding the workpiece in a toolmaker's universal vise (see Fig. 4-23). A compound angle being milled is shown in Fig. 6-26, with the workpiece clamped in a vise that is itself clamped to an adjustable angle plate. In this illustration, by clamping the workpiece in a vertical position, either directly against the adjustable angle plate or in a vise, as shown, an ordinary angle can be cut. Figure 4-25, in Chapter 4 illustrates how an angular surface is milled by tilting the spindle of a vertical milling attachment. Angular surfaces can also be milled by using angle milling cutters (Fig. 5-3), dovetail cutters (Fig. 6-32), and special form milling cutters (Fig. 5-4).

Another method of angular milling is illustrated in view A, Fig. 6-27

Courtesy of the Brown & Sharpe Mfg. Co.

Fig. 6-26. Adjustable angle plate and vise used to hold workpiece to slab mill compound angle.

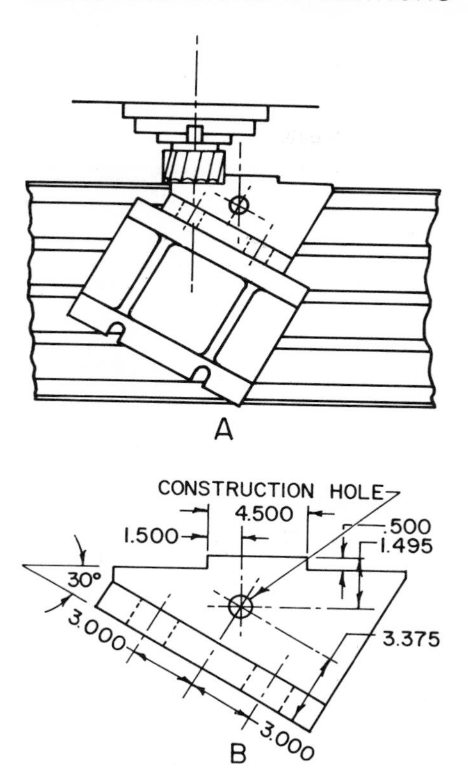

Fig. 6-27. A. Workpiece set up on angle plate to mill angular surface on horizontal spindle milling machine. For clarity, clamps are not shown. B. Workpiece to be machined.

and the workpiece is shown in view B. The workpiece has a construction hole, which serves only to provide a reference axis to which the dimensions of the angular holes and angular surfaces can be related. In the shop a pin is pressed into the construction hole to provide a convenient gaging surface from which to work.

The first step in doing this job is to align and clamp the angle plate to the milling machine table. For ordinary work the angle plate can be aligned with a bevel protractor, by holding the head against a convenient edge on the table, with the blade against the face of the angle plate. When the angle to be cut must be more precise, a sine bar (see Fig. 16-21) and precision gage blocks can be used as shown in view A, Fig. 6-28. With the sine bar and gage blocks held against the angle plate, the sine bar is indicated by means of a dial test indicator attached to the milling machine column (or spindle), moving the table back and forth longitudinally. If available, precision angle gage blocks (Fig. 16-22) can be used in place of the sine bar.

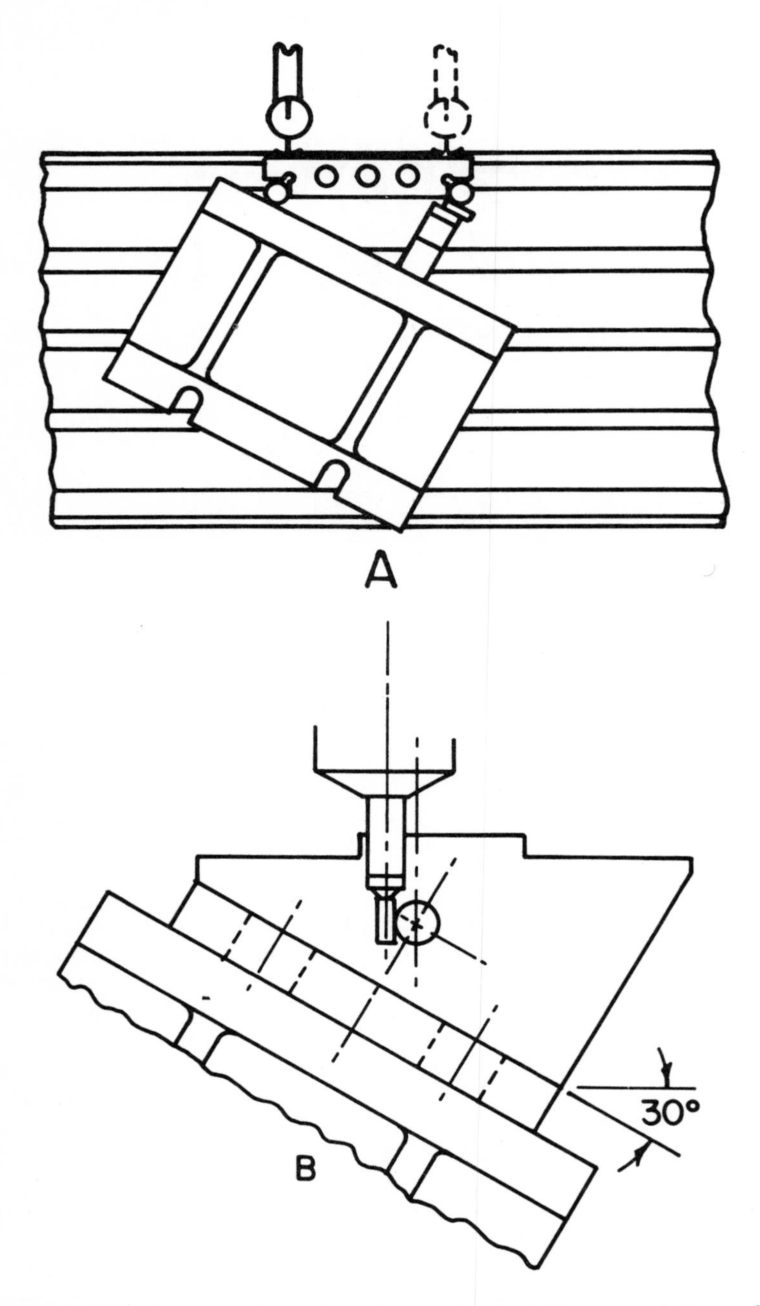

Fig. 6-28. A. Positioning angle plate with sine bar. B. Using edge finder and construction hole pin to align axes of spindle and construction hole.

The next step is to establish a zero reference position from which the table can be moved to mill the shoulder shown by the 1.500-inch dimension in Fig. 6-27. Since the 1.500-inch dimension is given from the axis of the construction hole, the zero reference position is the longitudinal table position where the axis of the spindle and the axis of the construction hole coincide. To find this position, mount an edge finder [1] in a chuck, preferably a collet chuck, that is held in the spindle, as shown in view B, Fig. 6-28. Offset the contact cylinder of the edge finder slightly and start the spindle. The contact cylinder will then rotate with a noticeable eccentricity. Carefully move the table longitudinally toward the left (from right to left) until the edge finder contacts the construction hole pin, stopping when the edge finder stops rotating eccentrically. The axis of the construction hole is now a known distance from the axis of the spindle, which is equal to one-half of the sum of the diameters of the construction hole pin and the contact cylinder of the edge finder. On most edge finders, the diameter of the contact cylinder is .200 inch and, in this case, the construction hole pin-diameter is .500 inch. The distance that the table must be moved to the left in order to align the axes of the spindle and the construction hole is then $\frac{.500}{2}+\frac{.200}{2}=.350$ inch. When moved this distance the table is at the longitudinal zero reference position. The table must be moved to this position by approaching it from right to left to eliminate the effect of the backlash. For this reason it will be very important to approach the 1.500-inch shoulder by also moving the table from right to left when this operation is performed. With the table at the zero reference position, the longitudinal feed screw micrometer dial is set to read zero; if the machine has a digital readout, that should also be set to read zero.

After mounting a four-inch diameter shell end milling cutter in the spindle, the first milling operation is to mill to size the surface defined by the 1.495-inch dimension in Fig. 6-27. The trial cut method is used and the necessary measurements are made with a depth micrometer from the 1.495-inch surface to the construction hole pin. One-half of the diameter of the construction hole pin must be subtracted from the 1.495-inch dimension to obtain the actual micrometer reading when the surface is to size; thus, the micrometer reading should be $1.495-\frac{.500}{2}=1.245$ inches.

The 1.500-inch shoulder can now be milled. First, however, the distance from the zero reference position to the axis of the cutter when it is positioned to take the finish cut on the shoulder must be found. To find this distance it is necessary to know the actual dimension of the shell end mill. This can be done by measuring the cutter diameter with a micrometer caliper, placing thin strips of good quality notebook paper between the cutter teeth and the micrometer anvil and spindle to protect these measuring surfaces. Of course, the thickness of the two strips of paper must be subtracted from the micrometer reading. In this case the cutter diameter

[1] Edge finders are described in Chapter 15, Vol. 1, K. H. Moltrecht, *Machine Shop Practice*, 2nd ed, (New York: Industrial Press Inc., 1980).

was found to be 3.994 inches. The table is then returned to the zero reference position, approaching this position by moving from right to left. The distance from this position to the position of the cutter when finish milling the shoulder is found by adding one-half of the cutter diameter to the 1.500-inch dimension, or $\frac{3.994}{2}+1.500=3.497$. After taking a rough cut the table is positioned 3.497 inches from the zero reference position to take the finish cut, moving to this position from right to left. The depth of these cuts can be established by the trial-cut method, taking the necessary measurements from the previously finished 1.495-inch surface.

The second shoulder can be milled by using the trial-cut method with the necessary measurements taken from the 1.500-inch shoulder. This shoulder can also be cut by positioning the table at a calculated distance from the zero reference position when taking the finish cut. The table setting for the finish cut is calculated by adding one-half of the cutter diameter to the distance to the shoulder from the zero reference position, or $\frac{3.994}{2}+(4.500-1.500)=4.997$ inches. This position also must be approached by moving the table from right to left.

Tramming the Spindle

The spindle on toolroom type vertical milling machines can be tilted at an angle to take various types of angular cuts. Similarly, the spindle on universal vertical milling attachments (Fig. 4-25) and on milling machines equipped with an overarm head (Fig. 4-29) can be tilted. After performing the angular cutting operation, the spindle must be brought back to the vertical position, with the spindle axis precisely perpendicular to the top of the milling machine table. This procedure, or operation, is called *tramming the spindle* and is illustrated in the four views shown in Fig. 6-29. The tools required to perform this operation are shown in these illustrations; they include two matched precision parallels, a dial test indicator, a tramming bar, and a chuck to hold the tramming bar on the spindle. The tramming bar in the illustration is counterbalanced by a thick knurled nut threaded onto the end opposite that to which the dial test indicator is attached.

The first step in performing this operation is to thoroughly clean the milling machine table top. All nicks, burrs, and small chips must be removed, and the clean condition should be verified by running a bare hand over the table-top surface. The setup consists of simply chucking the tramming bar, attaching the dial test indicator as shown, and placing the precision parallel bars on the table. The spindle is then lowered (or the knee raised) until the indicator touches a parallel and shows a reading. Before proceeding further, the knee must be clamped to the column; this detail is sometimes overlooked, thereby producing inaccurate results. The indicator can then be adjusted to read zero. Position a parallel at each end of the table, as shown in views A and B, Fig. 6-29, and take an indicator reading over the parallels as shown by turning the spindle to these posi-

Fig. 6-29. Procedure for tramming spindle of toolroom milling machine.

tions. Next, position a parallel below the spindle, as shown in views C and D, and indicate the parallel opposite each side of the table, again turning the spindle to these positions. The indicator readings will show the direction and the amount that the spindle head must be adjusted. Adjust the head accordingly and indicate over the parallels as before. Usually the spindle can be trammed most quickly by working in only one direction at a time; i.e., the direction shown by views A and B, or by C and D. When both indicator readings in one direction are zero, then tram in the other direction until both indicator readings are zero. When all four indicator readings are zero, the spindle is precisely perpendicular to the top of the table within the limits of the accuracy of the test equipment used.

Additional Milling-Machine Operations

Circular milling is performed on a circular milling attachment, also called a *circular table*. A great variety of work can be done on the circular milling attachment, or circular table. A typical circular milling process is shown in Fig. 6-30; in this operation a circular slot is being milled in the swivel table housing (see Fig. 4-5) of a universal knee and column milling machine, using a circular milling attachment having a power feed. Another type of job can be seen in Fig. 8-6, where a large spur gear is being milled while clamped onto a circular table. This attachment has an index plate, capable of indexing the circular table in equal increments or through a precise angle. A similar job is seen in Fig. 4-28, where the circular table is indexed for cutting the teeth of an internal gear. In this case the cutting tool is a single-point form tool, or slotting tool, which is given an up and down reciprocating motion by the slotting attachment mounted on the face of the milling machine.

T-slots, like those found on machine tool tables, can be machined on a milling machine. A T-slot milling operation is illustrated in Fig. 6-31. The first step in this operation is to mill a rectangular slot with an end milling cutter or a staggered-tooth side milling cutter. The bottom of the rectangular slot is then enlarged to the shape of an inverted T with the T-slot milling cutter, which is used like an end milling cutter. Also used like an end milling cutter is a dovetail milling cutter, which cuts dovetail slides as shown in Fig. 6-32.

Courtesy of the Brown & Sharpe Mfg. Co.

Fig. 6-30. Circular milling a slot using a power-actuated circular milling table.

In many shops the milling machine is also used to machine holes when it is convenient to do so. Drilling, boring, and reaming are operations readily performed on knee-and-column-type milling machines. The workpiece is always clamped firmly in a vise or to the table during these operations. The three directions of table movement, however, make it easy to align the drill, reamer, or boring tool with respect to the workpiece. On horizontal milling machines the workpiece is fed into the drill with the transverse table feed. The knee should be firmly clamped to the column when this operation is performed. The vertical head on heavy-duty vertical milling machines (see Fig. 4-9) should be used to feed the drill into the work so that the knee can be clamped to the column. Other vertical milling machines (see Fig. 4-10) have a quill-type spindle which is used to feed the drill into the workpiece. The hand feed of the quill on these machines is relatively sensitive, which is a real advantage

Courtesy of the Brown & Sharpe Manufacturing Company

Fig. 6-31. Milling a T-slot on an angular surface using a universal vertical milling head mounted on a sliding-head-type milling machine.

Courtesy of Cincinnati Milacron

Fig. 6-32. Milling a dovetail slide with a dovetail milling cutter.

when small holes are drilled. Some universal vertical-spindle milling heads that are mounted on horizontal milling machines have a quill-type spindle which can be used to feed the drill at an angle into the workpiece. Figure 6-33 illustrates a hole being drilled at an angle on a sliding-head-type milling machine with a quill-type universal vertical milling attachment.

The Coordinate Measuring Attachment

Vertical knee-and-column-type milling machines can be equipped with a coordinate measuring attachment, which is used to accurately position the table and thus the workpiece with respect to the axis of the spindle. Several different types of coordinate measuring attachments are available, one of the most common of which is shown in Fig. 6-34. A trough is attached to the table, and another trough is attached to the knee. Precision end measuring rods and a micrometer head placed end to end in these troughs act as a length gage to position the table. The rods are made in 1-inch increments of length. Lengths of less than 1 inch are produced with the micrometer head. It has a "tenth" vernier that enables increments of one-ten-thousandths (.0001) inch to be obtained.

An adjustable stop is located at one end of each trough. The adjustment is used only to zero the indicator located at the other end of the trough, after the reference axes of the workpiece have been located with respect to the spindle. No further adjustments are made to this stop. The dial test indicator provides a reference position in the trough for the rods and the micrometer. The table is always moved until the rods and

Courtesy of the Brown & Sharpe Manufacturing Company

Fig. 6-33. Drilling a hole at an angle on a sliding-head-type milling machine. The drill is fed into the work by the quill-type spindle of the universal milling attachment.

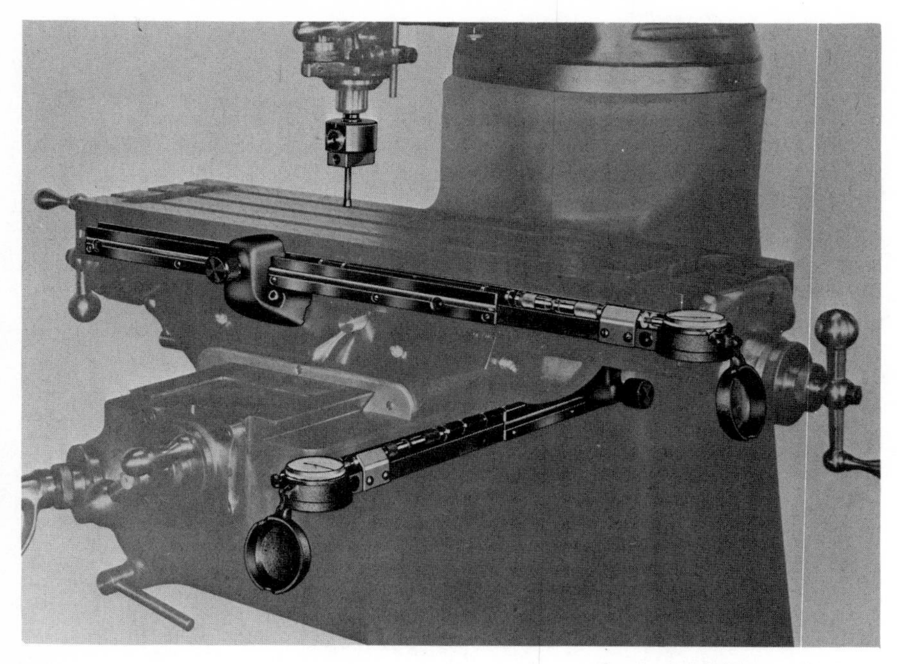

Courtesy of Bridgeport Machines

Fig. 6-34. Precision coordinate measuring attachment utilizing precision end measuring rods and a micrometer head placed in each of the troughs.

the micrometer in the trough cause the indicator to read zero. When the indicator reads zero, the table is in the desired position. The graduations on the indicator read in one-ten-thousandths inch; however; they should not be used to make "tenth" table settings, which are done by the setting of the micrometer head.

Care and cleanliness must be exercised in placing the rods and the micrometer in the troughs. The ends should be wiped clean before they are placed in the trough, and a check for cleanliness should be made after every table setting. This is done by slightly rotating each end rod and the micrometer while observing the hand of the dial test indicator. Any movement of the hand is an indication that there is dirt to be removed and that the table setting should be repeated. The table should always be backed off to release the rods and the micrometer before they are removed from the trough.

Precision Hole Location

Precision hole location operations are best performed on jig boring and jig grinding machines, as described in Chapter 15, Vol. 1 of the 2nd edition of this book. These machines are capable of producing the ultimate in precision machine work; however, they are not available in all shops. Moreover, not all precision hole location work will require the very highest degree of precision; therefore this work can be done on a milling machine using toolmakers' buttons and a procedure that is very similar to the method of using toolmakers' buttons on an engine lathe, which is also described in Chapter 15, mentioned above. Some toolroom milling machines are equipped with a coordinate measuring system, which was described previously in this chapter under, "The Coordinate Measuring Attachment." The method of precision hole location using the coordinate measuring system will now be described. The first step is always to tram the spindle as described earlier in this chapter, unless the spindle is known to be already in tram.

There are four basic steps which should be followed in sequence to locate and machine a hole on a milling machine equipped with a coordinate measuring attachment. They are: 1. align and clamp the workpiece to the milling machine table; 2. locate the two reference axes of the workpiece with respect to the spindle; 3. locate the hole to be machined; 4. drill and bore the hole to size. These steps will now be described.

1. *Align and Clamp the Workpiece.* Coordinate dimensions should be used on the drawings to specify the location of the holes to be machined. The reference axes should be to the left and at the top of the view on which they are shown. The workpiece should then be placed on the milling-machine table so that these reference axes will be to the left and toward the column of the machine. When placed in this position, the settings of the rods and the micrometer heads in the troughs will correspond exactly with the coordinate dimensions on the drawings.

Components of tools and dies are frequently made with two perpendicular edges from which the coordinate dimensions are given. Thus, they act as reference edges. An example is the Bushing Plate shown at A, Fig. 6-35. This plate must be placed on precision parallel bars to allow the cutting tools to clear the table. Before the clamps holding the plate to the table are firmly tightened, the upper edge—which is placed toward the column—should be indicated with a dial test indicator until this edge is parallel to the longitudinal table travel. It is then clamped in place.

2. *Locate the Two Reference Axes of the Workpiece with Respect to the Spindle.* Again using the Bushing Plate in Fig. 6-35 as an example, find the location of the two reference edges with respect to the axis of the spindle. Several methods can be used to do this (see Chapter 15, Volume 1). In this case these edges will be "picked up" with an edge finder, an instrument consisting of a cylindrical body on which a movable cylindrical contact is attached to one of the end faces. The diameter of this contact is .200 inch, which is less than the diameter of the body. The contact can be moved by applying a very light pressure.

With the edge finder held in the spindle by a chuck, the spindle is engaged. The contact is intentionally offset so that it wobbles as it rotates. The table is then moved longitudinally until the contact touches the left edge of the Bushing Plate. As this movement continues the wobble of the contact will decrease until it ceases altogether. When this occurs the movement of the table is stopped, and the axis of the spindle will be

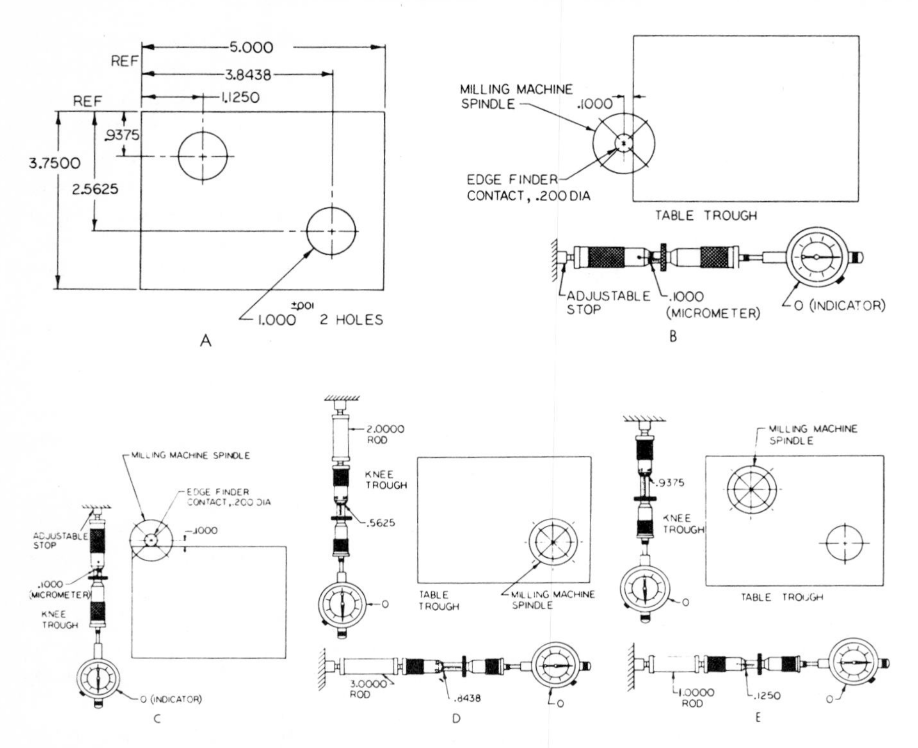

Fig. 6-35. Schematic drawings showing the micrometer head settings and precision end measuring rods required for jig boring two holes in a bushing plate. A. bushing plate; B. setting after picking up left-hand edge; C. setting after picking up upper edge; D. setting for boring first hole; E. setting for boring second hole.

one-half of the diameter of the contact, or .100 inch away from the left edge. The micrometer head is set to read .1000 inch and is to be placed in the trough located on the table. The adjustable stop is then adjusted until the dial test indicator reads zero. This position is shown at B, Fig. 6-35.

The reference edge which is placed toward the column is now picked up in the same manner with the transverse table movement used. A micrometer head set to read .1000 inch is placed in the trough attached to the knee, and the indicator is zeroed by the fixed stop. This position is shown at C, Fig. 6-35. After the two reference edges are located with respect to the axis of the spindle, the table can now be located to machine the holes from this position.

3. *Locate the Hole.* Set the micrometer head in the table trough to read .8438 inch and then place it in this trough. Also place a 3-inch precision end measuring rod in this trough. Move the table to the left until the dial test indicator on the table reads zero. Set the micrometer in the knee trough to read .5625 inch. Place it and a 2-inch precision end measuring rod in the knee trough. Move the table in the transverse direction until the indicator on the knee reads zero.

The table is now located for machining the first hole. This position and the settings are illustrated schematically at D in Fig. 6-35. The hole that is the greatest distance from the reference edges should always be machined first. *The other holes are machined by progressively moving the table so that the spindle approaches the reference edges.* In this case the second hole is located when the Bushing Plate is in the position shown at E in Fig. 6-35. The micrometer head in the table trough is set to read .1250 inch and a 1-inch precision end measuring rod is also placed in this trough. Only the micrometer head, which is set to read .9375 inch, is placed in the knee trough.

4. *Drill and Bore the Hole.* After locating each hole, follow the steps in the sequence below for the machining:

a. Drill a spot with a center drill. The center drill, being short and stiff, will not deflect as it is cutting and will produce a spot that is in line with the axis of the spindle.

b. Drill the hole through the plate with a ½-inch twist drill. The ½-inch drill will follow the spot made by the center drill closely. Enlarge the hole by drilling through the plate with a ¾-inch drill.

c. Bore the hole with a single-point boring tool held in an offset boring head. An offset boring head is shown in the spindle of the milling machine in Fig. 6-34. It has a micrometer dial that can be used to accurately offset the boring tool in order to bore the hole to the desired diameter.

   Before the hole is bored to the finish size, at least one and preferably two light boring cuts should be taken through the hole with a single-point boring tool, leaving the hole undersize. These cuts serve to correct any misalignment in the location of the hole caused by the previous drilling operations.

   The trial cut procedure is used to bore the hole. After a trial cut that leaves the hole undersize is taken for a short distance in the hole, the diameter of the hole is measured. The offset boring head is adjusted the amount required to bore the hole to the desired size. If the hole is close to the finish size, another trial cut is taken and a measurement is made to verify the previous setting of the offset boring head. If the size of the hole is correct, it is bored through the plate.

Milling-Machine Fixtures

Special work-holding fixtures are used on milling machines to increase their productivity by reducing the time required to position the part on the machine. Milling fixtures are also sometimes required to hold workpieces that are shaped in such a way that they cannot be held by any other method.

The casting in Fig. 6-33 is held on a plate which can be classified as a simple fixture. The bottom surface of the casting has been previously finish-machined and has a keyway cut into it. A key on the plate fits into this keyway. A second key on the lower side of the plate fits the T-slots of the milling machine and thereby aligns both the plate and the workpiece with respect to the table. The casting and the plate are clamped to the table by strap clamps which are supported by heel blocks as shown. The use of this simple fixture reduces the time required to set up the casting and provides a very effective method of positioning the bottom surfaces so that the upper surfaces will be machined in alignment with them.

A fixture mounted on a circular table is shown in Fig. 6-36. This fixture rapidly locates the part and holds it so that the circular slot can be cut by

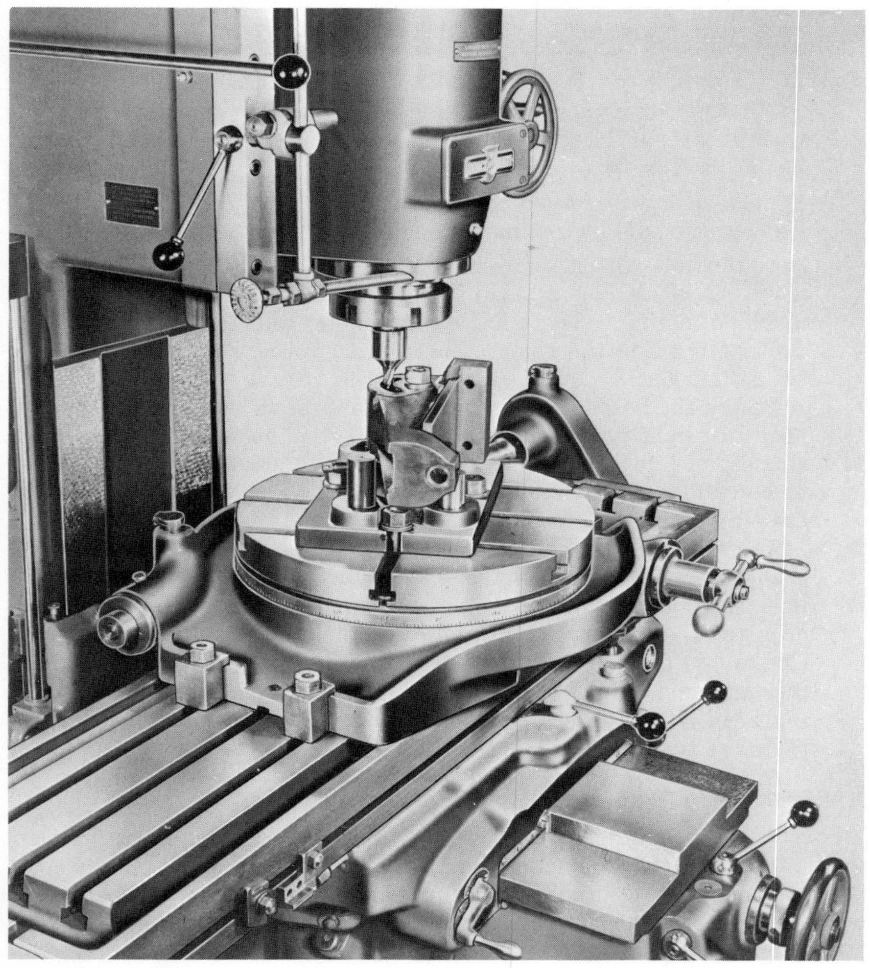

Courtesy of Cincinnati Milacron

Fig. **6-36.** End milling a circular slot in a part held by a milling machine fixture that is clamped on the circular table.

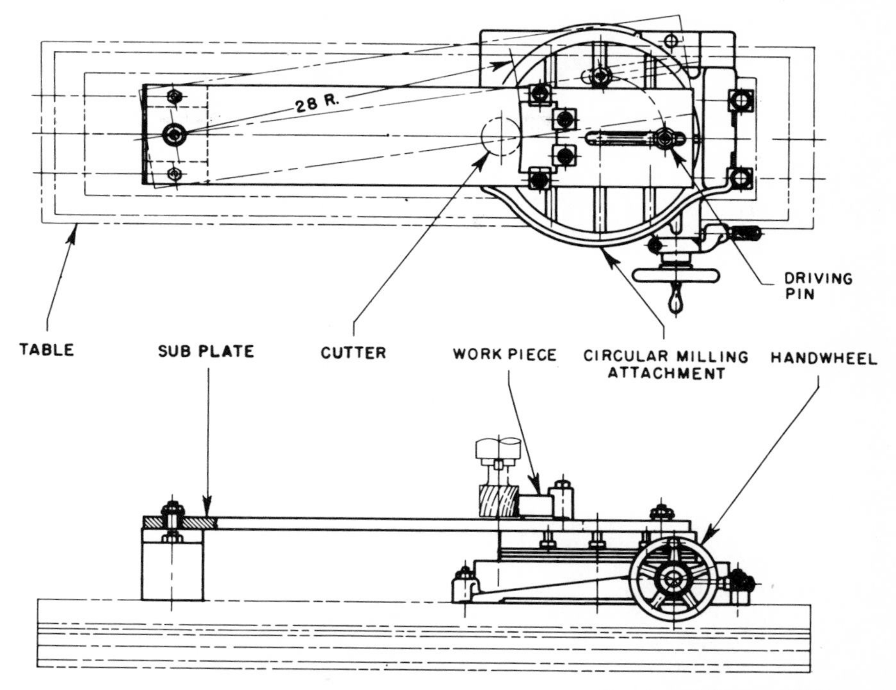

Courtesy of Cincinnati Milacron

Fig. 6-37. Setup for milling a large radius on a milling machine.

the end mill. A special fixture for milling large radii is shown in Fig. 6-37. The subplate on which the workpiece is mounted is pivoted by a stationary pin at one end and is driven by a driving pin fastened to the circular milling attachment at the other end. The driving pin is free to move in the slot in the subplate.

CHAPTER 7

Indexing

Indexing is the operation of rotating a workpiece an exact amount. The purpose may be to rotate the part an exact angle, or to space equally spaced divisions around a circumference, such as gear teeth or holes in a hole circle. Indexing is usually done with a mechanism called a *dividing head,* Fig. 7-1, which is also called an *index head.* Circular tables, also called rotary tables or circular milling attachments, Fig. 8-6, are also used to perform the indexing operations when they are equipped with an index plate.

Courtesy of Cincinnati Milacron

Fig. 7-1. Cincinnati Universal Dividing Head with descriptive keys.

Left: front view. A. spindle; B. direct index pin; C. housing; D. index plate; E. sector arms; F. index pin; G. index crank; H. index plate stop pin; I. base.
Right: rear view. A. drive shaft (for helical milling); B. spindle clamp lever; C. worm shaft housing and eccentric.

Indexing operations are most frequently performed on a milling machine because of its great versatility. Therefore, indexing is usually associated with milling-machine work. There are many occasions, however, when indexing is done on other types of machine tools. There are no restrictions to the use of the dividing head other than the ability of the machine tool to support it and the ability of the dividing head to support the workpiece. For example, dividing-head work is rarely, if ever, done on a lathe. Dividing heads are used occasionally, however, on shapers, planers, drilling machines, and often on jig borers. The dividing head is also used to index parts which are being inspected, this work being done on a surface plate.

This chapter will treat the calculations that are necessary in order to make practical use of the dividing head. Examples of the work done on the dividing head and the use of the dividing head for helical milling will be discussed in the following chapters.

The Cincinnati Universal Dividing Head

Dividing heads, or index heads, may vary somewhat in their design. However, the principle of operation of all dividing heads is the same. The main characteristics of a Cincinnati Universal Dividing Head are shown in Fig. 7-1. The dividing-head housing is held in a swivel block that is clamped to the machine tool table. Two keys attached to the bottom of the swivel block fit into the T-slots of the machine tool table and are used to align the dividing head. The keys may be removed if desired. The housing contains the mechanism of the dividing head. It can be tilted to position the spindle at any angle from 5 degrees below the horizontal position to 5 degrees beyond the vertical position. A scale in degrees which is supplemented by a vernier scale allows very accurate angular settings to be made. Brake-band-type clamps on the swivel block clamp the housing firmly in the desired position.

For indexing, the dividing-head mechanism is actuated by turning the crank at the side of the dividing head. A driving mechanism, shown in Fig. 9-5, is used to actuate the universal dividing head for helical milling operations. When the crank is used to actuate the dividing head, it turns a short shaft having a spur gear attached to its other end inside of the dividing-head housing. This gear engages a gear with an equal number of teeth which is located on a worm shaft; thus, the worm shaft rotates the same number of revolutions as the shaft to which the crank is attached. The worm shaft, illustrated in Fig. 7-2, engages a worm gear which is firmly mounted on the spindle of the dividing head.

The worm has a single thread and the worm gear has 40 teeth. Therefore, when the worm turns one revolution the worm gear and the spindle will turn $\frac{1}{40}$ revolution. In order for the spindle to make one complete revolution the worm must turn 40 revolutions. Since the gear ratio between the worm gear shaft and the shaft to which the crank is attached is 1 to 1, the crank must turn 40 revolutions to make the spindle turn one revolution; or, when the crank is turned one revolution, the spindle turns $\frac{1}{40}$ revolution. The dividing-head ratio, therefore, is 40 to 1. This is the ratio of most dividing heads. However, one prominent manufacturer, the Kearney & Trecker Corporation, builds a dividing head with a gear ratio of 5 to 1.

The worm shaft of the Cincinnati Universal dividing head, shown in Figs. 7-1 and 7-2, is mounted in a worm shaft housing. The bearings in this housing which support the worm shaft are eccentric with respect to the axis of the housing so that when the housing is rotated, the worm

Courtesy of Cincinnati Milacron

Fig. 7-2. View of spindle and worm-gear arrangement of a universal dividing head.

shaft moves toward or away from the worm gear in a circular path. Thus, the worm can be disengaged from the worm gear, thereby allowing the spindle to rotate freely. The worm is disengaged from the worm gear by turning the wormshaft housing that can be seen extending outside the dividing-head housing at C, Fig. 7-1. This feature can be used to perform direct indexing and to align workpieces held by the dividing head.

The spindle of the dividing head is hollow. The nose of the spindle of most modern dividing heads is identical in design to the spindle nose of modern milling machines. This permits chucks and collets to be used interchangeably between the nose milling-machine spindle and the dividing head. The spindle nose on dividing heads of older design do not have this feature. On older dividing heads the taper is usually a Morse or a Brown & Sharpe taper, while on the modern dividing heads the taper is the American Standard Milling Machine Spindle Nose Taper. Larger chucks and faceplates can be bolted to the nose and are driven by the driving key. Work-supporting centers and collet chucks are held in the inside taper of the spindle. The direct indexing pin engages a group of 24 equally spaced holes located on the outer cylindrical surface of the nose. These holes are used for direct indexing. The spindle can be clamped firmly in position by turning the spindle-clamping lever. The spindle should always be clamped whenever a cut is taken.

A basic component of the dividing head is the index plate which is mounted behind the crank. The index plate of the Cincinnati Universal Dividing Head has 11 concentric rows of holes called *hole circles,* each of which contains a different number of equally spaced holes. The crank can be positioned radially to allow the index pin, located at the outer end of the crank, to engage the holes in any hole circle. By placing the index pin in selected holes in a particular hole circle, the crank can be made to rotate a certain fractional part of a revolution. This, in turn, makes it possible to accurately rotate the dividing-head spindle a desired amount. The Cincinnati Universal Dividing Head is supplied with one standard index plate which has circles on both the front and back side. The hole circles in this plate are:

Side A: 24, 25, 28, 30, 34, 37, 38, 39, 41, 42, 43
Side B: 46, 47, 49, 51, 53, 54, 57, 58, 59, 62, 66

Although these hole circles will take care of most ordinary work, a greater range is sometimes required. This is obtained by using three additional index plates that can be furnished with the dividing head. These additional plates, called "High Number Index Plates," have the following hole circles:

Side A: 189, 177, 171, 147, 129, 117, 99, 91, 69, 48, 30
Side B: 199, 183, 169, 157, 141, 127, 111, 97, 81, 67, 36

Side A: 197, 181, 167, 153, 139, 123, 109, 93, 79, 46, 34
Side B: 193, 179, 163, 151, 137, 121, 107, 89, 77, 44, 32

Side A: 191, 175, 165, 149, 133, 119, 103, 87, 73, 42, 26
Side B: 187, 173, 159, 143, 131, 113, 101, 83, 71, 38, 28

One hole in each hole circle has a number stamped adjacent to it which designates the number of holes in the hole circle. The indexing operation should always start from this hole because it can be easily identified. Should an error occur in indexing, or be suspected of having occurred, it is always possible to return to the starting position from which a check can be made by repeating the previous indexing moves.

The index plate is held in position by the index plate stop pin (see H, Fig. 7-1). When plain indexing operations are performed, the stop pin is used to keep the index plate from rotating. When helical milling operations are performed, the index plate must rotate with the index pin engaged in order to drive the crank. In this case the stop pin must be disengaged. Notches are cut for a length of 2 inches along the circumference of the index plate and are spaced .060 inch apart. When the stop pin is disengaged and the index plate turned together with the crank a distance of one notch space, the spindle will rotate 1/18,460 of a revolution or an angle of 1.17 minutes. This adjustment is convenient for correcting slight misalignments that might occur during an indexing operation or the setting up of the workpiece on the dividing head.

The purpose of the sector arms, shown at E, Fig. 7-1, is to make it unnecessary to count the number of holes to be indexed during a number of repetitive indexing moves. This, of course, saves time, as well as avoids mistakes that can be made in counting. The sector arms are free to rotate over the face of the index plate, and the spread, or angle, made by the two sector arms can be adjusted to any desired setting. Each sector arm has one beveled edge which is used to locate the hole in the index plate. The total spread of the sector arms must include the hole in which the index pin is positioned before the start of each indexing movement. Failure to recognize this is sometimes a source of a mistake in setting up the dividing head. Thus, the spread of the sector arms must be equal to the number of holes to be indexed plus one.

The Brown & Sharpe Universal Spiral Index Center

The Brown & Sharpe Universal Spiral Index Center, shown in Fig. 7-3, has a 40 to 1 ratio. The worm can be disengaged from the worm gear in order to do direct indexing. A plate attached to the spindle nose has 24 equally spaced holes on a hole circle around its face and is used for direct indexing. A pin, located in the housing, engages the direct indexing holes. This index head is equipped with two index plates having the following hole circles:

Plate 1: 15, 16, 19, 23, 31, 37, 41, 43, 49
Plate 2: 17, 18, 20, 21, 27, 29, 33, 39, 47

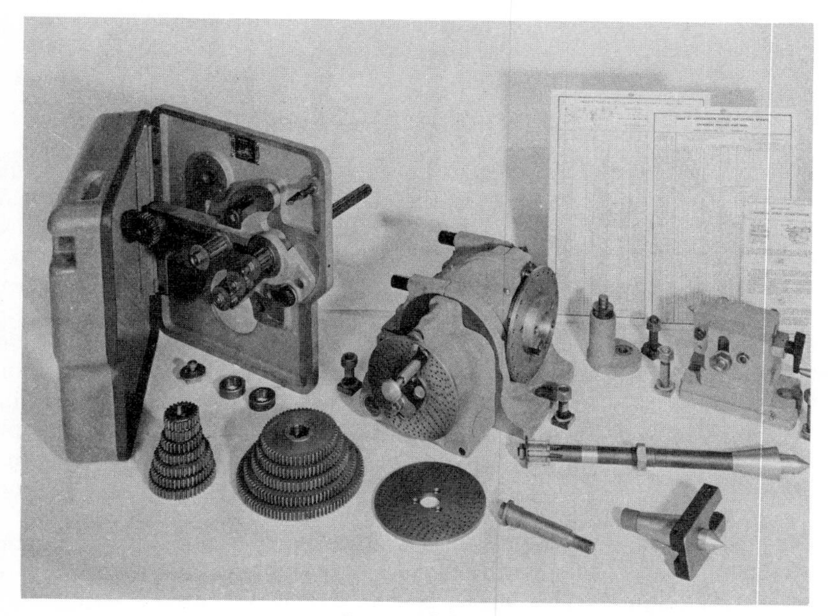

Courtesy of the Brown & Sharpe Manufacturing Company

Fig. 7-3. The Brown & Sharpe Universal Spiral Index Head.

A wider range of divisions can be obtained with the differential indexing method, which will be explained later in this chapter. For the present it is sufficient to say that this method of indexing involves coupling the gears and the driving unit to the index head. The same gears and driving unit are also used to perform helical milling operations. The change gears that are available have the following number of teeth: 24 (2 gears), 28, 32, 40, 44, 48, 56, 64, 72, 86, and 100.

Figure 7-3 also shows the footstock, or tailstock, which is used to support one end of a workpiece or a mandrel when it is being held between centers. The center of the footstock may be elevated to support tapered parts. The headstock center, shown in Fig. 7-3, has a driver plate attached to it. It is held in the spindle by means of a short draw-in-bolt shown lying to the left of the center. An extended center is shown lying behind the headstock center. This is used only for differential indexing. A spline is provided at the opposite end of the center for mounting a change gear. A center rest, used to support slender workpieces, is shown standing between the dividing head and the footstock.

Dividing-Head Operating Procedure

The procedure of operating a dividing head is quite simple. However, an indexing error will occur unless the details of this procedure are closely followed. The procedure starts with the initial setting of the crank. The first cutting operation should always be performed with the index pin in one of the holes adjacent to the number on the plate indicating the number of holes in the hole circle. In this way it is always possible to return to the initial dividing head setting because the first hole is identified by being a numbered hole. This is useful if a mistake in indexing has occurred. By returning to the initial position, the indexing movements can be retraced and the error found or corrected.

The index plate pin must be carefully positioned in each hole of the index plate which it is to enter. The pin can be kept in a retracted position when the crank is turned by giving the knob at the end of the crank a partial turn. The indexing operation is almost always performed by rotating the crank in a clockwise direction. Before the index plate pin reaches the hole in which it is to be inserted, it is allowed to come forward and touch the index plate. The crank is then gently "bumped" with the hand until the pin "drops" into the hole by the pressure of the spring behind it (see Fig. 7-3). The spindle is then locked and the sector arms rotated until the beveled edge of the trailing sector arm contacts the index plate pin. The cutting operation is performed, and the spindle is unlocked. The pin is retracted and the crank is turned the required number of complete turns plus a distance that would bring the pin just ahead of the hole in which it is to be placed. This hole is identified by the beveled edge of the leading sector arm. The pin is "bumped" into the hole as before. These operations are repeated until the job is done.

If the index plate pin is accidentally moved beyond the hole in which it should be positioned, the crank should be moved counterclockwise a partial turn and then moved clockwise again until the pin is just ahead of the hole. It is bumped toward the hole until the spring pressure causes it to "drop" into the hole. This procedure compensates for the backlash between the worm and the worm gear which, if ignored, could cause an error to occur.

Direct Indexing

Direct indexing is done by disengaging the worm from the worm gear so that the spindle can be rotated freely by hand. The direct index pin is inserted into one of the 24 equally spaced holes drilled into the spindle nose of the dividing head shown in Figs. 7-1 and 7-2. The dividing head in Fig. 7-3 has a 24-hole index plate mounted on the nose into which a direct index pin is inserted.

Any number of divisions that can be divided evenly into 24 can be indexed. These numbers are 2, 3, 4, 6, 8, 12, and 24. The number of holes that must be indexed to obtain the required number of divisions is calculated by dividing the total number of holes on the spindle nose by the number of divisions required.

Example 7-1:

The dividing head in Fig. 7-1 is to be used to drill eight equally spaced holes on a hole circle on a cover plate. Calculate the dividing-head movement if the direct indexing method is to be used.

$$\text{No. holes} = \frac{24}{8} = 3$$

Thus, the spindle should be moved three holes for each index. When counting the three holes, do not count the hole initially occupied by the direct index pin. The direct method of indexing is fast when only a few divisions are required because it saves the time necessary to turn the index crank a number of complete revolutions.

Plain Indexing

In *plain indexing*, perhaps the most common method of indexing, the dividing-head spindle is indexed a required amount with the crank and the index plate. The workpiece, being rotated by the dividing-head spindle, is indexed the same amount as the spindle. In order to do plain indexing, the worm and the worm gear of the dividing head must be engaged.

The information required to select the index plate and to make the correct indexing movement of the crank can be obtained by applying the following formula:

$$T = \frac{40}{N} \tag{7-1}$$

where: T = Number of complete and fractional parts of a turn of the crank
N = Number of divisions required on the workpiece

Formula 7-1 is applied by following the steps that are listed here:

1. Solve Formula 7-1 to obtain a whole number and a fraction.
2. The whole number, if it exists, is the complete number of turns that the crank must make in order to index the required divisions.
3. The remaining fraction is the fractional part of a turn that the crank must make. It is used to determine the hole circle and the number of holes that must be indexed by using the following procedure:
   a. Find a hole circle on an available index plate into which the denominator, or the lowest denominator, can be evenly divided and determine the resulting quotient.
   b. Multiply both the numerator and the denominator of the fraction resulting from Formula 7-1, or the reduced fraction having the lowest denominator, by the quotient obtained in Step 3a. The denominator of this product will be the hole circle that must be used, and the numerator will be equal to the number of holes to be indexed.

The procedure can perhaps best be learned by carefully following a few examples.

Example 7-2:

Determine the hole circle to be used and the indexing movement required to index 54 divisions using a Cincinnati Universal Dividing Head.

$$T = \frac{40}{N} = \frac{40}{54}$$

Thus, the crank must make 40/54 fractional part of a turn, and no whole number of turns is required. Since the standard index plate has a 54-hole circle, the crank can be made to rotate 40/54 of a turn by moving 40 holes in the 54-hole circle.

Example 7-3:

Determine the hole circle to be used and the indexing movement required to index 27 divisions using a Cincinnati Universal Dividing Head.

$$T = \frac{40}{N} = \frac{40}{27} = 1\frac{13}{27} = 1 + \frac{13}{27}$$

The crank must make one complete turn plus 13/27 fractional part of a turn. As a 27-hole circle is not available, a hole circle other than 27 must be found. An inspection of the hole circle on the standard index plate shows that a 54-hole circle is available and that $54 \div 27 = 2$. Therefore, the numerator and the denominator of the fraction 13/27 must be multiplied by 2—which does not change the value of the fraction.

$$T = 1 + \frac{13}{27} = 1 + \frac{13 \times 2}{27 \times 2}$$

$$T = 1 + \frac{26}{54}$$

Thus, to index 27 divisions the crank is made to turn one complete turn plus 26 holes in the 54-hole circle.

Example 7-4:

The workpiece is to be indexed 25 divisions using the Brown & Sharpe Universal Dividing Head. Determine the hole circle and the dividing-head movement required to do this job.

$$T = \frac{40}{N} = \frac{40}{25} = 1\,\frac{15}{25}$$

A 25-hole circle is not available, and there is no hole circle into which 25 can be divided. The fraction 15/25 must therefore be reduced to its lowest denominator. This is done by dividing the numerator and the denominator by the largest number which will divide evenly into both of these terms. In this example the numerator and the denominator are divided by 5.

Thus:

$$T = 1 + \frac{3}{5}$$

Now find a hole circle into which the numerator can be evenly divided and find the quotient. The numerator will divide evenly into 20, and $20 \div 5 = 4$. Multiplying both the numerator and the denominator by 4:

$$T = 1 + \frac{3 \times 4}{5 \times 4} = 1 + \frac{12}{20}$$

Thus, 25 divisions can be indexed by indexing one complete turn plus 12 holes in the 20-hole circle.

Formula 7-1 must be modified when the Kearney & Trecker dividing head, which has a 5 to 1 gear ratio, is used.

$$T = \frac{5}{N} \qquad \text{(7-1B)}$$

This dividing head is equipped with an index plate with the following hole circles:

Side A: 98, 88, 78, 76, 68, 58, 54
Side B: 100, 96, 92, 84, 72, 66, 60

Example 7-5:

Determine the hole circle and the indexing movement required to index 14 divisions with a Kearney & Trecker dividing head.

$$T = \frac{5}{N} = \frac{5}{14} = \frac{5 \times 6}{14 \times 6}$$
$$= \frac{30}{84}$$

Thus, the 14 divisions can be obtained by indexing 30 holes in the 84-hole circle.

Differential Indexing

Many divisions cannot be obtained by plain indexing because an index plate with the necessary hole circles is not available. The Brown & Sharpe Manufacturing Company has developed a method of indexing, called *differential indexing,* by which the range of their dividing head (see Fig. 7-4) can be extended to index all numbers between 1 and 382. In addition, many other numbers beyond 382 can also be indexed.

Courtesy of the Brown & Sharpe Manufacturing Company

Fig. 7-4. Index head geared to index 271 divisions by the differential indexing method.

By this method, the index crank is indexed a predetermined number of holes in a given hole circle. The procedure is like plain indexing except that the index plate moves in the same or opposite direction to the crank by the action of a train of gears shown in Figs. 7-4 through 7-6. The total movement of the crank at every indexing is, therefore, equal to its movement relative to the index plate plus the movement of the plate when the plate moves in the same direction as the crank; or minus the movement of the plate when the plate moves in the opposite direction to the crank.

Courtesy of the Brown & Sharpe Manufacturing Company

Fig. 7-5. Index head geared to index 319 divisions by the differential indexing method.

The spindle of the index head is connected to gear E, Fig. 7-4, by means of a special differential indexing center shown in Fig. 7-3. Gear E is connected to gear C by idler gear D. Gear C drives a shaft which rotates a worm and worm gear inside the index head causing the index plate to rotate. When differential indexing, the index plate must be disconnected by disengaging a stop pin that holds it in place. The idler gear D has no effect on the gear ratio between gears E and C except to change the direction of rotation of the index plate relative to the index crank. When one idler gear is used, as in Fig. 7-4, the index plate and the index crank rotate in the same direction. When two idler gears, D_1 and D_2, are used, as in Fig. 7-5, the crank and the index plate rotate in opposite directions.

It is necessary to make two calculations in order to determine the setup for differential indexing:

1. Calculate the number of holes and the hole circle required for the index head crank movement.
2. Calculate the change gear ratio.

The first step in calculating the index head setting for differential indexing is to select some arbitrary number of divisions that are either slightly greater or slightly less than the required number of divisions. The selected number of divisions must be such that they can be indexed with one of the standard index plates. It is suggested that the number selected be a number divisible by 10. The two calculations can then be made by applying the following formulas:

$$T = \frac{40}{A} \tag{7-2}$$

$$R = (A - N)\frac{40}{A} = \frac{\text{Gear connected to spindle or driving gear}}{\text{Gear on index plate drive shaft or driven gear}} \tag{7-3A}$$

$$R = (N - A)\frac{40}{A} = \frac{\text{Gear connected to spindle or driving gear}}{\text{Gear on index plate drive shaft or driven gear}} \tag{7-3B}$$

where: T = The number of complete turns and fraction of a turn of the index crank

N = Number of divisions actually required on the workpiece

A = Number of divisions which are arbitrarily selected and which can be obtained by available index plates

R = The gear ratio required to obtain the desired number of divisions (N) when the index head is set up to obtain the arbitrarily selected number of divisions (A)

Formula 7-3A is used when A is larger than N. In this case the index plate must move in the same direction as the crank. To do this one idler gear must be used in the gear train. Formula 7-3B is used when N is larger than A. When this occurs the index plate and the crank must turn in opposite directions, and two idler gears are required in the gear train.

Example 7-6 (See Fig. 7-4):

Calculate the index head movement and the gear ratio required to index 271 divisions.

Differential indexing is necessary since the required 271 divisions cannot be obtained by plain indexing. The number of divisions nearest to 271 that is selected is 280.

$$T = \frac{40}{A} = \frac{40}{280} = \frac{1}{7} = \frac{1 \times 3}{7 \times 3} = \frac{3}{21}$$

$$R = (A - N)\frac{40}{A} = (280 - 271)\frac{40}{280} = 9 \times \frac{1}{7} = \frac{9 \times 8}{7 \times 8}$$

$$= \frac{72}{56} = \frac{\text{Gear connected to spindle or driving gear}}{\text{Gear on index plate drive shaft or driven gear}}$$

The change gear arrangement for indexing 271 divisions is shown in Fig. 7-4. Gear E has 72 teeth and is connected to the index-head spindle. Gear C has 56 teeth and is connected to the shaft which causes the index plate to rotate. A single idler gear D is used to make the index plate rotate in the same direction as the index crank since Formula 7-3A was used. Idler gears such as shown have no influence on gear ratios except to determine the direction of rotation of the driven gear with respect to the driving gear. When the change gears are positioned as shown, the index head is indexed three holes in the 21-hole circle.

Example 7-7 (See Fig. 7-5):

Calculate the index head movement and the gear ratio required to index 250 divisions.

Let $A = 240$

$$T = \frac{40}{A} = \frac{40}{240} = \frac{1}{6} = \frac{1 \times 3}{6 \times 3} = \frac{3}{18}$$

$$R = (N - A)\frac{40}{A} = (250 - 240)\frac{40}{240} = 10 \times \frac{1}{6} = \frac{5}{3}$$

$$= \frac{5 \times 8}{3 \times 8}$$

$$= \frac{40}{24} = \frac{\text{Gear connected to spindle or driving gear}}{\text{Gear on index plate drive shaft or driven gear}}$$

The change gear arrangement for indexing 250 divisions is shown in Fig. 7-5. The 40-tooth gear, gear E, is connected to the spindle, and the 24-tooth gear, gear C, is connected to the index-plate drive shaft of the dividing head. Since Formula 7-3B was used, two idler gears are required. These gears are D_1 and D_2. The number of teeth on the idler gears does not matter as they in no way affect the gear ratio between the driver and the driven gear. They only affect the direction of rotation of the driven gear with respect to the driving gear. The crank is indexed three holes in the 18-hole circle.

Example 7-8 (See Fig. 7-6):

Calculate the index head movement and the gear ratio required to index 319 divisions.

Let $A = 290$

$$T = \frac{40}{A} = \frac{40}{290} = \frac{4}{29}$$

$$R = (N - A)\frac{40}{A} = (319 - 290)\frac{40}{290} = (29)\frac{4}{29} = \frac{4}{1}$$

Because a 4 to 1 ratio is unobtainable with available gears, a compound gear train, such as that shown in Fig. 7-6, must be used. On the compound gear train one of the idler gear shafts has two gears, G and F,

Courtesy of the Brown & Sharpe Manufacturing Company

Fig. 7-6. Index geared to index 319 divisions using a compound gear train.

mounted on it which rotate together at the same speed. These gears are called compound gears and, unlike other idler gears, they do affect the gear ratio. In compound gears a clear identification between driving and driven gears must be made. For example, in Fig. 7-6, E is connected to the spindle and is a driving gear which drives the smaller of the two compound gears, gear G. Thus gear G is a driven gear. The larger compound gear, gear F, drives the gear on the index drive plate shaft, gear C, through a single idler gear D. Gear F is then a driver gear, and the gear C on the index plate drive shaft is a driven gear. The single idler shaft has no effect on the gear ratio, changing only the direction of rotation. When the direction of rotation is considered, the compound gears are thought of as one idler gear. Thus, since Formula 7-3B was used to calculate the gear ratio, two idlers are required—which in this case are the two compound gears plus the idler.

To calculate the actual compound gears to be used, the gear ratio is expanded without changing its value.

$$R = \frac{4}{1} = \frac{4 \times 3}{1 \times 3} = \frac{12}{3} = \frac{3 \times 4}{1 \times 3} = \frac{3 \times 24}{1 \times 24} \times \frac{4 \times 16}{3 \times 16}$$

$$= \frac{72 \times 64}{24 \times 48} = \frac{\text{Driving gears}}{\text{Driven gears}}$$

It does not matter which driving gear is placed on the shaft connected to the index head spindle and which driving gear is on driving gear position on the compound gear shaft. Likewise either driven gear can be placed on the index plate drive shaft or in the driven gear position on the compound gear shaft. It is, however, necessary that driving gears be correctly placed in driving gear positions and that driven gears be placed in driven gear positions. The crank is indexed four holes in the 29-hole circle to do this job.

Indexing Angles

When a definite angle must be indexed, the calculations involved are somewhat different. The ratio of the crank and the spindle of the dividing head must be compared to the number of degrees, minutes, or seconds of an angle. It should be recalled that there are 360 degrees in a complete revolution or a full circle, there are 60 minutes per degree, and there are 60 seconds per minute. Furthermore, one revolution of the dividing-head crank causes the spindle to rotate 1/40 of a revolution. Expressed in degrees, this becomes $360° \div 40 = 9°$. Thus, one turn of the crank will cause the spindle to rotate 9 degrees. Therefore, to calculate the turns of the crank required to index an angle expressed in degrees, divide that angle by 9, or:

$$T = \frac{D°}{9} \tag{7-4A}$$

where: T = Number of complete and fractional parts of a turn of the crank
$D°$ = Angle to be indexed expressed in degrees

This formula can be used by expressing the angle in minutes or seconds. The total number of minutes in 9 degrees is $9 \times 60 = 540$, and the total number of seconds in 9 degrees is $9 \times 60 \times 60 = 32{,}400$. Therefore, Formula 7-4A can also be written as follows:

$$T = \frac{D'}{540} \tag{7-4B}$$

$$T = \frac{D''}{32{,}400} \tag{7-4C}$$

where: D' = Angle to be indexed expressed in minutes
D'' = Angle to be indexed expressed in seconds

When Formulas **7-4B** and **7-4C** are used, the entire angle must be expressed in minutes or seconds as will be demonstrated in the examples to follow.

Example 7-9:

A Brown & Sharpe Universal Index Head is to be used to index 35 degrees. Determine the index head movement required to perform the operation.

$$T = \frac{D^\circ}{9} = \frac{35}{9} = 3\,\frac{8}{9} = 3 + \frac{8 \times 3}{9 \times 3}$$

$$T = 3 + \frac{24}{27}$$

Thus, the crank must be turned three complete revolutions plus **24** holes in the 27-hole circle.

Example 7-10:

An angle of 14°20′ is required to be indexed with a Cincinnati Universal Dividing Head. Determine the index head movement required to perform this operation.

Formula **7-4B** must be used. In this equation the angle must first be entirely converted to minutes. This is done by multiplying the degrees by **60** and adding the remaining minutes.

$$D' = 14 \times 60 + 20 = 860'$$

$$T = \frac{D'}{540} = \frac{860}{540} = 1\,\frac{32}{54}$$

Thus, the crank must be turned one complete revolution plus **32** holes in the 54-hole circle.

Example 7-11 (See Fig. 7-7):

Four notches having a **90**-degree included angle are to be milled 50°47′20″ apart on a detent ring. A Cincinnati Universal Dividing Head is to be used, and the "high number" index plates are available. Calculate the dividing head setting required to perform this operation.

Courtesy of Cincinnati Milacron

Fig. 7-7. Setup for milling notches in detent ring.

Formula 7-4C is used. In order to use this formula, the entire angle must first be converted to seconds.

$$\begin{array}{lcr} \text{Converting degrees: } 50 \times 60 \times 60 & = & 180{,}000 \\ \text{Converting minutes: } 47 \times 60 & = & 2{,}820 \\ \text{Adding remaining seconds} & = & \underline{20} \\ \quad \text{Total angle in seconds} & = & 182{,}840 \end{array}$$

$$T = \frac{D''}{32{,}400} = \frac{182{,}840}{32{,}400} = 5\,\frac{521}{810}$$

The crank must, therefore, be turned five complete revolutions plus 521/810 fractional part of a revolution. The fractional part of a revolution could be obtained by indexing 521 holes in an 810-hole circle. However, no 810-hole circle is available. This situation is best resolved by finding a new fraction close to the ratio 521/810 which has a denominator equal to the number of holes in one of the hole circles available. First, determine the decimal equivalent of 521/810 by dividing the numerator by the denominator; thus, 521/810 = .643210. By the trial-and-error method find a fraction having a decimal equivalent as close as possible to .643210 and having as its denominator one of the available hole circles. The following fractions were found by this method:

$$\frac{110}{171} = .643275 \qquad \frac{128}{199} = .643216$$

The ratio 128/199 is selected because its decimal equivalent more nearly approaches the decimal equivalent of 521/810.

Thus:

$$T = 5 + \frac{128}{199}$$

The dividing head is indexed five complete revolutions of the crank plus 128 holes in the 199-hole circle. The error resulting from this procedure can be calculated by multiplying the index movement by the number of seconds per revolution of the crank, or 32,400 seconds.

$$\left(5 + \frac{128}{199}\right) \times 32{,}400 = 182{,}840.2''$$

Hence, the error will be 182,840.2 − 182,840.0, or .2 second which is very small.

Example 7-12:

Determine the indexing movement that will result in the nearest angle to 50°47′20″ that can be obtained if only the standard index plate for the Cincinnati Universal Dividing Head is available.

$$T = \frac{D}{32{,}400} = \frac{182{,}840}{32{,}400} = 5 + \frac{521}{810}$$

$$\frac{521}{810} = .643210$$

By trial and error find a fraction having a decimal equivalent as close as possible to .643210 having as its denominator one of the available hole circles. The following fractions were found by this method:

$$\frac{18}{28} = .642857 \qquad \frac{27}{42} = .642857$$

Hence:

$$T = 5 + \frac{27}{42}$$

The dividing head is indexed five complete revolutions of the crank plus 27 holes in the 42-hole circle. The error resulting from this procedure would be:

$$\left(5 + \frac{27}{42}\right) \times 32{,}400 = 182{,}828.57''$$

$$182{,}840.00 - 182{,}828.57 = 11.43''$$

The angular displacement error is then 11.43 seconds or $\frac{11.43''}{60} = .1905$ minute less than the required displacement. This cannot be corrected by the notches in the index plate because each notch rotates the spindle through an angle of 1.17 minutes.

The setup for milling the detent in Examples 7-11 and 7-12 is shown in Fig. 7-7.

The Cincinnati Wide-Range Dividing Head

The Cincinnati Wide-Range Dividing Head, shown in Fig. 7-8, provides a range of divisions from 2 up to 400,000—which is equivalent to an angle of 3.24 seconds. It consists of a standard universal dividing head

Courtesy of Cincinnati Milacron

Fig. 7-8. The Cincinnati Wide-Range Dividing Head.

with an additional gear ratio and indexing mechanism that is mounted on the large crank of the dividing head. The additional gear ratio is obtained through planetary gears mounted inside a housing that is built into the large index crank. A small index plate is mounted into the face of the planetary gear housing, and a small crank rotates concentrically with the large crank to drive the planetary gears. The entire wide-range mechanism rotates together with the large crank. The wide-range dividing head can be used as a standard dividing head to do direct indexing, or to do plain indexing simply by using the large crank only. The large index plate of the wide-range dividing head has the following hole circles:

Side A: 24-28-30-34-37-38-39-41-42-43-100
Side B: 46-47-49-51-53-54-57-58-59-62-66

The small index plate has only two hole circles, which are 100 and 54. The gear ratio of the planetary gears is 100 to 1.

The relationship of the gearing ratios with respect to the rotation of the spindle is as follows. One complete revolution of the large crank causes the spindle to rotate 1/40 revolution. One turn of the small crank is equivalent to 1/100 turn of the large crank or $1/100 \times 1/40 = 1/4{,}000$ turn of the spindle. Indexing the small crank 1 hole in the 100-hole circle results in 1/100 revolution of the small crank or $1/100 \times 1/4{,}000 = 1/400{,}000$ turn of the large crank. Thus, indexing the small crank one hole in the 100-hole circle will result in 400,000 divisions if repeated.

Of equal importance in understanding the principle of the wide-range dividing head is to understand the relationship of the rotation of the two cranks. Indexing the small crank one complete revolution is equivalent to indexing the large crank 1/100 revolution as a result of the planetary gear ratio. Indexing the small crank one hole in the 100-hole circle, or 1/100 revolution, is equivalent to indexing the large crank $1/100 \times 1/100 = 1/10{,}000$ revolution.

Formula 7-1 ($T = 40/N$) can be used in calculating the setting of the wide-range dividing head. When this formula is used for plain indexing, the answer can more conveniently be expressed as a common fraction. When intended for use with the wide-range dividing head, the answer should be expressed as a decimal fraction. When divisions, not angles, are indexed, the 100-hole circle on the large index plate and the 100-hole circle on the small index plate are the only two index plates that are used at any time.

Example 7-13:

Calculate the settings of the wide-range dividing head required to index 67 divisions.

$$T = \frac{40}{N} = \frac{40}{67} = .597015$$

The decimal fraction .597015 can be rewritten as follows:

$$T = .597015 = .59 + .0070 + .000015$$

or,

$$T = \frac{59}{100} + \frac{70}{10{,}000} + \frac{.15}{10{,}000}$$

Thus, the dividing-head crank must be rotated 59/100 + 70/10,000 turn if the last term can be neglected for the moment. In the relationship between the rotation of the two cranks, it was shown that each hole in the 100-hole circle of the small crank was equivalent to turning the large crank 1/10,000 revolution. Thus, if the 100-hole circle of the large index plate is used, the movement of the wide-range dividing head required to index 67 divisions is as follows:

Large Crank: 59 holes in the 100-hole circle.
Small Crank: 70 holes in the 100-hole circle.

The remainder of .15/10,000, which was previously neglected, is the indexing error. This error would amount to approximately ½ second too little in 360 degrees. Even this amount of error can be further reduced by indexing the small crank 71 holes instead of 70 holes every sixth index. This is based upon the fact that in six indexes the amount of the error is: $6 \times .15/10{,}000 = .90/10{,}000$. This is approximately 1/10,000, or one hole in the 100-hole circle of the small index plate.

Example 7-14:

Calculate the settings of the wide-range dividing head required to index 149 divisions.

$$T = \frac{40}{N} = \frac{40}{149} = .268456$$

$$T = \frac{26}{100} + \frac{84}{10{,}000} + \frac{.56}{10{,}000}$$

The movement of the wide-range dividing head required to index 149 divisions is:

Large Crank: 26 holes in the 100-hole circle.
Small Crank: 84 holes in the 100-hole circle.

The error is .56/10,000, which can be reduced by indexing the small crank 85 holes instead of 84 holes every second index:

$$\left(2 \times \frac{.56}{10{,}000} \cong \frac{1}{10{,}000}\right)$$

It is evident from the last two examples that the summation of the common fractions does not actually need to be rewritten. This was done in the examples in order to explain the principle involved. The setting of the wide-range dividing head can be determined by simply following the steps listed below:

1. Divide the number of divisions required into 40. The answer should be taken to six decimal places.
2. The first two numbers will be the number of holes to be indexed on the 100-hole circle of the large index plate.
3. The second two numbers will be the number of holes to be indexed on the 100-hole circle of the small index plate.

4. The last two numbers constitute the error. The error may be neglected, or it may be reduced by indexing one additional hole in the small index plate for each time that the total error approaches 1/10,000. An exception to this method of reducing the error is shown in Example 7-16 further on.

Example 7-15:

Calculate the settings of the wide-range dividing head required to index 103 divisions.

$$T = \frac{40}{N} = \frac{40}{103} = .388350$$

The dividing head movements are:

Large Crank: 38 holes in 100-hole circle.
Small Crank: 83 holes in 100-hole circle.

The error in this case is .50/10,000, which would be 1/10,000 when multiplied by 2. Thus, the small crank should be indexed 84 holes instead of 83 holes every second index if the error is to be reduced.

Example 7-16:

Calculate the settings of the wide-range dividing head required to index 209 divisions.

$$T = \frac{40}{N} = \frac{40}{209} = .191388$$

Note that in this case the error is approaching 1/10,000 for each index. In this situation the error will be significantly reduced by writing the answer in the following way:

$$T = .1914 - .000012$$

The dividing head movements would then be:

Large Crank: 19 holes in 100-hole circle.
Small Crank: 14 holes in 100-hole circle.

The error is reduced in this case by indexing the small crank 13 holes instead of 14 holes every eighth index movement:

$$\left(8 \times \frac{.12}{10{,}000} \cong \frac{1}{10{,}000}\right)$$

Indexing Angles on the Cincinnati Wide-Range Dividing Head

One advantage of the Cincinnati Wide-Range Dividing Head is the wide range of angles that can be indexed. When angles are indexed on this dividing head, the 54-hole circle of the large index plate and the 54-hole circle of the small index plate are always used.

It will be recalled that one complete revolution of the large crank causes the dividing-head spindle and the workpiece to rotate through an angle $360/40 = 9$ degrees or $9 \times 60 = 540$ minutes. Indexing the large crank one hole in the 54-hole circle, which makes it turn 1/54 of a revolution, causes the dividing-head spindle to rotate $1/54 \times 540 = 10$ minutes. One complete revolution of the small crank causes the large crank to make 1/100 of a revolution, thereby turning the dividing-head spindle $1/100 \times 540 = 5.4$ minutes. This is equal to $5.4 \times 60 = 324$ seconds. Indexing the small crank one hole in the 54-hole circle of the small index plate will cause it to make 1/54 of a revolution. Thus, if one revolution of the small crank indexes the dividing head spindle 324 seconds, 1/54 of a revolution of the small crank will index the spindle $1/54 \times 324 = 6$ seconds. Since this is the smallest movement normally made when angles are indexed on the wide-range dividing head, it establishes the limit of accuracy for this method. The limit of accuracy is actually one-half this amount, or 3 seconds. If the error is larger than 3 seconds, the small crank can be indexed an additional hole to reduce the error. This will be illustrated in a later example. It should be pointed out that an error of 3 seconds is far beyond the requirements of most practical jobs.

The following tabulation summarizes the results of the index movements used when angles are indexed.

Crank	Complete Turns	Holes	Hole Circle	Spindle Movement
Large	1	...	...	9° or 540′
Large	0	1	54	10′
Small	1	...	...	324″
Small	0	1	54	6″

The movement of the large crank is calculated by using Formula 7-4B. The movement of the small crank is calculated from the fact that one turn of this crank rotates the spindle 324 seconds.

Let: t = The number of complete turns and fraction of a turn of the small crank of the wide-range dividing head

D'' = The angle to be indexed by the small crank in seconds

$$t = \frac{D''}{324}$$

A more useful form of this equation is obtained if both the numerator and the denominator are multiplied by ⅙.

$$t = \frac{\frac{1}{6} D''}{\frac{1}{6} \times 324}$$

$$t = \frac{\frac{1}{6} D''}{54} \tag{7-5}$$

Formula 7-5 is applicable only for calculating the indexing movement of the small crank of the Cincinnati Wide-Range Dividing Head.

The procedure for calculating the required indexing movements for indexing angles on the Cincinnati Wide-Range Dividing Head can best be shown by the following examples.

Example 7-17:

Calculate the index movements of the Cincinnati Wide-Range Dividing Head that are required to index an angle of 50°47′20″.

1. Consider at first only that part of the angle to be indexed which is expressed in degrees and minutes, and convert this part of the angle entirely into minutes.

$$D' = (50 \times 60) + 47 = 3047'$$

2. Use Formula 7-4B to calculate the index movement of the large crank of the dividing head which would be required to index that part of the total angle expressed in degrees and minutes. Determine also the exact amount of this angle that cannot be indexed by the large crank.

$$T = \frac{D'}{540} = \frac{3047}{540} = 5\,\frac{347}{540}$$

The number 5(347/540) can be written in the following form without changing its value:

$$T = 5 + \frac{340}{540} + \frac{7}{540}$$

The value of the fractional parts of the term are not changed if both the numerator and the denominator are divided by 10, in which case the term appears as follows:

$$T = 5 + \frac{34}{54} + \frac{.7}{54}$$

The large crank must now be indexed five complete revolutions plus 34 holes in the 54-hole circle.

The remainder of .7/54 means that the large crank should be indexed an additional .7 part of the distance between two holes on the 54-hole circle. This, of course, cannot be done. The angle equivalent to this distance must be indexed by the small crank. If indexing the large crank one hole in the 54-hole circle causes the dividing-head spindle to rotate 10 minutes, indexing this crank seven-tenths (.7) of the distance between the holes of the large index plate causes the spindle to rotate $.7 \times 10 = 7$ minutes. Thus, the numerical value of the numerator of the remainder is always the number of minutes of the angle that cannot be indexed by the large crank.

3. Determine the amount of the angle remaining to be indexed by the small crank, and express this amount in seconds.

   There are 7 minutes remaining from Step 2 and 20 seconds which were not considered in Step 1.

$$D'' = (7 \times 60) + 20 = 440''$$

4. Calculate the index movement of the small crank using Formula 7-5.

$$t = \frac{\frac{1}{6}\,D''}{54} = \frac{\frac{1}{6} \times 440}{54} = \frac{73\frac{2}{6}}{54} = 1\,\frac{19\frac{2}{6}}{54}$$

   This may be written as follows:

$$t = 1 + \frac{19}{54} + \frac{\frac{2}{6}}{54}$$

   The index movement of the small crank is determined by the first two terms. Thus, index the small crank one complete turn plus 19 holes in the 54-hole circle.

   To index the entire angle of 50°47′20″ the small crank should be indexed an additional $\frac{2}{6}$ of the distance between the holes of the small index plate. This cannot be done and constitutes the indexing error. If the dividing-head spindle is indexed 6 seconds when the small crank is indexed one hole in the 54-hole circle, then a movement of the crank of $\frac{2}{6} \times 6 = 2$ seconds. This is then the amount of the error.

   A summary of the index movements required to index 50°47′20″ is given here:

Large Crank:	Complete turns	= 5
	Holes	= 34
	Hole circle	= 54
Small Crank:	Complete turns	= 1
	Holes	= 19
	Hole circle	= 54
Index Error:	2 seconds	

Example 7-18:

Calculate the index movements required to index an angle of 8°15′11″.

1. Convert 8°15′ into minutes.

$$D' = (8 \times 60) + 15 = 495'$$

2. Calculate the index movement of the large crank.

$$T = \frac{D'}{540} = \frac{495}{540} = \frac{490}{540} + \frac{5}{540}$$

$$T = \frac{49}{54} + \frac{.5}{54}$$

Thus, the large crank is indexed 49 holes in the 54-hole circle. The remainder is 5 minutes.

3. Convert the remaining angle to be indexed into seconds.

$$D'' = (5 \times 60) + 11 = 311''$$

4. Calculate the index movement of the small crank.

$$t = \frac{\frac{1}{6} D''}{54} = \frac{\frac{1}{6} \times 311}{54} = \frac{51\frac{5}{6}}{54}$$

$$t = \frac{51}{54} + \frac{\frac{5}{6}}{54}$$

If the small crank is indexed 51 holes in the 54-hole circle, the error would then be 5 seconds ($\frac{5}{6} \times 6 = 5$). Indexing one additional hole would, in this case, reduce the error to 1 second. The angle, in this case, would be 1 second too large. This can be shown by writing the answer to Formula 7-5 in the following manner:

$$t = \frac{51}{54} + \frac{\frac{5}{6}}{54} = \frac{52}{54} - \frac{\frac{1}{6}}{54}$$

Thus, the small crank should be indexed 52 holes in the 54-hole circle.

Summary:

Large Crank:	Complete turns	= 0
	Holes	= 49
	Hole circle	= 54
Small Crank:	Complete turns	= 0
	Holes	= 52
	Hole circle	= 54
Error:	1 second	

Any angle expressed in terms of only degrees and minutes can be indexed on the Cincinnati Wide-Range Dividing Head without error. Such angles occur much more frequently in shop practice than those which also include seconds. The following example will illustrate this point.

Example 7-19:

Calculate the index movement of the Cincinnati Wide-Range Dividing Head required to index an angle of 44°28′.

1. Convert 44°28′ into minutes.

$$D' = (44 \times 60) + 28 = 2668'$$

2. Calculate the index movement of the large crank.

$$T = \frac{D'}{540} = \frac{2668}{540} = 4\,\frac{508}{540}$$

$$T = 4 + \frac{50}{54} + \frac{.8}{54}$$

Index the large crank four complete turns plus 50 holes in the 54-hole circle. The remainder to be indexed by the small crank is equal to 8 minutes.

3. Convert the remaining angle to be indexed into seconds.

$$D'' = 8 \times 60 = 480''$$

4. Calculate the index movement of the small crank.

$$t = \frac{\frac{1}{6}\,D''}{54} = \frac{\frac{1}{6} \times 480}{54} = \frac{80}{54} = 1\,\frac{26}{54}$$

$$= 1 + \frac{26}{54}$$

Index the small crank one complete turn plus 26 holes in the 54-hole circle. There is no error.

Summary:

Large Crank:	Complete turns	= 4
	Holes	= 50
	Hole circle	= 54
Small Crank:	Complete turns	= 1
	Holes	= 26
	Hole circle	= 54

CHAPTER **8**

Dividing Head Work

The objective of this chapter is to show by example some basic operations that can be performed on a milling machine using a dividing head. Great stress will not be placed on the dividing head calculations which were treated in depth in the previous chapter.

Workpieces can be held on the dividing head by 3-jaw universal chucks, 4-jaw independent chucks, collet chucks, faceplates, or mounted on two centers, one held in the dividing-head spindle and the other fixed to the tailstock. Sometimes a special work-holding fixture must be made to clamp parts that cannot be held by conventional methods. The entire procedure for doing the job should be thought through in advance, and a method of overcoming any anticipated difficulty should be determined. Then the job can proceed with a minimum risk of an error being committed that could spoil the workpiece.

Cutting Spur Gear Teeth

Although most modern spur gears are cut on specialized gear-cutting machine tools, some gears are still cut on the milling machine. No special equipment other than an involute gear tooth cutter is required. These gear tooth milling cutters are standard form relieved cutters that can be obtained as stocked items. The milling machine is, therefore, quite adaptable for making spur gears in small quantities. A typical setup for milling a spur gear is shown in Fig. 8-1. The gear blank is pressed onto a standard mandrel held between the tailstock and the dividing-head center. The driving dog is clamped to the mandrel. The dog is driven by a driver which rotates with the spindle of the dividing head. The form relieved gear-cutting cutter is mounted on the milling-machine arbor. The gear blank must be located so that the cutter is centered with the gear blank in the transverse direction.

Certain dimensions of a spur gear are important when spur gears are milled. These dimensions, shown in Fig. 8-2, are defined here:

Outside Diameter. The outside diameter, O, is the diameter over the top of two opposing gear teeth if the number of teeth on the gear is an even number. If the number of teeth is an odd number, the outside diameter cannot be measured directly on the finished gear. It is usually measured on the gear blank before the gear teeth are cut. Compensations must be made for deviations from the theoretical outside diameter when the milling cutter is set to cut the required tooth depth and when the tooth size is measured with gear tooth verniers.

Pitch Diameter. In Fig. 8-2 the pitch diameter is the dimension D, or the theoretical mean diameter of the gear. The pitch diameters of two mating gears would contact each other at a point called the pitch point. Assuming for a moment that the pitch diameters of the two mating gears

Courtesy of Cincinnati Milacron

Fig. 8-1. Milling a spur gear.

are the diameters of two discs or rolls, and that these rolls rotate together without slipping, the speed ratio of the two rolls would be equal to the speed ratio of the two gears.

Standard Pitch Circle. The pitch circle is the circumference of the pitch diameter. The thickness of the gear tooth space and the thickness of the gear tooth are equal on the pitch circle. The thickness of the gear tooth is therefore measured on the pitch circle.

Diametral Pitch. The diametral pitch of a gear is an expression of the size of the gear teeth. The term *pitch* of a gear is sometimes used to mean the diametral pitch of the gear. The formula for the diametral pitch is given below:

$$P = \frac{N}{D}$$

where: P = Diametral pitch
N = Number of teeth on the gear
D = Pitch diameter

Small-sized gear teeth have a large diametral pitch, and large-sized gear teeth will have a small diametral pitch.

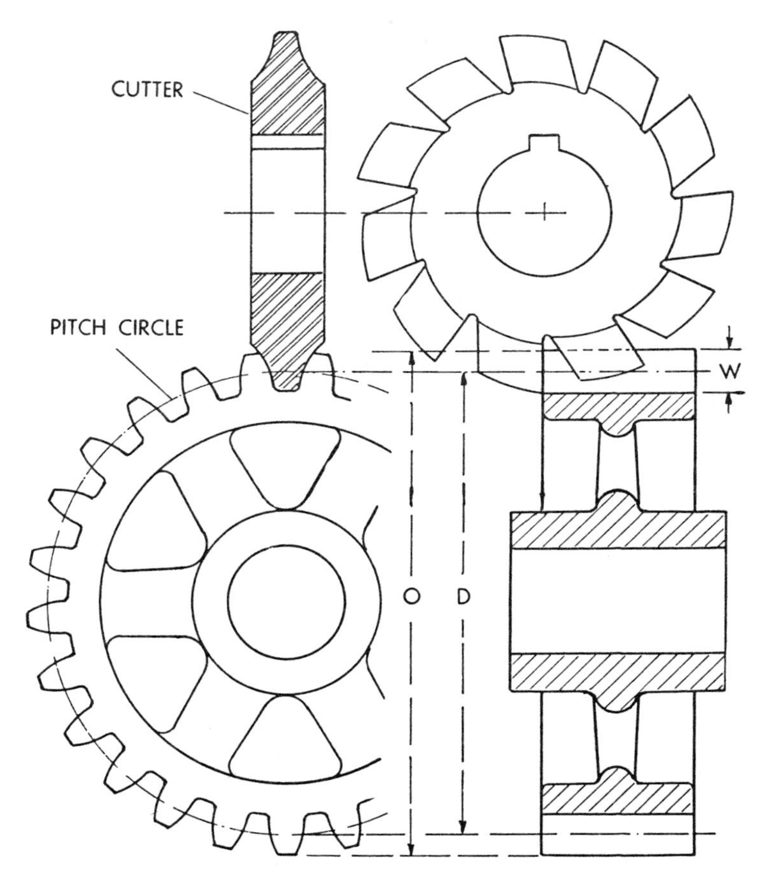

Fig. 8-2. Diagram showing important dimensions related to spur gear cutting.

Circular Pitch. The length of the arc along the pitch circle from the center of one gear tooth to the center of the next tooth is called the circular pitch. It is sometimes used to designate the size of teeth of very large gears.

Pressure Angle. The pressure angle is the angle at which the forces acting on the gear teeth react. It is measured from a line that is perpendicular to the centerline of the two meshing gear teeth. The pressure angle is determined by the profile of the gear teeth. Changes in the designed pressure angle of a gear tooth must accompany a change on the profile. Since gear teeth having different pressure angles will not work together, the pressure angle is an important gear tooth dimension. The American Standard Spur Gear Tooth Systems are:

American Standard 14½-Degree Involute Full-Depth Tooth
American Standard 20-Degree Involute Full-Depth Tooth
American Standard 20-Degree Involute Fine-Pitch Tooth
American Standard 14½-Degree Composite Tooth
American Standard 20-Degree Involute Stub-Tooth
Fellows Stub Tooth (20-Degree Pressure Angle).

The profile of the sides of the gear teeth is a special curve, which is called an *involute*. However, the pressure angle given for each system depends upon the orientation of the involute curve.

Whole Depth. The whole depth, W in Fig. 8-2, designates the whole of the gear tooth space. It is the depth to which the gear cutter must be positioned. This dimension is usually marked on the side of the gear tooth cutter.

The opening of the space between the gear teeth depends not only upon the diametral pitch of the gear but also upon the number of teeth on the gear. Theoretically a different milling cutter should be used for each gear having a different number of teeth. Because this would mean stocking a very large number of gear cutters, a practical compromise has been found by having eight different cutters for each diametral pitch. Each cutter will cut a number of teeth as listed below:

No. 1: 135 to rack	No. 5: 21 to 25
No. 2: 55 to 134	No. 6: 17 to 20
No. 3: 35 to 54	No. 7: 14 to 16
No. 4: 26 to 34	No. 8: 12 to 13

If greater accuracy of tooth shape is required in order to insure smoother and quieter operation of the gears, a series of cutters having half numbers is used. The half-numbered cutters made by the Brown & Sharpe Manufacturing Company are for the number of teeth listed here:

No. 1½: 80 to 134	No. 5½: 19 to 20
No. 2½: 52 to 54	No. 6½: 15 to 16
No. 3½: 30 to 34	No. 7½: 13
No. 4½: 23 to 25	

In preparation for setting up the gear blank it should first be mounted between the dividing-head centers without attaching the driving dog. A dial test indicator is placed against the periphery of the gear blank, and the gear blank is rotated to check for any eccentricity. If the peripheral surface of the gear blank is eccentric, a light cut should be taken on this surface on a lathe or cylindrical grinder in order to eliminate eccentricity. When it has been assured that the gear blank is concentric with its centers, or with the centers of the mandrel on which it is mounted, the driving dog is attached and the gear blank is remounted on the dividing-head centers. The dog is secured to the driver. The milling cutter is mounted on the arbor, and the machine is set up for the correct feed and speed.

The gear blank must now be located centrally with respect to the milling cutter. This procedure, illustrated in Fig. 8-3, will assure a very high degree of accuracy if it is followed with skill and care. Precise transverse table movements are required, with the micrometer feed screw dial used to position the table. In order to eliminate the error that can be caused by the clearance or lost motion between the feed screw and the feed screw nut, it is very important that all of the table settings be made by moving the table in one direction only. For locating the gear blank the transverse table settings can be made by moving the table toward or away from the column of the machine. However, once a direction has been selected, only that direction should be used to make the settings.

The procedure for locating the gear blank centrally with respect to the cutter is given below by referring to Fig. 8-3. In this case, all of the precise table settings will be made by moving the table away from the column of the milling machine, although, as just explained, the same results can be achieved by moving the table in the opposite direction. In Fig. 8-3, the column of the milling machine is at the left, and the operator would normally be standing at the right of the table.

A. *Establish a zero reference position.* Feed the table manually, using the transverse feed, in a direction toward the operator or away from the column of the machine until the position of the gear blank relative to the cutter is approximately as shown in view A. The final movement before coming to a stop must be toward the operator. When the table has stopped, set the micrometer dial of the transverse feed screw to zero.

B. *Establish the first transverse "touch up" position.* Feed the table manually toward the operator until the gear blank is located so that the milling cutter will touch the side shown in view B when the table is raised. The exact distance that the table has been moved from the zero reference position to this position must be determined by counting the number of turns of the feed screw and by reading the micrometer feed screw dial. Assume in this instance that this distance is 3.250 inches.

C. *Establish the vertical reference position.* First obtain a strip of good quality notebook paper approximately 1 inch wide by 8 or 10 inches long. Start the milling-machine spindle and place the paper feeler between the rotating cutter and the gear blank as shown in view C. The paper should be held on the side of the cutter where it will tend to pull the paper out of the hand. Be careful to keep the hand holding the paper from touching the cutter. Then carefully raise the table until the teeth of the cutter just graze the paper feeler without cutting through or tearing it. When the paper is held in the fingertips, this can readily be felt.

Fig. 8-3. Procedure for locating gear blank centrally with respect to formed tooth gear cutter.

A. Establish a zero reference position.
B. Establish the first transverse "touch up" position.
C. Establish the vertical reference position.
D. Return the table to the zero reference position.
E. Establish the second transverse "touch up" position.
F. Center the gear blank and the cutter.
G. Raise the table so that the cutter will cut the gear blank to size.
H. Cut the gear teeth.

D. *Return the table to the zero reference position.* Move the table a short distance with the longitudinal feed so that the gear blank will clear the cutter as it moves past to return to the zero reference position. Count the number of turns of the transverse feed screw and move the table slightly past the zero reference position. The table is then moved toward the operator until the micrometer dial reads zero to bring it to the zero reference position.

E. *Establish the second transverse "touch up" position.* Place the paper feeler between the rotating cutter and the gear blank again. The table is then manually fed toward the operator and the number of turns of the transverse feed screw counted. Continue to feed the table until the cutter just grazes the paper feeler. Then read the micrometer dial to determine the exact distance from the zero reference position to this position. Assume that this distance is 1.988 inches.

F. *Center the gear blank and the cutter.* First lower the table enough so that the gear blank can be positioned below the cutter as shown in view F. Next calculate the distance from the second transverse "touch up" position to the centered position. This distance is equal to one-half of the difference of the distances between the zero-reference position and the two "touch up" positions. In this case this distance is equal to:

$$\frac{3.250 - 1.988}{2} = .631 \text{ inch}$$

Using the micrometer dial, feed the table manually .631 inch toward the operator. The gear blank will now be located centrally with respect to the cutter as shown in view F.

The following two steps, shown in views G and H in Fig. 8-3, complete the setup for cutting the gear blank.

G. *Raise the table to position the gear blank so that the cutter will cut the gear blank to size.* The gear in Fig. 8-3 is a 30-tooth 8-diametral pitch American Standard 14½ Degree Full Depth Tooth Gear for which the whole depth of tooth is .2696 inch and the outside diameter is 4.000 inches. Since the tolerance specifications of the outside diameter of gear blanks permit them to be turned slightly undersize, the actual outside diameter must always be accurately measured with micrometer calipers before the depth of cut is set. The variation between the actual and the theoretical outside diameter must be considered in setting the machine to take the required depth of cut. A paper feeler gage is used, and the thickness of the paper must also be accounted for. Assume in this instance that the actual outside diameter of the gear blank is 3.997 inches and that the thickness of the paper feeler is .0035 inch. When the paper feeler is used to "touch up," as shown in view G, the vertical distance that the table is raised to allow the cutter to cut to size is then:

$$.2696 - \frac{4.000 - 3.997}{2} + .0035 = .2716 \text{ inch}$$

The depth of cut is set by starting the spindle and carefully raising the table until the cutter grazes the paper feeler. Move the table longitudinally until the gear blank is clear of the cutter. Then raise the table **.2716** inch, using the micrometer dial.

H. *Cut the gear teeth.* Engage the feed and cut the gear teeth as shown in view H. Index the dividing head one complete turn plus **11** holes in the 33-hole circle to cut all of the teeth. Most gear teeth are cut to depth in one cut; however, gears with teeth that are **6** or **7** diametral pitch and coarser are frequently milled in two cuts.

The size of a gear tooth can be measured by a special gear tooth vernier caliper shown in Fig. **8-4**. The instrument measures the chordal thickness of the tooth which is called the chordal addendum. Tables appearing in *Machinery's Handbook* provide the necessary data from which these dimensions can readily be calculated. The procedure is to cut one side of a gear tooth far enough to form a complete tooth surface for a short distance. The gear blank is indexed, and the adjacent tooth space is cut in

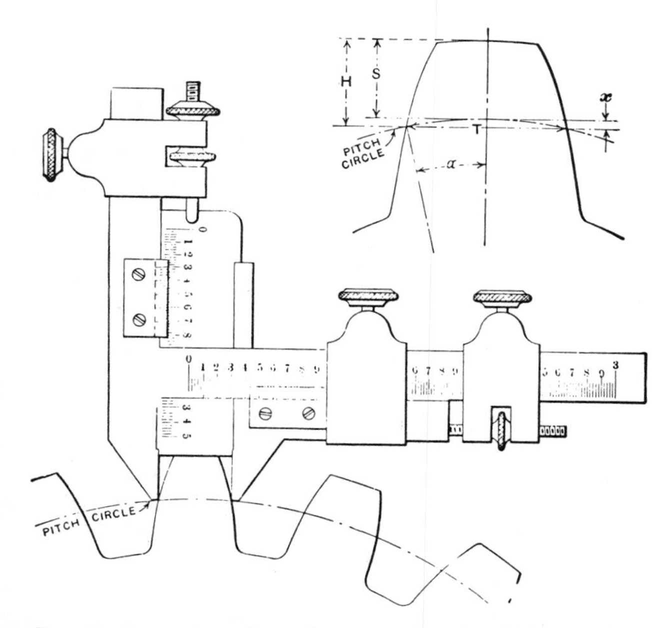

Fig. 8-4. Gear tooth vernier calipers for measuring the thickness of gear teeth at the pitch circle.

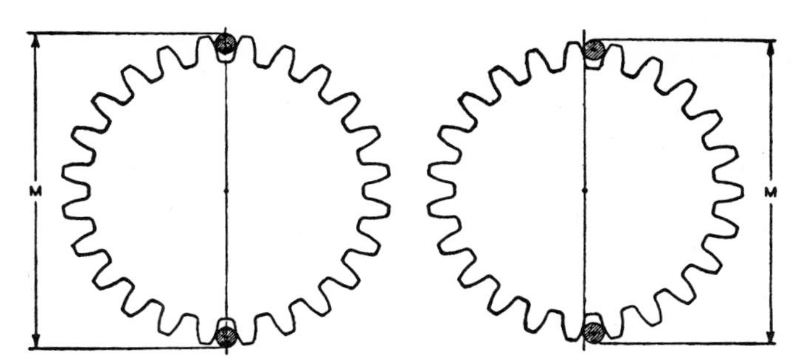

Fig. 8-5. Checking gear size by measuring over wires or pins.

the same manner. The thickness of the resulting gear tooth is then measured with the gear tooth vernier. Any necessary correction in the machine setting is made, and the procedure is repeated until the gear tooth size is correct.

One method of checking the finished gear is to measure over rolls or pins as shown in Fig. 8-5. This easily applied method is especially useful in shops with only a limited amount of inspection equipment. Two cylindrical rolls or wires with a predetermined diameter are placed in diametrically opposed tooth spaces. If the gear has an odd number of teeth, the rolls or wires are located as nearly opposite as possible, as shown by the diagram at the right in Fig. 8-5. The measurement, M, over the pins is made by any sufficiently accurate method of measurement. The required measurement is calculated from tables provided in *Machinery's Handbook.*

Milling a Large Spur Gear

Large spur gears can sometimes be milled by placing elevating or raising blocks between the table and the dividing head. Another method is to mill the gear on a circular table that is equipped with an index plate as shown in Fig. 8-6. The gear blank is placed on four parallel bars in order to raise it above the surface of the table and thereby provide clearance for the cutter. It is clamped to the circular table by bolts anchored in the T-slots of the table. The automatic vertical feed is used to take the cut.

The gear ratio of the circular milling attachment is 80 to 1; i.e., 80 turns of the index crank are required to make one revolution of the circular table. It is therefore necessary to modify Formula 7-1 in order to calculate the indexing movement. The number of teeth being cut on the gear shown is 96. The indexing movement required to cut this number of teeth is calculated in the following manner:

$$T = \frac{80}{96} = \frac{5}{6} = \frac{25}{30}$$

Thus the crank is turned 25 holes in the 30-hole circle for each index.

Cutting a Worm Gear

Worm gears may be cut on a universal milling machine. Two separate operations are required to do this job. The first operation is to cut gashes having the approximate shape of the worm gear teeth around the gear, and the second operation is to finish the gear teeth to size with a hobbing cutter. These operations are shown in Figs. 8-7 and 8-8 respectively.

Courtesy of Cincinnati Milacron

Fig. 8-6. Milling a large spur gear using a circular table equipped with an index plate.

The objective of the gashing operation is to provide teeth on the gear blank which can engage with the thread of the hob in order to drive the gear. The cutter used to gash the worm gear blank should preferably be an involute gear cutter of approximately the same size as the teeth of the worm gear. The cutter should be centered in both the crosswise and the lengthwise directions with respect to the gear blank. The table of the

Courtesy of Cincinnati Milacron

Fig. 8-7. Gashing the teeth of a worm gear in a milling machine.

universal milling machine must be set at the lead angle or helix angle of the worm thread (the tangent of the lead angle of the worm is found by dividing the lead of the worm thread by the circumference of the pitch circle).

Often the diameter of the milling cutter used to gash the worm gear blank is larger than the diameter of the hob to be used later. In this event the whole depth of the gear tooth should be marked on the side of the gear blank with a scribed line. The depth of the gashes must be less than the depth of the worm gear teeth. The gashes are cut by raising the table with the vertical feed until the cutter has reached the required depth for gashing as estimated by the layout line. The reading of the micrometer dial for the vertical feed is noted. The remaining gashes are then cut by indexing the dividing head and feeding the gear blank vertically into the cutter until the micrometer dial of the vertical feed has reached the same position noted in cutting the first tooth.

After the gashing operation is finished, the table of the universal milling machine is returned to a position perpendicular to the spindle without disturbing the longitudinal position of the table. The driving dog is removed from the work-holding mandrel to permit the work to have un-

Courtesy of Cincinnati Milacron

Fig. 8-8. Hobbing the teeth of a worm gear.

restrained rotation about the dividing-head centers. The gashing cutter is removed from the arbor and replaced by a hob, which has cutting teeth with the same shape as the worm thread. Flutes are cut in the lengthwise direction of the hob to form the faces of the cutting teeth and to provide space for the chips. The sides of the hob teeth are formed by a helical groove corresponding to the thread groove of the worm which is to work with the worm gear. The hob is the same diameter as the worm (except for a slight increase to provide a clearance for the top of the worm thread), and the lead of the hob thread groove is equal to the lead of the worm thread.

The worm gear is cut by raising the table until the gashes in the gear blank engage the thread on the hob. The machine spindle is then started so that the thread on the hob drives the worm gear and causes it to rotate.

As this occurs, the table is gradually raised, causing the worm gear to feed into the hob with the result that the hob cuts the teeth of the worm gear. Since the hob is a duplicate of the worm, except for those provisions necessary to convert it into a cutting tool, it is evident that the worm gear produced in this manner will also mesh with the worm.

It is advisable to cut the worm thread before hobbing the worm gear. The worm shaft can be used to an advantage for testing the center distance between the worm and the worm gear. In preparation for making this measurement the table must be lowered sufficiently to make room for the worm shaft beneath the hob. The lengthwise or longitudinal position of the table should not be disturbed. The worm shaft is placed on the top of the worm gear with their teeth engaged as if they were in operation. Turning the worm shaft slightly by hand will cause it to rotate around the worm gear. With the aid of a dial test indicator attached to a vernier height gage to indicate over each side of the cylindrical ends, the worm shaft is rotated until it is in a horizontal position on the worm gear. The vernier height gage and indicator are then used to measure the distance between the top of the mandrel on which the worm gear is mounted and the top of the worm shaft. The center distance is this measured distance plus one-half of the diameter of the mandrel and minus one-half of the worm shaft diameter. The calculated center distance is then compared to the specified distance and, if necessary, another cut is taken on the worm gear with the hob until the correct size is obtained.

Milling Hexagons

Nuts and the heads of bolts and cap screws are frequently machined to the shape of a hexagon. Figure 8-9 shows a hexagon being milled on a special nut which is held in a three-jaw universal chuck mounted on the dividing head. A round shim with its faces ground parallel is placed below the nut in order to seat it in the chuck so that the chuck jaws will clear the cutters. The dividing head is positioned with its spindle vertical. Either direct or plain indexing can be used to index the dividing head.

The cut can be taken with either one or two side milling cutters. If one cutter is used, six cuts will be required to cut the hexagon. If two side milling cutters are used, as shown in Fig. 8-9, to straddle mill the sides, only three cuts are required—since two sides are milled per cut. When a single-side milling cutter is used, the dimension across the flat of the hexagon is obtained by taking a trial cut along one side of the workpiece. The dividing head is indexed 180 degrees, a second trial cut is taken, and the distance across the flats is measured. The table is moved to bring the workpiece into the cutter an amount equal to one-half the distance between the measured size and the required size.

When straddle milling, the distance across the flats is determined by the spacing between the two side milling cutters. Appropriate collars should be placed between the two cutters so that they will cut the desired width. The straddle milling cutters, however, must be positioned so that the axis of the dividing-head spindle is centered with respect to the cutters. The

Courtesy of Cincinnati Milacron

Fig. 8-9. Straddle milling hexagon flats on nuts.

procedure is illustrated in Fig. 8-10. A trial cut is taken, and the distance M between the flat produced by the trial cut and the opposite cylindrical side of the work is measured. The table is then moved a distance S to center the work with respect to the straddle milling cutters. The distance S is calculated as follows:

$$S = \frac{D}{2} - (D - M) + W_c + \frac{W_F}{2} \tag{8-1}$$

where:
S = Distance moved by transverse feed to centralize straddle milling cutters
D = Original diameter of round blank
M = Measurement between flat produced by trial cut and opposite cylindrical surface
W_c = Width of cutter used to take trial cut
W_F = Width of flat on nut, or distance between straddle milling cutters

Milling Flutes in Taps

Flutes can be cut in taps on a milling machine, as shown in Fig. 8-11. The flutes are formed with a form milling cutter that is fed into the tap

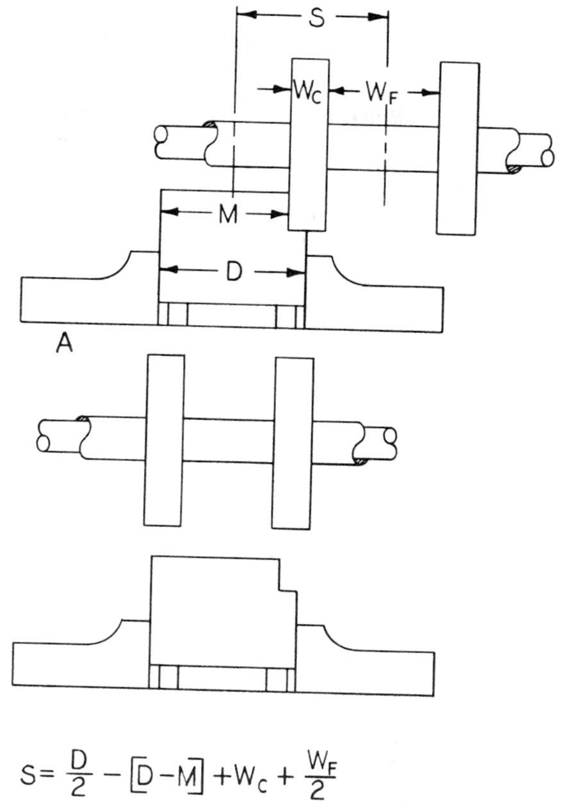

$$S = \frac{D}{2} - [D - M] + W_C + \frac{W_F}{2}$$

Fig. 8-10. Procedure for centering straddle milling cutters.

in an offset position. Although several different flute profiles are used on taps, the most common profile is a round profile, such as is illustrated being cut in Fig. 8-11. Since there are four flutes, the indexing operation can be performed by either direct or by plain indexing.

The tap blank can be positioned with respect to the cutter by trial and error methods; however, this can lead to somewhat uncertain results. It is better to calculate a predetermined position of the tap blank with respect to the cutter and to set the tap blank accurately in this position. This procedure can be seen in Fig. 8-12.

A straight edge or rule is held against that side of the cutter farthest away from the tap blank. The cutter, of course, should not be rotating. The table is moved in a transverse direction until the straight edge touches the side of the tap blank. A paper feeler may be used to obtain an accurate touch up. After the paper feeler has been removed, the table should be moved a distance equal to the thickness of the paper feeler. The transverse feed screw micrometer dial is set to read zero to establish

Courtesy of Cincinnati Milacron

Fig. 8-11. Milling flutes in a tap using a form milling cutter.

the transverse reference position. The tap blank is then centered with respect to the cutter by moving it a distance equal to the difference obtained when one-half of the width of the cutter is subtracted from one-half of the diameter of the tap blank. Next, the spindle is started and a long paper feeler held between the cutter and the tap blank as the table is carefully raised until the cutter just grazes the paper feeler. The table is then returned to the transverse reference position.

Before the table is offset, a corrective adjustment should be made in the vertical direction to compensate for the following: 1. the thickness of the paper feeler used to touch up the cutter; 2. the difference in diameter between the actual size of the tap blank and the finish size of the

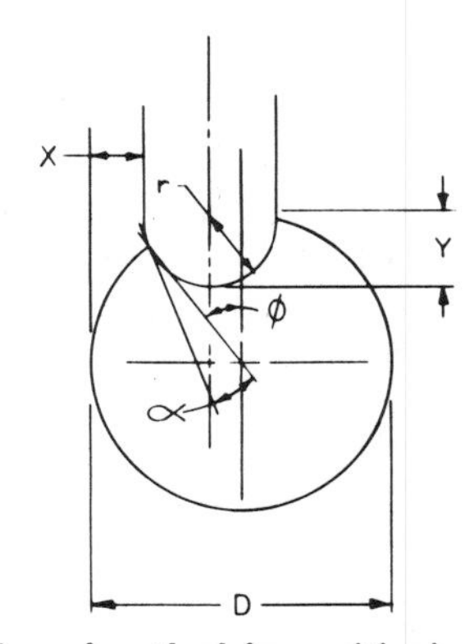

Fig. 8-12. Schematic view of method for positioning cutter for fluting taps.

tap blank. This is equal to the sum of the thickness of the paper feeler and one-half the difference between the tap blank diameter and the finish diameter of the tap. For example, if the thickness of the paper feeler is .003 inch, the diameter of the tap blank is 1.020 inches, and the finish outside diameter of the tap is to be 1.010 inches, the amount that the table must be moved upward is equal to

$$.003 + \frac{1.020 - 1.010}{2} = .008 \text{ inch}$$

The table is now in a position from which the predetermined offset movements can be made. It is moved a distance X, see Fig. 8-12, in the transverse direction, and it is raised a distance Y. Care must be taken to compensate for the lost motion in the transverse feed screw. The tap blank is now in the correct position to cut the flutes.

The distances X and Y in Fig. 8-12 can be calculated by the following formula:

$$X = \frac{D}{2}(1 - \sin\phi) + r\,[1 - \cos(\phi - \alpha)] \qquad (8\text{-}2)$$

$$Y = \frac{D}{2}(1 - \cos\phi) + r\,[1 - \sin(\phi - \alpha)] \qquad (8\text{-}3)$$

where: D = Actual outside diameter on the tap (not the tap blank diameter)

r = Radius of form milling cutter or radius of flute on tap

ϕ = One-half of angle between flutes;

on a four-fluted tap, $\phi = \frac{90}{2} = 45°$

α = Rake angle or tangential hook angle on tap, which is usually specified on drawing

If, for example, a tap is to be made for a 1-8-UNC-2B thread having four round profile flutes, assume that the radius of the flute is ¼ inch and that the tangential hook angle is to be 5 degrees. The outside diameter of the finished tap is to be 1.010 inches. The machine settings X and Y are calculated as follows:

$$X = \frac{D}{2}(1 - \sin\phi) + r\,[1 - \cos(\phi - \alpha)]$$

$$= \frac{1.010}{2}(1 - \sin 45°) + .250\,[1 - \cos(45° - 5°)]$$

$$= .505\,(1 - .70711) + .250\,(1 - .76604)$$

$$X = .20640 \text{ inch}$$

$$Y = \frac{D}{2}(1 - \cos\phi) + r\,[1 - \sin(\phi - \alpha)]$$

$$= \frac{1.010}{2}(1 - \cos 45^\circ) + .250\,[1 - \sin(45^\circ - 5^\circ)]$$

$$= .505\,(1 - .70711) + .250\,(1 - .64279)$$

$$Y = .23721 \text{ inch}$$

Milling Flutes in Reamers

The setup for milling flutes in reamers is similar to the setup for milling the flutes in the tap as shown in Fig. 8-11. Since the shape of the reamer flute differs from the shape of the tap flute, the formulas developed for positioning the work for milling circular tap flutes cannot be applied to reamers. Reamer flutes are generally cut using a double-angle form milling cutter with a radius in the corner, as seen in Fig. 8-13. Single-angle cutters are not recommended because their side teeth will leave feed marks on the face of the reamer teeth. Except when cutting a chip, the side teeth of the double-angle cutters will clear the work and therefore not scratch the face of the reamer teeth.

In general, the face of the reamer teeth for reaming steel or cast iron is made radial with respect to the axis of the cutter. Sometimes, however, a small radial rake angle is desirable, in which case the face of the cutter is not radial. One of the problems in setting up for cutting reamer teeth with a double-angle milling cutter is to position the workpiece relative to the cutter so that the flute will be correctly located on the reamer body.

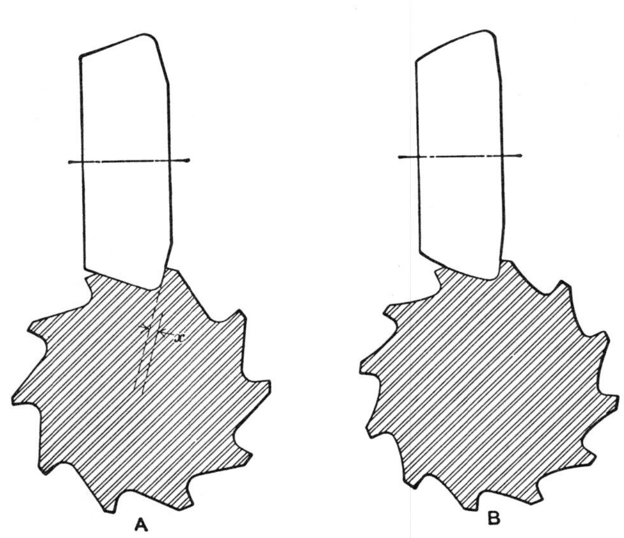

Fig. 8-13. Profile of reamer flutes and double-angle reamer fluting milling cutters.

Sometimes a trial and error procedure is used, and at other times an attempt is made to align the edges of the cutter with the layout lines on the workpiece. Trial and error procedures in this case produce uncertain results, and layout lines are difficult to see on small reamers. A better method is to place the largest diameter of the cutter over the highest point on the reamer blank when it is held between centers and then to move the milling-machine table a predetermined amount vertically and in the transverse direction. This procedure assures that the work will be in the correct relation to the cutter for milling the flutes.

The first step in aligning the reamer blank to the cutter when using the method involving exact table settings is to position the largest diameter of the cutter over the center of the workpiece. There are two ways of doing this. One method is to lay out a reference line on the side of the reamer blank and to rotate the reamer blank until this line is on the top of the blank. This can be done by carefully aligning the scriber point of a surface gage with the dividing head centers before the reamer blank is placed in the machine. With the reamer blank mounted on the dividing head centers and some blue layout paint painted on the side of the reamer, a line is scribed by the surface gage. The index head is indexed $\frac{1}{4}$ revolution or 90 degrees to bring this line to the top surface of the reamer blank. The largest diameter of the cutter is then positioned over the scribed line as judged visually.

A second method makes use of a small dial test indicator which is clamped to the reamer blank as shown in Fig. 8-14. With the reamer blank mounted on the dividing head centers and the indicator attached, the table is moved until the contact point of the indicator touches the bottom of one of the milling-machine arbor collars and a reading is obtained on the

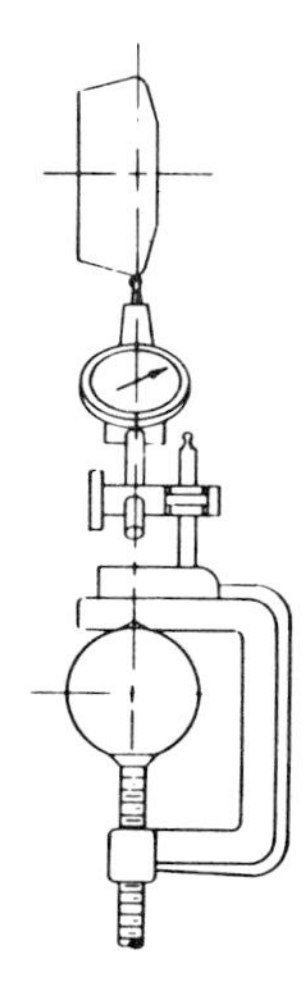

Fig. 8-14. Centering the largest diameter of reamer fluting cutter with respect to the axis of the dividing head.

indicator. The reamer blank and the indicator are then rocked back and forth slightly by turning the index crank of the dividing head. When the largest reading on the indicator is obtained, the dividing-head spindle is locked in position. The contact point of the indicator is now on a vertical plane passing through the axis of the cutter. Next, the table is moved so that the contact point of the indicator touches the cutter near its largest diameter. Using the transverse table movement until the largest indicator reading is obtained will place the largest diameter of the cutter on the center of the reamer blank. When this method is used, the indicator should be positioned with its contact point as close to the reamer blank as possible so that the contact point will be rotated through an arc with the smallest possible radius.

With the largest diameter of the fluting cutter located on the center of the reamer blank, the fluting cutter is touched up against the top of the blank. To do this the milling-machine spindle is engaged to cause the fluting cutter to rotate. The table is then carefully raised until the teeth of the fluting cutter just graze a long strip of paper that is held between the cutter and the reamer blank. After this procedure the table is moved longitudinally so that the fluting cutter will clear the end of the reamer blank in preparation for making the table offset adjustments.

Before these adjustments, however, a vertical corrective adjustment must be made. The purpose of such an adjustment is twofold: 1. to correct for the thickness of the paper feeler used to touch up the fluting cutter; 2. to correct for the difference in the actual diameter of the reamer blank and the finish diameter of the reamer. This adjustment is made by raising the table a distance equal to the thickness of the paper feeler plus one-half of the difference between the diameter of the reamer blank and the finish diameter of the reamer. For example, if the thickness of the paper feeler is .003 inch, the actual diameter of the reamer blank is 2.030 inches, and the finish diameter of the reamer is 2.000 inches, the distance that the table should be raised is:

$$.003 + \frac{2.030 - 2.000}{2} = .018 \text{ inch}$$

The reamer blank is now in a position to make the predetermined transverse and vertical offsets which are shown in Fig. 8-15. The table is moved a distance n in the transverse direction and raised a distance m to bring the reamer blank into position for cutting the flutes.

The method of calculating the offsets m and n will now be given. The terms in Formulas 8-4 and 8-5 are illustrated in Fig. 8-15.

$$n = \frac{D}{2} \sin(a + r) - d \sin a - R(\cos a - \sin a) \tag{8-4}$$

$$m = \frac{D}{2} [1 - \cos(a + r)] + d \cos a - R(\cos a + \sin a - 1) \tag{8-5}$$

where: n = Transverse offset; or the horizontal distance from the center of the reamer blank to the center of the radius on the cutting teeth of the fluting cutter, inches
m = Vertical offset; or, the vertical distance that the table must be raised, inches
D = Diameter of the reamer when finished, inches
d = Depth of the reamer flute, inches
a = Side angle of fluting cutter; or the angle on the side of the fluting cutter that will be used to cut the face surface on the teeth of the reamer, degrees
r = Rake angle to be cut on the reamer, degrees
R = Tooth radius at the largest diameter of the fluting cutter, degrees

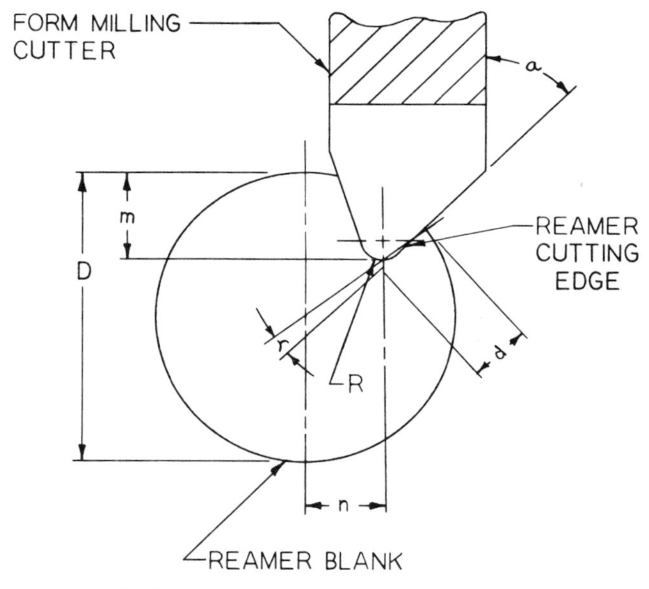

Fig. 8-15. Graphical relation between the fluting cutter angle, depth of flutes, radial rake angle, transverse set-over, and horizontal set-over which are used to position the workpiece in order to cut flutes in reamers and milling cutters.

When these formulas are used, the finish diameter of the reamer must be used instead of the actual diameter of the reamer blank; otherwise a very small error will occur on the resulting rake angle and a slightly larger error will occur in the depth of the flute. The correction for the difference between the actual diameter of the reamer blank and the finish diameter of the reamer should be made before the offset movements, m and n, are started, as explained earlier. If the rake angle on the finished reamer is to be negative, a minus (−) sign must be used ahead of the angle r in these formulas.

Example 8-1:

Flutes are to be milled in a reamer blank which is 2.030 inches in diameter. The finished size of the reamer after heat treating and grinding is to be 2.000 inches. The reamer teeth are to have a 5-degree negative rake angle and the flutes are to be .300 inch deep. The side angle of the fluting cutter is 10 degrees, and the radius of the arc connecting the angular cutting edges is .060 inch. Calculate the transverse and vertical offsets required to cut the flutes in the reamer.

$$n = \frac{D}{2} \sin (a + r) - d \sin a - R (\cos a - \sin a)$$

$$= \frac{2}{2} \sin (10° - 5°) - .300 \sin 10° - .060 (\cos 10° - \sin 10°)$$

$$= .08715 - .300 (.17365) - .060 (.98481 - .17365)$$

$$= .08715 - .05210 - .04867$$

$$n = -.0136 \text{ inch}$$

$$m = \frac{D}{2} [1 - \cos (a + r)] + d \cos a - R (\cos a + \sin a - 1)$$

$$= \frac{2}{2} [1 - \cos (10° - 5°)] + .300 \cos 10° - .060 (\cos 10° + \sin 10° - 1)$$

$$= 1 - .99619 + .300 (.98481) - .060 (.98481 + .17365 - 1)$$

$$= .00381 + .29544 - .0095$$

$$m = .2897 \text{ inch}$$

The negative answer for the transverse offset, n, means that the reamer blank must be moved toward the right instead of toward the left as shown in Fig. 8-15. This is due to the effect of the radius on the fluting cutter. If this radius had been zero, the offset would have been .08715 − .05210 = .035 inch toward the left.

Example 8-2:

Calculate the transverse and vertical offsets required to cut the reamer blank in Example 8-1 if the rake angle of the finished reamer is to be 5 degrees positive and all of the other dimensions are to remain unchanged.

$$n = \frac{D}{2} \sin (a + r) - d \sin a - R (\cos a - \sin a)$$

$$= \frac{2}{2} \sin (10° + 5°) - .300 \sin 10° - .060 (\cos 10° - \sin 10°)$$

$$= .25882 - .05210 - .04867$$

$$n = .1580 \text{ inch}$$

$$m = \frac{D}{2}[1-\cos(a+r)]+d\cos a-R(\cos a+\sin a-1)$$
$$= \frac{2}{2}[1-\cos(10°+5°)]+.300\cos 10°-.060(\cos 10°+\sin 10°-1)$$
$$= 1-.96593+.300(.98481)-.060(.98481+.17365-1)$$
$$= .03407+.29544-.00951$$
$$m = .3200 \text{ inch}$$

In this case the transverse offset, n, is made by moving the reamer blank toward the left, as shown in Fig. 8-15.

Example 8-3:

Calculate the transverse and vertical offsets required to cut the reamer blank in Example 8-1 if the rake angle of the finished reamer is to be zero degrees and all of the other dimensions are to remain unchanged.

$$n = \frac{D}{2}\sin(a+r) - d\sin a - R(\cos a - \sin a)$$
$$= \frac{2}{2}\sin(10° + 0°) - .300\sin 10° - .060(\cos 10° - \sin 10°)$$
$$= .17365 - .05210 - .04867$$
$$= .0729 \text{ inch}$$

$$m = \frac{D}{2}[1-\cos(a+r)]+d\cos a-R(\cos a+\sin a-1)$$
$$= \frac{2}{2}[1-\cos(10°+0°)]+.300\cos 10°-R(\cos 10°+\sin 10°-1)$$
$$= (1-.98481)+.300(.98481)-.060(.98481+.17365-1)$$
$$= .01519+.29544-.0095$$
$$m = .3011 \text{ inch}$$

The transverse offset is made by moving the reamer blank toward the left, as shown in Fig. 8-15.

The flutes of reamers may be spaced unevenly to prevent the reamer from chattering. The flutes of half of the reamer may be spaced irregularly but made to correspond with the other half of the reamer, the opposite cutting edges being diametrically opposite each other. (See Fig. 8-16.) The advantage of this method is that the diameter of the reamer can be measured directly. Another method is to space the cutting edges around the whole reamer irregularly so that no two cutting edges are diametrically opposite each other. These reamers should be measured when mounted on bench centers with the aid of a vernier height gage and a dial test indicator.

It is desirable to have the widths of the lands of the reamers equal. When the reamer teeth are unequally spaced, the land widths will not be equal if the depth of the cutter setting remains constant. It is, therefore,

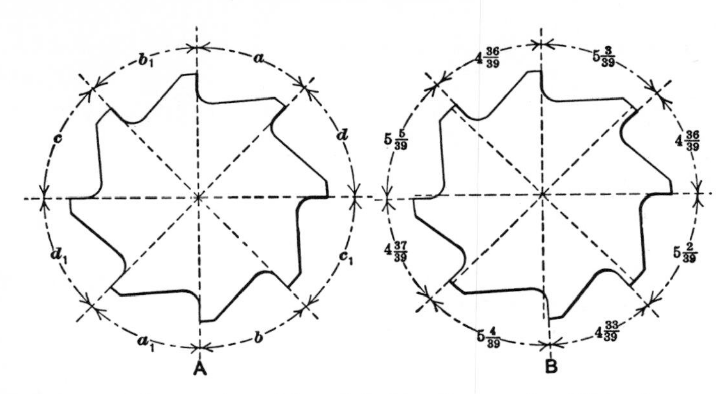

Fig. 8-16. A. Irregular spacing with each half uniform and cutting edges opposite. B. Flutes so spaced that cutting edges are not exactly opposite.

necessary to raise the table during the cut in order to produce a uniform land width on the reamer.

When the flutes of the reamer have unequal spacing—which is alike on each half so that the opposite cutting edges are in line, the indexing may be done by milling the flutes in pairs; that is, after a flute is milled, the dividing head is turned half a revolution and the corresponding flute on the opposite side of the reamer is cut. Then, after the adjacent flute is milled, the dividing head is again turned half a revolution, and so on. Two cuts should be taken in each flute. The first series of cuts around the reamer should be slightly less than the required depth. When the second cut in each flute is made, the depth should be varied slightly so that the widths of the lands will be equal.

To illustrate how the indexing movements for irregular spacing are determined, suppose that a reamer is to be cut having eight flutes with the spacing of each half equal. Assume that a 20-hole-circle index plate is to be used. The total number of holes on the index plate that the crank passes in making one revolution of the dividing-head spindle is $20 \times 40 = 800$. The number of holes for eight equal divisions would be $800 \div 8 = 100$. The next decision to make is the amount of irregularity to have in the spacing. The difference should be slight and need not exceed 2 degrees, although it is often made 3 or 4 degrees. Assuming that the difference is to be 2 degrees, the movement of the index crank necessary to give this variation must be determined. As 800 holes represent a complete revolution of the dividing-head spindle, or 360 degrees, a movement of one hole $= 360 \div 800$, or nearly ½ degree. Therefore, the number of holes required for a movement of 2 degrees is approximately equal to $2 \div 0.5 = 4$ holes. If the divisions were equal, the eight flutes would be cut by turning the crank five turns or 100 holes. However, by varying the movement four holes one way or the other as nearly as can be arranged, an irregularity of approximately 2 degrees is obtained. Thus the successive movements could be 96, 100, 103, and 101 holes, or 4 turns 16 holes,

5 turns, 5 turns 3 holes, and 5 turns 1 hole. In diagram A, Fig. 8-16, flutes a and a_1 would be milled first, diametrically opposed; then by indexing 96 holes, the work would be located for milling flute b. After milling flute b_1 on the opposite side, another movement of 100 holes would locate flute c. Flutes c and c_1 would be milled, and 103 holes indexed to locate d. After fluting d and d_1 the cutter could be aligned with flute a by a movement of 101 holes. The maximum amount of spacing between adjacent flutes is that represented by the spacing of flutes a and b. This is equal to 101 − 96, or 5 holes, which is approximately 2¼ degrees. When selecting the number of holes by which the indexing movements are to be varied, remember that the total sum must equal one-half the number of holes representing a complete revolution when each half of the reamer is spaced alike and indexed as described. Thus:

$$96 + 100 + 103 + 101 = 400$$

When all of the flutes are to have irregular spacing, the indexing movements may be obtained from Table 8-1. To illustrate its application, suppose that a reamer is to have eight flutes. If the spacing were equal, five turns of the crank would be required ($40 \div 8 = 5$). In order to obtain the irregular spacing of the flutes the indexing move for the second flute should be five turns minus three holes; the third flute five turns plus five holes, etc. The last movement will complete the 40 turns of the crank in order to make one revolution of the spindle. In conclusion, it should be emphasized that the irregularity in spacing can be obtained by variations in indexing other than those that have been described.

Table 8-1. Indexing Movements for Milling Irregular Spacing of Teeth on Reamers

Number of flutes in reamer......	4	6	8	10	12	14
Index circle to use...........	39	39	39	39	39	49
Before cutting...	Move index crank the number of holes below more or less than for regular spacing.					
2d flute.....	8 less	4 less	3 less	2 less	4 less	3 less
3d flute.....	4 more	5 more	5 more	3 more	4 more	2 more
4th flute....	6 less	7 less	2 less	5 less	1 less	2 less
5th flute....	....	6 more	4 more	2 more	3 more	4 more
6th flute....	....	5 less	6 less	2 less	4 less	1 less
7th flute....	....	....	2 more	3 more	4 more	3 more
8th flute....	....	....	3 less	2 less	3 less	2 less
9th flute....	....	....	....	5 more	2 more	1 more
10th flute....	....	....	....	1 less	2 less	3 less
11th flute....	....	....	....	....	3 more	3 more
12th flute....	....	....	....	....	4 less	2 less
13th flute....	....	....	....	....	......	2 more
14th flute....	....	....	....	....	......	3 less

Milling a Tang and a Center Key Slot

The procedure for milling a center-key slot in a sleeve adaptor is shown in Fig. 8-17. The sleeve is held between the tailstock and the headstock centers. A driving dog is placed over the sleeve which is not shown in the illustration. A two-fluted end mill is held in the spindle of the machine by a suitable chuck. The spindle is started and the end mill is carefully brought up to the cylindrical portion of the work. (If a vertical milling machine is used, the table movement should be toward the column and touched up on the side away from the column so that the graduations on the transverse feed screw micrometer dial will read directly. On a horizontal milling machine the direction of the movement is vertically up and the touch up is on the top of the workpiece.) When the end mill can be felt to graze a long paper shim held between the work and the cutter, the table feed is stopped and the cutter is moved to clear the work. The table is moved a distance equal to the thickness of the paper feeler plus one-half of the cutter diameter plus one-half of the workpiece diameter against which the touch up was made. Next, the table is positioned so that the cutter is in the correct lengthwise position. The sleeve is then fed into the cutter until it has penetrated to a depth of approximately 0.250 inch, and a lengthwise cut is taken until the slot is 1$\frac{3}{16}$ inch long. Repeat this operation until the depth of the slot is about one-half the diameter of the sleeve. Index the sleeve 180 degrees, and cut the opposite half of the key slot until it is completely through the sleeve. Remove the two-fluted end mill, and replace it with a four-fluted end mill that is long enough to extend through the key slot. Start the spindle and position the end mill in the slot as shown in Fig. 8-17. Move the table and take a very light cut along one side of the key slot. Remove the end mill from the hole and index the sleeve 180 degrees. With the spindle running, feed the end mill through the hole, and then take a light cut along the other side

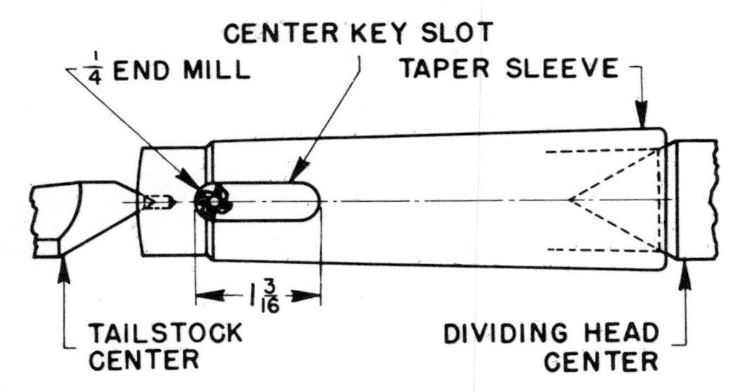

Courtesy of Cincinnati Milacron

Fig. 8-17. Milling a center key slot on a tapered sleeve adaptor.

of the key slot. After measuring the width of the slot, adjust the table a distance equal to one-half of the difference between the measured dimension and finished dimension to which it must be cut. Take another cut on each side of the slot as before, and measure the slot again. Repeat until the slot is finished to size.

Because the tang on this sleeve is designed to be 90 degrees with respect to the key slot, the dividing head should be indexed 90 degrees. (See Fig. 8-18.) A $\frac{5}{8}$-inch four-fluted end mill is placed in the spindle. The spindle is started, and the cutter is fed into the end of the sleeve in order to take a trial cut that will leave the tang oversize. Since the length of the tang is only $\frac{21}{32}$ inch, the trial cut is made to this length. With the cutter clear of the work the sleeve is indexed 180 degrees, and a trial cut is taken on the opposite side with the depth of cut setting of the cutter undisturbed from that used to take the first cut. Measure the thickness of the tang with a micrometer caliper. When the work has been moved the required amount into the cutter, take a cut on both sides of the tang as before to bring it to size.

Machining the Opening in a Blanking Die

The dividing head is used in conjunction with a slotting attachment in Fig. 8-19 in order to machine the opening in a blanking die. The die is accurately centered in a four-jaw independent chuck that is mounted on the nose of the dividing-head spindle. The opening in the die has been previously drilled and bored on an engine lathe. There are 10 triangular-shaped internal notches that must be cut to size. The slotting tool is ground to the shape of the notches. It is centered on the vertical diameter of the bore, and the notches are cut by using the hand feed to feed the knee in a down direction. The notches are machined a uniform distance from the axis of the die by using the micrometer dial on the hand adjustment of the knee.

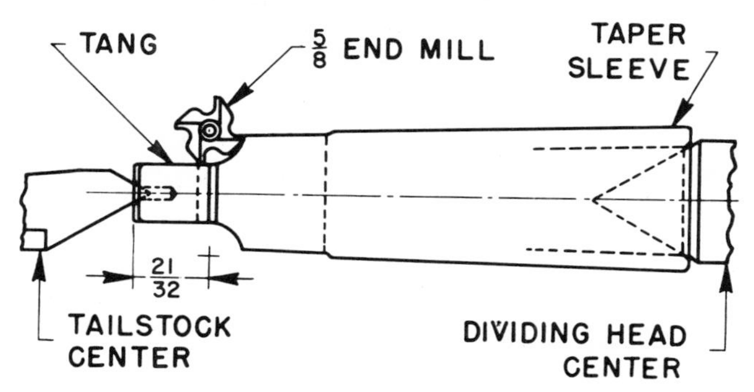

Courtesy of Cincinnati Milacron

Fig. 8.18. Milling the tang on a tapered sleeve adaptor.

Courtesy of Cincinnati Milacron

Fig. 8-19. Machining the opening of a blanking die.

Graduating a Micrometer Dial

Micrometer dials can be accurately graduated on a dividing head as shown in Fig. 8-20. The micrometer dial is mounted on a special mandrel that is held in the dividing-head spindle. A single-point high-speed steel tool is held in a boring bar which is fastened to the slotting attachment. The cutting tool is ground to an included angle of 60 degrees. If the depth of the cut is .010 inch, the width of each graduation will be approximately .005 inch. In this case the depth of cut is established by the longitudinal feed because the graduations are on a 45-degree conical surface. The amount of longitudinal feed required to obtain a .010 depth of cut is:

$$\begin{aligned}\text{Longitudinal Feed} &= .010 \times \csc 45^\circ \\ &= .010 \times 1.4142 \\ &= .014 \text{ inch}\end{aligned}$$

There are to be 250 graduations, and every fifth graduation is to be made longer than the others. The stroke of the slotting attachment is adjusted for cutting the longer graduations. The dividing-head setting is calculated as follows:

$$T = \frac{40}{N} = \frac{40}{250} = \frac{4 \text{ holes}}{25\text{-hole circle}}$$

Courtesy of Cincinnati Milacron

Fig. 8-20. Graduating a micrometer dial using a dividing head and slotting attachments.

However, instead of indexing four holes in the 25-hole circle, the long graduations are cut first by indexing $4 \times 5 = 20$ holes in the 25-hole circle. After the long graduations are cut, the length of the slotting attachment stroke is adjusted to cut the short graduations. The short graduations are cut by indexing four holes in the 25-hole circle, but every fifth graduation is skipped since it has already been cut. The graduations should not be cut to depth in one stroke of the ram in order to avoid an excessive burr. Several strokes should be used with the table fed by hand a small amount during each stroke. The depth can be accurately controlled by using the micrometer feed dial.

If a slotting attachment is not available, the cutting tool may be clamped in a fly cutter holder, or it may be clamped between the collars directly on the milling-machine arbor. When the cutting tool is to be clamped directly on the arbor as explained above, two tool bits should be used. One tool bit is the actual cutting tool, and the other is clamped on the opposite side of the arbor. This prevents the clamping force exerted when the arbor nut is tightened from bending the arbor. When this method is used instead of the slotting attachment, the graduations are cut by hand with the table fed longitudinally and the depth of cut obtained by raising the knee. The length of the graduations can be determined by using the longitudinal feed micrometer dial. Graduations on conical surfaces, such as in Fig. 8-20, are cut by positioning the dividing head at the required angle.

Milling End Teeth on End Milling Cutters and Side Teeth on Side Milling Cutters

When the end teeth of an end mill or the side teeth of a side mill are being cut, the dividing head must be set at an angle as shown at A and B in Fig. 8-21. This is necessary in order to mill the lands or tops of the teeth to a uniform width. The angle of elevation to which the dividing head must be set is determined by the following formula:

$$\cos \alpha = \tan \frac{360°}{N} \times \cot \beta \tag{8-6}$$

where: α = Angle of elevation of dividing head
N = Number of teeth on the cutter
β = Cutter angle (Fig. 8-21, B)

Example 8-4:

An end mill is to have ten teeth. A 70-degree fluting cutter is to be used. Calculate the angle at which the dividing head should be set.

$$\cos \alpha = \tan \frac{360°}{N} \times \cot \beta$$

$$= \tan \frac{360°}{10} \times \cot 70° = .72654 \times .36397$$

$$\cos \alpha = .26444$$

$$\alpha = 74°40'$$

The angle of elevation for cutting the side teeth of a side milling cutter is determined in exactly the same way. Sketch B shows a dividing head set for milling the side of a side mill with a 70-degree cutter, the angle α being approximately 85½ degrees.

Milling the Teeth of an Angular Milling Cutter

The flutes of angular milling cutters must be cut so that the teeth have a uniform width at the top. Referring to Figs. 8-22 and 8-23, the procedure for calculating the angle of elevation required to mill angle cutters is given here:

a = Angle of elevation of dividing head

B = Angle of cutter blank to be milled

C = Angle of fluting cutter

$$T = \text{Tooth angle} = \frac{360}{N}$$

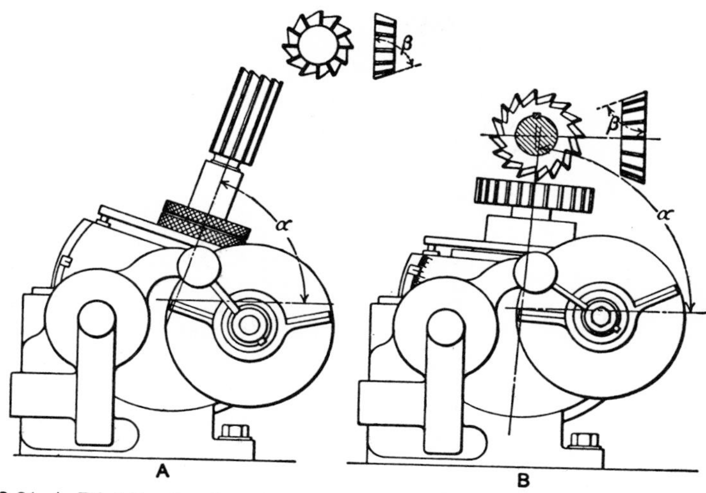

Fig. 8-21. A. Dividing head set for milling teeth of end mill. B. Dividing head set to mill the side teeth of a side milling cutter.

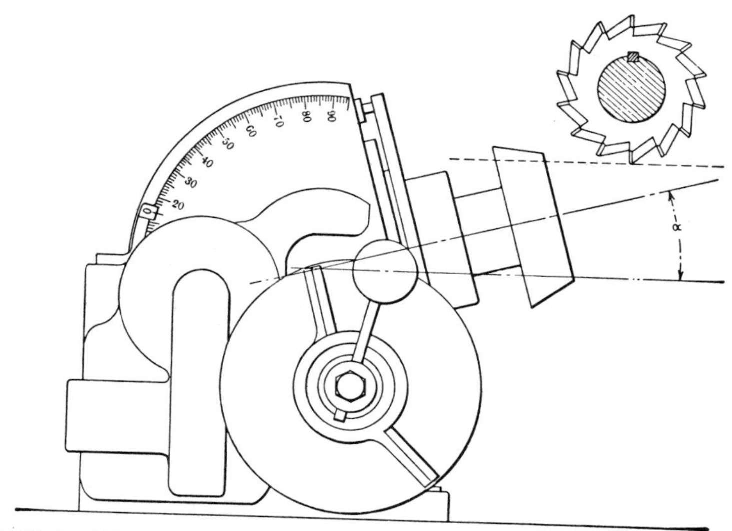

Fig. 8-22. Dividing head set at angle of 15 degrees for cutting teeth in 70-degree angle cutter blank with a 60-degree cutter.

N = Number of teeth to be milled

D and E = Angles indicated on Fig. 8-23

Then:

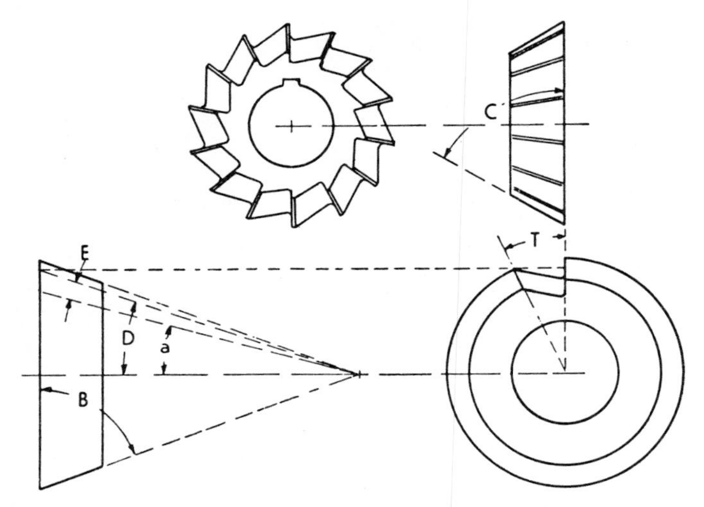

Fig. 8-23. Angles involved in calculation for determining position of dividing head when milling teeth on angle cutter.

$$\tan D = \frac{\cos T}{\tan B}$$

$$\sin E = \tan T \cot C \sin D$$

$$a = D - E$$

Example 8-5:

The flutes of a 70-degree milling cutter with 18 teeth are to be milled with a 60-degree single-angle cutter. Calculate the angle a to which the dividing head must be elevated in order to obtain lands of uniform width.

$$T = \frac{360°}{N} = \frac{360°}{18} = 20°$$

$$\tan D = \frac{\cos T}{\tan B} = \frac{\cos 20°}{\tan 70°} = \frac{.93969}{2.7475}$$

$$= .34202$$

$$D = 18°53'$$

$$\sin E = \tan T \cot C \sin D$$

$$= \tan 20° \cot 60° \sin 18°53'$$

$$= .36397 \times .57735 \times .32364$$

$$= .068009$$

$$E = 3°54'$$

$$a = D - E = 18°53' - 3°54'$$

$$a = 14°59'$$

Drilling and Boring Hole Circles

Figure 8-24 shows the setup for boring 19 holes in a cast-iron plate. The holes are patterned in concentric circles around a hole located in the center of the plate. The dividing head is mounted on special elevating blocks which elevate it above the level of the table in order to clear the face plate. The spindle of the dividing head is positioned parallel to the spindle of the milling machine. A face plate is mounted on the dividing-head spindle nose, and the workpiece is clamped to the face plate with the axis of the central hole coinciding with the axis of rotation of the dividing-head spindle. This may be done by means of a layout, or, if the holes have been machined previously, by indicating the sides of the central hole when rotating the dividing-head spindle. The central hole is then aligned with the axis of the milling-machine spindle by using layout lines as a reference or by placing a dial test indicator on the spindle of the milling machine and indicating around the sides of the central hole. The dividing-head spindle must then be turned so that the first hole in the inside hole

Courtesy of Cincinnati Milacron

Fig. 8-24. Boring 19 holes in a cast iron plate.

circle can be aligned with the milling machine by moving the table longitudinally. The dividing-head crank should be positioned so that the index plate pin is in one of the numbered holes on the index plate. This may require releasing the plate stop and rotating the index plate a small amount in order to fit the index plate pin into the numbered hole of the 24-hole circle.

The central hole is then bored to the required size with a single-point boring tool that is rotated by the milling-machine spindle while the work is fed with the transverse table feed. An offset boring head, with a micrometer dial to allow the boring tool to be offset an exact amount, would be very helpful in boring the holes. After the central hole has been finished to size, the table is moved longitudinally a distance equal to the radius of the second hole circle. This hole circle has six holes equally spaced around its circumference. After the first hole of this hole circle has been bored to size, the dividing head is indexed six turns of the crank plus 16 holes in the 24-hole circle. This movement is determined from the following calculation:

$$T = \frac{40}{N} = \frac{40}{6} = 6\frac{4}{6} = 6 + \frac{16}{24}$$

All of the holes in the second hole circle are bored to size, after which the table is moved longitudinally to the position for boring the outside hole circle. There are 12 holes in the outside hole circle which are indexed by setting the dividing head to index three complete revolutions of the crank

Courtesy of Cincinnati Milacron

Fig. 8-25. Drilling holes around a hole circle in a flanged cap.

plus eight holes in the 24-hole circle. The movement is determined as follows:

$$T = \frac{40}{N} = \frac{40}{12} = 3\frac{4}{12} = 3 + \frac{8}{24}$$

A similar procedure can be used to drill holes in a hole circle such as the one shown in Fig. 8-25. The operation illustrated is drilling and counterboring 42 holes for clamping screws on a flanged cap. The indexing movement for each successive hole is to rotate the crank 40 holes in the 42-hole circle. In performing the operation, every hole is started with a drill and countersink, or center drill, held in the spindle of the machine. Then, after all of the holes have been drilled to the required size, they are countersunk to the correct depth.

CHAPTER 9

Helical and Cam Milling

A helical surface is produced on a milling machine by rotating the workpiece while at the same time feeding it in the direction of the axis of rotation. This motion is accomplished using a universal dividing head which is driven from the table feed screw through change gears (see Fig. 9-5). The change gears permit varying the ratio between the table feed rate and the rate of rotation of the dividing head on which the

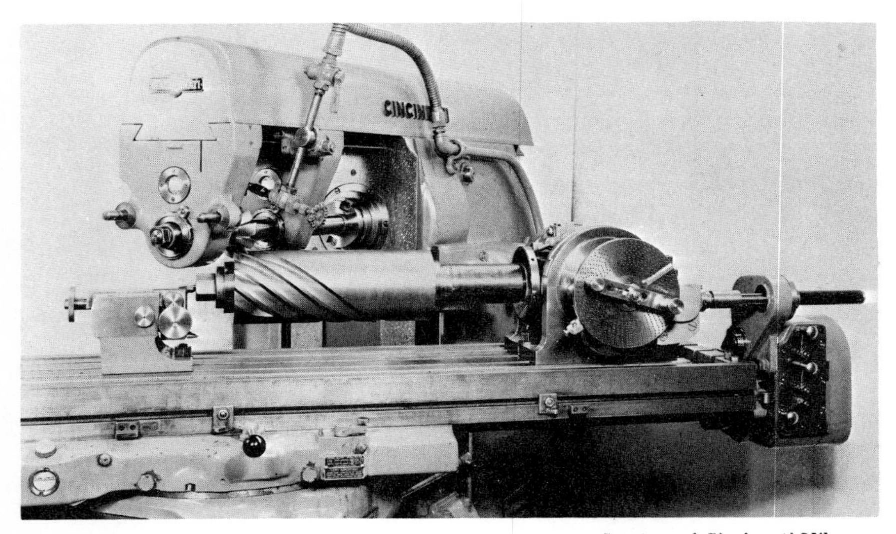

Courtesy of Cincinnati Milacron

Fig. 9-1. Cutting helical flutes in a helical milling cutter blank.

workpiece is mounted. The dividing-head indexing mechanism can be used independently of the helical milling mechanism in order to space the helical grooves around the periphery of the workpiece as required. An exception to this is differential indexing, which cannot be done during helical milling; however, this method of indexing is seldom required for helical milling. A typical helical milling setup and operation is shown in Fig. 9-1, which shows helical flutes being milled in a helical milling cutter blank. The operation is being performed on a universal milling machine which has its table swiveled at an angle in order to cut the helix. The dividing head is driven by enclosed gears located behind it.

The contours of most cams are generated by taking a series of incremental cuts which combine the rotation and the displacement of the cam blank. On a radial cam and on a face cam the movement of the cam blank would be rotation and radial displacement; on a drum cam the movement would be rotation and longitudinal displacement parallel to the axis of the cam. There are some cams, however, that can conveniently be cut by a method similar to helical milling. A vertical milling attachment is used together with a universal dividing head for cutting these cam surfaces.

Helical Curves and Helical Surfaces

A helical curve is a curve which winds around a cylinder and advances at a uniform rate in the direction of the axis. Such a curve is shown in

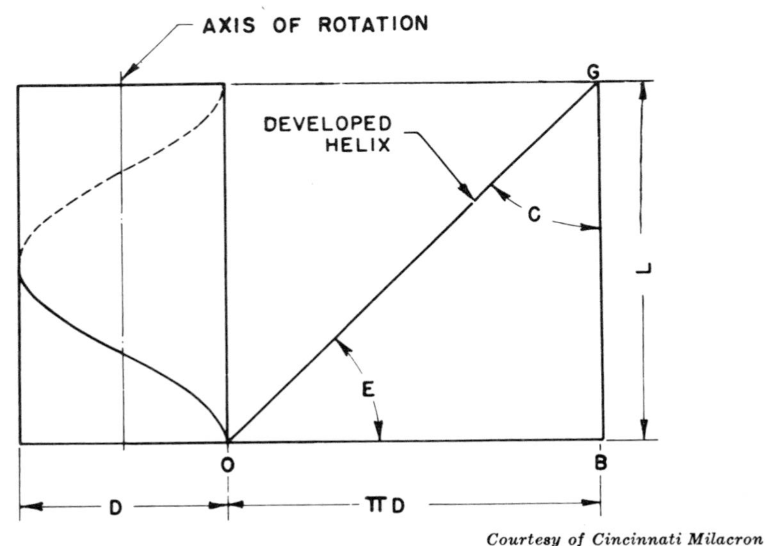

Courtesy of Cincinnati Milacron

Fig. 9-2. Development of the helix.

Fig. 9-2. If the surface of the cylinder in this illustration is unfolded into a plane, the circumference of the cylinder would be represented by the line OB which would be equal in length to πD, where D is the diameter of the cylinder. The helix, when unfolded, becomes the straight line OG. The helix on the cylinder is shown having made one complete revolution. The length along the cylinder required for the helix to make the complete revolution is called the *lead*. When unfolded, the lead is represented by the distance BG, which is also given the dimension L.

The helix has two angles, E and C, which are called helix angles and which are used in practice. It is, therefore, necessary to indicate clearly which helix angle is meant. The helix angle of screw threads is commonly specified by the helix angle E, while, on the other hand, the helix angle on helical gears and milling cutters is specified by C.

The following equations are obtained from the geometry of Fig. 9-2.

$$L = \pi D \tan E \tag{9-1A}$$

$$L = \pi D \cot C \tag{9-1B}$$

where: E and C = Helix angles in degrees
L = Lead of helix, inches
D = Diameter of the cylinder, inches

A number of equally spaced helices all with the same lead and helix angle are shown in Fig. 9-3. When the cylinder on which they are marked

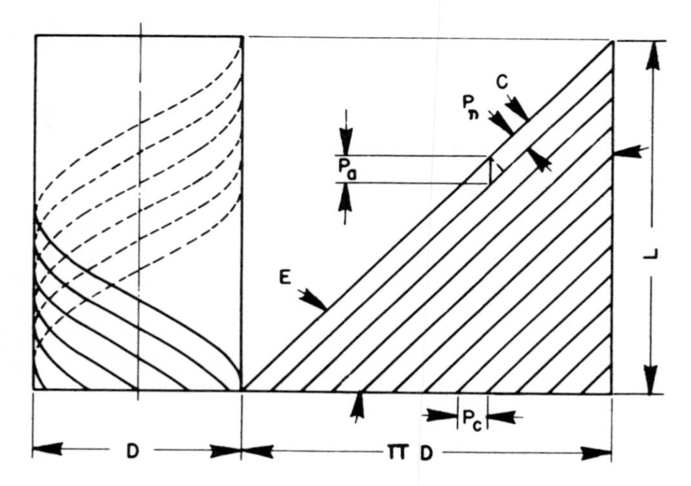

Courtesy of Cincinnati Milacron

Fig. 9-3. Axial (P_a), circular (P_c), and normal (P_n) pitch of equally spaced helices.

is unfolded, the helices form a series of parallel lines. The pitch of the helix is defined as the distance between each helix. Figure 9-3 illustrates three pitch distances, each of which must be clearly distinguished:

P_a = Axial pitch—the distance between consecutive helices when measured parallel to the axis of the cylinder
P_n = Normal pitch—the distance between consecutive helices when measured perpendicular, or normal, to the helices
P_c = Circular pitch—the distance between consecutive helices when measured in a direction perpendicular to the axis of the cylinder
N = Number of equally spaced helices

From the geometry of Fig. 9-3:

$$P_c = \frac{\pi D}{N} \tag{9-2}$$

$$L = P_a N \tag{9-3}$$

$$P_n = P_c \cos C \tag{9-4}$$

$$P_n = P_c \sin E \tag{9-5}$$

$$P_a = P_c \tan E \tag{9-6}$$

$$P_a = P_c \cot C \tag{9-7}$$

Helical surfaces are formed by the sides of the flutes of helical milling cutters, reamers, etc. Since the diameter along these surfaces varies from top to bottom, their helix angle varies and is dependent upon the diameter at which it is measured. This is shown in Fig. 9-4, where the helices corresponding to the diameters D_1, D_2, and D_3 are developed. The helix angles C_1, C_2, and C_3, corresponding to the diameters D_1, D_2, and D_3 along the helical groove, are not equal to each other. It is therefore necessary to select a diameter on which the helix angle is measured. In the case of helical gears, the diameter is the pitch diameter of the gear. In the case of the helical milling cutters the helix angle is generally given on the outside diameter of the flutes.

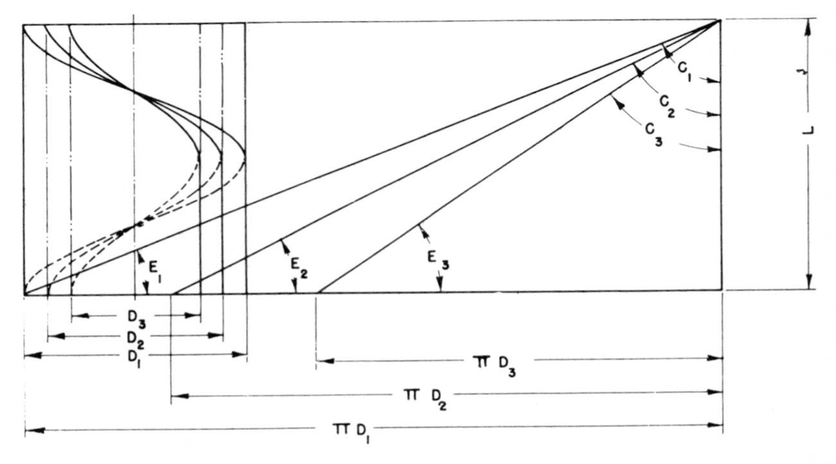

Courtesy of Cincinnati Milacron

Fig. 9-4. Development of a helical surface.

Change Gears for Helical Milling

When milling a helix on a milling machine the dividing head is used to rotate the workpieces as the table feeds in the longitudinal direction.

The rotation of the dividing head is derived from a gear train which is driven by the longitudinal feed screw or lead screw. Figure 9-5 illustrates two dividing-head drives: at the left it can be seen how the gears on an older machine are connected to the dividing head. At the right a newer machine has a cover over the gears which acts as a shield. In the gear trains illustrated, gears D and B are driving gears, and C and A are driven gears. Gears B and C are compound gears since they are located on the same shaft and rotate together at the same speed. Because these gears have different numbers of teeth, they affect the speed ratio between gears A and D. Single idler gears, on the other hand, affect only the direction of rotation and not the speed ratio of the driver and driven gears. The standard change gears available on the Cincinnati Universal Dividing Head have the following number of teeth: 17, 18, 19, 20, 21, 22, 24 (2 gears), 27, 30, 33, 36, 39, 42, 45, 48, 51, 55, 60. On the Brown & Sharpe Universal Index Head the gears used for differential indexing are also used for helical milling. These were listed in Chapter 7.

The lead of the milling machine must be known or calculated before the change gears required to cut a given helix can be calculated. The lead of the milling machine is the distance that the table advances by the longitudinal feed when the dividing-head spindle makes one revolution while being driven by change gears having a one-to-one ratio. If the number of threads per inch on the longitudinal feed or lead screw and the ratio of the dividing head are known, the lead can be easily calculated in the following manner:

$$\text{Lead of Milling Machine} = \frac{\text{Dividing Head Ratio}}{\text{No. Threads per Inch on Lead Screw}} \tag{9-8}$$

Most dividing heads have a 40 to 1 ratio, and if, for example, the lead screw of a given machine has four threads per inch, the lead would be:

$$\text{Lead of Milling Machine} = \frac{40}{4} = 10$$

In other words, if the change gear ratio were 1 to 1, the table would advance 10 inches while the dividing-head spindle made one revolution. This would result in a helix with a lead of 10 inches being cut on a workpiece held on the dividing head.

The change gears for helical milling are calculated by using Formula 9-9:

$$\frac{\text{Driven Gears}}{\text{Driving Gears}} = \frac{\text{Lead of Helix to Be Cut}}{\text{Lead of Machine}} \tag{9-9}$$

The application of this formula will be illustrated by the following examples.

Example 9-1:

A helix with a lead of 48 inches is to be cut on a milling machine. A Cincinnati Universal Dividing Head is to be used. The milling-machine

Courtesy of Cincinnati Milacron

Fig. 9-5. Driving mechanism for driving the dividing-head spindle when performing a helical milling operation. Left: older machine. Right: new machine with cover for gears. A. driven gear; B. driving gear; C. driven gear; D. driving gear; E. leadscrew.

lead screw has four threads per inch. Calculate the required change gears, using only those gears available with the Cincinnati Universal Dividing Head.

$$\text{Lead of Milling Machine} = \frac{40}{4} = 10$$

$$\frac{\text{Driven Gears}}{\text{Driving Gears}} = \frac{\text{Lead of Helix to Be Cut}}{\text{Lead of Machine}} = \frac{48}{10}$$

$$= \frac{6 \times 8}{2 \times 5} = \frac{(6 \times 10)\,(8 \times 6)}{(2 \times 10)\,(5 \times 6)}$$

$$= \frac{60 \times 48}{20 \times 30}$$

Referring to Fig. 9-5, the driving gears (20 and 30) must be placed in positions *D* and *B*. It does not matter which gear is on *D* or which gear is on *B*. The driven gears (60 and 48) are placed in positions *C* and *A*. Again, it does not matter which of the driven gears is in position *C* or in position *A*. For example, the correct lead (48 inches) would be cut if the gears were positioned in any of the following sequences:

Position	Sequence Number			
	1	2	3	4
Driving Gear *D*	20	30	20	30
Driving Gear *B*	30	20	30	20
Driven Gear *C*	60	60	48	48
Driven Gear *A*	48	48	60	60

Gear Ratios for Complex Leads

It is usually not possible to obtain the change gears necessary to cut all leads. Leads which cannot be cut exactly by existing change gears may be classified as complex leads. On certain classes of work, such as helical gears, it is necessary to find a gear ratio for which change gears are available and which will cut a lead as close as possible to that required. Many instruction books supplied by dividing-head builders as well as *Machinery's Handbook* have tables listing the change gears for cutting a wide range of leads; however, it is very useful to be able to calculate these gear ratios.

Although Formula 9-9 is used for calculating change gears in helical milling, special methods must be employed to calculate gear ratios for complex leads. There are two basic methods for calculating the gear ratios for complex leads: the logarithm method and the method of continued division. Since the first method requires a special table of logarithms of gear ratios, it will not be treated in this book; however, excellent information on this method and the actual tables needed, can be found in *Ma-*

chinery's Handbook.[1] The continued division method, on the other hand, can best be described by carefully studying the following examples:

Example 9-2:

Calculate the change gears required to cut the flutes of a 10-tooth, 4-inch diameter, plain helical milling cutter which has a 25-degree helix angle. This job is to be performed using a Cincinnati Universal Dividing Head on a milling machine on which the lead is 10 inches.

As a first step the lead corresponding to this helix angle must be calculated.

$$\begin{aligned} L &= \pi D \cot C \\ &= \pi\ (4)\ \cot 25° = \pi\ (4)\ (2.1445) \\ &= 26.949 \text{ inches} \end{aligned}$$

The required gear ratio can now be calculated by following the steps below in the exact order given:

1. Apply Formula 9-9.

$$\frac{\text{Driven Gears}}{\text{Driving Gears}} = \frac{\text{Lead of Helix to Be Cut}}{\text{Lead of Milling Machine}} = \frac{26.949}{10}$$

2. Expand the gear ratio so that it becomes a whole number.

$$\frac{\text{Driven Gears}}{\text{Driving Gears}} = \frac{26.949 \times 1000}{10 \times 1000} = \frac{26{,}949}{10{,}000}$$

3. Start the continued division process.

   a. If the numerator of Formula 9-9 is larger than the denominator, divide the denominator into the numerator.

   b. If the denominator is larger than the numerator, divide the numerator into the denominator. Thus:

$$\frac{\text{Driven Gears}}{\text{Driving Gears}} = \frac{26{,}949}{10{,}000}$$

$$26{,}949 \div 10{,}000 = 2\,\frac{6949}{10{,}000} \quad \text{(A)}$$

4. Continue the continued division process. This is done by taking the remainder or fractional term of the previous quotient and dividing the denominator of this remainder by its numerator. The process is repeated until a quotient is obtained that has no remainder. Start-

[1] Eric Oberg, Franklin D. Jones, and Holbrook L. Horton, *Machinery's Handbook*, 21 ed. (New York: Industrial Press, Inc., 1979), pp. 1439-1465.

ing with the remainder of the first quotient obtained, the continued division process is carried to completion below.

$$10{,}000 \div 6{,}949 = 1\,\frac{3051}{6949} \quad \text{(B)} \qquad 337 \div 173 = 1\,\frac{164}{173} \quad \text{(G)}$$

$$6{,}949 \div 3{,}051 = 2\,\frac{847}{3051} \quad \text{(C)} \qquad 173 \div 164 = 1\,\frac{9}{164} \quad \text{(H)}$$

$$3{,}051 \div 847 = 3\,\frac{510}{847} \quad \text{(D)} \qquad 164 \div 9 = 18\,\frac{2}{9} \quad \text{(I)}$$

$$847 \div 510 = 1\,\frac{337}{510} \quad \text{(E)} \qquad 9 \div 2 = 4\,\frac{1}{2} \quad \text{(J)}$$

$$510 \div 337 = 1\,\frac{173}{337} \quad \text{(F)} \qquad 2 \div 1 = 2 \quad \text{(K)}$$

5. Tabulate the results of the continued division.

			A	B	C	D	E	F	G	H	I	J	K
Quotients (Whole Numbers)			2	1	2	3	1	1	1	1	18	4	2
Driven Gears	0	1	2	3	8	27	35	62	97	159	2,959	11,995	26,949
Driving Gears	1	0	1	1	3	10	13	23	36	59	1,098	4,451	10,000
Quotients (Whole Numbers)			2	1	2	3	1	1	1	1	18	4	2

The results of the continued divisions must be tabulated as shown. The whole number part of each quotient obtained, including that of the first continued division, must be entered from left to right in the table in the exact order in which it was obtained in the continued division process and in the appropriate row. Two of the horizontal rows represent the gear ratios or the driven gears/driving gears. There are four spaces in the gear ratio rows that are ahead of the vertical quotient columns. These spaces must be filled in according to the rules given below:

a. If the original gear ratio is greater than unity or one:

Driven Gears	0	1
Driving Gears	1	0

b. If the original gear ratio is less than unity or one:

Driven Gears	1	0
Driving Gears	0	1

In the example at hand the gear ratio is 26,949/10,000, which is greater than one.

The actual gear ratios in columns *A, B, C,* etc., can then be calculated by the following procedure.

Driven Gears. Starting from left and going to right, successively multiply each number in the quotient row by the number that is one space to the left in the row below and add to this product the number that is two spaces to the left in this lower row. Enter the result in the space for the driven gear which is below the quotient.

Driving Gears. Starting from left and going to right, successively multiply each number in the quotient row by the number that is one space to the left in the row above and add to this product the number that is two spaces to the left in this upper row. Enter the result in the space for the driving gear which is above the quotient.

The final gear ratio must always be equal to the gear ratio used at the start of the continued divisions. It is important to note that it is the *ratio* which must be equal to the original ratio, although the exact numbers in the numerator and denominator may be different. The original ratio can always be obtained by multiplying the final ratio in the table by a number that is equivalent to one such as 2/2, 3/3, 4/4, etc.

The mathematical procedure for calculating the gear ratios has just been described. In order to present an example of these calculations they are shown below for each space in the gear ratio rows. By comparing these calculations to the numbers in the tables the procedure for making the calculations can be learned.

Column	*Driven Gear*	*Driving Gear*
A	$2 \times 1 + 0 = 2$	$2 \times 0 + 1 = 1$
B	$1 \times 2 + 1 = 3$	$1 \times 1 + 0 = 1$
C	$2 \times 3 + 2 = 8$	$2 \times 1 + 1 = 3$
D	$3 \times 8 + 3 = 27$	$3 \times 3 + 1 = 10$
E	$1 \times 27 + 8 = 35$	$1 \times 10 + 3 = 13$
F	$1 \times 35 + 27 = 62$	$1 \times 13 + 10 = 23$
G	$1 \times 62 + 35 = 97$	$1 \times 23 + 13 = 36$
H	$1 \times 97 + 62 = 159$	$1 \times 36 + 23 = 59$
I	$18 \times 159 + 97 = 2959$	$18 \times 59 + 36 = 1098$
J	$4 \times 2959 + 159 = 11{,}995$	$4 \times 1098 + 59 = 4{,}451$
K	$2 \times 11{,}995 + 2{,}959 = 26{,}949$	$2 \times 4{,}451 + 1{,}098 = 10{,}000$

The last gear ratio is Driven Gears/Driving Gears = 26,949/10,000, which in this instance is exactly equal to the original gear ratio. This is proof that the gear ratio calculations are correct.

6. Select the best gear ratio. A gear ratio is now selected from the table for which change gears for the dividing head will be available and which will be as close as possible to the original gear

ratio. The original ratio is expressed as a decimal by dividing the denominator into the numerator so that it can be compared to the selected gear ratio—which in this example is 2.6949. The gear ratio in the vertical column E (step 5), Driven Gear/Driving Gear = 35/13, is selected. The decimal equivalent of 35/13 is 2.6923. All of the gear ratios to the left of E are rejected because their deviation from the original ratio is greater. The ratios to the right of E are rejected because they contain numbers that are larger than the available gears.

7. Calculate the required change gears by factoring and expanding the gear ratio selected in Step 6.

$$\frac{\text{Driven Gears}}{\text{Driving Gears}} = \frac{35}{13} = \frac{7 \times 5}{6.5 \times 2} = \frac{7 \times 6}{6.5 \times 6} \times \frac{5 \times 12}{2 \times 12}$$

$$\frac{\text{Driven Gears}}{\text{Driving Gears}} = \frac{42 \times 60}{39 \times 24}$$

The actual lead that can be cut with the gear ratio of 35/13 can be calculated by manipulating Formula 9-9 as follows:

$$\frac{\text{Driven Gears}}{\text{Driving Gears}} = \frac{\text{Lead of Helix to Be Cut}}{\text{Lead of Machine}} \qquad (9\text{-}9)$$

$$\text{Lead of Helix to Be Cut} = \text{Lead of Machine} \times \frac{\text{Driven Gears}}{\text{Driving Gears}}$$

$$= 10 \times \frac{35}{13} = 10 \times 2.6923$$

$$= 26.923 \text{ inches}$$

Thus, the total error of the lead that will be cut by the 35/13 gear ratio will be equal to $26.949 - 26.923 = .026$ inch, or about .001 inch per inch.

Example 9-3:

A helical gear is to be cut on a milling machine. The lead of this machine is 10 inches, and the lead to be cut on the gear is 7.882 inches. The dividing head to be used is a Cincinnati Universal Dividing Head. Calculate the change gears required to cut the teeth on this gear using the continued division method.

$$\frac{\text{Driven Gears}}{\text{Driving Gears}} = \frac{\text{Lead of Helix to Be Cut}}{\text{Lead of Milling Machine}} \qquad (9\text{-}9)$$

$$= \frac{7.882}{10} = \frac{7{,}882}{10{,}000}$$

When the continued divisions are performed:

$$10{,}000 \div 7{,}882 = 1\,\frac{2118}{7882} \quad \text{(A)}$$

$$7{,}882 \div 2{,}118 = 3\,\frac{1528}{2118} \quad \text{(B)}$$

$$2{,}118 \div 1{,}528 = 1\,\frac{590}{1528} \quad \text{(C)}$$

$$1{,}528 \div 590 = 2\,\frac{348}{590} \quad \text{(D)}$$

$$590 \div 348 = 1\,\frac{242}{348} \quad \text{(E)}$$

$$348 \div 242 = 1\,\frac{106}{242} \quad \text{(F)}$$

$$242 \div 106 = 2\,\frac{30}{106} \quad \text{(G)}$$

$$106 \div 30 = 3\,\frac{16}{30} \quad \text{(H)}$$

$$30 \div 16 = 1\,\frac{14}{16} \quad \text{(I)}$$

$$16 \div 14 = 1\,\frac{2}{14} \quad \text{(J)}$$

$$14 \div 2 = 7 \quad \text{(K)}$$

In tabulating the quotients of the continued division, care must be exercised to place the first four numbers correctly in the first gear ratio spaces. Since the required gear ratio, 7.882/10 is less than unity, the following tabulation is used:

			A	B	C	D	E	F	G	H	I	J	K
Quotients (Whole Numbers)			1	3	1	2	1	1	2	3	1	1	7
Driven Gears	1	0	1	3	4	11	15	26	67	227	294	521	3,941
Driving Gears	0	1	1	4	5	14	19	33	85	288	373	661	5,000
Quotients (Whole Numbers)			1	3	1	2	1	1	2	3	1	1	7

As a check on the calculations, the final gear ratio appearing in column K is equal to the original gear ratio $3941/5000 \times 2/2 = 7882/10000$. The calculations for the gear ratios appearing in columns A, B, C, etc., are given below in order to illustrate once more how these calculations are made.

Column	*Driven Gears*	*Driving Gears*
A	$1 \times 0 + 1 = 1$	$1 \times 1 + 0 = 1$
B	$3 \times 1 + 0 = 3$	$3 \times 1 + 1 = 4$
C	$1 \times 3 + 1 = 4$	$1 \times 4 + 1 = 5$
D	$2 \times 4 + 3 = 11$	$2 \times 5 + 4 = 14$
E	$1 \times 11 + 4 = 15$	$1 \times 14 + 5 = 19$
F	$1 \times 15 + 11 = 26$	$1 \times 19 + 14 = 33$
G	$2 \times 26 + 15 = 67$	$2 \times 33 + 19 = 85$

Column	*Driven Gears*	*Driving Gears*
H	$3 \times 67 + 26 = 227$	$3 \times 85 + 33 = 288$
I	$1 \times 227 + 67 = 294$	$1 \times 288 + 85 = 373$
J	$1 \times 294 + 227 = 521$	$1 \times 373 + 288 = 661$
K	$7 \times 521 + 294 = 3{,}941$	$7 \times 661 + 373 = 5{,}000$

The required gear ratio is:

$$\frac{\text{Driven Gears}}{\text{Driving Gears}} = \frac{7.882}{10} = .7882$$

The ratio selected from the table is 26/33 which is equal to .7878. The required change gears are:

$$\frac{\text{Driven Gears}}{\text{Driving Gears}} = \frac{26}{33} = \frac{2 \times 13}{3 \times 11} = \frac{2 \times 9}{3 \times 9} \times \frac{13 \times 3}{11 \times 3}$$

$$\frac{\text{Driven Gears}}{\text{Driving Gears}} = \frac{18 \times 39}{27 \times 33}$$

The actual lead cut by the gears selected is:

$$\begin{aligned}\text{Lead of Helix to Be Cut} &= \text{Lead of Machine} \times \frac{\text{Driven Gears}}{\text{Driving Gears}} \\ &= 10 \times \frac{26}{33} \\ &= 7.879 \text{ inches}\end{aligned}$$

Thus, the total error in the lead when the selected gears are used is $7.882 - 7.879 = .003$ inch.

Milling Helical Flutes

Milling the helical flutes on a plain milling cutter is a typical example of a helical milling operation. This job will be used as an example in order to present some of the details involved in actually doing a job of helical milling. An end view of the plain milling cutter is shown in Fig. 9-6. The flute is to have a right-hand cut (see Fig. 9-12) and a right-hand helix with a 25-degree helix angle. The cutter blank is made from an M2 high-speed steel that has been carefully annealed to a hardness of 240 Bhn. It has been turned and bored with some stock left on each surface for finish grinding after it has been hardened.

Before the job is started each step should be planned and the necessary calculations should be made in advance. Starting with the calculations, the procedure for milling the flutes on a universal milling machine is described in the following steps:

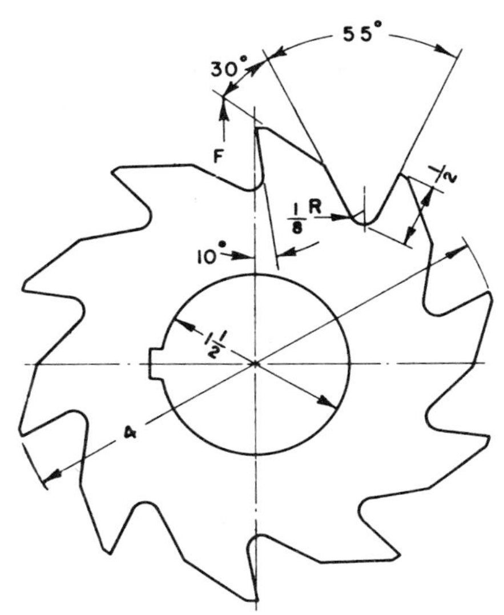

Courtesy of Cincinnati Milacron

Fig. 9-6. Dimensions of a plain milling cutter.

1. Make all of the necessary calculations.
2. Set up the cutter blank on the universal milling machine and mount the fluting cutter on the arbor.
3. Position the cutter blank with respect to the fluting cutter.
4. Swivel the table of the universal milling machine to the required swivel angle.
5. Cut the flutes.

1. Make All of the Necessary Calculations. This is one of the most important and difficult of the steps. The following calculations should be made before the job is set up on the machine so that it will not be idle while the calculations are made.

a. Calculate the change gears required to cut the desired helix angle.
b. Calculate the hole circle and index plate to be used; and determine the required movement of the dividing head.
c. Determine the included angle of the fluting cutter.
d. Calculate the corrected angle of table swivel.
e. Calculate the transverse and vertical cutter offsets.
f. Calculate the cutting speed and the feed rate.

a. *Calculate the change gears.* As these calculations have already been shown in Example 9-2; they will not be repeated.

b. *Calculate the hole circle and dividing-head movement.* Since there are 10 teeth on the cutter, the indexing movement will consist of making four complete turns of the index crank which can be done by using any index plate. Sometimes the dividing head is used in making a layout on the end face of the gear blank and in positioning this layout, once it has been made, in correct orientation to the cutter. This requires indexing the cutter blank in degrees of an angle. It is, therefore, an advantage to have an index plate attached to the dividing head with a hole circle which can be used to index angles. On the Cincinnati dividing head this would be the 54-hole circle while on the Brown & Sharpe dividing head it would be the 27-hole circle.

c. *Determine the included angle of the cutter.* The 55-degree included angle in the base of the flute is measured on the end face, which is a plane perpendicular to the axis of the cutter blank. This angle is represented by angle f_c in Fig. 9-7. The cutter blank will be swiveled at an angle b when the flutes are being cut. Thus, the included angle of the fluting cutter must be angle f, which is called the normal flute angle, in order to produce the required included angle, f_c, in the plane on the end face of the cutter blank. The following formula can be used to calculate the included angle of the cutter.

$$\tan f = \tan f_c \cos b \tag{9-10}$$

where: f = Normal flute angle and also the included angle of the cutter, degrees

f_c = The flute angle perpendicular to the axis of the cutter blank which is measured on the end face, degrees

b = Angle of swivel of the milling-machine table, degrees

For the moment it can be assumed that the angle of swivel of the milling-machine table is equal to the helix angle to be cut, or 25 degrees.

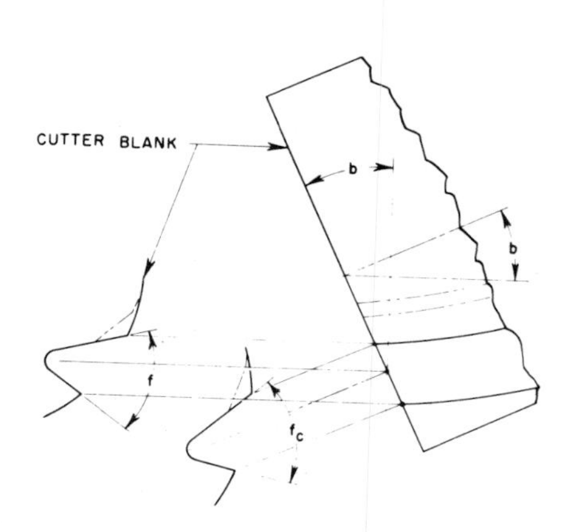

Courtesy of Cincinnati Milacron

Fig. 9-7. Relation between normal flute angle (f) and flute angle at right angle to cutter blank (f_c).

$$\tan f = \tan f_c \cos b = \tan 55° \cos 25° = 1.4281 \times .90631$$
$$= 1.2943$$
$$f = 52° \, 19'$$

A fluting cutter which is ground to this included angle will cut an angle of 55 degrees on the end face of the cutter blank. On the other hand, if a fluting cutter with a 55-degree included angle is available it would cut the following included angle on the end face of the cutter blank.

$$\tan f_c = \frac{\tan f}{\cos b} = \frac{\tan 55°}{\cos 25°} = \frac{1.4281}{.90631}$$
$$= 1.5757$$
$$f_c = 57° \, 36'$$

It is now a matter of deciding whether to accept the small error **(2°36′)** in the included angle on the cutter blank caused by using a standard 55-degree fluting cutter or to grind a fluting cutter to have an included angle of 52°19′. In this case it will be assumed that the fluting cutter will be ground.

d. *Calculate the corrected angle of table swivel.* The table must be swiveled in order to cut the helix as shown in Fig. 9-8. When the helix is

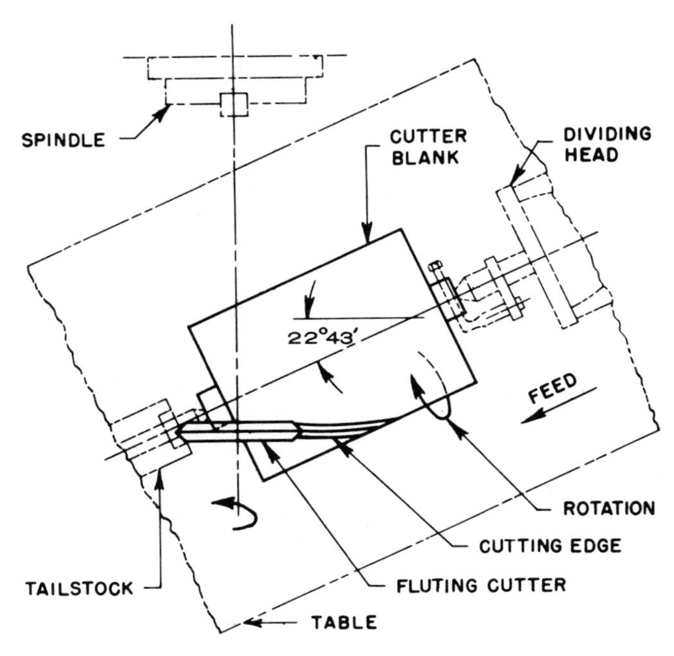

Courtesy of Cincinnati Milacron

Fig. 9-8. The table of the milling machine must be swiveled to 22°43′ for milling the right-hand helix.

cut, the fluting cutter will be offset or set over. This offset will cause an error in the helix angle cut on the cutter blank if the table is swiveled to the designated helix angle—in this case, 25 degrees. The cause of the error is shown in Fig. 9-9. As a result of the offset of the fluting cutter, the helix is produced in the plane *GG* while at the same time the table is swiveled in the horizontal plane *HH*. View A illustrates this condition for a single-angle fluting cutter while view B shows a double-angle fluting cutter.

The correct angle of swivel can be calculated, or it can be determined from a layout on a drawing board, as in Fig. 9-9. The tooth form of the cutter is laid out, and a projection is made above the tooth form on which the helix angle is to be cut. Angle c is laid off for some convenient length d (here d is not the depth of the tooth). A second projection is made vertically above the tooth form from the line *GG* as shown. On the second projection the distance d is laid off, and the angle b is determined as shown. Careful and accurate drafting practice to an enlarged scale must be used to make this layout.

The corrected angle of swivel can be calculated by the following formula:

$$\tan b = \tan c \cos (r + a_c) \tag{9-11}$$

where: b = Angle of swivel of the milling-machine table, degrees
c = Helix angle to be cut, degrees
r = Radial rake angle to be cut on the cutter blank, degrees

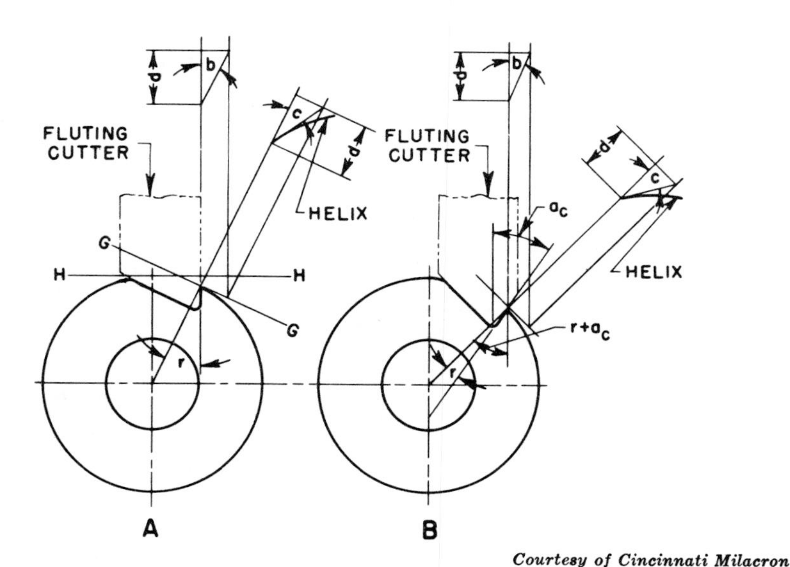

Courtesy of Cincinnati Milacron

Fig. 9-9. Graphic determination of the angle of swivel.

a = Side angle on that side of the fluting cutter which will cut the face of the cutting edge on the cutter blank, degrees (Fig. 8-15)

a_c = The angle of the fluting cutter (angle a) projected to the axis of the cutter blank (Fig. 9-9)

When a single-angle fluting cutter is used, the term a is zero and Formula 9-11 becomes:

$$\tan b = \tan c \cos r$$

A double-angle fluting cutter is recommended over a single-angle fluting cutter because the single-angle cutter will tend to back cut, leaving the face of the flute marred with cutter marks. When double-angle fluting cutters are used to mill flutes with a large helix angle, some back cutting may occur unless a special fluting cutter is used.

Before Formula 9-11 can be used to calculate the angle of swivel, the value of the angle a_c must be determined. To avoid some laborious calculations, the angle a_c can be assumed to be equal to the angle a because their difference is usually small. If this is done, the angle of swivel, b, can be calculated from Formula 9-11 with the knowledge that a small error will be present.

In the present example, the radial rake angle to be produced is 10 degrees (Fig. 9-6), and the side angle of the fluting cutter is 15 degrees.

$$\begin{aligned} \tan b &= \tan c \cos (r + a_c) = \tan 25° \cos (10° + 15°) \\ &= (.46631)(.90631) = .42262 \\ b &= 22° \, 55' \end{aligned}$$

It is possible to calculate angle a_c; however, the calculations involve a mathematical process known as the method of successive approximations. This procedure for calculating angle a_c is shown in Appendix 1.

e. *Calculate the transverse and vertical cutter offsets.* The procedure for calculating the transverse and vertical cutter offsets is given in Appendix 2. It is shown in the Appendix that in this example:

$$\text{Transverse offset } n = .6575 \text{ inch and}$$

$$\text{Vertical offset } m = .6554 \text{ inch}$$

A significant difference occurs in the answer for m and n if 16°10′ is used for the angle a_c instead of 15° in making the calculations. It is thus necessary to calculate the angle a_c as shown in Appendix 1, if this method of offsetting the cutter is selected.

An alternative method of offsetting the cutter, which does not involve these calculations, requires that a layout be made on the end of the cutter blank. This method, which will be described later, is practical when flutes are cut in cutter blanks having a large diameter because the layout will be readily visible.

f. *Calculate the cutting speed and the feed rate.* This is done in the conventional manner also making use of Tables 5-4 and 5-8. The fluting cutter is a high-speed steel form-relieved-type milling cutter, and the cutter blank is made from high-speed steel (tool steel) which has been annealed to have a hardness of approximately 240 Bhn. The diameter of the fluting cutter is 4 inches, and it has ten teeth.

$$V = 50 \text{ fpm} \qquad \text{(from Table 5-4)}$$

$$f_t = .003 \text{ ipt} \qquad \text{(from Table 5-8)}$$

$$N = \frac{12\,V}{\pi\,D} = \frac{12 \times 50}{\pi \times 4}$$

$$N = 48 \text{ rpm}$$

$$f_m = f_t\, n_t\, N = .003 \times 10 \times 48$$

$$f_m = 1.4 \text{ ipm}$$

2. Set Up the Cutter Blank and Mount the Fluting Cutter. The cutter blank is mounted on a mandrel and held in place by a nut. The mandrel is then mounted between the centers of the dividing head. Although the dividing head center acts as a live center, the tailstock center is a dead center and must be provided with lubrication. Usually a mixture of white lead and oil serves as a lubricant. It is necessary to disengage the stop pin of the dividing head because the index plate rotates when the helix is cut.

A form-relieved shaped-profile fluting cutter is mounted on the arbor. Here the fluting cutter is a right-hand cutter which is mounted on the arbor with its largest diameter opposite the spindle, as shown in Fig. 9-10. In this way the profile cutter will mill the flute by the conventional or up milling process. A left-hand fluting cutter could be used as shown in Fig. 9-14.

A double-angle milling cutter with the desired radius on the end of the teeth and with a 55-degree included angle could also be used. This type of cutter will require taking two separate cuts through each flute. The first cut is used to form the 55-degree profile in the base of the flute. A second cut must be taken through each flute to form the 30-degree profile located behind the cutting edge (Fig. 9-6).

3. Position the Blank with Respect to the Fluting Cutter. In most instances when setting up to perform a helical milling operation, the workpiece and the milling cutter should be aligned with respect to each other before the table is swiveled. When the axis of the workpiece and the milling cutter are perpendicular to each other, it is relatively easy to center the workpiece and the cutter or to obtain a desired amount of offset. When they are not perpendicular to each other, this procedure becomes very difficult to follow without an attendant loss of accuracy—unless, of course, the workpiece is positioned by the use of layout lines.

Two methods can be used to align the fluting cutter and the cutter blank. One method is to use a layout which will be described later. The second method is to offset the workpiece a predetermined amount after the center of the cutter blank has been positioned below the largest diameter of the fluting cutter. Since this method of offsetting the cutter blank

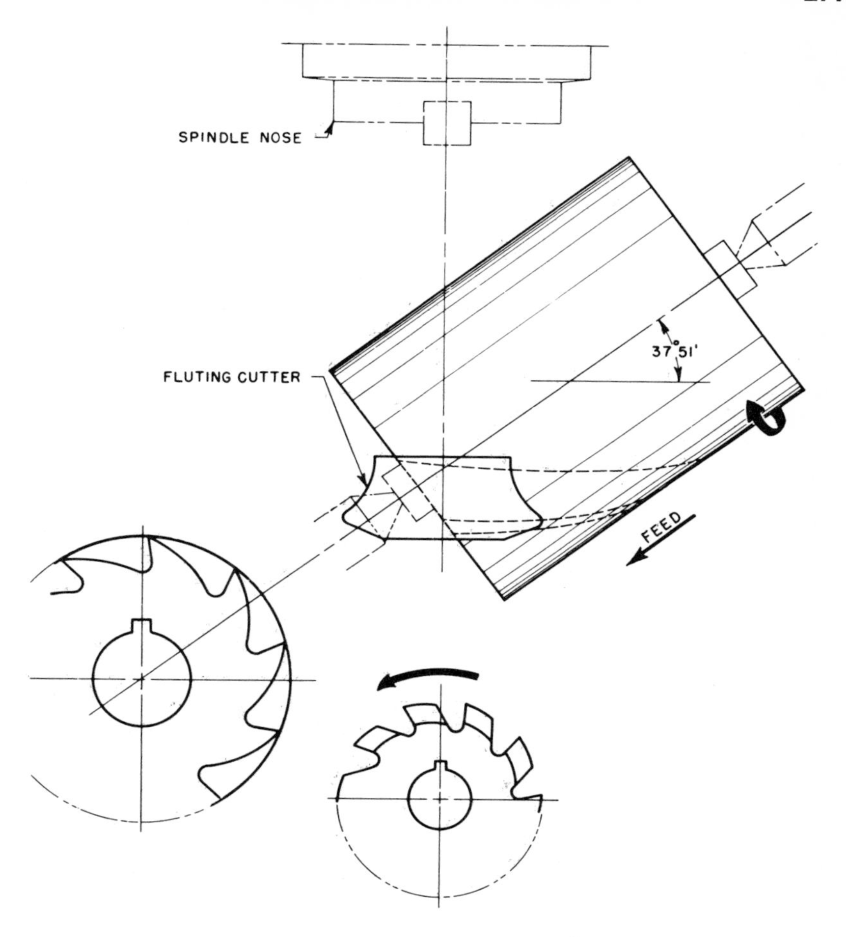

Courtesy of Cincinnati Milacron

Fig. 9-10. Plan view showing relation between fluting cutter and cutter blank for milling a right-hand helix with a right-hand cut.

is the same as explained in Chapter 8 for the reamer, it will not be given in detail here. The four basic steps are:

1. Align the center of the cutter blank with the largest diameter of the fluting cutter.
2. Touch up the cutter blank against the rotating fluting cutter using a paper feeler.
3. Adjust the table to compensate for the thickness of the paper feeler as well as for the difference in diameter of the cutter blank and its diameter when it has become a completely finished plain milling cutter.
4. Offset the table .6575 inch in the transverse direction and raise the table .6554 inch to obtain the vertical offset m.

If a roughing cut is to be taken through each flute first, the vertical offset is reduced about .030 inch to .625 inch on the first cut.

When the layout method is used to align the cutter blank and the fluting cutter, the layout must be made before the table is swiveled. The first step is to remove temporarily one of the change gears that drive the dividing head in order to cut the helix. When this is done, the gear train will not cause the dividing-head spindle to rotate as the table is moved in the longitudinal direction. The layout is now made with the aid of the dividing head which is used to index the required angles. (It is not for this reason, however, that the gear must be removed. The reason will be explained later on.)

The end face of the cutter blank should be painted with a layout blue so that the layout lines will be clear and sharp. In the layout procedure as shown in view A, Fig. 9-11, line *a* is first scribed through the center of the blank. The cutter blank is then indexed 10 degrees, and line *b* is scribed as shown in view B. Then the cutter blank is indexed 90 degrees, and line *c* is scribed ½ inch below the top of the blank as seen in view C. The layout is now complete; however, the cutter blank must be indexed 15 degrees to the position shown in view D. This angle is calculated from the geometry of the layout lines so that line *b* will be parallel to the 15-degree side cutting edge of the fluting cutter. This position can be checked by placing a straight edge or rule against the side of the cutter as shown in view D.

The reason for removing the change gear will now become apparent. The table must be positioned longitudinally until the end face of the cutter blank is below the axis of the fluting cutter, or in a vertical plane passing this axis. The cutter blank must not rotate while the table is moved longitudinally to this position. Views E and F in Fig. 9-11 illustrate this procedure. Place a machinist's square or a straight edge against the end face of the cutter blank and move the table longitudinally until the blade of the square is against the collar of the milling-machine arbor as shown in view E. Insert a paper feeler between the collar and the blade of the square to check this position. When the correct position is attained, it should be possible to pull the paper feeler from between the arbor and the blade of the square; however, a definite drag should be felt. Using the longitudinal feed micrometer dial, move the table a distance equal to the sum of the thickness of the paper feeler and one-half of the diameter of the collar on the milling-machine arbor. The cutter blank will then be in the position shown in view F.

4. Swivel the Table to the Required Swivel Angle. The swivel angle in this case was calculated to be 22°55′. After the table has been swiveled, the change gear that was temporarily removed should be replaced. However, before the table is swiveled, the direction in which it is to be swiveled must be known. Since there are several possibilities when flutes are cut in a cutting tool such as a milling cutter, some thought and attention should be given to this matter.

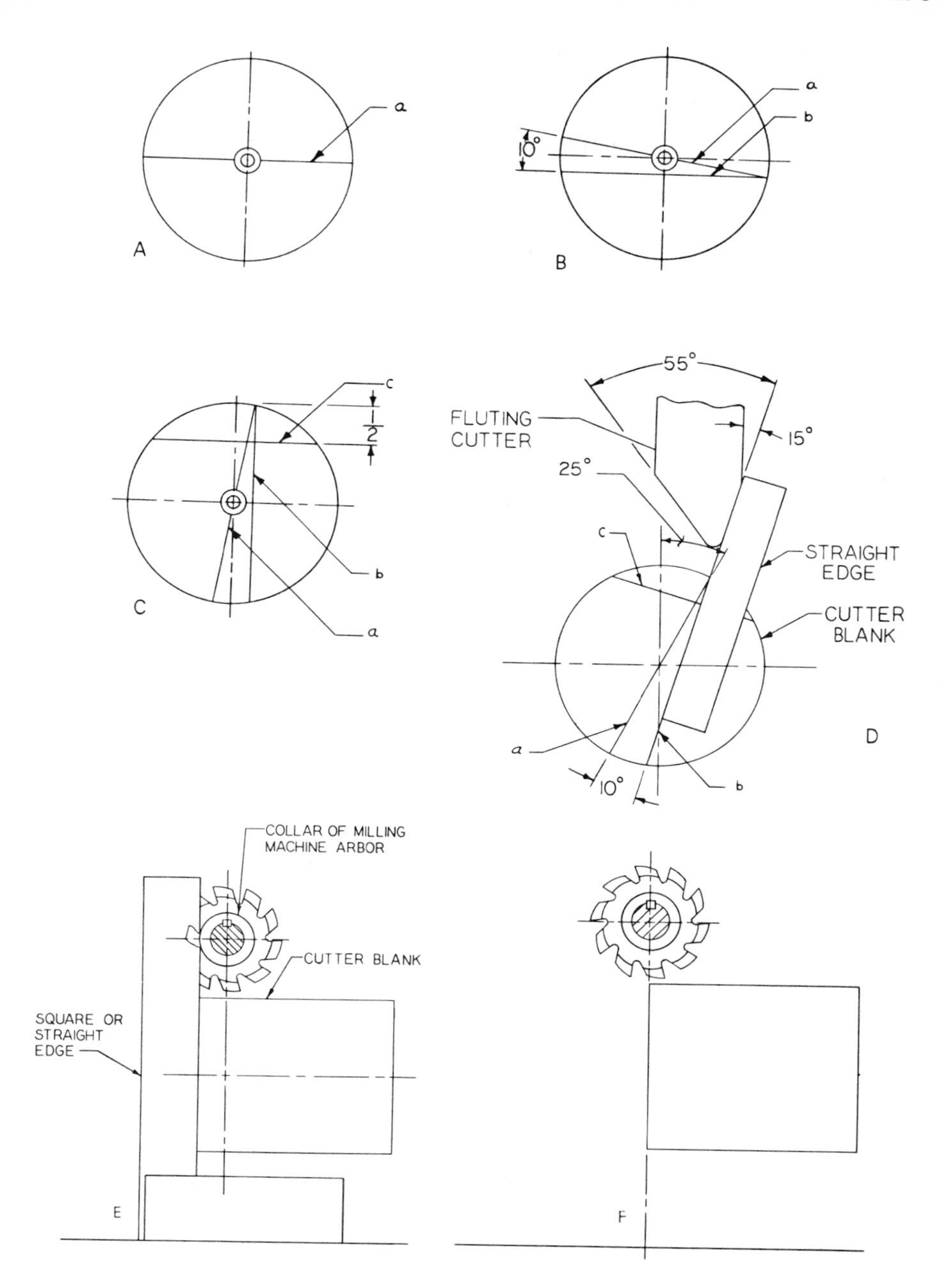

Fig. 9-11. Steps in the layout method for aligning the cutter blank and the fluting cutter.

The four possibilities are shown in Fig. 9-12, with an end milling cutter used as an example. At A a right-hand cut with a right-hand helix is shown. To cut, this cutter must be rotated counterclockwise. The teeth on the end face will cut with an effective positive rake angle which is formed by the flute. The cutter at B has a right-hand cut and a left-hand helix. It will also cut while rotating counterclockwise when viewed from the end; however, the teeth on the end face will have an effective negative rake angle. The cutter at C, with a left-hand cut and a right-hand helix, will cut while rotating clockwise when viewed from the end with the teeth. The effective rake angle in this case will be negative. A cutter with a left-hand cut and a left-hand helix, shown at D, will cut when rotating

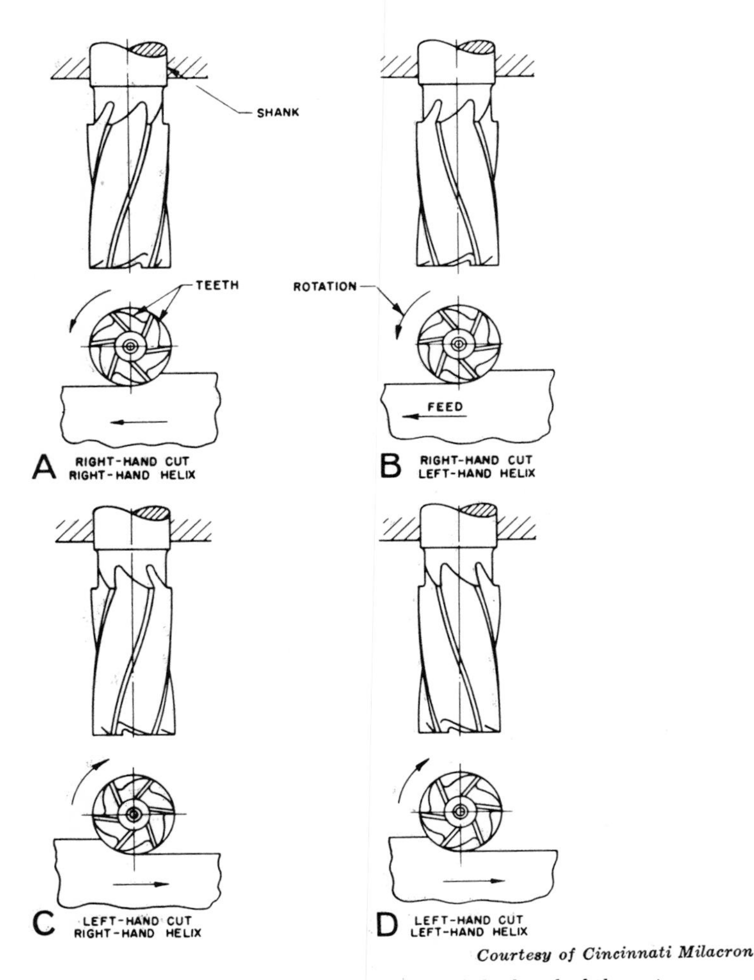

Courtesy of Cincinnati Milacron

Fig. 9-12. Illustration showing the relationship of the hand of the cut and the hand of the helix on end milling cutters.

clockwise; and the teeth on the end face will have an effective positive rake angle as a result of the helix. Similarly, plain milling cutters have a specified right- or left-hand cut and a right- or left-hand helix.

Cutters that have a right-hand cut are usually cut with a right-hand fluting cutter, and those that have a left-hand cut are usually cut with a left-hand cutter. The setup for cutting a right-hand helix with a right-hand cut is shown in Fig. 9-11. The fluting cutter is cutting by the conventional or up milling procedure. Figure 9-13 shows the possible combinations for helical milling. In this illustration cutter A is a right-hand cutter and cutter B is a left-hand cutter. Climb or up milling should be avoided unless the milling machine has a backlash eliminator. Whenever possible the direction of the table feed should be such that the load

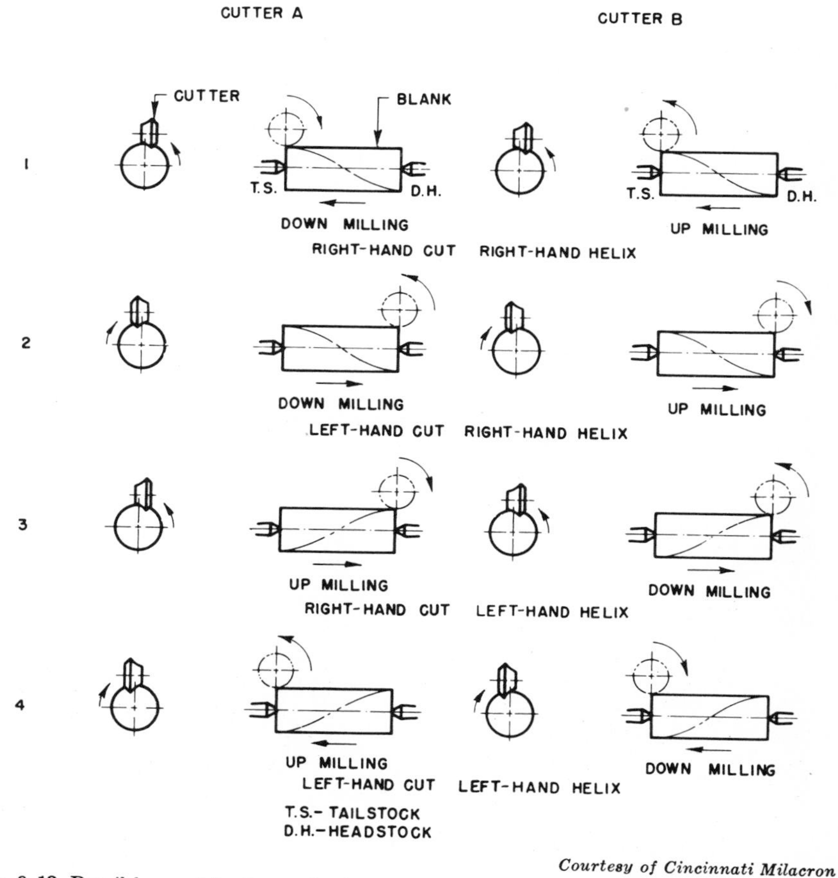

Courtesy of Cincinnati Milacron

Fig. 9-13. Possible combinations of relationships between types of fluting cutters, their positions, the directions of blank rotation, and table feed when cutting teeth for right- or left-hand cut and on right- or left-hand helix.

caused by the cutting action is directed against the dividing-head headstock, because the headstock center rotates with the workpiece. The tailstock center, on the other hand, is a dead center, and the cutting load can impose a heavy load on it which can result in excessive wear. In Fig. 9-13, the dividing-head headstock is shown mounted on the right side of the table. On some milling machines the dividing head is mounted on the left side of the table. In either case, the direction of the cutting load should be against the dividing-head headstock center.

5. Cut the Flutes. If the cutter blank has been offset by moving the table the offset distances n and m, the cut can be started without any further work. Usually two cuts, a roughing and a finishing cut, are taken through each flute. The vertical offset, m, is made approximately .030 inch less than the calculated dimension for taking the roughing cut. It is set to the calculated dimension before the finish cuts are taken.

If the layout method has been used to offset the cutter blank, additional table adjustments must be made to position the cutter blank in the correct relation to the fluting cutter. The table is moved until the cutter blank is in the position shown in Fig. 9-10. Then by a combination of transverse and vertical table movements made while the cutter is rotating, the blank is moved to allow the cutter to cut into the flute until the cut is about .030 inch away from the layout lines b and c (Fig. 9-11). This positions

Courtesy of Cincinnati Milacron

Fig. 9-14. Position of cutter blank relative to the left-hand fluting cutter when milling a right-hand helix with a left-hand cut.

Courtesy of Cincinnati Milacron

Fig. 9-15. Milling right-hand flutes with a right-hand cut in shell end mills using right-hand cutter and the up milling method.

the cutter blank for the roughing cut. After the rough cut is completed on all flutes, a similar table adjustment is made until the cutter touches or "splits" the layout line. The finishing cut can then be taken through all of the flutes. Lower the table when returning to the starting position.

Two helical flute milling operations are shown in Figs. 9-14 and 9-15. In Fig. 9-14 a left-hand fluting cutter is in a position to mill a left-hand cut on a right-hand helix in a flute in a plain milling cutter blank. An interesting operation is shown in Fig. 9-15, where a right-hand helix with a right-hand cut is being milled in the flutes of four shell end milling cutters that are mounted together on the same arbor. The notches seen in one of the bottom helices show the position of each shell end mill on the arbor. These notches are caused by the keyseats located on the back face of each cutter. A right-hand fluting cutter is used, and the setup is the same as shown in Fig. 9-10.

Helical Milling Operations

Several additional helical milling operations are illustrated in Figs. 9-16 through 9-19. In Fig. 9-16 the flutes of a tapered milling cutter are being milled. In this case the flutes are cut with an end mill. The dividing-head spindle is tilted at an angle equal to one-half of the included angle of

Courtesy of Cincinnati Milacron

Fig. 9-16. Setup for milling teeth on tapered milling cutter.

the taper. The tailstock spindle is also tilted as shown in the illustration so that it is supported at the correct angle. This job can be done on either a plain or a universal milling machine since it is not necessary to swivel the table—because an end milling cutter instead of an arbor-mounted cutter is being used. Whenever a helix is cut with an end milling cutter the table does not need to be swiveled.

A universal milling machine can be used to mill the teeth on the helical gears, as shown in Fig. 9-17. Helical gears can also be milled on plain knee-and-column-type milling machines. The cutter is swiveled to the helix angle required instead of the milling-machine table. One attachment with which this can be done is the universal milling attachment shown in Fig. 9-18. It is preferable, however, to use a universal milling machine for milling helical gears if one is available. The cutter is shown being held on a stub arbor while the workpiece is mounted between the dividing-head centers in the conventional way.

Milling Worm Threads

The helical grooves of a worm thread are shown being cut on a milling machine in Fig. 9-19. One advantage of milling worm threads is that a rather wide variety of leads can be obtained, and another advantage is that multiple worm threads can be accurately indexed to provide a precise spacing of threads.

In Fig. 9-19, the worm is being cut with a form-relieved milling cutter with an included angle equal to the angle of the worm thread. It is held on a stub arbor which is mounted on a universal vertical milling attachment. This attachment is swiveled to the helix angle of the worm threads.

Since the lead of the worm is relatively small, a short-lead attachment is used to drive the table and the dividing head—otherwise the table feeding mechanism would be overloaded and possibly damaged. The milling machine must be modified by the builder before this attachment can be used, because a special splined shaft, not installed on standard milling machines, drives the gears in the gear train. These gears are located in the housing attached to the dividing-head end of the table (see Fig. 9-19). When the low lead attachment is engaged, the table lead screw is driven by the gear train instead of by the splined shaft and gears that are nor-

Courtesy of the Brown & Sharpe Manufacturing Company

Fig. 9-17. Milling teeth on a helical gear on a universal milling machine.

mally used. A train of gears inside this attachment connects the lead screw and the dividing head. Another attachment similar to the low-lead attachment is the long-and-short lead attachment that can be seen attached to the table of the milling machine in Fig. 9-20. A range of leads from .100 inch to 1,000 inches can be cut with this attachment.

Although short leads can be cut by feeding the table manually with a hand crank such as the one in Fig. 9-5, an attachment is necessary when any extensive helical milling involving low leads is done. Since by this method the power table feeding mechanism is not used, it will not be overloaded and damaged. The use of the hand feed to feed the table is very tedious and is not recommended except in an emergency.

Cam Milling Constant-Rise Cams

Cams with a constant rise can be generated in a milling machine by holding the cam blank in an angular position on the dividing head (Fig. 9-20) and combining the longitudinal feed of the table with the rotation of the dividing head. The rotation of the dividing-head spindle is synchronized with the table feed through the gears that are used for helical milling. Since the leads involved in cam milling are generally rather short, a short-lead attachment or a long-and-short lead attachment is useful. In Fig. 9-20, where a long-and-short lead attachment is shown, the cam is cut with an end milling cutter held in a vertical milling attachment.

Courtesy of Cincinnati Milacron

Fig. 9-18. Milling a helical gear on a plain knee and column milling machine using a universal milling attachment.

Courtesy of Cincinnati Milacron

Fig. 9-19. Milling a worm with a universal milling attachment with a wide-range dividing head and a short lead attachment.

The vertical milling attachment is tilted to the same angular position as the dividing head.

If, in Fig. 9-20, the axes of the dividing-head spindle and the spindle of the vertical milling attachment were both horizontal, the combined motion of the table feed and the dividing spindle rotation would cause a cylindrical surface to be cut on the "cam"—since the distance between the axis of the dividing-head spindle and the vertical milling attachment would not change. On the other hand, if the axes of the dividing head and vertical milling attachment were vertical, the rise cut on the cam would be equal to the lead for which the milling machine was geared. If the axes of the dividing-head spindle and the vertical milling attachment spindle are inclined, as shown in Fig. 9-20, any rise can be cut on the cam profile, providing it is less than the lead for which the machine is geared. The cam lead, then, can be varied within certain limits by simply changing the angle of inclination, i, of the dividing head and the vertical milling attachment.

Formula 9-8 and the methods previously described in this chapter can be used to calculate the lead of the milling machine and therefore will not be repeated. The following formulas are used to calculate the angle of inclination of the dividing-head spindle, assuming that the milling machine is geared for a given lead.

Courtesy of Cincinnati Milacron

Fig. 9-20. Setup for milling a uniform rise cam on a universal milling machine equipped with a long and short lead attachment.

$$H = \frac{360\ h}{a} \tag{9-12}$$

$$\sin i = \frac{H}{L} \tag{9-13A}$$

$$\sin i = \frac{360\ h}{aL} \tag{9-13B}$$

$$L_c = h \cot i + w \tag{9-14}$$

where: i = Angle to which the dividing-head spindle and the vertical milling attachment are set, degrees
h = The rise of the cam in a given part of a circumference, inches
H = Lead of the cam, or the rise if the rise is considered to continue at a given rate for one complete revolution, inches
L = Lead for which the milling machine is geared, inches
a = The included angle of the cam profile corresponding to the cam rise h, degrees
L_c = The minimum length of the flute on the end of the milling cutter required to cut the cam rise, inches
w = thickness of cam

The relationship of these quantities is shown in Figs. 9-21 and 9-22. Formulas 9-12 and 9-13 are generally used by first calculating the amount of cam lead H required in order to obtain a given rise h in an angle a. A convenient rule to follow is: *Select a lead L that is equal to the number nearest to twice the cam lead H*. The angle of inclination can then be calculated. The minimum length of flute on the end milling cutter required to cut the cam rise, L_c, can be calculated when there is doubt about the length of the cutter that is available.

Example 9-4:

A cam with a constant rise of .750 inch in 90 degrees is to be cut on a milling machine. Determine the lead to which the milling machine should be geared and the angle of inclination to which the dividing head and the vertical milling attachment should be positioned.

$$H = \frac{360° \, h}{a} = \frac{360° \times .750}{90°} = 3.00 \text{ inches}$$

$$L = 2H = 2 \times 3 = 6 \text{ inches}$$

$$\sin i = \frac{H}{L} = \frac{3}{6} = .500$$

$$i = 30°$$

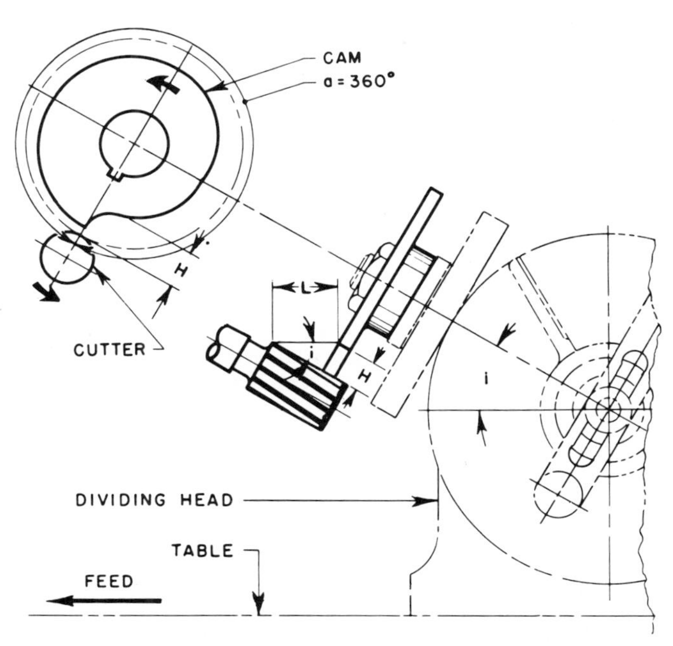

Courtesy of Cincinnati Milacron

Fig. 9-21. Relation of the cam and the table lead.

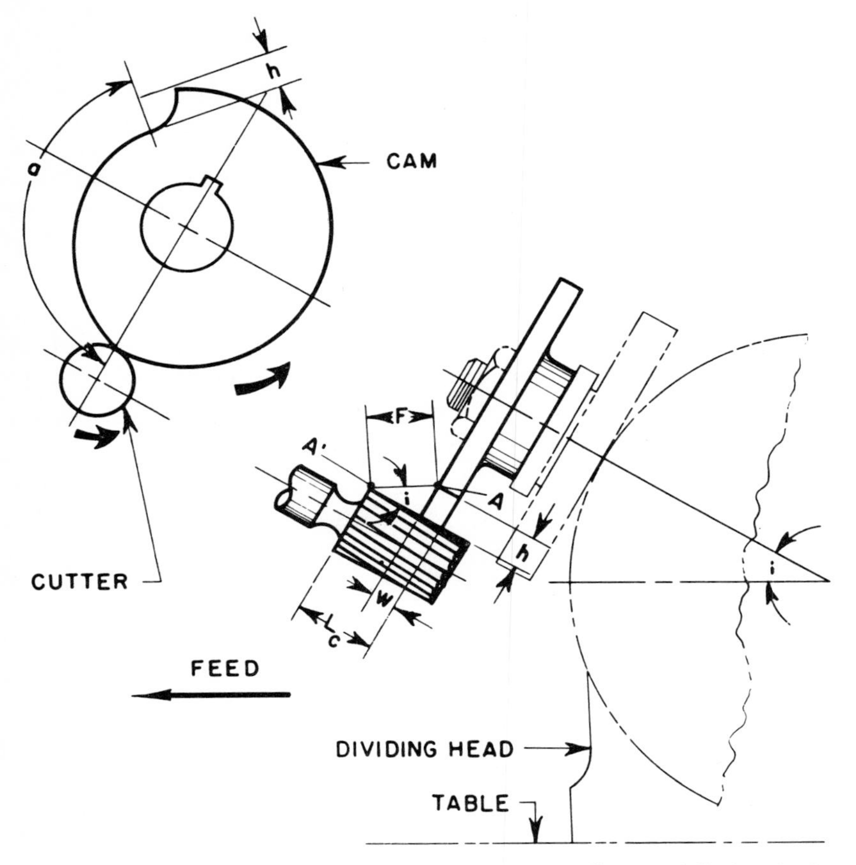

Courtesy of Cincinnati Milacron

Fig. 9-22. Angle of inclination of dividing head and milling cutter spindle in relation to cam rise and machine lead.

Example 9-5:

A ¼-inch-thick cam is to be cut with a constant rise of .125 inch in 300 degrees. Determine the lead to which the milling machine is to be geared and the angle of inclination to which the dividing head and the vertical milling attachment should be positioned.

$$H = \frac{360^\circ\, h}{a} = \frac{360^\circ \times .125}{300^\circ} = .150 \text{ inch}$$

$$L = 2\,H = 2 \times .150 = .300 \text{ inch}$$

A lead of .300 inch cannot be cut by using the standard gearing mechanism available for helical milling; however, this lead can be cut exactly using a long-and-short lead attachment. Therefore, two solutions will be offered: one for standard gearing and the other for use when a long-and-short lead attachment is available.

For a long-and-short lead attachment:

$$\sin i = \frac{H}{L} = \frac{.150}{.300} = .500$$

$$i = 30°$$

$$\begin{aligned} L_c &= h \cot i + w = .125 \cot 30° + .250 \\ &= .125 \times 1.7320 + .250 \\ &= .467 \text{ inch} \end{aligned}$$

The standard leads that can be cut on the Cincinnati Universal Milling Machine equipped with the Standard Enclosed Driving Mechanism range from 2½ to 100 inches. It is assumed below that the machine will be geared for the minimum available lead.

Thus, for standard gearing:

$$\sin i = \frac{H}{L} = \frac{.150}{2.5} = .06000$$

$$i = 3° \, 26'$$

$$\begin{aligned} L_c &= h \cot i + w = .125 \cot 3° \, 26' + .250 \\ &= .125 \times 16.668 + .250 \\ &= 2.334 \text{ inches} \end{aligned}$$

This answer is unsatisfactory, because a rather long end milling cutter would be necessary to cut the required rise. Although the standard driving mechanism is rated to cut a minimum lead of 2½ inches, a shorter lead can be cut if the hand feed is used. The minimum lead is obtained by using the smallest available gears as driven gears and the largest available gears as driving gears.

$$\frac{\text{Driven Gears}}{\text{Driving Gears}} = \frac{\text{Lead to Be Cut}}{\text{Lead of Milling Machine}}$$

$$\begin{aligned} \text{Lead to Be Cut} &= \text{Lead of Milling Machine} \times \frac{\text{Driven Gears}}{\text{Driving Gears}} \\ &= 10 \times \frac{17 \times 18}{55 \times 60} = \frac{3{,}060}{3{,}300} \end{aligned}$$

$$\text{Lead to Be Cut} = .927 \text{ inch}$$

With this lead, the angle of inclination and the length of flute on the milling cutter become:

$$\sin i = \frac{H}{L} = \frac{.150}{.927} = .1618$$

$$i = 9° \; 19'$$

$$L_c = h \cot i + w = .125 \cot 9° \; 19' + .250$$
$$= .125 \times 6.0955 + .250$$

$$L_c = 1.012 \text{ inches}$$

This answer is satisfactory since the length of flute on the end milling cutter required to cut the cam rise is only slightly more than 1 inch.

Cam Milling—Incremental Cut Method

Most cams are cut by means of the incremental cut method. The cam blank is indexed a very small increment, and the table is moved a small increment. By making such numerous and fine incremental movements the profile is developed. The cam is then finished by hand filing, hand honing, or hand polishing the small ridges that are left by the incremental cuts.

The size of the ridges is determined by the number of increments used to cut the profile and the relationship between the cutter diameter, the angle indexed per increment, and the distance from the cam center. This relationship can be seen in a greatly exaggerated form in Fig. 9-23. The

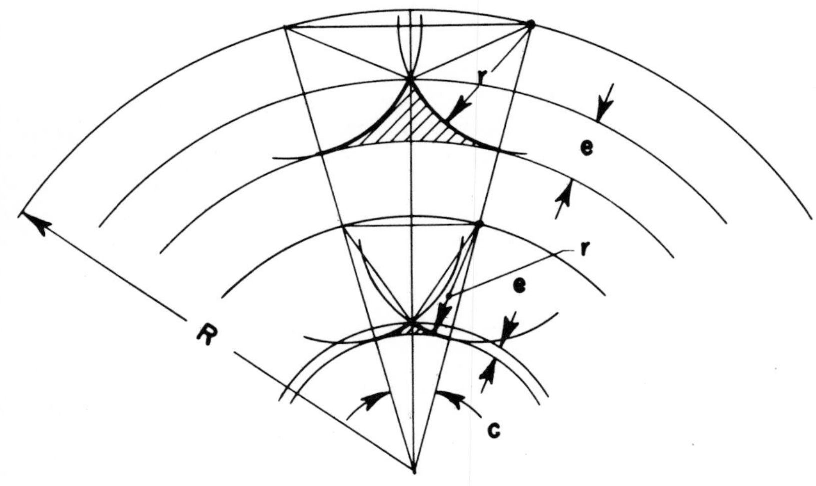

Courtesy of Cincinnati Milacron

Fig. 9-23. Greatly exaggerated geometric relation between the center distance of the cam profile, height of ridges, and angular indexing.

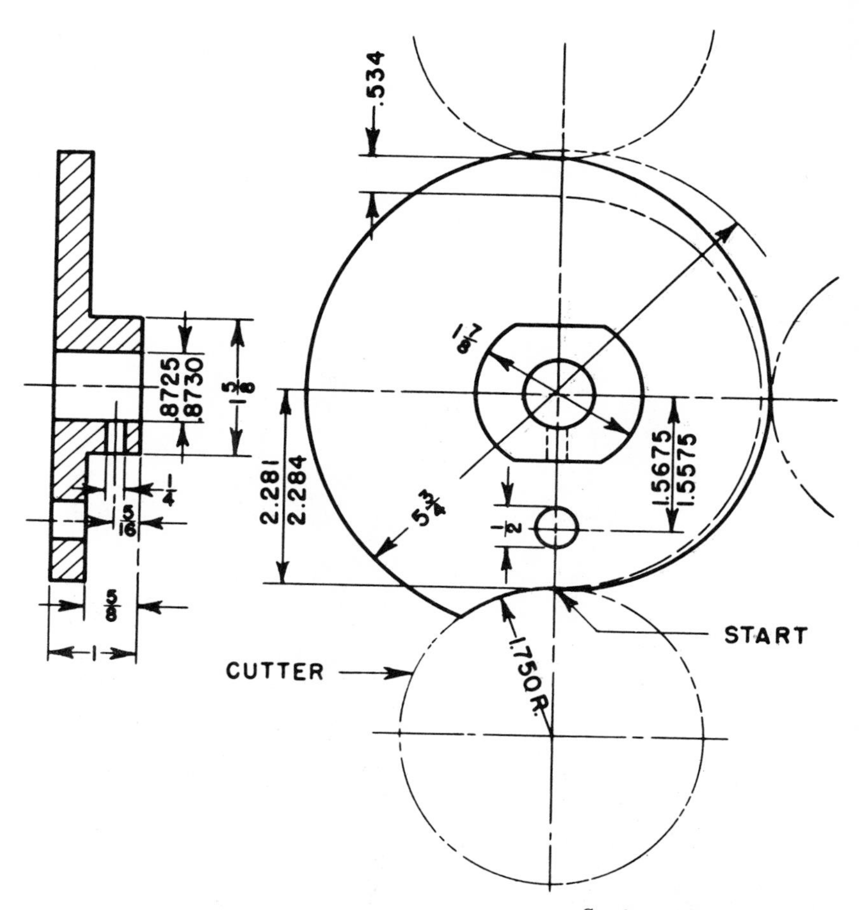

Courtesy of Cincinnati Milacron

Fig. 9-24. Radial cam with cam rise arranged in a geometric progression.

height of each ridge produced, *e,* increases as the distance from the cam center, *R,* is increased, as the angle indexed, *c,* is increased, and as the radius of the milling used, *r,* is decreased. No firm rule can be given regarding the best combination of these variables. Each job must be analyzed and an estimate made of the combination that will most effectively do the job.

An example of the incremental cut method of cam milling is the cam illustrated in Fig. 9-24. Figure 9-25 shows the setup for milling this cam. The cam rise here is .534 inch in an angle of 180 degrees. During this interval the cam rise, which is to be in accordance with a geometric progression, requires a 3-degree angular displacement in relation to the radial distance from the cam center for each index. Sixty indexes are, therefore,

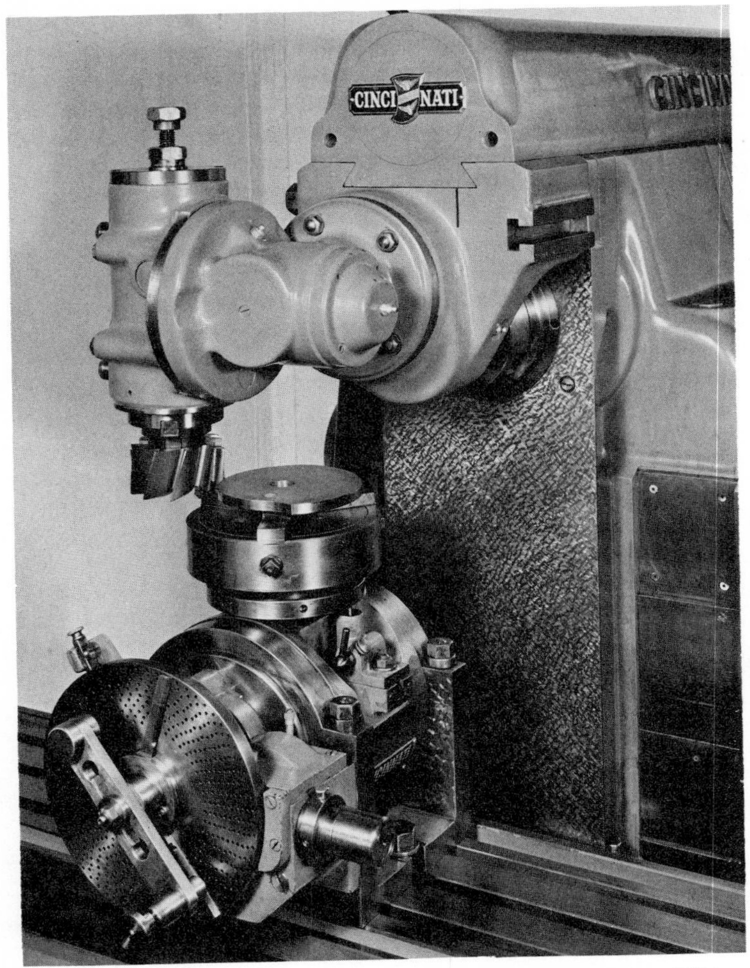

Courtesy of Cincinnati Milacron

Fig. 9-25. Setup for milling the radial cam using the incremental method.

required to complete the cam profile. The rise required for each 3-degree angular displacement has been calculated and tabulated in Table 9-1. The cam blank has a ½-inch diameter construction hole, as in Fig. 9-24, which is used to locate cam contour. The distance between the center of the cam and the periphery of the cutter must be kept within 2.281 and 2.284 inches (Fig. 9-24).

The milling machine is set up by first mounting the vertical milling attachment and tramming the spindle of this attachment until it is per-

Table 9-1. Micrometer Dial Adjustment for Each Increment in Milling Cam

Setting	Dial Adjustment	Setting	Dial Adjustment
Start		31	.0817
1	.004	32	.0871
2	.009	33	.0928
3	.0133	34	.0988
4	.0142	35	.105
5	.0152	36	.112
6	.0162	37	.120
7	.0173	38	.128
8	.0184	39	.137
9	.0197	40	.146
10	.0210	41	.155
11	.0224	42	.166
12	.0238	43	.177
13	.0254	44	.189
14	.0271	45	.202
15	.0290	46	.216
16	.0309	47	.231
17	.0330	48	.247
18	.0352	49	.263
19	.0375	50	.280
20	.0400	51	.298
21	.0427	52	.307
22	.0456	53	.338
23	.0486	54	.361
24	.0518	55	.385
25	.0553	56	.411
26	.0590	57	.438
27	.0630	58	.467
28	.0673	59	.499
29	.0718	60	.534
30	.0765		

Courtesy of Cincinnati Milacron

pendicular to the top of the table. A 1-inch diameter aligning bar is placed in the spindle and a dial test indicator is attached to this bar. With the knee clamped to the column of the milling machine, the tabletop is indicated by rotating the spindle. Better results can usually be obtained by indicating over matched parallels that are placed on the table. When the indicator reading is the same in all positions, the spindle must be vertical with respect to the tabletop.

The dividing head can now be mounted on the tabletop and an independent jaw chuck attached to its spindle. The dividing head is positioned with its spindle vertical, as shown in Fig. 9-25. A 1-inch diameter aligning bar is then chucked in the dividing head. The indicator attached to the vertical head is used to true the aligning bar held in the chuck while the chuck is rotated. The vertical alignment of the dividing head is then checked by indicating the aligning bar in the chuck while the knee is traversed up and down. Any vertical misalignment is corrected. Next, the axes of the vertical attachment and the dividing head are made to coincide by indicating around the aligning bar that is held in the chuck. The dial test indicator can then be removed. When in this position, the table can be moved in the longitudinal direction to cut the cam; however, it should be locked in the transverse direction and not moved in this direction until the job is completed.

There are several methods of positioning the table for cutting the cam. If the milling machine is equipped to handle precision end measuring

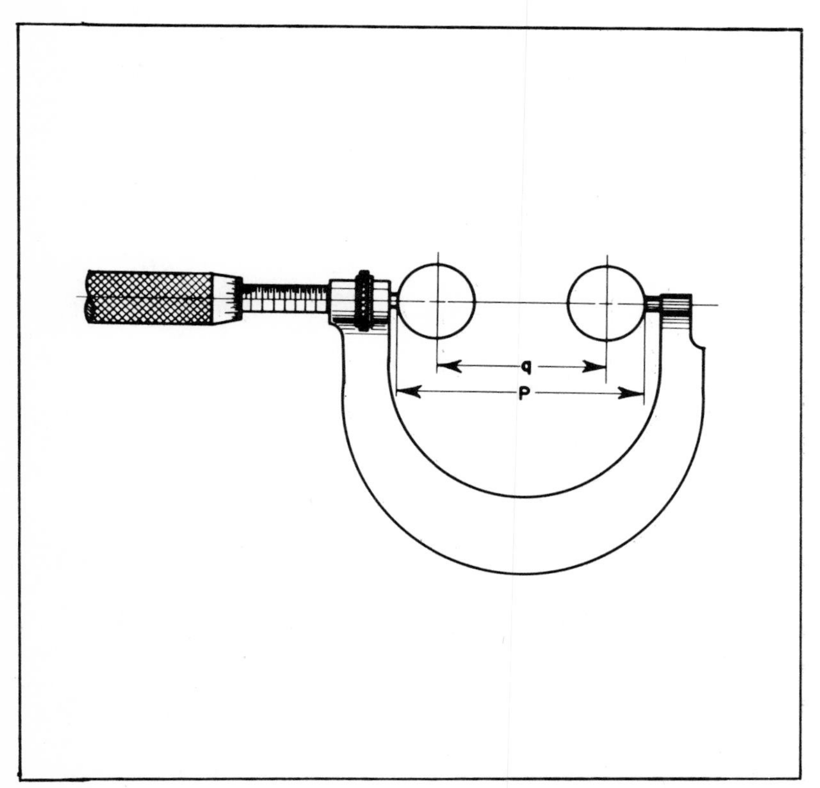

Courtesy of Cincinnati Milacron

Fig. 9-26. Method of aligning the spindle of the dividing head with the spindle of the machine preliminary to milling the radial cam.

rods, these can be used. On ordinary milling machines the feed screw should not be relied upon to move the table to an accuracy of .003 inch in a distance of about 2¼ inches. When moving shorter distances, however, an accuracy of about .001 inch can be relied upon if the machine is in good condition. One method that can be used on ordinary milling machines will now be described.

The cam is cut from an initial zero reference position and this position must now be established. This can be done by measuring over the alignment bars with vernier calipers or with a micrometer caliper as shown in Fig. 9-26. One alignment bar is placed in the milling-machine spindle, and the second alignment bar is on the dividing head. First the distance q must be determined. Measure the diameter of the milling cutter with micrometer calipers by placing pieces of good notebook paper between the teeth of the milling cutter and the two contact surfaces of the micrometer. The combined thickness of the two strips of paper must be subtracted from the micrometer reading in order to obtain the true diameter of the cutter. Assume that in this case the milling cutter is exactly 3.500 inches in diameter. From Fig. 9-24, the dimension of the low point on the cam, which is now considered the reference dimension, is 2.284 max. and 2.281 min. The distance q and the measurement over the 1-inch aligning bars p (see Fig. 9-26) are then calculated as follows:

$$q \text{ max} = 2.284 + 1.750 = 4.034 \text{ inches}$$
$$q \text{ min} = 2.281 + 1.750 = 4.031 \text{ inches}$$
$$p \text{ max} = 4.034 + .500 + .500 = 5.034 \text{ inches}$$
$$p \text{ min} = 4.031 + .500 + .500 = 5.031 \text{ inches}$$

The table is then moved by using the longitudinal hand feed until the micrometer reading shows that the two aligning bars are the correct distance apart, as seen in Fig. 9-26. This position must be obtained by feeding the table from right to left when the position of the dividing head and the vertical milling attachment is as shown in Fig. 9-25. In this manner the lost motion in the feed screw and feed screw nut will not affect the table settings when the cam is cut. When the table has been positioned as described, the longitudinal feed screw micrometer dial at the end of the table is set to read zero. Thus, it is always easy to return to this zero reference position.

Now the aligning bar can be removed from the dividing head, and the cam blank can be inserted in the chuck. The cam blank is then centered in the chuck by indicating its outside diameter with a dial test indicator that can be attached to the aligning bar in the vertical attachment spindle. This indicator is also used to align the ½-inch construction hole in the cam blank. A ½-inch pin is inserted in the construction hole with a light press fit. By adjusting the table longitudinally and by rotating the dividing head, the pin is indicated until it is in perfect alignment with the spindle. The cam blank is now in the correct position for the first cut.

Courtesy of Cincinnati Milacron

Fig. 9-27. Setup for finish grinding the surface of a hardened cam using the incremental method.

The 3½-inch shell end mill is placed in the spindle of the vertical milling attachment, and the machine is set up for the correct cutting speed. After the table is moved away from the zero reference position, it is raised to bring the cutter in the position shown in Fig. 9-25. The first cut is now taken by feeding the cam toward the cutter using the hand longitudinal feed. The cam is fed into the cutter until the micrometer dial reaches the zero reading that was established by measuring over the aligning bars. The cam blank is now moved away from the cutter, and the dividing head is indexed 3 degrees. The table is then moved longitudinally to move the cam blank into the cutter until the micrometer dial reading is .004 less than the zero reading. This value is obtained from Table 9-1, which lists the distances from the zero position for each step. Each increment is then cut by indexing the dividing head 3 degrees and feeding the cam blank into the cutter, stopping short of the zero setting by the distance specified in the table. If a large amount of stock must be removed from the cam, a rough cut should first be taken to remove the bulk of the metal so that a small amount of stock is left for finishing.

A similar procedure is used to finish grind the hardened cam shown in Fig. 9-27. The cam surface is produced by an incremental setting in the vertical direction for each degree indexed. The cut is taken by feeding the table longitudinally with a rapid power feed.

The face cam in Fig. 9-28 is cut with a combination of methods used. This cam has a uniform rise portion which is milled with the high-and-low lead attachment in a manner similar to milling a helix. The dwell portion of the cam is cut by turning the cam blank manually with the index crank while the table is in the stop position. The method for cutting the drum cam, shown in Fig. 9-29, is similar to the method for cutting the face cam. The uniform rise portion of the drum cam can be cut by helical milling. Complex contours can be cut in both face cams and drum cams by means of the incremental method. The best procedure is to rough out the channel, leaving about $\frac{1}{32}$ to $\frac{1}{16}$ stock on each side. A two-fluted end milling cutter with a diameter equal to the width of the channel is used to finish the cam. The incremental movements are made by rotating the workpiece

Courtesy of Cincinnati Milacron

Fig. 9-28. Setup for milling a face cam.

Courtesy of Cincinnati Milacron

Fig. 9-29. Setup for milling a drum cam.

with the dividing head and feeding the table longitudinally. The cutter is fed into the work during each increment with the vertical feed. The cams are then finished by hand methods.

CHAPTER **10**

The Horizontal Boring Machine

The horizontal boring machine is used to machine holes and plane surfaces primarily on larger workpieces. It is especially indispensable for machining large castings and weldments. Figure 10-1 shows a horizontal boring machine with the large table surface area available for the clamping of workpieces. Horizontal boring machines are interesting and challenging to operate because of the variety of different workpieces that can be machined on them and the variety of operations which they perform. They are sometimes called horizontal boring mills or horizontal boring, drilling, and milling machines.

Courtesy of the Giddings & Lewis Machine Tool Company

Fig. 10-1. A horizontal boring machine machining a horizontal boring machine headstock.

Courtesy of the Giddings & Lewis Machine Tool Company

Fig. 10-2. Line boring a weldment on a horizontal boring machine.

Horizontal Boring-Machine Construction

There are three basic types of horizontal boring machines: the table type, the planer type, and the floor type. A table-type horizontal boring mill is shown in Figs. 10-1 and 10-2. The principal parts of the table-type horizontal boring machine are the base or runway, the saddle, the table, the end support column, the headstock column, the headstock, and the spindle. The base, or runway, has precision machined slides or ways on which the saddle slides parallel to the spindle. The saddle carries the table, which can slide perpendicular to the axis of the spindle. The headstock column is firmly bolted to the bed. It has ways machined on its face upon which the headstock slides and against which the headstock can be clamped in position. The headstock is raised and lowered on the column by an elevating screw. Counterweights inside of the headstock column are attached to the headstock by chains to reduce the work required in raising and lowering the headstock. Cuts can be taken with the face milling cutter by feeding the headstock vertically as well as with the two directions of the table feed. The spindle has a holding type of internal taper which is used to hold boring bars and other tools. Two key slots, shown in Fig. 10-3, intersect the spindle taper. One key slot is provided for the insertion of a draw key, which draws the tools firmly into the boring bar and prevents their accidental release. The other key slot at the end is provided for the insertion of a cross key, which drives the tools through their tang. Drift pins or drift keys are also placed in the end key slots in order to drive the tools loose from the taper. The drift pins are struck sharply with the hammer to release the tools. The spindle can be moved in and out of the headstock manually or by power feed, or it can be clamped in place. Long boring bars, called *line* boring bars, must be

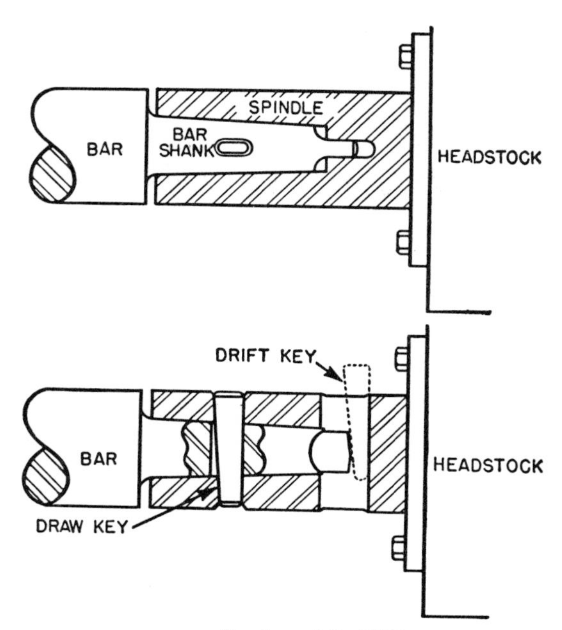

Courtesy of the Giddings & Lewis Machine Tool Company

Fig. 10-3. Method of holding boring bars and other tools in a horizontal boring machine spindle.

supported in their outer end in a bushing which is held on the end support column. A line boring bar supported in an end support bushing is shown in Fig. 10-2. Here holes are being bored in a large weldment.

The table of the planer-type horizontal boring machine can move in only one direction: perpendicular to the axis of the spindle. The headstock column is mounted on a slide upon which the entire column and headstock can be set in motion parallel to the spindle axis. When cutting, however, the feed parallel to the axis of the spindle is made by feeding the spindle. The end support column is mounted on a slide on the side of the table opposite the headstock column. The planer-type horizontal mill is designed to provide exceptional rigidity in machining long heavy workpieces.

Floor-type horizontal boring machines are designed to machine workpieces so large and heavy that it would not be practical to handle them on a table. The headstock column is mounted on a slide which provides a feeding movement perpendicular to the axis of the spindle. The spindle feed provides the second feeding direction, while the vertical feed of the headstock is the same as with all horizontal boring machines. Heavy floor plates upon which the workpieces are clamped are embedded in a fixed position in the foundation of the machine.

Setting Up the Workpiece

An important step, and the first step, in horizontal boring machine work is to set up the machine correctly. It must be set up so that the surfaces to be machined will clean up and will have the correct dimensional relationship to the other surfaces on the workpiece. The setup, of course, should be planned in advance.

The nature of the surfaces on the workpiece that are available for aligning it on the table of the horizontal boring machine table should

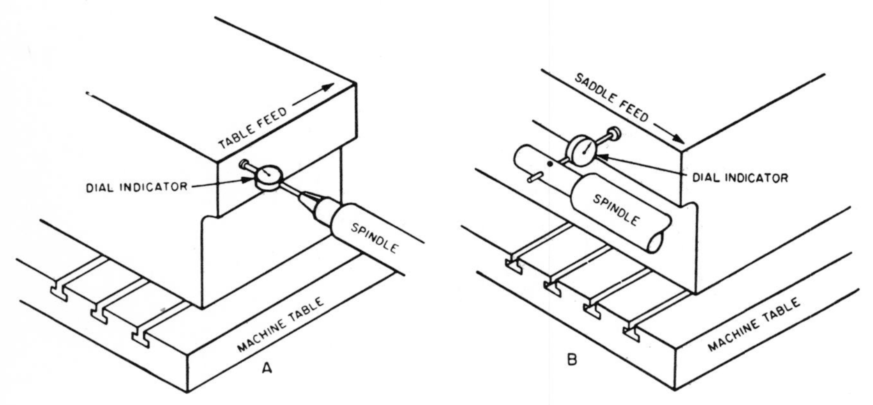

Courtesy of the Giddings & Lewis Machine Tool Company

Fig. 10-4. A. Checking the alignment perpendicular to the axis of the spindle with a dial test indicator. B. Checking the alignment parallel to the axis of the spindle with a dial test indicator.

be understood. The best surfaces are always those that have been machined in a previous operation, because machined surfaces form accurate seating surfaces for placing the workpiece on the table or on parallel bars. Precise measuring tools such as precision squares, protractors, and dial indicators can be used against a machined surface to align the part. For example, a dial test indicator can be used against a machined surface to align it parallel or perpendicular with respect to the spindle, as shown in Fig. 10-4. The workpiece can be aligned by the finished surfaces of a bored hole by indicating lengthwise along the surfaces of the bore. The spindle can be aligned on the axis of the bore by indicating around the circumference of the bore as shown in Fig. 10-5. A finished surface can be placed against a slot block, as in Fig. 10-6, in order to align it parallel to the T-slots of the table.

Unfinished surfaces that are rough cannot be used in the same manner to align the workpiece as finished surfaces; yet, it is frequently necessary to use rough surfaces. Of course, if there is a choice, machined surfaces are always preferred. The tools used to measure and align rough surfaces should, if possible, average out the errors on these surfaces. For example, the bent point of a surface gage can be used to test an unmachined surface for parallelism with the tabletop. The surface gage is placed on the table or on a parallel bar, and the bent point is moved around over the top of

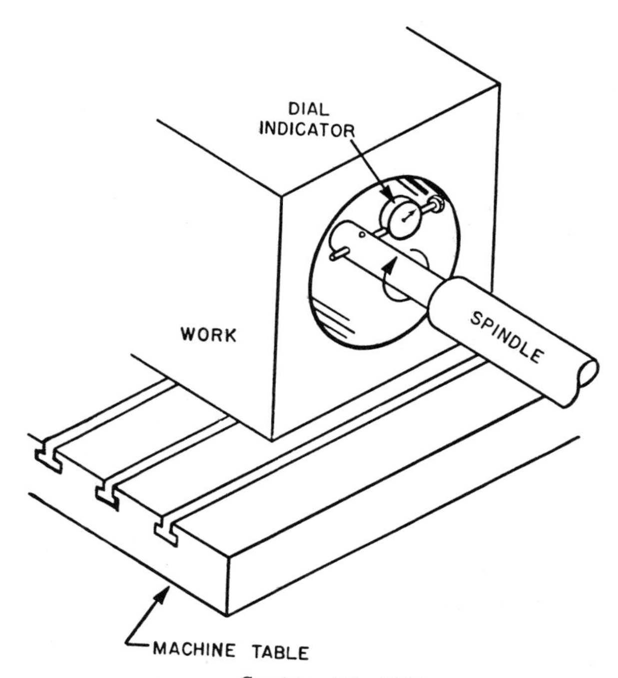

Courtesy of the Giddings & Lewis Machine Tool Company

Fig. 10-5. Aligning the spindle with a finish machined bore using a dial test indicator.

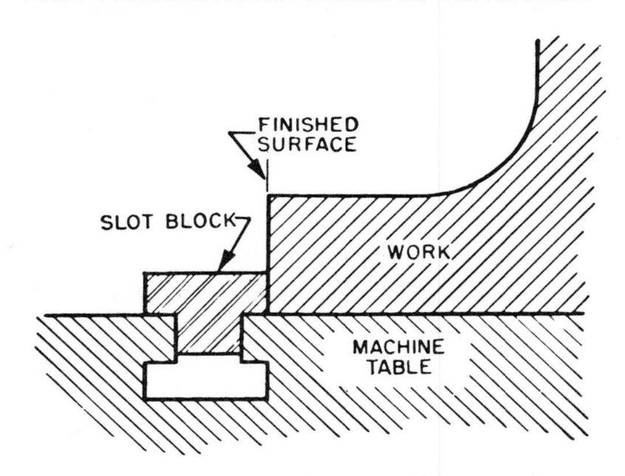

Courtesy of the Giddings & Lewis Machine Tool Company

Fig. 10-6. Using a slot block placed in T-slot to align workpiece with the T-slot.

the unmachined surface of the workpiece. High and low spots can be detected by feel and the workpiece positioned until it is as level as possible. This could not be done using a dial test indicator over the rough surface because the unevenness would cause the hand to jiggle so much that no readings would be certain. A good combination square or an equivalent tool should be used on rough surfaces, such as castings, instead of a precision machinist's square. In most instances it is best to make a layout on rough workpieces before they are brought to the horizontal boring machine for machining.

The most difficult setup to make on any machine tool, including a horizontal boring machine, is the setup for the first machining operation of a casting, forging, or weldment on which no previous machining operations have been performed, and hence with all the surfaces rough. In such cases the layout greatly simplifies the procedure of setting the part up. Figures 10-7 and 10-8 illustrate the method for setting up a rough casting from layout lines. These layout lines have identified the center of the casting and the center of the bore. The casting is to be set up to machine the raised surfaces, or pads, on the base with a face milling cutter. Placed on its side as shown in Fig. 10-7, the casting is first positioned to lie parallel to the table feed direction with the pads overhanging the side of the table enough to allow the cutter to clear the table when milling the pads. The T-slots of the table, which are accurately machined parallel to the direction of the table feed, can be used as reference surfaces. In this case a measurement is made with a rule from the edge of the T-slot to the vertical layout or setup line (view A, Fig. 10-7). This measurement is made at each end of the workpiece and when it is equal at each end, the workpiece is aligned with the direction of the table travel.

The next step in making this setup is to level the casting. This procedure is illustrated at Fig. 10-7. The distance of the horizontal layout line from the top of the table is checked at a number of positions by bringing the

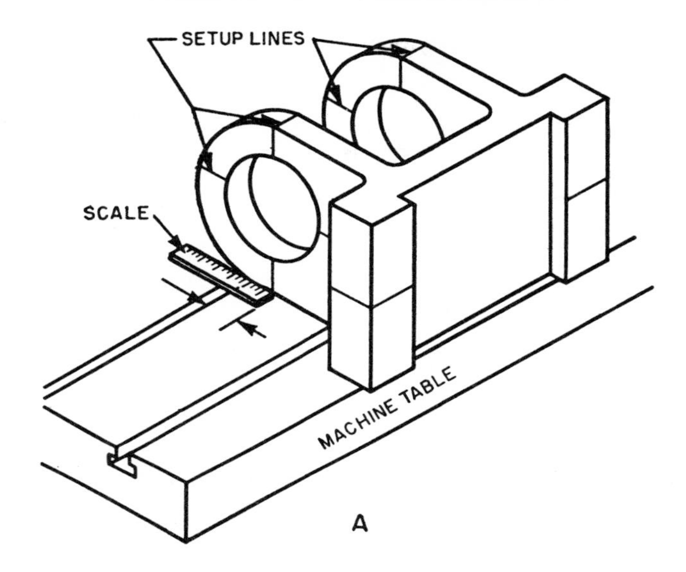

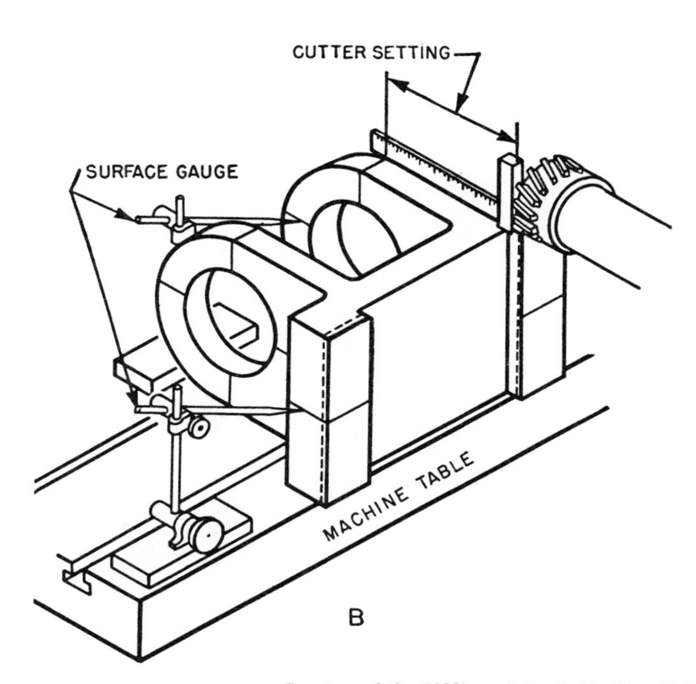

Courtesy of the Giddings & Lewis Machine Tool Company

Fig. 10-7. Using setup lines to align a workpiece upon which no previous operations have been performed.

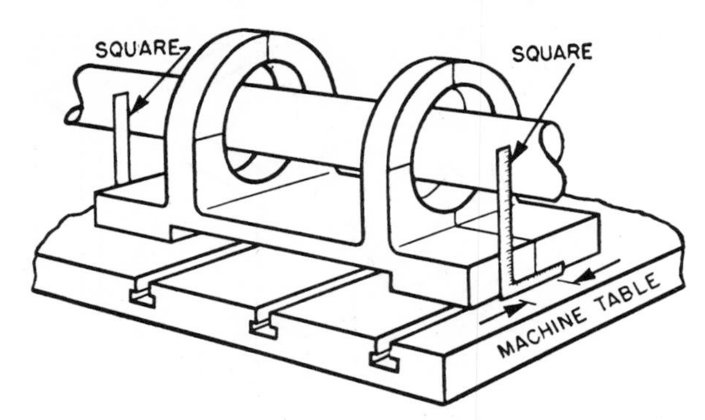

Courtesy of the Giddings & Lewis Machine Tool Company

Fig. 10-8. Using setup lines to align workpiece parallel to a line-boring bar.

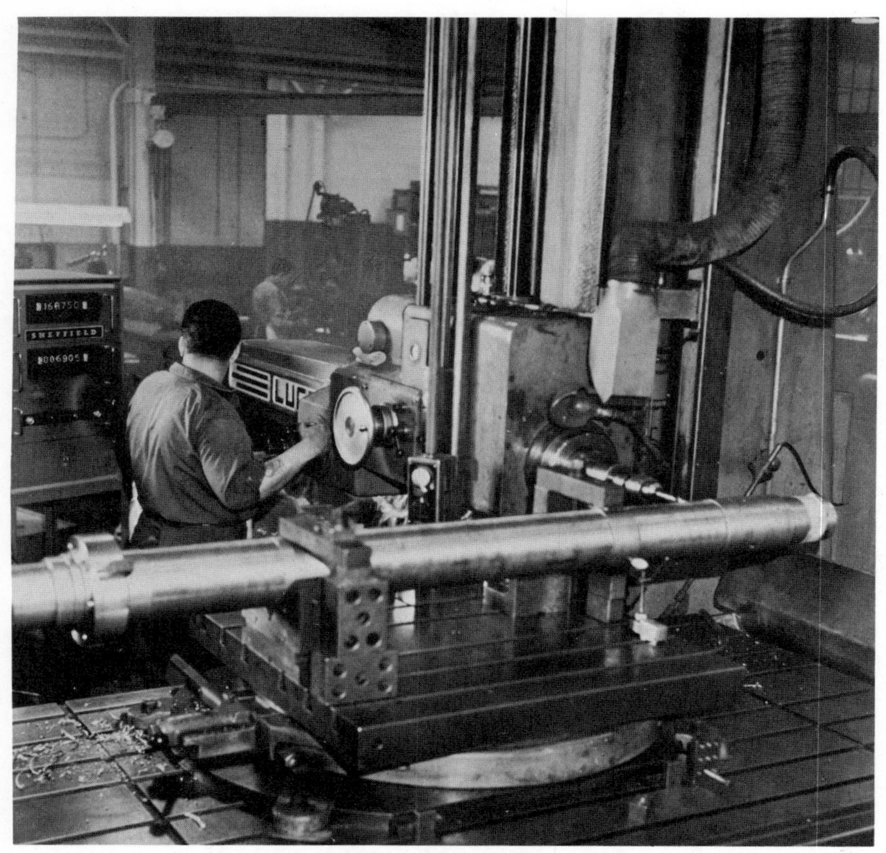

Courtesy of The Bendix Corporation—Automation & Measurement Division

Fig. 10-9. Machining a large shaft on a horizontal boring machine equipped with a Sheffield Cordex Digital Readout System.

Courtesy of The G. A. Gray Company

Fig. 10-10. Face milling on a large planer-type horizontal boring machine.

scriber point of the surface gage up to the layout line. If necessary, shims are used to lift the casting until it is level. The casting is level when the scriber point can touch the layout line in any position without disturbing the adjustment of the surface gage. All of the settings should be rechecked, and then the casting is clamped in place. As a precautionary measure, a final check of the alignment should be made after the casting has been clamped in position. The face milling cutter is set in position to take the cut by measuring from the layout line to the face of the cutter with a combination square, as shown at B, Fig. 10-7.

After the pads have been face milled, the clamps are removed from the casting and the table is cleaned up in preparation for the second operation. This will be line boring of the two bores with a line-boring bar. The line-boring bar is held in the bushing in the support column—but not in the spindle—so that the workpiece can be placed on the table without interference from the boring bar. The workpiece is placed on the table with the previously machined pads against the surface of the table, as shown in Fig. 10-8. The line-boring bar is then pulled through the cored holes in the casting and fastened in the spindle of the horizontal boring machine. The casting is aligned parallel with the line boring bar by placing a square against the side of the boring bar and measuring the distance from the side of the square to the vertical layout line at the end of the workpiece, as seen in Fig. 10-8. This must be done at each end of the workpiece, and the square must be placed on the same side of the boring bar at each end. When the measurement from the square to the layout line is the same at each end, the workpiece is parallel to the boring bar and can be clamped in place in this position.

The table is then moved until the boring bar is in the center of the cored hole. This is checked by placing a square against the side of the boring bar and measuring the distance from the square to the layout line as before; however, in this case the square is placed against the two opposite sides of the boring bar and the measurement need only be made at one end of the workpiece. After the boring bar has been positioned laterally in the center of the cored holes of the casting, it is positioned so that its axis will coincide with the axis of the cored hole by raising or lowering it as required. The vertical position of the boring bar is obtained in some instances by measuring from the top and bottom of the boring bar to the top and bottom sides of the cored hole, respectively, or by measuring to layout lines made adjacent to the cored hole to define its position when it is machined. More frequently, the vertical position of the boring bar is determined by finding the dimension from the finished surfaces of the pads to the axis of the finished holes. In this case the position

Courtesy of The New Britain Machine Company—Lucas Machine Division

Fig. 10-11. Line boring a blower housing using a boring head mounted on the boring bar.

Courtesy of The New Britain Machine Company—Lucas Machine Division

Fig. 10-12. A stub boring operation using a boring head.

of the boring bar is obtained by a measurement taken from the top of the table to the boring bar. This can be done by adding one-half of the diameter of the boring bar to the required measurement and measuring over the top of the bar with the aid of a dial test indicator. With the line-boring bar in position the two holes can now be bored to size.

Horizontal Boring Machine Work

The horizontal boring machine is adaptable to machining surfaces and holes on many different sizes and shapes of workpieces. As already stated, these workpieces are often too large to be machined on most other machine tools. A few of the many types of jobs done on the horizontal boring machine are shown in Figs. 10-9 through 10-15.

The horizontal boring machine in Fig. 10-9 is equipped with a positional readout measuring system that displays the exact position of the table and the spindle relative to a reference position. The table and spindle positions are displayed by the digits that appear on the cabinet behind the operator. The workpiece, which is a large shaft, is placed on a circular table that can be rotated 360 degrees to enable machining operations to be performed on both sides of the shaft. One end of the shaft is clamped on two parallel blocks which are of such size that the axis of the shaft is level. An angle plate at this end of the shaft helps to position it so that its axis is parallel to the T-slots of the circular table. This setup allows the side of the shaft viewed by the camera in this illustration to be free from obstructions along a large portion of its length so that the machining operations can be performed on these surfaces without interference. In the drilling operation of Fig. 10-9 the drill is fed into the shaft by the spindle feed. A dial test indicator, which is mounted on a magnetic base, is positioned against the side of the shaft as shown in order to make certain that the shaft is not moved by the drill thrust or other cutting forces. The dial test indicator is not needed when the circular table has been rotated around for the other side to be machined since the angle plate then prevents the shaft from moving.

A large planer-type horizontal boring machine in Fig. 10-10 is shown performing a face milling operation. The workpieces are clamped to a large angle plate which is mounted onto the table of the machine. The large diameter of the spindle permits a heavy milling cut to be taken even though the spindle is extended so that the cutter can reach the workpiece. The spindle rotates inside the large quill extended from the headstock.

A large rotary blower housing is shown being bored in Fig. 10-11. The line-boring bar is placed in the spindle and supported at the opposite

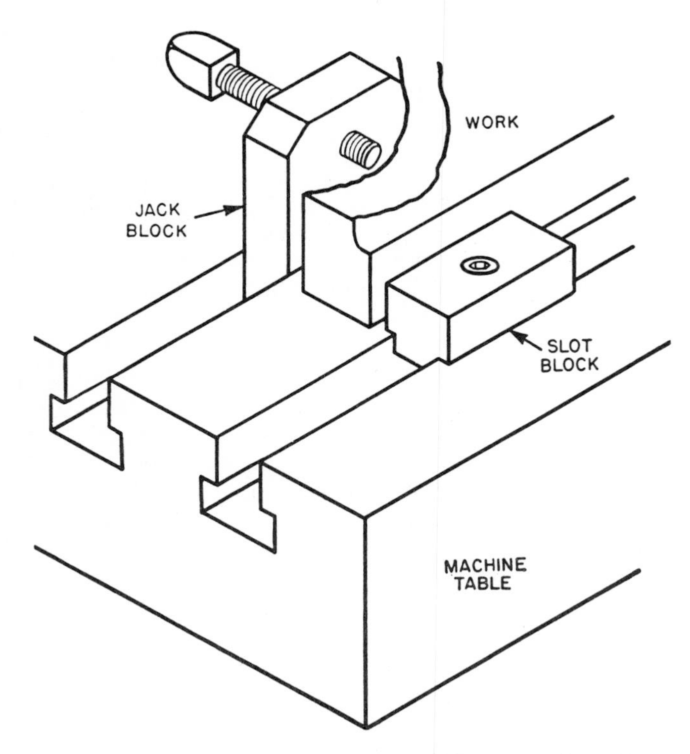

Courtesy of the Giddings & Lewis Machine Tool Company

Fig. 10-13. Using a jack block to block a workpiece.

end by the end of the support column. A boring head is clamped to the boring bar upon which one or two single-point boring tools can be fastened. The job actually consists of boring approximately half of the two adjacent holes. Two rotors with conjugate surfaces are mounted side by side in the finished bore. The blower housing is clamped in place with strap clamps. Hardwood blocks are used as heel blocks; and bracing jacks, or screw jacks, are placed inside extension tubes below the clamps. On jobs like this the table feed should be used instead of the spindle or bar feed. Long boring bars such as the one shown deflect downward somewhat because of their own weight. If the lengthwise position of these bars is changed significantly, the amount of deflection will change and result in an error in the position of the bore. If the length of the bore is large, the table feed should also be used because the table can feed a greater distance than the spindle.

Holes shorter in length are bored with stub-boring bars, as shown in Fig. 10-12. The boring tool may be fed through the hole by using either the table feed or the spindle feed. Usually the table feed is preferred, as the

Courtesy of the Giddings & Lewis Machine Tool Company

Fig. 10-14. The continuous feed facing head can be used to feed the cutting tool toward the center of the part while it is rotating. Simultaneously a boring cut is taken using the spindle feed of the machine.

Courtesy of the Giddings & Lewis Machine Tool Company

Fig. 10-15. Using the continuous facing head to turn an outside diameter while simultaneously boring.

deflection of the boring bar will not vary by being extended farther at the end of the cut than at the start of the cut. The workpiece is clamped in place with a large strap clamp on top and jack blocks in front. The method of using jack blocks is shown in Fig. 10-13.

A continuous feed facing head is shown in Figs. 10-14 and 10-15. A single-point cutting tool mounted in the head can be rotated and at the same time continuously fed toward the center of rotation in order to take a facing cut. It may also be fed continuously in an outward direction while rotating, which is done to cut a face inside a large bore. In Fig. 10-14 a boring cut is taken by feeding the offset boring tool with the spindle feed while simultaneously taking the facing cut with the continuous feed facing head. This head can also be used to take a turning cut, as shown in Fig. 10-15. The radial infeed of the facing head is used to adjust the single-point tool to the proper depth of cut. A boring cut and a turning cut are being taken simultaneously in Fig. 10-15 by using the saddle feed.

One of the most useful horizontal boring mill attachments is the large rotary table, such as shown in Figs. 10-9, 10-11, and 10-16. Many different surfaces around the workpiece can be machined in a single setup on the rotary table by indexing the table to bring these surfaces opposite the machine spindle; holes running in different directions can be machined in a like manner. As an example, the elevator housing casting shown in Fig. 10-16 has surfaces and holes on four sides that are machined in a single setup. Obviously, both time and effort are saved in this way by reducing the number of setups required.

Two or more holes that are not in close proximity must sometimes be machined to produce a common axis. When these holes are large enough to allow passage of a long boring bar, they can be machined most accurately and conveniently by line boring, illustrated in Fig. 10-16. The boring bar passes through all of the holes to be machined and is supported at each end; additional intermediate supports are sometimes provided. Single-point boring tools are clamped in the boring bar at different lengthwise positions for boring the holes to size. This method is recommended and should be used whenever possible.

There are some jobs, however, where this method cannot be used; because the holes are frequently not large enough or are so far apart that a single line boring bar of sufficient size and rigidity cannot be used. When this occurs other methods must be employed, and one of these is illustrated in Fig. 10-17. The workpiece is clamped to a rotary table and is aligned so that the common axis of the two holes to be machined passes through the center of the rotary table; i.e., the workpiece is set up in such a way

Fig. 10-16. Machining all surfaces on four sides of the casting by indexing it with a rotary table. Operation shown is line boring the bearing holes.

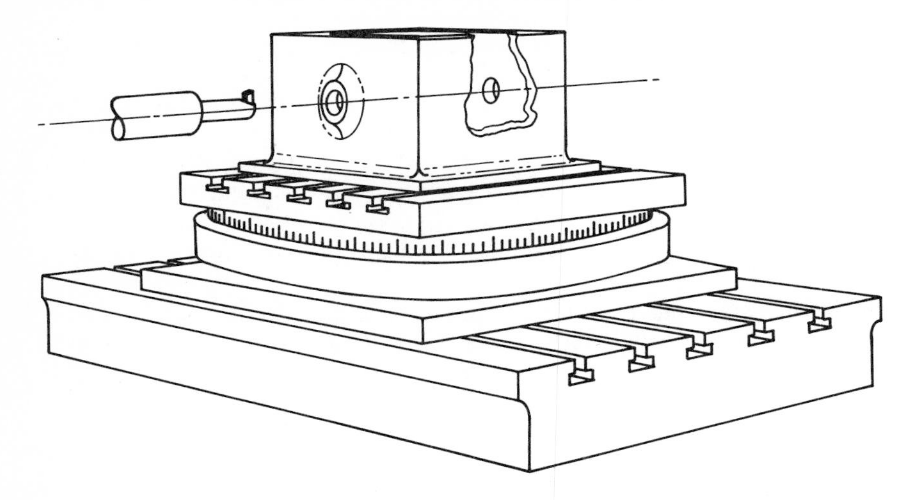

Fig. 10-17. Using a rotary table and a stub boring bar to bore two holes that are a large distance apart and have a common axis. Rotary table is indexed 180° to bore the second hole.

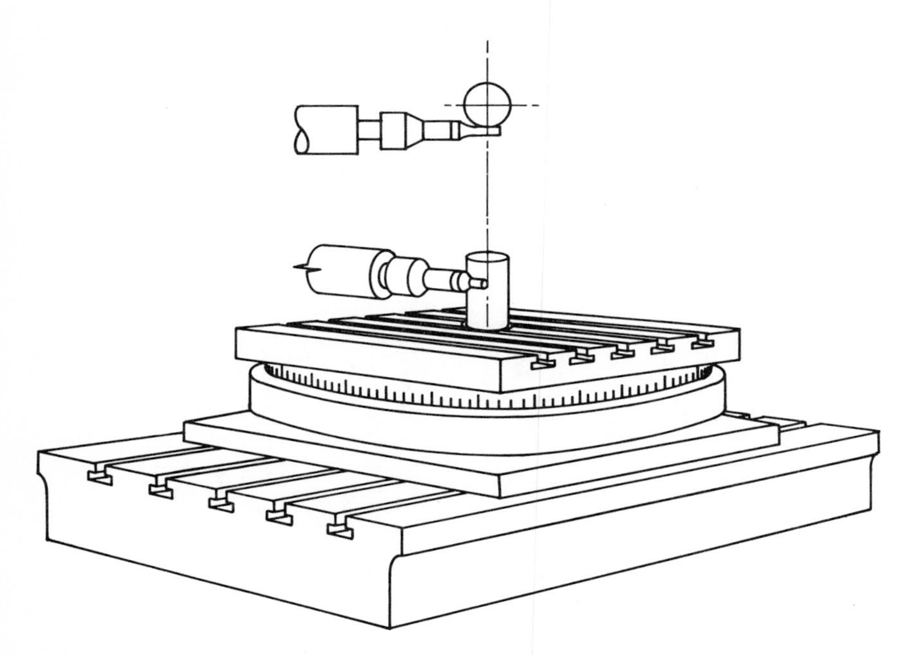

Fig. 10-18. Method of aligning axes of machine spindle and rotary table by use of edge finder.

that the axes of the two holes and that of the rotary table intersect. The table of the horizontal boring machine must then be positioned so that the axis of the machine spindle and that of the rotary table also intersect, and the spindle is raised to the required height for machining the holes. Each hole is then separately machined; one hole is first machined to size and the rotary table is then indexed 180 degrees for the second hole to be machined to size. Boring operations in these holes are performed by a stub boring bar, as shown in Fig. 10-17. Holes parallel to the first set can be machined by moving the machine table to an offset position. If the holes are on opposite sides and must, like the first pair of holes, have a common axis, the machine table offset will have to be made in opposite directions when the rotary table is indexed 180 degrees.

For smaller workpieces this method can also be used on a milling machine, or on a jig borer by mounting the rotary table on an angle plate.

An essential requirement of this method of machining holes having a common axis is to position the machine table so that the axis of rotation of the rotary table will intersect the axis of the machine spindle. There are several methods by which this can be done, one of which is demonstrated in Fig. 10-18. A centering pin is placed in the centering hole of the rotary table and an edge finder is chucked in the machine spindle. With the spindle rotating and the contact cylinder offset so that it wobbles, the machine table is moved bringing the contact cylinder against the centering pin; when the contact cylinder ceases to wobble, the machine table movement is stopped. From this position the table is moved a distance equal to one-half of the sum of the diameters of the contact cylinder and the centering pin, which will result in the required alignment.

CHAPTER **11**

Grinding Wheels

The grinding wheel is a cutting tool which utilizes a multitude of abrasive particles as cutting edges. These abrasive particles are held together in the general shape of a disc by a bonding material to form the grinding wheel. On the surface of the wheel abrasive particles form many very small chips which in their sum total can represent a significant rate of metal removal. A representative group of grinding wheels used in the machine shop is shown in Fig. 11-1.

Grinding wheels are used for many purposes, which include snag grinding of castings in foundries and billets of steel in steel mills, hand

Courtesy of Cincinnati Milacron

Fig. 11-1. Variety of typical grinding wheel shapes and sizes.

grinding of welds, and serving as cut-off wheels to make saw cuts in metal parts. In the machine shop, however, grinding wheels are used primarily on machines. These machines can be precision grinding machines which are used to grind tools and machine parts to close dimensional tolerances, or they can be pedestal grinders which are used to hand-grind single-point cutting tools. One advantage of the grinding process is that workpieces that are too hard to machine by other methods can be ground to a required shape and an accurate size. Precision grinding machines, in addition to producing close dimensional tolerances on soft or hardened metals, also produce an excellent finish on the surface of the work.

Abrasives

For several thousand years abrasives have been used to shape materials and to produce keen cutting edges. Sandstone is a natural abrasive which is used to this day. For many centuries sandstone grinding wheels were used to sharpen cutlery, tools, agricultural implements, and weapons. Emery and corundum are also natural minerals, which are harder and have a better abrasive action than sandstone. Corundum and emery both consist of variable size crystals of aluminum oxide. Emery contains a considerable amount of iron oxide and other impurities. Corundum also contains impurities. The first man-made grinding wheels were made from these materials during the middle of the nineteenth century. The term *emery wheel* is still synonymous with grinding wheel, although grinding wheels made from natural abrasives have been replaced by artificial abrasive wheels because the natural abrasives have nonuniform properties.

All modern abrasives used in making grinding wheels are man-made, except for natural industrial diamonds. The man-made abrasives are: aluminum oxide, silicon carbide, synthetic diamonds, and cubic boron nitride; the latter is available under the trade name, Borazon™. An important property of an abrasive is its hardness, or its ability of resistance to penetration. The Knoop hardness values of abrasives used in grinding, honing, and lapping are given below, with the hardness value of hardened tool steel, the last item, included for comparison:

Diamond	7000
Cubic Boron Nitride	4700
Silicon Carbide	2480
Aluminum Oxide	2050
Hardened Tool Steel (60 HRC)	740

On the majority of grinding wheels, aluminum oxide or silicon carbide is used as the abrasive. Aluminum oxide abrasives are recommended for grinding high tensile strength material. Both hardened and unhardened plain carbon steels, alloy steels, and tool steels should be ground with aluminum oxide grinding wheels. Other materials that are ground with aluminum oxide wheels are: ferritic and pearlitic malleable iron, certain titanium alloys, high-temperature alloys, and martensitic and ferritic stainless steels. Several different grades of aluminum oxide are manufactured: regular, white, modified, monocrystalline, microcrystalline, and sintered. Each grade has certain unique properties making it more suitable for certain applications. For specific recommendations, the grinding wheel manufacturers should be consulted. In addition to these grades, alumina-zirconia grades are available; they are used primarily for heavy duty grinding operations, such as snag grinding in foundries and steel mills.

Silicon carbide abrasives are harder than aluminum oxide and the abrasive grains are jagged and sharp. Vitrified silicon carbide grinding wheels are cooler cutting than comparable aluminum oxide wheels. In general, however, silicon carbide wheels should not be used for grinding steel. As an exception, however, vitrified bonded black silicon carbide wheels having a grain size of 240, or finer, can be recommended for finish grinding the cutting edges on high-speed-steel cutting tools. The cool cutting action of these wheels reduces the tendency to damage the cutting edges, and the sharp abrasive particles produce a keen cutting edge. Black silicon carbide wheels are recommended for grinding gray cast iron, austenitic stainless steel, copper alloys, certain high-strength aluminum alloys, certain titanium alloys, many refractory materials, and rubber, stone, and marble. Another form of silicon carbide, green silicon carbide, is used to grind cemented carbides.

The hardest known substance is the diamond, whether it is natural or synthetic. It can penetrate all other materials. In precision grinding, diamonds are used for two purposes: 1. as an abrasive on grinding wheels, and for honing and lapping; 2. to true and dress grinding wheels. Cemented carbide, ceramic, and cermet cutting-tool materials should always be ground with a diamond grinding wheel; this is the principal field of application of these wheels. They are not recommended for grinding steel.

Cubic boron nitride, or CBN, is used to grind hardened plain carbon and alloy steels and certain high-temperature alloys. It is especially recommended for grinding all tool and die steels having a hardness of 50 R_c, or harder. High-speed-steel cutting tools, including those made from the difficult-to-grind types of high-speed steel, are ground with great efficiency using CBN grinding wheels on cutter and tool grinding machines. CBN grinding wheels are very cool-cutting and hold their size and shape exceptionally well. This allows heavier cuts to be taken without damage to the cutting edges. An amount equal to the infeed of the wheel will usually be removed from each cutting edge that is ground, resulting in a more precise and uniform geometry on the finished tool. CBN wheels used for tool grinding should have a somewhat finer grain size and they should be mounted accurately on the grinding spindle. It is most important to properly condition the face of CBN grinding wheels by careful truing and dressing before they are used. CBN is not recommended for grinding cemented carbides and non-metallic materials.

The Cutting Action of Abrasives

The individual abrasive particles form very small chips which can vary in shape and size over a wide range. In rough snag grinding operations the chips can be relatively thick and wide, whereas in precision grinding operations they are usually thin and narrow. The speed of the abrasive particle when it is used on a grinding wheel is usually very great, being approximately 60 to 70 miles per hour for vitrified wheels. This speed creates a very high temperature at the surface being ground. On the other hand, honing and lapping are abrasive operations which are conducted at low speeds and result in relatively low cutting temperatures. The advantage of the high speed of the grinding wheel is that the great number of chips formed in a short time result in a relatively high metal cutting rate. The advantage of honing and lapping is their cool cutting action.

The individual abrasive particle may act in different ways on the surface of the workpiece. The abrasive grain may simply plow through the surface of the work as shown in the upper view of Fig. 11-2. The metal is pushed aside, and some of it breaks off the work as a highly distorted particle. The abrasive particle can also produce a chip in a manner similar to the formation of a chip with a single-point tool. This is shown in the

Courtesy of Cincinnati Milacron

Fig. 11-2. Top: Plowing action of abrasive grain. Bottom: Chip formation action of abrasive grain.

lower view in Fig. 11-2. The shape of the abrasive grains and the amount of interference between the work surface and the grain largely determine if cutting or plowing takes place. Some of the abrasive grains will be in such a position that no more than rubbing can take place, and no metal is removed.

The chips produced by a grinding wheel are very hot. The temperature at the grinding contact surface is not known exactly, but it may be as high as 2,000 to 3,000 degrees Fahrenheit. When the chip leaves the workpiece it is surrounded by the atmosphere where it reacts with the oxygen to produce the characteristic sparks associated with grinding. In other words, the chips are ignited by the oxygen in the air and burn or oxidize. The grinding fluid, if used, will quench many of the sparks. The heat generated by the grinding action also increases the temperature of the surface of the metal significantly to a depth which can be as great as several thousandths of an inch. Stresses will be developed in the surface which are called residual stresses. The residual stresses in ground surfaces are tensile stresses and act like a rubber band stretched around the workpiece. When these stresses are large enough, they can cause the formation of cracks on the surface. They also reduce the strength of the workpiece when it is subjected to repetitive loads. Residual stresses can cause relatively thin parts to warp and to bend.

The rapid heating and cooling of the surface of the workpiece can also cause it to be heat-treated—as though it were heated in a furnace and quenched. This causes a change in the structure of the metal on the surface and changes its properties. For example, the surface of hardened tool steel can be softened by this action; or it can be softened and immediately rehardened. A third type of injury to ground surfaces may be *burn*. Burn is recognizable by the visible discoloration caused by the extremely thin oxide film formed by the momentary exposure of the surface to the high temperature. The discoloration itself is not objectionable, except for its appearance, because the thickness of the oxidized layer is only a few millionths of an inch. The discoloration, however, is a visible indication that the surface below the oxide film has reached a high temperature and is probably damaged.

Grinding wheel wear occurs for several reasons. One of the prime causes is the attrition of the abrasive particles as a result of chemical reactions. Chemical reactions between the abrasive and the work material can occur very rapidly at the high temperature produced while they are in intimate contact as the abrasive is forming a chip or plowing through the work material surface. For example, at this high temperature, silicon carbide will react rapidly with the iron in steel to form iron carbide and certain silicon compounds, which is the reason silicon carbide grinding wheels dull so rapidly when used on steel. Since aluminum oxide is more resistant to such chemical attack by the iron in steel, aluminum oxide grinding wheels are generally recommended instead for grinding steel. At the high temperature of grinding the abrasive may also react chemically with the atmosphere, the coolants, or even the bond material that holds the wheel together. Attrition resulting from chemical reactions causes the abrasive particles to diminish somewhat in size; the sharp edges and sharp corners become rounded and thus the abrasive particles get dull. When this occurs extensively over the face of the wheel, the wheel is said to be *glazed.*

Another cause of grinding-wheel wear is the impact force occurring when the abrasive particles strike the surface of the work. In precision grinding on machine tools these forces are usually, but not altogether, small enough so that they have very little effect on the abrasive particle. In some circumstances, however, the impact forces can cause some of the abrasive particles to fracture and others to be torn from the surface of the wheel. While this results in wheel wear, it also exposes new, sharp edges and new, sharp abrasive particles on the wheel face. As an abrasive particle becomes more dull, the magnitude of the grinding forces will increase, thereby increasing their tendency to fracture or to tear out the dull abrasive particles from the wheel. The "ideal grinding wheel" was thought then to be self-sharpening by this action. However, this usually occurs only to a very limited extent and grinding wheels must normally be sharpened from time to time. When this self-sharpening action does occur in precision grinding, the wheel will be too soft and it will not hold to size on the part. In this event, the remedy would be to use a harder grinding wheel or to change the grinding conditions so that it will act harder. Self-sharpening can occur to a greater extent when off-hand grinding on a pedestal grinder or with a portable snag grinder.

Loading is the name of the condition when chips of work material are trapped in the pores of the grinding wheel or when work material adheres to the abrasive particles on the face. While this does not cause grinding wheel wear in itself, it does require wheel sharpening because a loaded wheel will not grind freely, if at all. Loading can occur to some extent when grinding most materials; it is, however, much more prevalent when grinding soft materials, such as soft steel, brass, and aluminum. When grinding oxidation-resistant materials such as stainless steel, the grinding chips do not readily burn in the atmosphere and some of them will weld to the surface of the workpiece. In order to prevent loading and rewelding, chemical grinding additives may be incorporated into the grinding wheel bond or be added afterwards, in which case they fill the pores of the wheel.

Grain Size

The size of the grains or particles of abrasives is expressed by the size of the screen opening through which they are sifted. Such screens have a certain number of openings per linear inch. For example, a 46-mesh screen has 46 openings per lineal inch. Grains which pass through this screen and do not pass through the next finer screen are classified as 46-grain size or 46-grit size. The size of these grains is approximately 1/46 inch. The commercial grit sizes are given in Table 11-1. The most commonly used grit sizes for cylindrical and surface grinding range from 36 to 100. For

Table 11-1. Commercial Abrasive Grit Sizes

Coarse Range		Medium Range		Fine Range	
6	14	30	70	120	280
8	16	36	80	150	320
10	20	46	90	180	400
12	24	54	100	220	500
		60		240	600

tool and cutter grinding this range is 36 to 120, although sometimes grit sizes as small as 320 are used to obtain a fine finish on a cutting edge.

The Bonding Materials

The types of bonds used in making grinding wheels are called vitrified, resinoid, rubber, shellac, silicate, and oxychloride. The function of the bond is to hold the abrasive particles in place and to allow them to break away after they have become dull. A brief description of the various bonds follows:

Vitrified. Bonds made from vitrified clay are called vitrified bonds. About 75 percent of the grinding wheels manufactured are vitrified grinding wheels. The bond material consists of feldspar and clays. After measured amounts of bond and abrasive have been mixed together, they are pressed to shape and fired in a kiln. The vitrified wheels have high strength and good porosity which exposes many abrasive particles and provides a chip space. The rigidity and strength of the vitrified bond helps in attaining precise dimensions on the parts being ground. The vitrified bond is unaffected by water, acid, oils, and ordinary temperature variations.

Resinoid. Resinoid-bonded wheels are made from a mixture of abrasive particles, synthetic resins, and a plasticisor. Molded either hot or cold and then baked in an electric oven, resinoid-bonded wheels are very strong and can withstand rough usage. They are, therefore, used extensively in foundries, steel mills, and in welding shops. Some large diameter cut-off wheels are resinoid bonded. In the machine shop resinoid-bonded diamond wheels are extensively used to grind cemented-carbide and ceramic cutting tools. They are also used in the machine shop for grinding threads and for other operations requiring an accurate wheel shape.

Rubber. Rubber-bonded wheels are used for snag grinding wheels in foundries, welding shops, and steel mills because of their strength. Cut-off wheels are also made with a rubber bond. In the machine shop the principal use for rubber-bonded wheels is to make regulating wheels for centerless grinding machines.

Shellac. Shellac grinding wheels are cool cutting and produce an excellent surface finish on steel. They are used to grind crankshafts, camshafts, paper mill rolls, etc. They are also used to grind the cutting edges on knives because they will not overheat the workpiece when properly used. Shellac-bonded wheels are recommended for grinding hardened tool steels and thin sections that are apt to overheat.

Silicate. Silicate-bonded wheels are made from sodium silicate, commonly known as "water glass." This bond releases the abrasive grains readily which gives the wheel a mild and cool cutting action. They are used in shops where rather large slow-speed wheels are used to grind fine cutting edges on cutlery and on edged tools.

Oxychloride. The oxychloride bond is made from a mixture of magnesium oxide and magnesium chloride. It is considered to be a very cool cutting bond and is sometimes used on large disc grinding wheels. Because most grinding fluids attack this bond, the wheel is always used dry.

Metal. Although metal is not normally considered as a bonding material, it is used extensively as a bond for diamond wheels. Diamond wheels are commercially made with three types of bonds: metal, resinoid, and vitrified. Metal-bonded diamond wheels are usually made in two parts. The center or body of the wheel is made from plastic, steel, or bronze, and the portion of the wheel holding the diamond particles is usually made from bronze.

Grade

The grade of a grinding wheel refers to the degree of strength with which the bond holds the abrasive particle in place. Actually the strength of the bonding material does not vary; it is usually quite uniform. The amount of the bonding material surrounding each abrasive particle and filling the space between the particles can be varied to alter the firmness with which the abrasive particle is held in place. The bond forms a series of "posts" between each abrasive particle which holds the particles in place as shown in Fig. 11-3. If less bonding material is used, the bonding posts will be weak and the grinding wheel will be designated as a soft wheel. Increasing the amount of bonding material strengthens the bonding posts so that the abrasive will be held more firmly in place, and the wheel

Courtesy of The Carborundum Company

Fig. 11-3. "Posts" of bonding material hold the grains of a grinding wheel in place and determine its hardness or grade. From left to right are weak posts, medium posts, and strong posts.

will be designated as a medium wheel. If still more bonding material is used, the bonding posts will be further strengthened and the wheel will be classified as a hard wheel. The grade of grinding wheels is designated by capital letters, starting with A as the softest wheel and ending with Z as the hardest wheel. The majority of the wheels used for precision grinding on machine tools will have a grade ranging between F and N. Rough-grinding and snag-grinding wheels usually range in grade from M to Z.

Structure

The structure of a grinding wheel refers to the spacing of the abrasive particles. An open structure, as in Fig. 11-4, provides ample space between adjacent abrasive particles for the chip formed in grinding to clear itself from the wheel. An open spacing also provides fewer abrasive particles for each unit area (say each square inch) on the face of the wheel. Since there are fewer abrasive particles, each particle in contact with the work will exert a greater pressure on the workpiece, thus penetrating more deeply and removing a larger chip. Open wheels will tend to cut freely. However, because of there being fewer particles, such wheels will not produce as smooth a surface finish on the workpiece. A denser spacing will provide more abrasive particles per unit area of wheel face. Then, because the load between the work and the wheel will be carried by more abrasive particles, the pressure on each particle will be less and it will not penetrate as deeply into the surface of the work. Thus, the chip formed will be smaller. In order for the abrasives in the denser structure to penetrate as deeply as in the more open structure, a greater force must be exerted between the work and the wheel—which is usually undesirable since it will generate more heat and increase the temperature of the ground surface. The denser structure has an advantage in that it generally produces an improved surface finish on the workpiece. The structure of a grinding wheel is designated from 1 to 15, 1 being a dense structure and 15 being an open structure. The basis of the selection of a structure is to find the best compromise between the surface finish produced and the free cutting ability of the wheel. In some cases only one structure, which is determined by the grinding wheel manufacturer, is available for certain grinding wheels, while in other cases several structures may be available from which a choice can be made.

Grinding Wheel Marking

Aluminum-oxide and silicon-carbide grinding wheels are marked in a somewhat uniform manner by all grinding wheel manufacturers. Although there are some differences in the markings, all manufacturers follow the basic pattern of identification so that a person who understands this pattern can read and interpret the marking on any grinding wheel. It should be emphasized, however, that two different manufacturers' grinding wheels may have the same designation but may not necessarily perform in the same manner.

Courtesy of The Carborundum Company

Fig. 11-4. The structure of the grinding wheel defines the spacing of the abrasive particles. From left to right are dense spacing, medium spacing, and open spacing.

The basic pattern for marking aluminum-oxide and silicon-carbide grinding wheels is given below:

Prefix	1 Abrasive Type	2 Grain Size	3 Grade	4 Structure	5 Bond Type	6 Manufacturer's Record
51 —	A —	36 —	L —	5 —	V —	23

1. *Abrasive Letters.* The letter (A) is used for aluminum oxide and (C) for silicon carbide. The manufacturer may designate some particular type in either of these broad classes by using his own symbol as a prefix (Example, 51A).
2. *Grain Size.* The grain sizes commonly used and varying from coarse to fine are indicated by the following numbers: 10, 12, 14, 16, 20, 24, 30, 36, 46, 54, 60, 70, 80, 90, 100, 120, 150, 180, 220. The following additional sizes are used occasionally: 240, 280, 320, 400, 500, 600. The wheel manufacturer may add to the regular grain number an additional symbol to indicate a special grain combination.
3. *Grade.* Grades are indicated by the letters of the alphabet from A to Z in all bonds or processes. Wheel grades from A to Z range from soft to hard.
4. *Structure.* The use of a structure symbol is optional. The structure is indicated by Nos. 1 to 15 (or higher if necessary) with progressively higher numbers indicating less density and a wider grain spacing or a "more open" structure.
5. *Bond or Process.* Bonds are indicated by the following letters: V, vitrified; S, silicate; E, shellac or elastic; R, rubber; RF, rubber reinforced; B, resinoid (synthetic resins); BF, resinoid reinforced; O, oxychloride.
6. *Manufacturer's Record.* The sixth position may be used for the manufacturer's private factory records; this is optional.

Diamond and Cubic Boron Nitrate Wheel Marking

A group of typical diamond-grinding wheels is shown in Fig. 11-5. A typical American National Standard identification symbol for designating a diamond wheel is illustrated in Fig. 11-6. A brief explanation of the symbols follows. For more detailed data, see *Machinery's Handbook.*

Basic Core Shape. This portion of the symbol indicates the basic shape of the core on which the diamond abrasive section is mounted. The shape is actually designated by a number preceded by the letter D.

Diamond Cross-Section Shape. This, the second component consisting of one or two letters, denotes the cross-sectional shape of the diamond abrasive section.

Diamond Section Location. The third component of the symbol consists of a number which gives the location of the diamond section, i.e., periphery, side, corner, etc.

Modification. The fourth component of the symbol is a letter designating some modification such as drilled and counterbored holes for mounting or special relieving of a diamond section or core. This modification position of the symbol is used only when required.

Some diamond wheel manufacturers use their own marking instead of

Courtesy of the Norton Company

Fig. 11-5. Typical diamond grinding wheel shapes.

the American Standard marking. For example, the Norton Company uses its own system, which is explained below:

Abrasive	Grit Size	Grade	Concen-tration	Bond	Bond Modification	Depth of Diamond Section	Manufacturer's Identification Symbol (Omitted)
D	100—	N	75	B	56	1/8	

Abrasive. D designates a natural diamond. SD designates a manufactured diamond. CB designates cubic boron nitride, or CBN.

Grit Size. The grit size appears in the second position. The following grit sizes are available: 36, 46, 68, 80, 90, 100, 105, 120, 150, 180, 200, 240, 320, 400, 500, 600.

Grade. The available grades ranging from soft to hard are H, J, L, N, P, and R. These grades should not be compared to those shown on the wheel markings of other bonded abrasive products, since the diamond wheel grade is considerably harder. For each wheel type and bond type there is a standard or preferable grade.

Concentration. The concentration of the diamond is shown in the fourth position. In Norton diamond products the concentrations from lowest to highest are given by the numbels 25, 50, 75, and 100.

Bond. Bonds are indicated by the following letters: B, resinoid; M, metal; V, vitrified.

Bond Modification. The bond modification is shown in the sixth position. In the case of the example, the B56 bond is for wet or dry grinding of tungsten carbide.

Depth of Diamond Section. The seventh position indicates the depth of the diamond section of the wheel. Standard depths available are 1/16″, 1/8″, and 1/4″.

Manufacturer's Identification. The use of this symbol is optional.

Standard Grinding Wheel Shapes

The United States Department of Commerce and Grinding Wheel Institute in cooperation with the builders of grinding machines have established nine standard grinding wheel shapes. The dimensions in which these wheels are available have also been standardized. Most of the grinding wheels used are standard wheels, but there are a large number of special

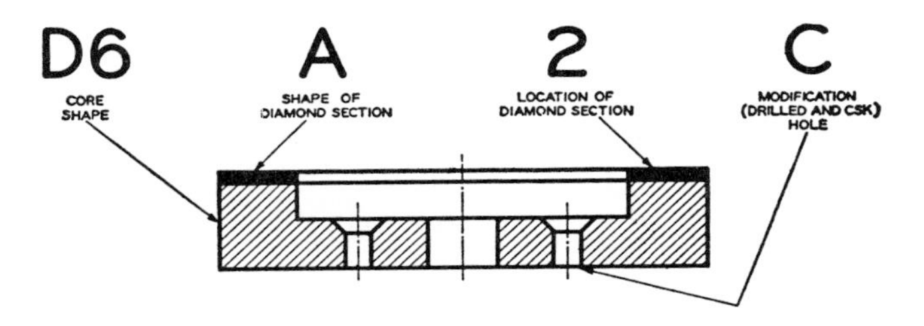

Fig. 11-6. Typical American National Standard diamond wheel designation symbol.

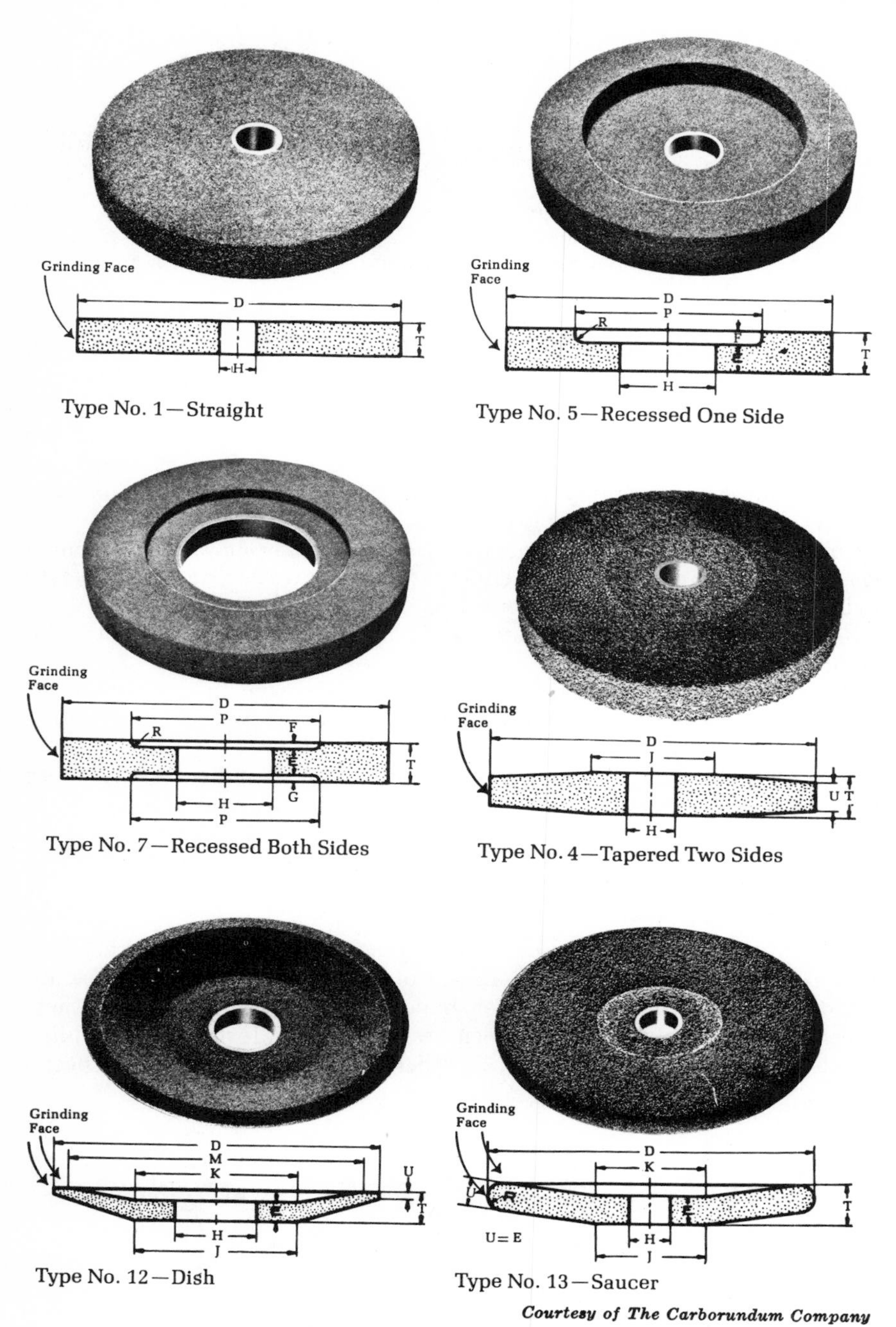

Courtesy of The Carborundum Company

Fig. 11-7. Standard grinding wheel shapes.

wheel shapes that are used less frequently for the less common grinding operations.

The nine standard grinding wheel shapes are shown in Figs. 11-7 and 11-8. The following letters designate the dimensions of the grinding wheels in the illustrations.

D—Diameter (overall)
E—Center or back thickness
F—Depth of recess (See types 5 and 7)

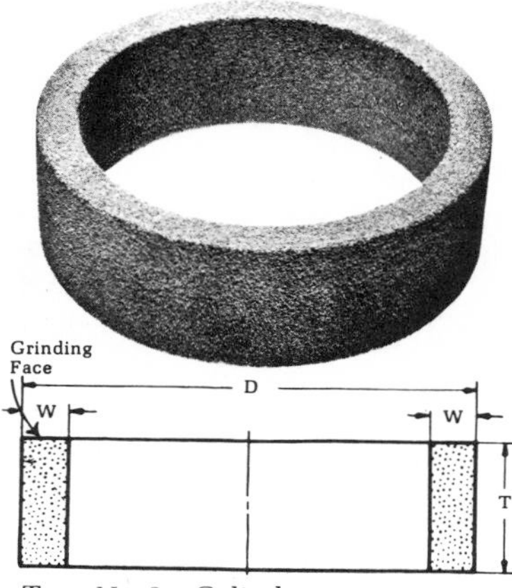

Type No. 2—Cylinder

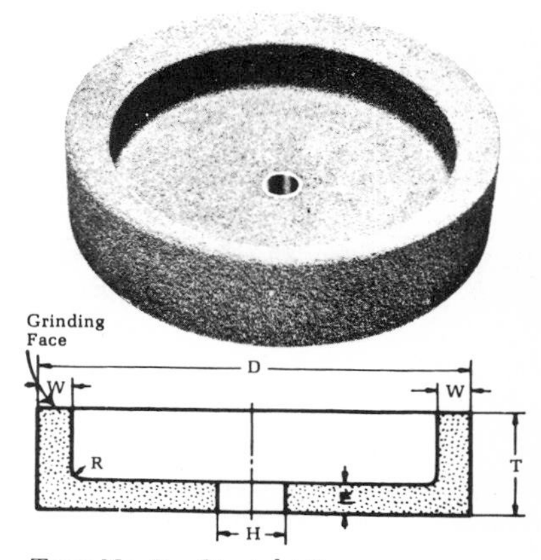

Type No. 6—Straight Cup

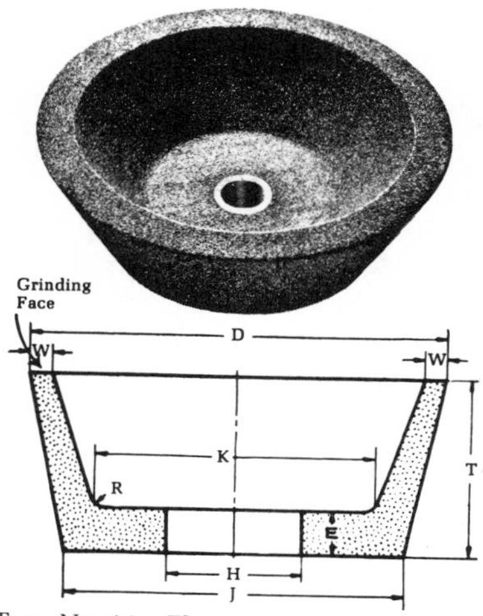

Type No. 11—Flaring Cup

Courtesy of The Carborundum Company

Fig. 11-8. Standard grinding wheel shapes.

G—Depth of recess (See type 7)
H—Hole diameter
I—Diameter of outside flat
K—Diameter of inside flat
M—Large diameter of bevel
P—Diameter of recess
R—Radiused corner
T—Thickness (overall)
U—Width of edge
W—Wall thickness at grinding face

Wheel types Nos. 1, 5, and 7 are used for cylindrical grinding, surface grinding, off-hand grinding, and snag grinding. Thin cut-off wheels ranging in thickness from .005 to ⅛ inch are also specified as No. 1 wheels. The No. 7 and No. 5 wheels are provided with recesses to give clearance for mounting flanges. This group of wheels is classified as *Straight Wheel Types*.

Type No. 4 is classified as *Tapered Wheel Type*. Tapered wheels are generally used for snag-grinding operations. Wheel Type No. 2 is classified as *Cylinder Wheel Type,* which is used on either horizontal or vertical spindle surface grinding machines. Either the peripheral surface or the face of the wheel is used as the grinding surface. A *Straight Cup Wheel Type* is designated as Type No. 6, for use on horizontal and vertical spindle surface grinding machines. It is also used for off-hand grinding utilizing the flat surface on the wheel face. The wheel face may be either plain or beveled. Type No. 13 is a *Saucer Wheel* or a *Saw Gummer* which is used primarily for resharpening saws. Wheel Type No. 11, designated as a *Flaring Cup Type Wheel,* and Wheel Type No. 12, designated as a *Dish Wheel Type,* are used for tool and cutter grinding. The Type No. 11 wheel, which may have a plain or beveled face, is also used in conjunction with a resinoid bond as a snag-grinding wheel. The thinness of the Type No. 12 wheel permits the grinding edge of the grinding wheel to fit into narrow places.

In addition to these grinding wheel shapes there are a large number of shapes available on mounted wheels. A mounted wheel is a very small diameter grinding wheel permanently fastened to its own spindle or mandrel. Mounted wheels are used for small hole precision internal grinding, such as jig grinding, and for machine grinding mold cavities. They are also used for off-hand grinding to finish mold cavities, to remove tool marks in molds, for deburring, and to sharpen tools. Mounted wheels can be obtained having any of the four basic types of abrasives.

Mounting Grinding Wheels

Before any grinding wheel is mounted on the spindle of a grinding machine it should be carefully checked to be sure that it is not fractured. The wheel should be carefully examined visually for cracks. Vitrified wheels, where size permits, may be suspended by slipping one finger through the hole and tapping with a light instrument, such as the handle of a screwdriver, which should produce a clear metallic ring. If a clear ring is not produced it may be evidence of a crack in the wheel; however, oil- or water-soaked wheels do not ring clearly. Often grinding wheel manufacturers impregnate grinding wheels with various resins and greases which deaden the tone of the wheel to modify the cutting action. Larger wheels may be suspended and tapped with a wooden mallet. The bushing in the center of the wheel should be checked for evidence of looseness. It should not extend beyond the side of the wheel and it should slip over the grinding-machine spindle or the wheel-mounting arbor without binding. If the wheel does bind it should be carefully hand-scraped or reamed to provide an easy sliding fit.

The correct and incorrect method of mounting grinding wheels with small holes directly onto the spindle of the grinding machine is shown in Fig. 11-9. The inner flange should be keyed or otherwise fastened to the wheel spindle, and the face of this flange must run true without any endwise eccentricity. A blotter is placed between the inner flange and the wheel, and the wheel is pushed snugly against the flange. Blotters are sometimes glued to the face of new grinding wheels or are provided separately. The blotting paper provides an even bearing surface on the face of the wheel which prevents damage to the surfaces on the wheel clamped by the flanges. The blotters help to drive the grinding wheel by increasing the friction between the flanges and the wheel. The outer flange is then placed against the wheel with a blotter between the wheel and this flange. The outer flange should have an easy sliding fit over the spindle so that it will provide a uniform bearing against the wheel and the blotter. Finally, the nut is placed on the spindle and tightened enough to hold the wheel firmly in place, i.e., tight enough to prevent it from slipping and to transfer the driving torque. It should not be too tight as it would then set up excessive strains in the wheel.

Grinding wheels with a large-diameter hole are mounted on an arbor as shown in Fig. 11-10. The method of mounting the wheel on the arbor is similar to mounting small hole wheels directly on the spindle. The wheel is usually, though not always, mounted on the arbor first. The outer flange of larger-diameter wheels is held in place by means of a

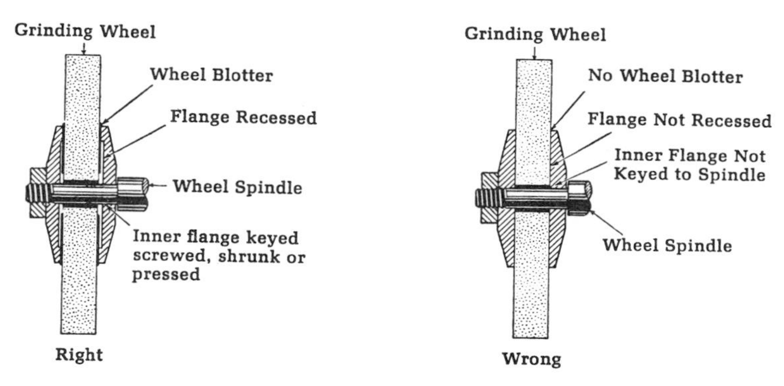

Courtesy of The Carborundum Company

Fig. 11-9. Correct and incorrect methods of mounting grinding wheels with small holes.

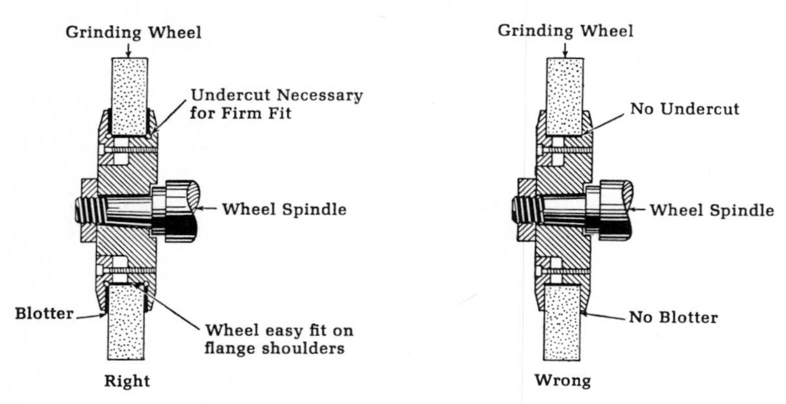

Courtesy of The Carborundum Company

Fig. 11-10. Correct and incorrect methods of mounting grinding wheels with large holes.

series of bolts. First the bolts should be made finger tight. Then after one bolt is tightened slightly with a wrench, the diametrically opposed bolt is tightened slightly. All of the bolts are tightened in this manner until they are tight enough to hold the grinding wheel firmly in place. The arbors have a tapered hole which fits onto a taper on the grinding-machine spindle. Both tapers must be inspected to be sure that they are clean and free of nicks. The arbor and the wheel are then placed on the spindle and tightened firmly against the spindle by means of the nut.

The grinding wheels must be balanced in order to prevent the strains caused by unequal centrifugal forces that can occur if the wheel is out of balance. Most new wheels are reasonably concentric with the arbor hole and inherently in balance. Experience may show that smaller wheels do not require balancing. Small wheels can be balanced by determining the heavy side before the wheel is mounted. The heavy side is then placed in the top or highest position when the wheel is on the spindle. The tolerance in the hole then causes the wheel to contact the side of the hole which is on the heavy side of the wheel. This positions the heavy side slightly closer to the axis of the spindle, which tends to balance the wheel.

Large grinding wheels must be balanced more precisely. The wheel is mounted on the arbor, and the arbor and the wheel together are mounted on a balancing spindle. The balancing spindle is then placed on special parallel ways or balancing discs. The heavy side of the wheel tends to come to rest in the lower position. The flanges of the large arbors will have two, three, or four adjustable balance weights which are adjusted until the wheel and the arbor are in true balance. The Cincinnati cylindrical grinding machines have an automatic balancing mechanism which is built into the machine. The wheel is balanced in a few seconds while rotating on the spindle of the machine by the mere turn of a handle. Large diameter wheels should be checked for balance from time to time because they can become unbalanced when their diameters decrease as the result of wear. Care should be exercised in rebalancing wheels that have been used with a grinding fluid to make certain that the grinding fluid retained by the wheel has not settled in one portion of the wheel to cause an unbalance. This condition is temporary and can lead to an unbalanced condition of the wheel after it has been eliminated. The moisture in the wheel can be eliminated by rotating it for a few minutes dry or without using the grinding fluid. Before a machine that uses a grinding fluid is shut down, it is good practice to stop the flow of coolant and allow the wheel to run at full speed for several minutes. The centrifugal force will eliminate most of the moisture in the wheel, thus preventing the formation of heavy unbalanced regions due to moisture settling when the wheel is standing still for a prolonged period of time.

As a safety precaution, the operator should stand to one side when first starting up a new wheel or one that has been remounted and allow it to run at full operating speed for at least one minute. If the wheel is damaged in any way it is most likely to fail during this period. It is also advisable to stand aside when the newly mounted wheel makes contact with the workpiece for the first time.

Dressing and Truing Grinding Wheels

Dressing is the operation performed on the face of the grinding wheel which is intended to sharpen the abrasives or to change the nature of its grinding action. Truing is the operation that is performed to create concentricity or parallelism, or to alter the shape of the grinding wheel.

There are several different kinds of dressing tools available for dressing a grinding wheel. Star dressers have pointed discs that are loosely mounted on a pin. The pin is held in place by a frame which has a handle on one end. The dresser is held against the rotating grinding wheel which causes the pointed discs to spin. The discs remove some of the abrasive from the wheel and tend to pick metal out of the wheel which has become loaded on the face. This type of dresser is used primarily to dress coarse-grained wheels on pedestal grinders and snag grinders. It is sometimes used to dress segmental surface grinding wheels. Abrasive sticks are used for dressing and truing smaller grinding wheels and thin grinding wheels. They are especially useful in forming a profile and in dressing grinding wheels used on cutter- and tool-grinding machines.

Abrasive wheels mounted on a holder with precision antifriction bearings are used to dress grinding wheels mounted on cylindrical or surface grinding machines. The grinding wheels which act as the dressing wheel are made from silicon carbide. These dressers will impart a smooth, clean cutting face on the surface of the wheel which leaves no dressing marks on the work. The dressing wheel is set at a slight angle to the axis of the grinding wheel and is driven by contact with the wheel. The dressing wheel is traversed across the grinding wheel face; however, at least two-thirds of the dressing wheel should always remain in contact with the grinding-wheel face.

The crush-form dressing method, for dressing profiles on the face of grinding wheels, is particularly useful when doing repetitive work. The crush-form rolls are power driven and made from hardened high-speed steel, cemented carbide, or boron carbide. The rolls are 4 to 6 inches in diameter and are made to the profile to be produced on the workpiece. Driven at a speed of 150 to 300 feet per minute, the roll is brought in contact with the face of the grinding wheel. The grinding wheel is rotated by the crush rolls, and the wheel is fed into the rolls gradually until the desired profile has been produced on the wheel.

Perhaps the most common method of dressing and truing grinding wheels on precision grinding machines is to use an industrial quality diamond. The diamonds are set in a variety of holders which can be firmly held in the grinding machine. In the diamond dressing operation shown in Fig. 11-11, the diamond is held at an angle of 10 to 15 degrees in the direction of the wheel rotation. It may be held on center or 1/8 to 1/4 inch below center. In the crossfeed direction the diamond may be positioned perpendicular to the face of the grinding wheel, or 10 to 15 degrees with respect to the direction of the crossfeed, as shown in Fig. 11-11. When the diamond is positioned so that it is perpendicular to the face of the wheel it is traversed back and forth across the face of the grinding wheel using a slow traverse feed. The amount of infeed of diamond into the wheel per pass across the face should not exceed .001 inch, and the infeed should be decreased to less than .001 inch per pass for the last few finish dressing passes across the wheel face. When a very fine finish must be ground on the workpiece, the diamond should be passed across the face of the work several times using no infeed and a slow traverse. A fast traverse of the diamond past the wheel will give a fast cutting wheel but one that may produce diamond marks on the surface of the work. The diamond can be seriously damaged by overheating, and care must be exercised to prevent this from occurring by not taking too deep a cut. Grinding wheels can be diamond-dressed dry; however, if a grinding fluid is available it is best to use it while dressing. On surface grinding machines the diamonds can be placed in a holder held by the magnetic chuck which is usually used to hold the work on the table.

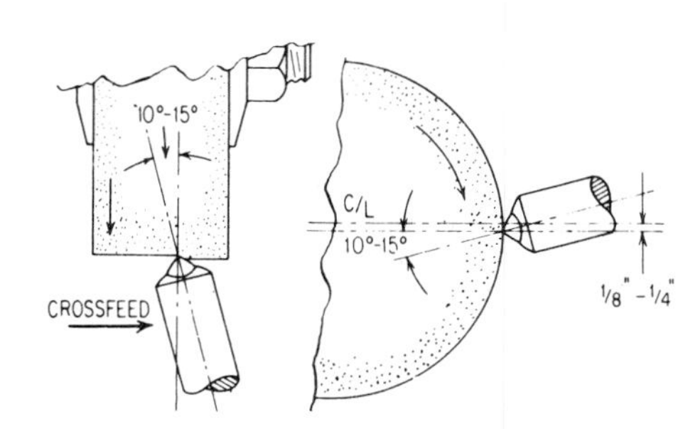

Fig. 11-11. Position of diamond for dressing and truing grinding wheel.

In addition to the single-point diamond truing and dressing tool, multiple-point dressers are available. Another type of truing and dressing tool, which is used like the single-point diamond tool, consists of a block of concentrated diamond abrasive grit mounted on the end of the holder.

Diamond and cubic boron nitride (CBN) grinding wheels require a different truing and dressing procedure. These wheels are precision made and, if correctly used do not require frequent truing or dressing. Whenever possible each wheel should be mounted on its own adaptor, which should not be removed for the life of the wheel. As a first step in setting up a new wheel, and later on when required, the wheel must be accurately mounted on the adaptor and on the machine spindle. The spindle and the flange face of the grinding machine should be indicated with a dial test indicator to make certain that it is running true. Then the adaptor is mounted and tightened on the spindle, with the wheel held only lightly by the adaptor. Using a dial test indicator to find the high spot on the wheel, lightly tap the high spot with a soft wooden block and repeat until the wheel runs true, after which it is tightened onto the adaptor. Cup wheels should be mounted to run true within .001 in. (0.025 mm), resinoid bonded peripheral wheels within .0005 in. (0.013 mm), and metal bonded wheels within .0002 in. (0.005 mm).

When metal bonded diamond and CBN wheels require additional truing in order to rotate within the prescribed limits, the wheel together with

Courtesy of the General Electric Co., Specialty Materials Dept.

Fig. 11-12. Truing a Borazon(TM) grinding wheel on a surface grinder, using a steel bonded, diamond-impregnated tool. The traverse speed should be 3 to 5 fpm and the infeed .0005 in. per pass.

the adaptor should be mounted on a special mandrel that is held between centers on a lathe, cylindrical grinder, or preferably between the centers of a cutter and tool grinder. Using a 60- to 100-grit aluminum oxide or silicon carbide grinding wheel operating at 5000 to 6000 fpm, slowly grind the face of the metal bonded wheel until it is true. The metal bonded wheel should be rotating slowly, at approximately 100 to 200 fpm. Do not rotate this wheel manually because the uneven motion may cause flat spots to be ground. Metal bonded wheels having the abrasive on the face, such as cup wheels, can also be trued by lapping on a flat glass or cast iron plate using a figure 8 motion. The lapping compound is a slurry composed of water and 120 grit silicon carbide.

Resinoid bonded diamond and CBN wheels may be trued with a metal bonded diamond stick which is held in a holder and used like a single-point diamond truing and dressing tool, as shown on a surface grinder in Fig. 11-12. The diamonds are held in a metal bond and the diamond concentration is high to make a matrix that is more abrasion resistant than the wheel. Another method is to use a brake controlled truing attachment. This attachment has a silicon carbide wheel that is slowed down by a brake while the resinoid bonded wheel, operating at full speed, is slowly fed into the silicon carbide wheel. Resinoid bonded wheels are sometimes trued by carefully grinding a soft steel block.

After the diamond or CBN wheel has been trued the grinding face is shiny and smooth. Before it can be used it must be dressed. Dressing is accomplished by holding an aluminum oxide or a silicon carbide dressing stick against the grinding face of the wheel while it is rotating. The dressing stick may be held in a holder, in which case the operation is performed very much like dressing with a single point diamond. It may also be held against the wheel by hand, as shown in Fig. 11-13, in which case the wheel can be jogged to make it run more slowly while scrubbing the face with a stick. A small amount of coolant should be used to create a slurry or paste that rolls like grains of sand between the stick and the wheel to abrade the bonding material from between the abrasive grains. The wheel will feel smooth before it is dressed; after properly being dressed it will have a rough texture. As the wheel becomes sharp there will be a rapid wear of the dressing stick, which is a good indicator that the wheel is being opened. Diamond and CBN wheels should not require frequent dressing if used correctly; when dressing is required, the method described should be used.

Grinding Wheel Speeds

The operating speed of a grinding wheel depends upon many variables, among which are: shape and size of the wheel; bond from which it is made; abrasive used on the wheel; material to be ground; type of grinding operation; type of coolant used, if any; and spindle speed range for which the machine is designed. No hard and fast rule for selecting the spindle speed can be given except one: In all cases the maximum speed specified by the grinding wheel manufacturer must not be exceeded. This speed is

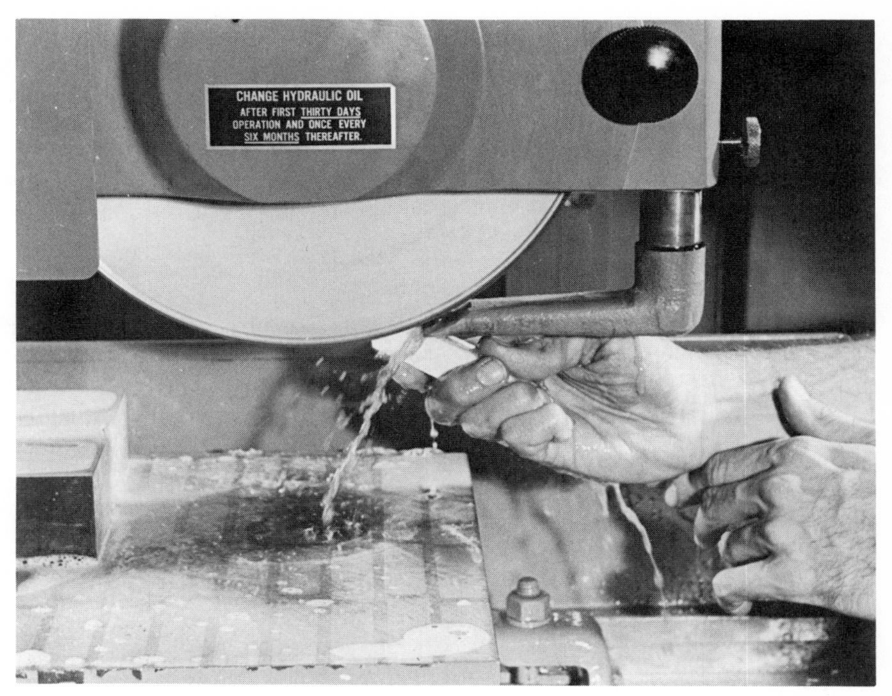

Courtesy of the General Electric Co., Specialty Materials Dept.

Fig. 11-13. Dressing a Borazon (TM) grinding wheel by hand, using a 220 grit, or finer (G hardness), aluminum oxide dressing stick.

usually marked on the wheel. A speed that is below the maximum allowable speed is often used. Normally, a speed that is too slow is not used because the wheel will act softer and will wear away more rapidly. In certain cases, however, a very slow grinding wheel speed is used in order to improve the surface integrity of the ground surface; i.e., in order to reduce the damage done to the surface by grinding. At the lower speed the grinding wheel will generally cut cooler and more freely. Very high speed grinding, sometimes called abrasive machining, is done, but only on special machines and with special wheels designed for this purpose.

A few representative examples of grinding wheel speeds are given below, which, however, may have to be qualified by the variables listed above.

Type of Wheel	fpm	m/min
Vitrified bonded	4000-6500	1220-1980
Metal bonded	5000-12,000	1525-3660
Resinoid bonded		
Al_2O_3 or SiC abrasive	5000-9500	1525-2900
Diamond or CBN abrasive (dry)	3000-5000	915-1525
Diamond or CBN abrasive (wet)	3500-8500	1070-2590

Most cylindrical grinding operations using an aluminum oxide (Al_2O_3) wheel are performed at a wheel speed of about 5000 to 6000 fpm (1525 to 1830 m/min); most surface grinding operations are performed at 4000 to 6000 fpm (1220 to 1830 m/min). Cutter and tool grinding operations are also performed at 4000 to 6000 fpm when an aluminum oxide or a silicon carbide wheel is used. The grinding wheel speeds listed above should be used for cutter and tool grinding when metal or resinoid bonded diamond or cubic boron nitride (CBN) wheels are used.

Grinding Wheel Performance and Selection

Although a particular grinding wheel may perform better than others on a given job, most grinding wheel operations can be performed satisfactorily by a number of different grinding wheels. By adjusting the work speed, the traverse rate, the infeed, and the method of truing and dressing, most wheels can be used for a rather wide range of applications. It is usually not necessary to have a large number of different grinding wheels available for each machine.

The following paragraphs will describe how some of the basic grinding conditions affect the performance of the wheel. Caution should be exercised, however, in attempting to predict the change in grinding wheel performance by varying only one of the basic grinding conditions. Many of these conditions are so interrelated that a change in one brings about a change in the other.

Arc of Contact. The arc of contact, Fig. 11-14, is that portion of the circumference of the grinding wheel that is in contact with the work. It is dependent upon the type of grinding operation being performed, being greatest for internal grinding and least for cylindrical grinding. The arc of contact is also affected by the depth of the infeed, the diameter of the grinding wheel, and the diameter of the workpiece. It is increased by increasing the diameter of the grinding wheel and by increasing the depth of the wheel infeed. In cylindrical grinding the arc of contact will be increased when the diameter of the work is increased, and in internal grinding when the diameter of the hole is decreased.

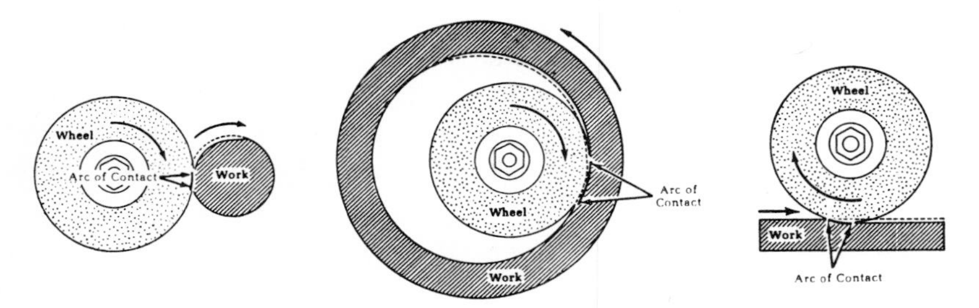

Courtesy of The Carborundum Company

Fig. 11-14. The arc of contact is the portion of the circumference of the grinding wheel that is in contact with the work. The arc of contact is shown for: left, cylindrical grinding; center, internal grinding; right, surface grinding.

The arc of contact is important because of its effect upon the area of contact. Increasing the arc of contact will increase the area of contact if other grinding conditions remain constant.

Area of Contact. The area of contact is that portion of the surface of the grinding wheel in contact with the surface of the workpiece at any one instant. It is equal to the length of the arc of contact multiplied by that portion of the width of the wheel which is in contact with the work. For cylindrical and internal grinding it is the arc of contact times the rate at which the work traverses the face of the wheel per revolution of the work. For surface grinding it is the cross-feed rate times the arc of contact.

It is necessary to force the grinding wheel and the workpiece together in order to grind. The force on the wheel will be equal to the force on the work. This force will be distributed over all of the particles of abrasive that are in contact with the workpiece at any one instant. Each abrasive grain will then carry a portion of the exerted force in order to grind material from the workpiece. This portion of the grinding force will be referred to as the grinding pressure, although technically this is not quite correct.

If there is a constant force between the grinding wheel and the workpiece, an increase in the area of contact will cause more abrasive particles to be in contact with the surface of the work and there will be a consequent decrease in grinding pressure on each particle because the force is divided by more particles. When the pressure on each particle is decreased, the abrasive particles will not penetrate into the work as deeply and each particle will grind off less metal. Since the pressure on the particle is decreased, it is less likely to be torn from the surface of the wheel. It is also less likely to be fractured; therefore the grinding wheel will tend to perform like a harder wheel. Thus, the increasing area of contact tends to make the wheel act harder if the force between the wheel and the workpiece is not changed. If, however, the magnitude of this force is increased in order to make the wheel act as before, more power will be required and more heat will be generated. This might cause the surface of the work to overheat and be damaged. Furthermore, the increased force will cause the wheel and the workpiece to deflect away from each other.

Conversely, if the area of contact is decreased, the force will be distributed over fewer abrasive particles and the pressure on each particle will increase. Because each particle will penetrate more deeply into the surface of the work, it will be more likely to be torn from the surface of the grinding wheel or to fracture, thereby creating new sharp cutting edges on the surface of the grinding wheel. Thus, decreasing the area of contact will tend to make the grinding wheel act softer.

When a large area of contact must be used, a softer grade of grinding wheel with a more open structure should be used. The open structure provides more chip space and decreases the number of abrasive particles that are in contact with the surface of the work, thereby increasing the pressure on each particle. A coarser wheel dressing action with a diamond will also help to produce a better grinding action.

Traverse Speed. The traverse movement is the movement of the work past the face of the wheel, and the traverse speed is the rate of this movement. For surface grinding the traverse speed is equivalent to the cross-feed or the distance that the wheel is fed across the surface of the work per stroke of the table. Increasing the traverse speed increases the area of contact, which tends to make the grinding wheel act somewhat harder. In practice, however, increasing the traverse speed will also have the effect of increasing the force between the work and the wheel so that the grinding pressure on each abrasive particle is increased and more heat is generated. Furthermore, increasing the traverse speed will cause more abrasive particles to cut into the work, and as a result more of the abrasive particles on the face of the grinding wheels will be subject to wear. Since worn particles will fracture or be separated from the surface, the grinding wheel will act softer. A soft grinding wheel can be made to act harder by decreasing the traverse speed, which affects the surface finish produced on the work. A slow traverse speed should be used to obtain a good finish.

Wheel Infeed. The wheel infeed is the depth of cut or the depth of penetration of the wheel into the work. Increasing the infeed results in a larger arc of contact, thereby increasing the area of contact. A greater force must be exerted to cause the grinding wheel to penetrate into the work. Such force tends to offset the effect of the larger area of contact. Since increased pressure causes each abrasive particle to penetrate more deeply into the work and cut more deeply, the effect of increasing the wheel infeed is to make the wheel act softer. The heat generated will also increase to the point where the amount of infeed used is limited; otherwise the surface of the work will be seriously damaged.

Work Speed. The work speed for all grinding operations, generally expressed in feet per minute, is usually in the range of 15 to 100 feet per minute. Some operations such as cam grinding and thread grinding are performed using a work speed as low as 2 to 6 feet per minute. Some nonferrous alloys and soft metals may be ground as fast as 200 feet per minute.

The effect of the work speed is to increase the length of the chip that is produced. Since each abrasive particle does more work, the amount of heat generated will also increase. Furthermore, each abrasive particle will tend to get dull more rapidly, and the wheel may act somewhat softer. In practice the work speed is sometimes decreased to permit a deeper cut or more infeed to be used. In general, the work speed for rough grinding and for finish grinding is the same. Because a slower work speed will usually produce a somewhat better surface finish, the speed is sometimes reduced for finish grinding except when a deep, slow-speed rough grind precedes the finish grind. If the wheel acts too soft when rough or finish grinding, reducing the work speed will tend to make it act harder.

Grinding Wheel Speed. The maximum speed of the grinding wheel is determined by the grinding wheel manufacturer and should never be exceeded. Providing that the maximum safe speed is not exceeded, increasing the speed will make the wheel act harder because each abrasive particle will be in contact with the work for a shorter length of time and the length of the cut made by the abrasive will be shorter. The generation of more heat is due to the faster cutting speed. As the grinding wheel wears and is dressed, it will become smaller in diameter, thereby reducing the surface speed of the wheel. This will cause the wheel to act softer and to generate less heat. Many grinding machines are equipped with different sets of pulleys so that the wheel speed can be varied in a stepwise manner. It is the common practice on these machines to increase the wheel speed when the diameter of the wheel is reduced to a certain limit. Also, by reducing the speed of the wheel, a harder grade grinding wheel than is required for the work can be used.

Grinding Wheel Diameter. A small-diameter wheel will always act softer than a larger wheel, even though their hardnesses are the same and they are operating at the same surface speed in terms of feet per minute. The smaller diameter wheel has a smaller arc of contact. However, it has fewer abrasive particles on the surface, and each abrasive particle cuts into the work more frequently in a given period of time. The abrasive particles will therefore get dull more rapidly and the resulting increase in the grinding force will cause them to fracture or to be separated from the surface of the wheel more rapidly. Inevitably, grinding wheels will act softer as they are used up and their diameter becomes smaller. It should also be kept in mind that large-diameter wheels should generally be softer than small-diameter wheels for grinding a given material.

Work Diameter. The work diameter affects the arc of contact, the larger diameter having the greater arc of contact. The resulting increase in the area of contact associated with the larger work diameter will tend to cause the grinding wheel to act slightly harder. Thus, theoretically the surface speed of the work should be slightly increased or the traverse speed reduced in order to decrease the area of contact. In practice, however, it is sometimes better to decrease the surface speed of the work slightly and to maintain a reasonably fast traverse rate when larger diameter workpieces are ground. Furthermore, the grinding wheel used on such workpieces should have a more open structure so that more chip space is provided and each abrasive particle can penetrate more deeply into the work.

The Material to Be Ground. The material to be ground determines the type of abrasive to be used on the grinding wheel. Silicon carbide is used to grind very hard and brittle materials and very soft materials that can easily be penetrated. Among the materials ground with silicon carbide are hardened tool steels, high-speed steel, aluminum, gray cast iron, brass, rubber, plastics, marble, and other stones. Aluminum oxide is used to grind materials that are not easily penetrated and are tough. This includes carbon steels, alloy steels, malleable iron, wrought iron, and tough bronzes. Hardened tool steels and high-speed steel are also ground with aluminum oxide. For general grinding operations on these materials, aluminum oxide is preferred. However, a fine cutting edge can be ground with a fine-grained silicon carbide wheel. Diamonds are used to grind cemented carbides and oxides. Cubic boron nitride is used to grind hard tool steel.

Amount of Material to Be Removed. When a large amount of material is to be ground from the surface of the workpiece, a grinding wheel with a coarse grain size and an open structure should be used. As the amount of stock to be removed is decreased, the grain size can be decreased and a less-open structure used. Most precision grinding work should be done with abrasives having a medium size grain and a medium structure even if the amount of stock to be removed is small. Cutter and tool grinding operations generally require the use of a finer grain size and a more closed structure.

Surface Finish. The finish required on the workpiece affects the selection of the grinding wheel. In general, fine-grained wheels will produce a finer surface finish. Where a very high degree of surface finish is required—such as in roll grinding—shellac-bonded wheels are used. The surface finish obtained is also affected by the procedure used in dressing the grinding wheel with the diamond. For a fine finish on the workpiece, several light cuts should be taken across the face of the wheel with the diamond. A good surface finish can be obtained with a fairly coarse-grained wheel by skillfully truing and dressing the wheel. The application of a synthetic chemically active grinding fluid or a grinding oil will also assist in improving the surface finish on the work. Water-soluble oil-type cutting fluids seem to have little effect on the finish obtained.

Condition of the Grinding Machine. The performance of the grinding wheel is affected by its condition. Heavy and rigidly constructed grinding machines that are not subject to vibrations can use softer grinding wheels. A machine subject to vibrations due to wear or light construction usually requires the use of somewhat harder grinding wheels.

Grinding Fluids. Grinding fluid is always helpful and often essential in performing precision grinding operations. Most wet grinding operations are performed by using a water-soluble oil, which acts as a coolant. The cooling action of any liquid coolant, including soluble oil, has little if any effect upon the rate of wear of the grinding wheel. It is even more important that the coolant has very little influence on the maximum temperature generated in grinding and on the total heat generated. Dry grinding does not produce larger instantaneous temperatures than wet grinding. The principal function of the coolant in grinding is to prevent the gradual rise of the temperature of the workpiece. In this respect the coolant performs a very important function in precision grinding operations.

Other grinding fluids are used to improve the action of the grinding wheel. Synthetic chemically active grinding fluids which are sold under trade names can reduce grinding forces, provide a better surface finish on the work, and increase the life of the grinding wheel. Chemically active fluids may also be good coolants, although their primary function is that of a chemical agent. Mineral-base oils that have been blended with other agents such as active sulphur or chlorine are also used as grinding fluids. Grinding oils can sometimes reduce the amount of heat generated by the grinding action, but their action as a coolant is poor. The application of a grinding oil can result in a marked improvement of the surface finish obtained on the workpiece. Because of their poor cooling ability, mineral-base oils should not be used when a high level of heat generation by the grinding action cannot be avoided. During all operations with grinding oils, care must be exercised to prevent the occurrence of an explosion. The mist caused by the action of the grinding wheel can be ignited by a grinding spark from the wheel when the conditions are just right.

The revolving grinding wheel is surrounded by a rapidly moving layer of air caused by the friction between the air and the surface of the wheel. The most rapid movement of the air occurs adjacent to the peripheral surface of the wheel. When the flow of this layer of air is disrupted by the surface of the workpiece, a region of turbulence and higher pressure is created. As this region occurs where the coolant is applied, some of the coolant is deflected away by the turbulence and pressure. This limits the quantity of coolant that can reach the interface of the wheel and the work, where the metal is actually being ground. More effective use of the coolant can be obtained by placing a piece of sheet metal a short distance above the nozzle and as close to the peripheral face of the wheel as possible. This air diverting sheet will disrupt the rapidly moving layer of air before it reaches the interface of the wheel and the work, thereby allowing more coolant to enter the region where grinding actually occurs.

Grinding Wheel Width. In some circumstances narrow face grinding wheels will act softer than wheels having a wider face, other conditions being equal. This will occur when plunge grinding with the entire face of the grinding wheel. In this case the narrow face grinding wheel will act softer because it will have a smaller area of contact than a wheel having a wider face. Plunge grinding is the method by which the grinding wheel is fed into the workpiece without any traverse of the wheel or the workpiece. Theoretically, when traverse grinding, only the leading portion of the face of the grinding wheel that is within the traverse length per revolution (or per stroke) will actively grind the workpiece. While traverse grinding does occur in this manner in many instances, there are other times when grinding takes place beyond the traverse length on the wheel face as a result of wheel wear, or small surface deflections caused by the grinding pressure. Although this should make the grinding wheel act harder as a result of the increased area of contact, any increase in hardness that might occur is negligible. It may, however, cause the heat generated by the grinding action to increase noticeably. When grinding cutting tools on a cutter and tool grinding machine, it is common practice to true and dress the grinding wheel so that it will have only a narrow face width in contact with the workpiece, thereby limiting the area of contact. This prevents any undue increase in the grinding heat and results in a cooler grinding action. It will, however, usually cause the grinding wheel to act softer than when the full width of the wheel is used.

CHAPTER **12**

Cylindrical Grinding

The term *cylindrical grinding* designates the precision grinding of true cylindrical surfaces, tapered or conical surfaces, and flat shoulders. Although generally performed on a cylindrical grinding machine, sometimes these operations can be performed on cutter and tool grinding machines or even on engine lathes. *Internal grinding* is a term that designates the grinding of internal surfaces in a hole. Internal grinding is often done on special internal grinding machines as well as on a cutter and tool grinding machine, an engine lathe, and rather frequently on a universal cylindrical grinding machine. For this reason, internal grinding will be treated in this chapter on cylindrical grinding.

Cylindrical and internal grinding operations are performed on workpieces for several reasons. Surfaces can be ground to very close tolerances with greater ease and rapidity than they can be machined with a single-point tool in a lathe. Also, precision ground surfaces usually have a better surface finish than can be produced by a single-point tool in a lathe. Finally, hardened surfaces that cannot be economically machined otherwise can be machined by grinding. Parts are frequently machined close to size in a lathe. If need be they are hardened. The final operation is to finish these parts to size in a cylindrical grinding machine.

Cylindrical Grinding Machines

There are two basic types of cylindrical grinding machines: the plain cylindrical grinding machine and the universal cylindrical grinding machine. In general the plain cylindrical grinding machine is somewhat more rigidly constructed, while the universal cylindrical grinding machine is constructed to provide greater flexibility and thereby allow a wider range of work to be done. The principal differences in construction are in the wheel spindle head slide, the headstock, and the table. These differences will be discussed in the description of the universal cylindrical grinding machine.

A universal cylindrical grinding machine is shown in Fig. 12-1. The bed of this machine contains the controls for its manual and automatic operation. The controls vary in detail on different grinding machines and will, therefore, not be examined in detail. They can be learned by studying a specific machine in the shop or by studying literature supplied by the grinding machine builder. Nevertheless, all cylindrical grinding machines do have controls mounted on the bed which perform the following specific functions:

1. Table traverse handwheel for manually traversing the table past the grinding wheel.
2. Automatic table traverse engagement for engaging the automatic table traverse past the grinding wheel.
3. Table traverse speed selector for regulating the speed of the automatic table traverse past the grinding wheel.

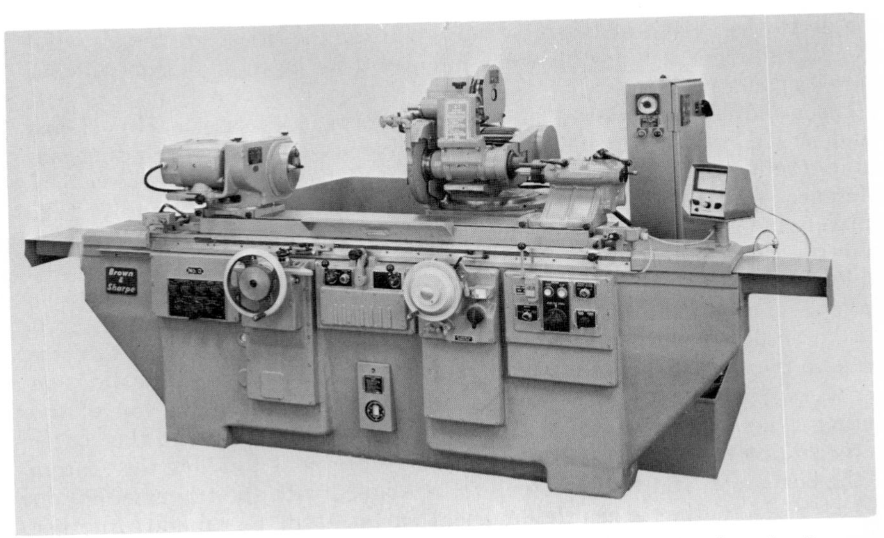

Courtesy of the Brown & Sharpe Manufacturing Company

Fig. 12-1. Universal cylindrical grinding machine.

4. Table reverse lever for changing the direction of the automatic table traverse. This lever may be engaged manually. When not engaged manually, it is moved by two dogs, or stops—one of which is mounted at each side of the table. The position of these dogs determines the length of stroke of the automatic table traverse.
5. Table dwell controls which regulate the amount of time that the table dwells or pauses at the end of each stroke before moving in the opposite direction.
6. Grinding-wheel infeed, or cross-feed, handwheel which is used to feed the grinding wheel toward or away from the workpiece manually. On some machines this control is mounted on the grinding-wheel spindle head.
7. Automatic grinding-wheel infeed, or cross-feed, engagement which engages the automatic infeed of the grinding wheel.

8. Automatic grinding-wheel infeed, or cross-feed, control which regulates the distance that the wheel feeds into the work at the end of each stroke of the table traverse.
9. Headstock speed control for regulating the speed of the headstock spindle.
10. Headstock-spindle start-stop control.
11. Grinding-wheel spindle start-stop control.
12. Grinding-machine start-stop control.

Many cylindrical grinding machines have additional controls which are unique to the design of the machine. For example, most universal cylindrical grinding machines have an internal grinding spindle for which a start-stop control is provided.

The grinding machine table slides upon ways that are machined and hand scraped on the top of the bed. On universal cylindrical grinding machines the top of the table is in a horizontal plane to provide a convenient reference surface. Plain cylindrical grinding machines usually take heavier cuts and require a more copious supply of grinding fluid. In order to drain the grinding fluid toward the wheel-spindle head, the top of the table on these machines is usually slanted toward this head. The table traverse can be manually actuated by the hand wheel on the bed, or it can be actuated automatically by a hydraulic mechanism. The table is made in two parts, as can be seen in Fig. 12-9. The sliding table slides on the ways of the bed and the swivel table is mounted on the top of the sliding table. The swivel table pivots around a pin attached to the center of the sliding table. The purpose of the swivel table is to allow the centers of the headstock and the footstock to be aligned with the traverse motion of the table so that a true cylindrical surface will be ground. Another purpose of the swivel table is to offset the centers so that a tapered surface can be ground as shown in Fig. 12-9.

The headstock of the grinding machine is mounted at the left side of the top of the table. Ways machined on the top of the table are used to align the headstock in place. The headstock can be clamped in different positions along the ways in order to accommodate different lengths of workpieces. The headstock spindle of cylindrical grinding machines is of unique construction. When centers are mounted in the taper inside the spindle, the center may be made to rotate, or the center and the spindle may be held stationary while a driving plate revolves around them. The driving plate drives the dog which is attached to the work causing it to rotate. This feature permits the workpiece to be ground while mounted on two dead centers, thus eliminating any error in the center and in the spindle bearings and resulting in a more accurately ground surface. Chucks and faceplates can be mounted on the universal grinding machine headstock, which can be swiveled at an angle (Figs. 12-12 and 12-13) in order to grind tapers, angles, and faces. On plain cylindrical grinding machines the headstock cannot be swiveled.

The footstock is mounted at the opposite end from the headstock. Used to support one end of the workpiece when grinding between centers, the footstock can be clamped in different positions along the length of the table. It is aligned to the headstock by the ways on the tabletop. The footstock spindle, which does not rotate, is pressed forward by a spring used to keep the footstock center in the center hole of the workpiece at all times. The spindle can be locked in position lengthwise. This is done when it is necessary to bring a grinding pressure against the footstock spindle, such as when shoulders are ground.

A major difference in the construction of plain and universal cylindrical grinding machines is the design of the grinding-wheel spindle head. The grinding-wheel spindle head moves along by sliding on a slide. On plain cylindrical grinding machines this slide is in a fixed position, per-

Courtesy of the Brown & Sharpe Manufacturing Company

Fig. 12-2. Grinding-wheel spindle head.

mitting the wheel spindle head to move only back and forth in a direction perpendicular to the longitudinal table movement. The wheel spindle head of a universal cylindrical grinding machine is shown in Fig. 12-2. The slide on which this head moves is not constructed in a fixed position and can be positioned at an angle with respect to the table. Thus, the wheel spindle head can be moved at an angle in order to grind steep tapers as shown in Fig. 12-10. In addition, the wheel spindle head of the universal cylindrical grinder can be rotated independently of the wheel spindle head slide and clamped in an angular position in order to grind shoulders (Figs. 12-7 and 12-8) or tapers (Fig. 12-11).

On both plain and universal cylindrical grinding machines the speed of the grinding wheel can be varied by changing the sheaves, or V-belt pulleys, on the motor and on the spindle. In Fig. 12-2 the guards are removed in order to show these sheaves. On universal cylindrical grinding machines the position of the wheel and the pulleys can be reversed. This feature facilitates the performance of shoulder grinding operations such as those shown in Figs. 12-7 and 12-8.

Universal cylindrical grinding machines are generally equipped with an internal grinding head like the one seen above the external grinding wheel in Fig. 12-2. The internal grinding head is clamped in this position when performing external grinding operations in order not to interfere with these operations. When it is to be used, it is simply pivoted down and clamped into the position shown in Figs. 12-15 and 12-16.

Cylindrical Grinding Machine Accessories

The standard accessories that are furnished with the universal cylindrical grinding machine in Fig. 12-1 are shown in Fig. 12-3. These parts are named below:

A. Universal back rests and adjustable bronze shoes (2 furnished)
B. Table-type wheel-truing fixture
C. Steady rest
D. Set of wrenches
E. Wheel sleeve and sheave pullers
F. Face plate
G. Grinding wheel
H. Footstock-type wheel-truing fixture
I. Wheel sleeve
J. Motor sheaves
K. Instruction booklet and repair parts booklet
L. Set of work-driving dogs
M. Table splash guards
N. Turret-setting bar
O. Four-jaw independent chuck

Additional attachments are available but are not regularly furnished with the machine. A partial listing of accessories is given here:

Permanent magnet rotary chuck
Three-jaw universal chuck
Collet chucks
Radius wheel-truing attachment
Angle wheel-truing attachment
Wheel-truing fixture for internal grinding wheels

Courtesy of the Brown & Sharpe Manufacturing Company

Fig. 12-3. Standard equipment for universal cylindrical grinding.

Back Rests (Steady Rests)

Back rests are also called steady rests. A pair of back rests is shown at A, Fig. 12-3, and an application of back rests is shown in Figs. 12-4 and 12-5. The purpose of the back rest is to support the workpiece by resisting its tendency to deflect away from the grinding wheel as a result of the grinding pressure. Thus, they help to prevent any inaccuracy being ground into the workpiece. By resisting the tendency of the workpiece to deflect, they serve another very useful function—helping avoid the occurrence of chatter. Back rests are especially useful when long slender shafts are ground, as shown at A, Fig. 12-4. The number of back rests to be used is a matter of judgment. For an approximation, the diameter is multiplied by a number between six and ten, and the product will be the distance between the steady rests. For example, if a workpiece is 30 inches long and 1.250 inches in diameter, the distance between the back rests for maximum support should be $6 \times 1.25 = 7.500$ inches. Thus, three back rests should be used. Two of the back rests are placed about 7.500 inches from each end, and the third is placed in the center.

Back rests can also be used to an advantage during the grinding of workpieces with relatively larger diameters which are stiff enough to resist serious deflection by the grinding wheel. When a heavy, coarse, rough-grinding cut is taken on such a part, the grinding-wheel pressure can cause the tailstock to back out unless the spindle is clamped. If the part is susceptible to heat and the tailstock spindle is clamped, the result will be that a heavy load on the centers will cause wear and that the workpiece will readily distort from too much center pressure. These conditions could actually cause the spindle to be forced slightly back, with the workpiece becoming loose on the centers during the finish grinding cuts. The application of one or two back rests can prevent this difficulty from occurring. When a large and heavy workpiece is ground, such as the roll shown at B, Fig. 12-4, the heavy weight would impose a large load on the centers. In this case the back rests shown in the illustration are used to help support the workpiece and relieve the load on the centers.

Although the design of back rests varies somewhat, they are always provided with two shoes. One shoe is called the horizontal shoe and the other shoe is called the lower shoe. In Fig. 12-3, both shoes are constructed in one piece and are shown lying flat in front of the back rest. Other types of back rests have two separate shoes. In either case, both shoes can be adjusted independently. The surface on which the shoes rest should be rotating true before the shoes are brought in contact. This may require taking a light truing cut across the workpiece, or if this cannot be done, as in the case of very slender parts, a true spot should be ground by plunge grinding using a very slow infeed. The first spots should be ground near the ends of the slender workpiece where it is supported by the centers.

Courtesy of Cincinnati Milacron

Fig. 12-4A. Using back rests to grind a slender workpiece.

Courtesy of Cincinnati Milacron

Fig. 12-4B. Using back rests to grind a heavy roll.

When the surface against which the shoes rest is true, the back rest is applied by first bringing the lower shoe against the work with just enough pressure to provide support. The horizontal shoe is then moved against the workpiece with a very light but firm pressure. If too much pressure is used on the horizontal shoe, it will spring the work toward the wheel causing it to be ground undersize at this point. The spring pressure in the back rests will keep the shoes in contact with the workpiece. However, if much stock must be removed, repeated settings of the shoes are often necessary. As the workpiece approaches the finish size, the wheel is withdrawn and the work stopped. The diameter is measured in several locations, including the position opposite each steady rest. The measurements are compared and the deviations corrected by adjusting the offending back rest. After the first workpiece has been ground to size, the stops can be set on the back rests so that they will not move forward beyond this setting. Subsequent workpieces can be placed in the machine by depressing the springs in the back rests, and the shoes will follow the workpiece until it reaches its finish size.

The face of the horizontal shoe should be vertical, and the face of the lower shoe should be at an angle of about 120 degrees with respect to the face of the horizontal shoe. Care should be taken to make sure that this angle is not less than about 120 degrees; otherwise the workpiece can be ground out of round. The shoes will wear and occasionally require redressing when the wear has become excessive.

Basic Methods of Cylindrical Grinding

The two basic methods of performing precision grinding operations on cylindrical grinding machines are called *plunge cut grinding* and *traverse grinding*. In plunge cut grinding the grinding wheel is continuously fed into the workpiece which rotates but does not traverse back and forth. Thus, the ground surface on the workpiece will be a reflection of the shape dressed onto the periphery of the grinding wheel. The length of the surface ground by this method cannot be greater than the thickness of the grinding wheel. Cylindrical and tapered surfaces can be ground in this manner. On production cylindrical grinding machines even stepped surfaces with shoulders can be ground by dressing the grinding wheel accordingly or by mounting more than one wheel on the grinding-machine spindle. Some production-type cylindrical grinding machines can be equipped with a spindle reciprocating arrangement which causes the spindle to move the grinding wheel back and forth a short distance parallel to the workpiece while it is plunge cutting. This action improves the surface finish on the workpiece by eliminating the marks caused by the diamond wheel dresser.

Courtesy of the Brown & Sharpe Manufacturing Company

Fig. 12-5. Universal cylindrical grinding machine set up for traverse grinding.

The most common method of cylindrical grinding is traverse grinding. Here the workpiece is traversed back and forth past the grinding wheel by the table. At the end of each stroke the table is usually made to stop or to dwell for a short period of time before the motion continues. The infeed of the wheel, which may be manual or automatic, usually occurs at the end of a stroke. A traverse grinding operation being performed on a universal cylindrical grinding machine is shown in Fig. 12-5.

Performing a Cylindrical Grinding Operation

Cylindrical grinding is a precision machining operation which requires that careful attention be given to every detail of the setup and operation. A step-by-step procedure for setting up and performing a cylindrical grinding operation will be given below. Figure 12-5 shows the operation to be described.

1. Select the best grinding wheel. Before starting the grinding operation, true and dress the grinding wheel. If the grinding wheel must be changed, it should be balanced before being mounted on the spindle.
2. Set the grinding wheel to operate at the correct speed. Measure the diameter of the wheel, and calculate the revolutions per minute at which the wheel should operate. Mount the required pulleys on the end of the spindle and on the motor (Fig. 12-2). The tension of the belt should not be excessive to avoid damaging the spindle bearings.
3. If a universal cylindrical grinding machine is used, the wheel slide base should be adjusted to feed the grinding wheel perpendicularly into the workpiece.
4. Position the headstock and the footstock about the same distance from each end of the table and far enough apart to hold the workpiece approximately in the center of the table.
5. Lock the headstock spindle. The workpiece is to be ground on dead centers or with both centers stationary.
6. Inspect the headstock and footstock (or tailstock) centers to make sure that they are in good condition. If the centers are scored or not at the correct angle, they must be replaced.
7. Set the swivel table to the zero position. This alone should not be relied upon to position the table for grinding a true cylindrical surface.
8. Inspect the center holes in the end of the workpiece. They must be clean and in good condition; otherwise accurate cylindrical surfaces cannot be ground. Place a lubricant in the center holes at each end of the work. If a proprietary lubricant is not available, white lead mixed with a small amount of lubricating oil makes an excellent lubricant.
9. Attach the correct size driving dog to the left end of the workpiece.

10. Place the workpiece between centers in the machine. First place the workpiece on the headstock center while holding the tailstock spindle back. Then slowly release the tailstock spindle and allow the tailstock center to come forward into the center hole of the workpiece. The spring tension in the tailstock spindle should be enough to hold the workpiece securely between centers.
11. Select a suitable work speed and adjust the headstock so that the drive plate will rotate at the required speed.
12. Select the table traverse speed and adjust the machine to operate at this table speed.
13. Select the desired table dwell time and adjust the machine accordingly. In general, the length of time that the table should dwell (or pause) at the end of each stroke is related to the speed of the table traverse. When the table traverse is relatively slow, a long dwell period should be used. When the table traverse speed is increased, the dwell period can be shortened correspondingly.
14. Adjust the length of the table traverse. The two dogs attached to the front face of the table are positioned to give the required length of table traverse. The table traverse should be long enough to permit the face of the grinding wheel to extend over the end of the workpiece a distance of about one-quarter to one-half of the width of the grinding wheel. At the other end of the workpiece, the table should stop when the grinding wheel is approximately .010 to .030 inch away from the shoulder. Whenever the grinding wheel approaches a shoulder, the length of the table traverse at the shoulder must be checked before starting to grind each new workpiece that is placed in the machine—even if identical parts are being ground. Differences in the depth to which center holes are drilled will cause the lengthwise position of each workpiece in the machine to vary.
15. Select the desired automatic grinding wheel infeed rate and set the machine accordingly. For average conditions this should be .001 to .0015 inch per stroke for rough grinding and .0005 inch per stroke or less for finish grinding. Sometimes the infeed for the final finishing cuts are made by hand. Likewise, the infeed for the first cuts taken when the workpiece is set up should be made manually.
16. Place the back rests in position on the table, but do not place the back rest shoes against the workpiece at this time.
17. Adjust the coolant nozzle to the grinding wheel. Place the splash guards in position at the front of the table. The splash guards are not shown in Fig. 12-5.
18. Start the grinding-wheel spindle. Stand to one side until the grinding wheel has reached its maximum speed. Turn on the flow of the grinding fluid, and allow it to flow over the grinding wheel for several minutes before starting to grind.

19. Start the driver plate of the wheel spindle to rotate, thereby rotating the work.
20. Bring the grinding wheel carefully up to the workpiece until a few grinding sparks can be seen to indicate that the wheel has just touched the workpiece. Contact between the grinding wheel and the workpiece can also be detected by sound.
21. Engage the table traverse and allow the wheel to pass over the entire surface to be ground.
22. Grind the workpiece until it is entirely cylindrical, using an infeed rate of not more than a .0005 inch per stroke. The infeed may be made by hand at the end of each table traverse stroke.
23. Measure both ends of the ground surface with micrometer calipers (Fig. 12-6). This is done to check if the machine is grinding a taper.
24. If necessary, adjust the angular setting of the swivel table in order to position the headstock and tailstock centers so that they will not cause a taper to be ground.
25. Repeat Steps 21 through 24 until the measurements at each end of the workpiece indicate that the machine is no longer grinding a taper.

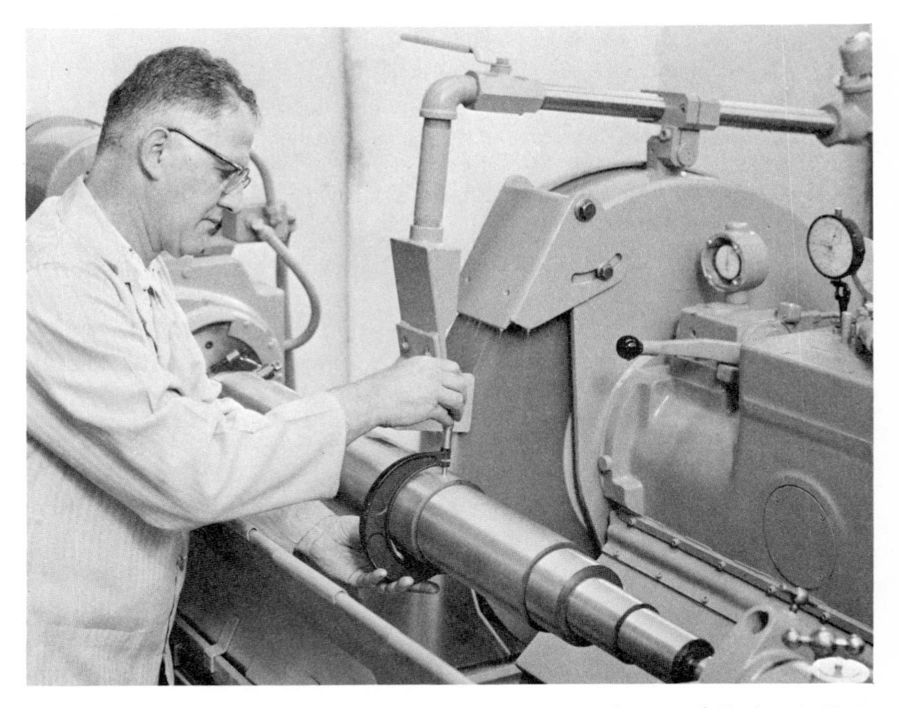

Courtesy of Cincinnati Milacron

Fig. 12-6. Measuring the diameter of a surface on a plain cylindrical grinding machine.

26. Set the back rest shoes in position against the workpiece.
27. Rough grind the workpiece until the diameter of the workpiece is within .002 to .005 inch of the required size. When this size is reached the grinding-wheel infeed should be stopped, but the table traverse is allowed to continue until the absence of grinding sparks indicates that the grinding wheel is no longer cutting. In shop terminology this is called "sparking out."
28. Withdraw the grinding wheel and stop the rotation of the workpiece.
29. Measure the diameter of the workpiece with micrometer calipers as shown in Fig. 12-6. Calculate the amount that must be removed from the diameter of the workpiece to bring it to the required size.
30. Set the automatic grinding-wheel infeed mechanism to feed the wheel into the work at the required rate for finish grinding.
31. Finish grind the workpiece to size. Feed the grinding wheel into the workpiece the necessary distance to grind it to the required diameter. Allow the wheel to spark out and then withdraw the wheel. Stop the rotation of the workpiece and measure its diameter with micrometer calipers. If a few more "tenths" must be ground off in order to obtain the required diameter, feed the wheel carefully up to the workpiece and then feed it manually into the workpiece to bring the workpiece to size.
32. Grind the other surfaces of the workpiece.
33. After the last surface has been ground on the final part, allow the grinding wheel to rotate at full speed for several minutes with the grinding fluid shut off. This allows the wheel to spin out most of the grinding fluid that has been absorbed during the course of the grinding operation.
34. Clean up the grinding machine. Precision work cannot be done on dirty equipment.

If the part to be ground has the same diameter for its entire length, one end of the workpiece should be ground almost up to the driving dog. The workpiece is then turned around in the grinding machine, and the remaining end is ground to size. The remaining end is short and well-supported as it is close to the footstock. Therefore, it is relatively easy to grind the second end to match the first end. Parts with a number of shoulders should also be ground by grinding as many surfaces on one end as possible before the second end is ground.

Shoulder Grinding

Shoulder grinding is an operation that is frequently performed on a cylindrical grinding machine. In some cases the shoulders are machined square on a lathe and are not machined in the cylindrical grinding machine. In this case there is usually a neck or undercut of the cylindrical surfaces adjacent to the shoulder which makes it unnecessary to bring the grinding wheel against the shoulder. There are several methods of grinding shoulders, and the method selected will depend upon the nature of the workpiece, the design of the grinding machine, and the experience of the operator.

One method of grinding shoulders is to grind the cylindrical surface adjacent to the shoulder to the finish size. The traverse is stopped so that the grinding wheel is .010 to .030 inch away from the shoulder. After the last cut on the cylindrical surface, the table traverse is stopped with the grinding wheel adjacent to the shoulder. By carefully turning the table traverse handwheel, the table is moved toward the shoulder to grind away the excess metal remaining on the cylindrical surface until the side of the grinding wheel touches the shoulder. The shoulder is then ground to size with the side of the grinding wheel.

A second procedure for grinding shoulders is to grind the cylindrical surface to finish size as before. When this surface is ground, the shoulder is brought to within .010 to .030 inch from the face of the grinding wheel. The grinding wheel is withdrawn, and the finished diameter is measured. When this diameter is known to be to size, the grinding wheel is brought forward until it is about .0005 to .001 inch away from the finished surface. The table is then moved by hand to grind off the metal remaining between the shoulder and the finished surface. The table traverse handwheel is carefully turned until the side of the grinding wheel contacts the shoulder and grinds it to size. The grinding wheel is then carefully moved forward by manually turning the infeed handwheel until the shoulder blends perfectly with the adjacent cylindrical surface. Micrometer graduations on the infeed handwheel can be used by noting the reading used to finish the cylindrical surface to size and then by bringing the wheel back to this reading when finishing the shoulder.

Still another procedure is to grind the shoulder to size first. The grinding wheel is fed into the cylindrical surface immediately adjacent to the shoulder by plunge grinding, and this surface is ground to size. Then the remaining portion of the cylindrical surface is ground to size.

When shoulders are ground with the side of the grinding wheel, a cross-hatched pattern of grinding marks will appear on the shoulder. Often this is not objectionable. A better appearing surface and a better shoulder, however, can be ground by positioning the grinding wheel at an angle of 45 degrees as shown in Fig. 12-7. Some plain cylindrical grinding machines are built with an angular wheel slide which causes the wheel to be held in the angular position permanently. On universal cylindrical grinding machines the wheel slide base is positioned so that the infeed of the grinding wheel will be perpendicular to the axis of the workpiece. The wheel spindle head is positioned at 45 degrees. The grinding wheel is mounted on the right-hand side of the spindle, and the pulleys are mounted on the left-hand side.

The advantage of this method of grinding shoulders is that the shoulder is ground with the face instead of the side of the grinding wheel. Thus,

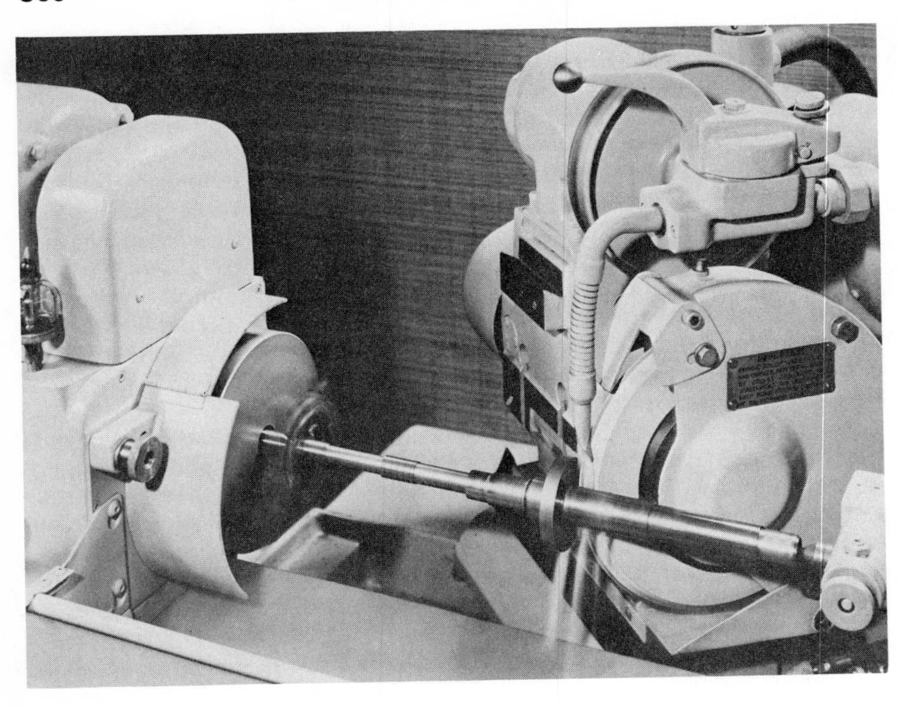

Fig. 12-7. Grinding a shoulder on a universal cylindrical grinding machine with the grinding wheel positioned at an angle of 45 degrees.

the crosshatched grinding wheel marks are avoided, and the shoulder has a scratch-free surface. The shoulder and the cylindrical surfaces are ground with the wheel positioned in this manner. Figure 12-8 shows how large shoulders can be ground by this method.

An important factor in grinding shoulders by this method is the correct truing and dressing of the grinding wheel. The diamond is mounted in the table wheel truing fixture (see B, Fig. 12-3), and the fixture is fastened to the grinding-machine table. With the grinding wheel positioned at the 45-degree angle, the grinding wheel is trued by moving the diamond across the face of the wheel using the traverse movement of the table. Slightly more than one-half of the width of the face of the grinding wheel should be trued in this manner. This surface of the grinding wheel is used to grind the cylindrical surface of the workpiece. The surface of the grinding wheel that grinds the shoulder is trued next by turning the diamond in the holder toward the left side of the wheel. The left side of the face of the grinding wheel is then trued by moving the grinding wheel in and out across the diamond with the infeed handwheel. A V-shaped form is thus trued on the grinding wheel. The two surfaces on the face of the wheel are 90 degrees with respect to each other, and the width of the two faces is approximately equal.

Either of the first two methods of grinding shoulders can be used when the width of the face of the shoulder does not exceed the width of the surface of the grinding wheel that is in contact with the shoulder. Larger shoulders are ground by moving the grinding wheel in and out across the face of the shoulder with the infeed handwheel. This procedure is illustrated in Fig. 12-8. Regardless of which procedure is used, the shoulders and the cylindrical surfaces produced by grinding with the grinding wheel at an angle will have an excellent surface finish.

Whenever possible all shoulders should be ground by having the pressure of the grinding wheel directed toward the headstock. Sometimes, however, it is necessary to grind a shoulder with the grinding-wheel pressure directed toward the footstock. In this case the footstock spindle must be clamped firmly in place, as the pressure of the grinding wheel will tend to move the footstock spindle backward and thus cause the workpiece to become loose.

Taper and Angle Grinding

The most accurate method of finish machining external tapers is by grinding on a cylindrical grinding machine. This is also usually the most

Courtesy of the Brown & Sharpe Manufacturing Company

Fig. 12-8. Grinding the face of a large shoulder with the grinding wheel positioned at 45 degrees.

rapid method of finishing external tapers. Tapers are ground by positioning the swivel table at the required angle, thereby positioning the workpiece at an angle with respect to the motion of the table traverse as shown in Fig. 12-9. The sliding table traverses back and forth in the usual manner. In Fig. 12-9 the taper on a milling-machine arbor is shown being ground. A graduated scale at the end of the table can be used to set the

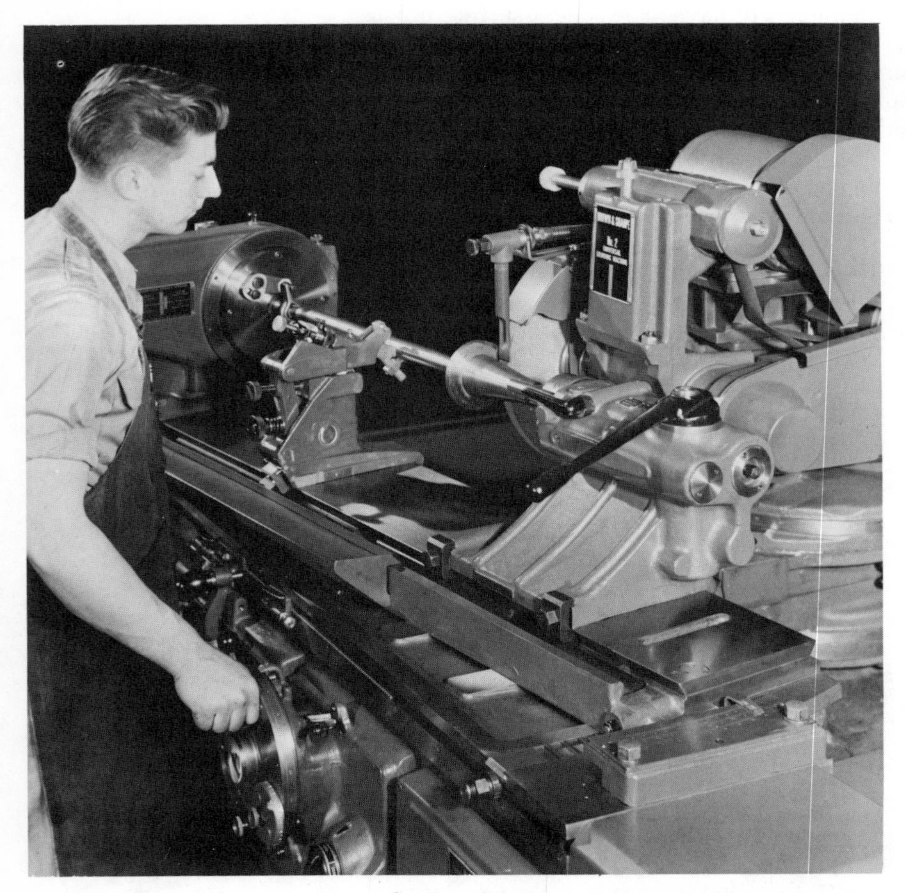

Courtesy of the Brown & Sharpe Manufacturing Company

Fig. 12-9. Grinding a taper on a milling-machine arbor by offsetting the swivel table.

table to the correct angle. This scale should not be relied upon to obtain extremely accurate settings. It is graduated in degrees and in taper per foot.

Accurate external tapers, such as in Fig. 12-9, are ground to fit a taper ring gage or to fit the taper in a mating part. Sometimes, when a taper ring gage or a mating part is not available, the taper must be measured on a sine plate or a sine bar. A step-by-step procedure for grinding a taper follows:

1. Set the swivel table to the angle of the taper as accurately as possible using the scale at the end of the swivel table.
2. True and dress the grinding wheel.
3. Mount the workpiece between dead centers in the same manner as for grinding a cylindrical surface.
4. Adjust the table traverse speed, the length of the table traverse, the dwell, and the grinding-wheel infeed to the desired settings.

Courtesy of Cincinnati Milacron

Fig. 12-10. Grinding a steep taper on a universal cylindrical grinding machine using the wheel slide to traverse the grinding wheel over the taper.

5. Start the rotation of the grinding wheel, and allow it to rotate for a few minutes with the coolant turned on.
6. Start the spindle driver plate, thereby rotating the workpiece.
7. Bring the wheel forward carefully by turning the infeed handwheel until the grinding wheel just touches the workpiece.
8. Engage the automatic table traverse and take a light cut across the taper, feeding the wheel into the work by hand.

9. When the entire surface of the taper has been ground, allow the grinding wheel to spark out. Then withdraw the wheel from the workpiece.
10. Dry the taper, removing all traces of the coolant, and rub a thin layer of bluing over the surface of the taper. If no bluing is available, ordinary board chalk can be used as a substitute—although the use of bluing is preferred. Make four chalk marks 90 degrees apart lengthwise along the taper.
11. Remove the workpiece from the grinding machine, and place a ring gage over the taper. Wring the gage on the taper and then remove it.
12. "Read" the fit of the taper ring gage on the taper. If the bluing or chalk is rubbed off more at the large diameter of the taper than at the small diameter, the taper is too steep. If more is rubbed off on the small diameter, it is not steep enough. When an even amount is rubbed off along the entire length of the taper, the taper is correct. When holding-type tapers are correct, the gage will stick firmly to the workpiece and must be driven off with a soft hammer.
13. Make the necessary adjustment to the swivel table in order to correct any error in the taper as determined in Step 12.
14. Repeat Steps 7 through 13 until the angle of the taper is correct.
15. Grind the taper to size. The size of the taper is determined by the

Courtesy of the Brown & Sharpe Manufacturing Company

Fig. 12-11. Grinding a steep taper and a slight taper in one setup on a universal cylindrical grinding machine.

distance that the ring gage will fit onto the taper. This will require the workpiece to be removed several times to gage the diameter by fitting the ring gage over the taper. Subsequent pieces can be ground by eliminating Steps 7 through 14. The size of the taper should, however, be checked before it is finish ground to size.

A steep taper can be ground on a universal cylindrical grinding machine by positioning the grinding-wheel slide at the required angle as shown in Fig. 12-10. The grinding wheel is moved back and forth across the taper with the wheel slide by means of the infeed handwheel. The workpiece is fed into the grinding wheel by moving the table slightly with the table traverse handwheel. By using both the wheel slide and the offset table, it is possible to grind a steep taper and a slight taper in one setup as shown in Fig. 12-11. The wheel slide must be set at the angle that is the complement of the difference between half of the included angle of one taper and half of the included angle of the other. For example, if the included angle of the steep taper is 50 degrees and the included angle of the slight taper is 10 degrees, the wheel slide must be set at the following angle:

Courtesy of the Brown & Sharpe Manufacturing Company

Fig. 12-12. Grinding an angle or steep taper on a part by setting the headstock at the required angle.

$$\text{Wheel Slide Setting} = 90° - \left(\frac{50°}{2} - \frac{10°}{2}\right)$$
$$= 70°$$

The grinding wheel is trued with one face parallel to the surface of the steep taper and the other face parallel to the table traverse motion.

Angular surfaces and tapers can also be ground on a universal cylindrical grinding machine by positioning the headstock at the required angle, as shown in Figs. 12-12 and 12-13. In Fig. 12-12 the workpiece is clamped to a face plate which is mounted on the headstock. Centers for grinding machines and lathes can be conveniently and rapidly ground by the set-up shown in Fig. 12-13. The center is placed in the taper of the headstock spindle which is made to rotate as the cone point of the center is being ground.

Flat surfaces can be ground on the universal cylindrical grinding machine as seen in Fig. 12-14. The workpiece in this case is mounted on a magnetic chuck and the headstock is positioned 90 degrees with respect to the table ways. Parts can be ground in a similar manner while held on a faceplate or in a chuck. This operation is called face grinding.

Internal Grinding

Straight and tapered holes can be ground on a universal grinding machine as shown in Figs. 12-15 and 12-17. The internal grinding spindle is

Courtesy of the Brown & Sharpe Manufacturing Company

Fig. 12-13. Grinding a center on a universal cylindrical grinding machine.

lowered and clamped into position. The workpiece is held in a chuck or on a faceplate. Long workpieces must be given additional support by a steady rest or center rest as shown in Fig. 12-16. For grinding straight holes, the swivel table and the axis of the headstock spindle must be set parallel with the longitudinal table traverse. The length of the table traverse should be adjusted to allow the internal grinding wheel to pass partly out of the hole at each end unless the inside end of the hole is not open. The distance that the face of the wheel should pass out of the hole should be about one-fourth to one-half of the width of the face of the wheel. If the inside end of the hole is not open or "blind," the dwell at that end of the hole should be made slightly longer than at the open end of the hole. A relatively short dwell is usually used at open ends of holes. The headstock should be set for the correct work speed, and the grinding wheel should be trued using a table-type truing fixture (B, Fig. 12-2) to make sure that the face of the wheel is parallel to the bore.

When all of the preparations have been completed, start the work to rotate, and feed the grinding wheel into the hole by hand. Then engage the power table traverse and feed the wheel into the work by hand. When the wheel has started to grind, the automatic infeed can be engaged or

Courtesy of Cincinnati Milacron

Fig. 12-14. Face grinding on a universal cylindrical grinding machine.

Courtesy of Cincinnati Milacron

Fig. 12-15. Internal grinding on a universal grinding machine using an internal grinding head permanently attached to the machine.

the hole can be enlarged by infeeding manually. When the cut is finished, allow the wheel to spark out and then withdraw the wheel from the work by hand. The size of the hole is obtained by taking trial cuts and measuring the diameter. If several holes with the same diameter are to be ground, the wheel infeed stop can be used. The infeed stop is set to disengage the automatic infeed when the diameter of the hole is to size. Allowance should be made for wheel wear and for the dressing of the wheel.

One problem encountered in grinding internal holes is that the hole sometimes becomes larger at each open end. Holes in this condition are called "bell-mouthed" holes. The problem is caused by the difference in pressure of the grinding wheel against the side of the hole when the wheel approaches the end of the hole. It can be prevented by taking relatively light cuts in order to reduce the wheel pressure, by reducing the length of time that the table dwells at each reversal, by reducing the distance that the wheel is allowed to pass out of the hole at the end of each stroke, and by allowing the wheel to spark out before it is entirely withdrawn from the hole. If, for some reason, the wheel cannot be allowed to spark out, it should be backed away from the side of the hole before it is withdrawn out of the hole. Another cause of bell-mouthed holes can be looseness in the grinding-wheel spindle.

Courtesy of the Brown & Sharpe Manufacturing Company

Fig. 12-16. Internal grinding a hole in a long workpiece which is supported by a steady rest.

Tapers and internal conical surfaces or angles can be ground in much the same manner as external tapers or angles. Slight tapers are usually ground by setting the swivel table to the required taper or angle. The angle of the taper and the size of the taper are checked with a taper plug gage. Steep tapers can be ground by positioning the headstock at the required angle or by positioning the wheel spindle head and the wheel slide base in the same manner as for grinding external tapers. It is possible to grind a straight and a tapered hole in one setup as shown in Fig. 12-17. Two operations are, however, required. The wheel spindle head is set so that the internal spindle is parallel to the longitudinal traverse, and the wheel stand is set parallel to the side of the tapered holes. The grinding wheel is dressed so that the leading edge is parallel to the side of the tapered hole, and the back portion is parallel to the side of the straight hole. The straight hole is ground with the power longitudinal table traverse used, and the wheel infeed is made by hand or automatically. The tapered hole is ground by the manual cross-slide movement of the wheel head, and the infeed is accomplished by a slight longitudinal table movement.

Form Grinding

Contoured surfaces can be ground on cylindrical parts by truing the grinding wheel to a contour that reflects the required shape on the part.

Courtesy of the Brown & Sharpe Manufacturing Company

Fig. 12-17. Internal grinding a straight and tapered hole in one setup.

Form grinding is performed by plunge grinding the formed grinding wheel into the workpiece, as shown in Fig. 12-18. On this job the workpiece is held in a collet chuck that is mounted in the headstock of the grinding machine.

Centerless Grinding

Many cylindrical parts can be ground on centerless grinding machines like the one shown in Fig. 12-19. The work is supported on a work rest blade between a grinding wheel and a regulating wheel. The grinding wheel does the grinding by operating at a surface speed of 5,500 to 6,000 feet per minute. The regulating wheel, made from a rubber-bonded abrasive, operates at a relatively slow speed of 50 to 200 feet per minute. This regulating wheel acts as a brake to counteract the tendency of the workpiece to spin rapidy because of the action of the grinding wheel. The speed of rotation of the work is regulated by the regulating wheel. The regulating wheel is set at a slight angle, which imparts a lateral motion to the workpiece and thereby causes it to feed through the machine. The centerline of the workpiece is generally held above the centerline of the grinding and regulating wheels by the work rest blade. Although sometimes the centerline of the work is held below the centerline grinding or regulating wheels, the centerline of the work is never at the same height as the centerline of the two wheels.

An advantage of the centerless grinding method is that long slender workpieces are supported between the two wheels and therefore do not deflect. Another advantage is that the machine can be very rapidly loaded and unloaded. In many cases a continuous flow of parts can be made to pass through the machine so that high production rates can be achieved. Of course, the time required for making the setup on centerless grinding machines is generally greater. The three methods of centerless grinding, called *through-feed, in-feed,* and *end-feed,* will be described in the following section:

Through-feed Method. The through-feed method can be used when the

Courtesy of Cincinnati Milacron

Fig. 12-18. Form grinding by plunge grinding the formed grinding wheel into the workpiece, which is held in a collet chuck.

Courtesy of Cincinnati Milacron

Fig. 12-19. Centerless grinding the outside diameter of tubing.

parts can be passed completely through the machine, as shown in Fig. 12-19. Other parts that can be ground by the through-feed method include automotive wrist pins, mortar shells, and a drill rod. The work is moved through the machine by the angularity of the regulating wheel. The rate of feed depends upon the speed of the regulating wheel and its angular adjustment. It may be necessary to pass the work through the machine more than once, the number of passes depending on the amount of stock to be removed, the roundness and straightness of the unground work, and the limits of accuracy required. Sometimes several centerless grinding machines are placed in line, and the workpieces are made to pass from one directly into another machine.

In-feed Method. When parts have shoulders, heads, or some larger part than the ground diameter, the in-feed method usually is employed. This method is similar to plunge-cut grinding on a cylindrical grinding machine. The length of the surface that can be ground is limited to the width of the wheel. There is no axial feeding movement, and the regulating wheel is positioned with its axis approximately parallel to that of the grinding wheel—there being a slight inclination to keep the work tight against the end stop. The regulating wheel is moved toward the workpiece and feeds it into the grinding wheel to reduce its size.

End-feed Method. The end-feed method is applied to taper work. The grinding wheel, regulating wheel, and the work rest blade are set in a fixed relation to each other, and the work is fed in from the front mechanically or manually to a fixed stop.

Cylindrical Grinding Technology

Certain variables encountered frequently in cylindrical grinding will be given some further consideration in this section. The cylindrical grinding machine operator must understand these factors because he must control them when he operates the machine. The performance of the grinding wheel was treated in detail in Chapter 11 and should be reviewed. Although this performance will not be considered in detail in this section, it, too, has a significant influence on the cylindrical grinding operation.

Work Speed. The work speed for cylindrical grinding should be approximately 15 to 100 feet per minute. Increasing the work speed will make the grinding wheel act softer and will usually result in an increased rate of production. The surface finish produced at a higher work speed is somewhat inferior to the surface produced at a slower work speed. For this reason the work speed is sometimes reduced to a very slow speed for taking the final finishing cuts on the work. Excellent surface finishes can be obtained by using a very slow work speed. Fast work speeds can sometimes be the cause of chatter and vibration, which must be eliminated by reducing the speed. In general the work speed should be as fast as possible for rough grinding and much slower for finish grinding.

Traverse Length. The traverse length should be slightly more than the length of the surface being ground so that the wheel extends beyond the ends of the surface about one-quarter to one-half of the width of the wheel. If there is a shoulder at one end of the cut, this, of course, is impossible. If the grinding wheel does not overlap the end of the work, there is a tendency for the ends to grind large. If the wheel overlaps too much, the end may be ground too small.

Traverse Speed. The traverse speed affects the rate of grinding and the surface finish obtained on the workpiece. The traverse speed should not exceed three-quarters of the width of the wheel face per revolution of the work. Usually it is best not to exceed one-half of the width of the wheel face per revolution of the work for rough grinding and one-eighth of the width of the wheel per revolution or less for finish grinding. In general, the depth of cut or infeed can be increased when the traverse speed is reduced. A slow traverse speed will result in a better surface finish than a fast traverse speed—which can cause the work to chatter, especially on long slender workpieces. A faster traverse speed combined with a slow infeed rate can be used to grind slender workpieces when it is possible to use a wide grinding wheel, which tends to stabilize the workpiece. Another technique used to grind slender workpieces is to true the grinding wheel so that it will have a slight taper at one end and to traverse the workpiece into the tapered end of the grinding wheel using a very heavy infeed per pass and a very slow traverse speed. The infeed is made at only one end of the workpiece, which can be ground to size by this method in a few passes only, often in a single pass. To use this method, however, the part configuration must allow the grinding wheel to overtravel at both ends.

Dwell. Dwell is the time that the table pauses at the end of each traverse before it reverses. The purpose of the dwell is to lessen the shock of the table reversal and to grind the diameter to a uniform size. Depending upon the traverse speed, the leading surface of the face of the grinding wheel does the bulk of the grinding. The trailing surface on the face of the wheel cleans up the work and reduces it to size. Since the wheel does not completely pass over the end of the work, the trailing surface of the wheel does not clean up and reduce the surface of the work to size at each end of the traverse. The dwell allows the wheel to pause long enough at each end of the traverse for the wheel to finish the ends to size.

Grinding-Wheel Infeed. The amount of infeed of the grinding wheel per table traverse is dependent upon many factors. Among these factors are the surface finish required on the work, the type of grinding wheel used, the size and rigidity of the workpiece, and the traverse speed. No hard and fast recommendations can be made that will fit every situation. Other factors to be considered in selecting an infeed rate are the required accuracy on the workpiece, the use of coolants, and the support provided to the workpiece by back rests. The infeed must be reduced when grinding dry in order to prevent the workpiece from overheating, which could cause metallurgical damage on the surface of the work. Depending on these factors the infeed for rough grinding may be from .001 to .005 inch (0.025 to 0.13 mm) per pass and for finish grinding .0001 to .0005 inch (0.002 to 0.013 mm) per pass. In general, deeper cuts are taken for rough grinding and the infeed rate is decreased for finish grinding. If the workpiece has a shoulder at one end, the infeed should take place only at this end together with a suitable dwell period to give the grinding wheel the required time for stock removal in the absence of overtravel.

Centers and Center Holes. The condition of the grinding-machine centers and the center holes in the workpiece have a great effect on the accuracy of the workpiece. Accurate cylindrical grinding is impossible when either the machine centers or the workpiece center holes are in poor condition or are dirty. Since the work rotates on the centers, any unevenness or roughness in the rotation will be immediately reflected on the surface being ground.

The centers should contact the conical surface center holes for the entire length of the 60-degree cone surface in the hole. In other words, the included angle in the center hole and on the center must be exactly 60 degrees. If this is not the case, the workpiece will not be held in place as rigidly as possible by the centers. This can cause chatter and increase the chance for the work to deflect away from the grinding wheel. Furthermore, the center and the center hole will wear very rapidly, which results in inaccuracy on the surface being ground. In many modern machine shops the center holes are finished by grinding on special center hole grinding machines before the part is ground on a cylindrical grinding machine. In all cases the condition of the centers and the center holes must be checked, and any deficiencies corrected before proceeding with the cylindrical grinding operation.

Warming Up. The cylindrical grinding machine should be allowed to run for a few minutes before precision work is attempted. This allows the temperature of the moving parts and the oil to stabilize. It is particularly important that the grinding wheel be allowed to run with the coolant flowing over the wheel, because the grinding wheel will operate differently depending upon whether it is dry or wet. Thus, if the grinding is started with the wheel dry and the coolant is then turned on, it could cause a variation in the diameter of the workpiece being ground. It is equally important that the grinding wheel be allowed to run for a few minutes with the coolant shut off before shutting down the grinding machine for a longer period of time. This is done to allow the grinding wheel

Courtesy of Cincinnati Milacron

Fig. 12-20. A hand caliper gage for continuously measuring the diameter of the workpiece as it is being ground.

to expel as much of the absorbed coolant as possible by the centrifugal action of the wheel. If this is not done and the wheel is standing still for a long period of time, the absorbed coolant will settle in the lower part of the wheel and cause it to be unbalanced when it is started up the next time. Of course, the wheel will regain the balanced condition when it has run for some time with the coolant on.

Grinding Machine Gaging Systems

There are a number of cylindrical grinding machine gaging systems available which continuously measure the diameter of the workpiece as the part is being ground. A hand caliper gage is shown in Fig. 12-20. The hand on the dial indicates the size of the workpiece as it is being ground. This gage can be used to measure the outside diameter continuously during plunge grinding, as shown, or during traverse grinding. It can be used over interruptions on the work circumference, and it can measure the diameter of a taper from a shoulder.

CHAPTER **13**

Surface Grinding

The grinding of plane or flat surfaces is known as surface grinding. A variety of sizes and types of surface grinding machines are used to grind both large and small workpieces. As an example, the faces of some large steam turbine casings are rough and finish machined on a surface grinding machine with an abrasive wheel used to perform both the rough and finishing operation. Hardened machine tool ways are finish machined by grinding as shown in Fig. 13-3. Perhaps the most extensive application of the surface grinder is found in the toolroom, where tools, dies, and gages are made. Since many of these parts are hardened, the only practical method of finishing them to close dimensional tolerances and a good surface finish is by grinding. The surface grinder is almost indispensable in the toolroom.

Surface Grinding Methods and Machines

The basic methods of surface grinding are shown in Fig. 13-1. Surface grinding machines are constructed to grind surfaces by one of these methods. The most common method of surface grinding is seen at A, Fig. 13-1 and in Fig. 13-3. The table moves back and forth longitudinally below the grinding wheel. At the end of each longitudinal movement, the saddle, upon which the table is mounted, moves an increment across the workpiece usually from .010 to .050 inch. This feed is called the cross-feed. On some surface grinding machines, such as shown in Fig. 13-3, the cross-feed is obtained by moving the wheel instead of the table. The wheel can be moved up and down vertically to establish the depth of cut and to grind the workpiece to the required size. This motion is called the down feed.

The plunge cut method of grinding is illustrated in Fig. 13-1, view B. This method is primarily used to grind narrow surfaces that are less wide than the grinding wheel. The wheel is fed a small increment downward into the workpiece at the end of each longitudinal stroke of the workpiece.

Figure 13-1, C, shows a method of surface grinding which is used to grind large castings such as crankcase covers, crankcases, gear cases, and similar work. The face of the grinding wheel is brought into contact with the work, although most of the stock is removed by the edge of the wheel. In many instances the grinding wheel is made from a number of smaller segments held on a disc-shaped wheel head. On some machines of this

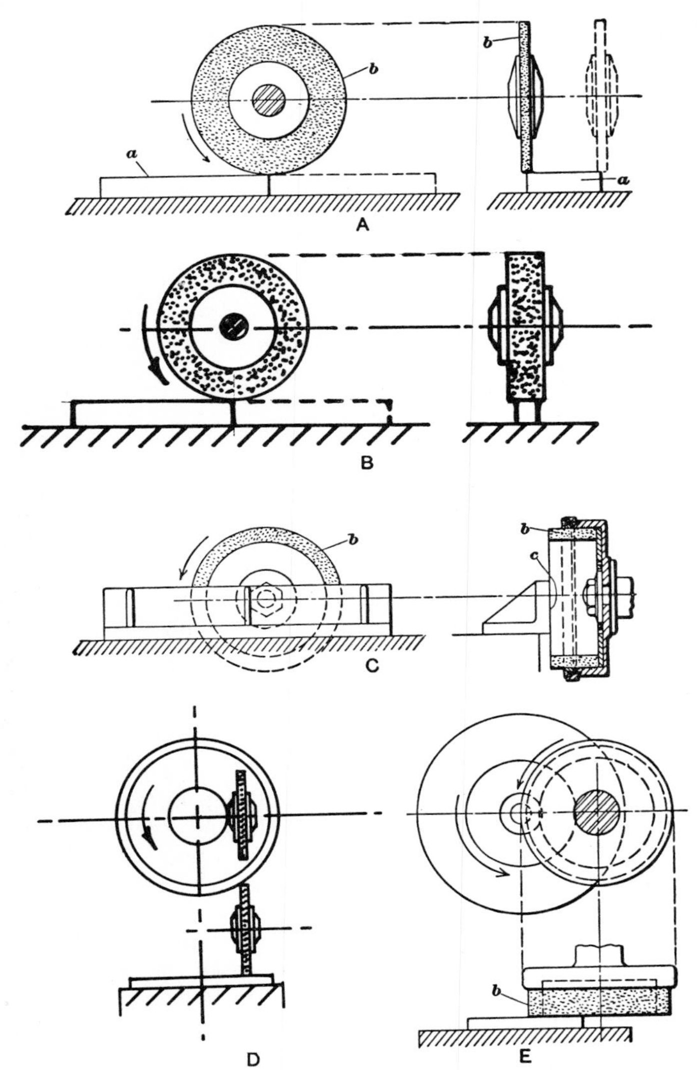

Fig. 13-1. Different methods used to grind plane surfaces.

type the table is reciprocated back and forth past the grinding wheel while on others the column supporting the grinding-wheel spindle is moved back and forth. An advantage of this method of grinding is the large surface area of the workpiece that can be ground in one pass of the work or the wheel. Some surface grinding machines of this general type have a vertical spindle instead of a horizontal spindle. Rotary surface grinding is shown at D, in Fig. 13-1; the table of a rotary grinding machine rotates while the grinding wheel slowly moves toward and away from the center of the rotary table. This method is used to grind the sides of disc-shaped workpieces.

The method shown at E, Fig. 13-1 is frequently used to grind both large and small workpieces. A machine of this type is illustrated in Fig. 13-4. The work table has a rotary movement, and the wheel head is fed down a certain amount for each revolution of the table. The grinding wheel may be cylindrical, as shown, or it may be of a segmental type. The wheel head may be positioned vertically or given a slight tilt. Tilting the head permits increasing the grinding wheel penetration which results in faster rates of stock removal. The grinder in Fig. 13-4 has a wheel head that can be tilted for rough grinding so that the maximum horsepower is concentrated on the leading edge of the grinding wheel to allow deeper penetration. After the work is ground to within a few thousandths of an

Courtesy of the Gallmeyer & Livingston Company

Fig. 13-2. Modern surface grinding machine equipped with digital readout of the wheel head.

Courtesy of the Mattison Machine Works

Fig. 13-3. Large double-spindle surface grinding machine used to grind ways.

inch of the final size, the wheel can be quickly returned to the vertical position for finish grinding. Finish grinding with a flat wheel removes all concavity produced by the elliptical generating plane of the wheel. Surface grinding machines of this type are used to grind flat plates as well as large castings. Also, it is possible to grind a large number of smaller parts simultaneously on these machines. The work-holding surface is usually a circular magnetic chuck. On some machines designed for mass production two or more wheel heads are mounted around the circular table so that the workpiece can be ground to finish size in one revolution of the table.

The surface grinding machine shown in Fig. 13-2 is equipped with a digital readout which displays the position of the grinding wheel relative to the working surface of the machine. After a workpiece is ground to a convenient size and measured, the display is adjusted to show this size. The wheel head can then be moved up or down, and the display will always show the relationship between the grinding wheel and the work-holding surface.

The large surface grinding machine in Fig. 13-3 has two spindles mounted between three massive box-type columns. The axis of one of the spindles is horizontal. The second spindle is a vertical spindle which can be swiveled up to 55 degrees on either side of the vertical position. Horizontal and angular surfaces of ways, dovetails, and V-blocks, as well as other machine parts, can readily be ground.

Table Speed, Cross-Feed, and Down Feed

The table speed—the velocity of the lengthwise or longitudinal travel of the table—is usually specified in feet per minute, although on many machines this speed cannot be set precisely. In this case, the grinding-machine operator must judge whether the table is moving at a slow or fast speed. The table speed available varies for different makes of surface grinding machines, and on some machines the longitudinal or lengthwise travel of the table can only be actuated manually. Often the automatic longitudinal table movement is actuated by a hydraulic mechanism with the speed generally ranging from 0 to 100 or 150 feet per minute.

The cross-feed on many machines can be actuated automatically while on others it can only be actuated manually. The automatic cross-feed can be accurately controlled ranging from 0 to .250 inch or more per table reversal. Micrometer dials make it possible to obtain very accurate movements of the cross-feed. Some machines are equipped with an automatic down-feed mechanism, although on most surface grinders the down feed of the grinding wheel is done manually. Precise adjustments of the down

Courtesy of the Mattison Machine Works

Fig. 13-4. Vertical-spindle rotary-table-type surface grinding machine.

feed can, however, be made. Vertical adjustments of the down feed can be made by reading the micrometer dial of the elevating or down-feed handwheel which is usually marked in increments of .0001 or .0002 inch per graduation.

It is not possible to give recommendations for the table travel, cross-feed, and down feed (or depth of cut) to be used except in some specific instances. These three cutting conditions for surface grinding are interrelated with each other. For example, increasing the depth of cut generally requires a reduction in the cross-feed; decreasing the depth of cut and the table speed permits an increase in the cross-feed. Of equal importance in determining the cutting conditions are the size and rigidity of the machine, the condition of the machine (especially the spindle), the type of grinding wheel used, the workpiece material, and whether a roughing or a finished cut is being taken. In general, the depth of cut for finishing cuts should be .0001 to .0002 inch; the table speed should be fast, and large cross-feed increments should be used. The wheel should be allowed to spark out on the final finishing cut by passing it over the workpiece one or more times without any down feed.

The Magnetic Chuck

Magnetic chucks are used extensively on surface grinding machines because they provide a convenient and accurate method of holding steel and cast-iron parts. One of the major advantages of the magnetic chuck is the speed of loading and unloading the workpieces. The need for special clamps and other holding devices is eliminated or reduced. The workpiece is held onto the face of the chuck by the magnetic field; however, it is the friction between the workpiece and the chuck surface that prevents the workpiece from sliding as a result of the grinding pressure. Only magnetic materials can be held on the magnetic chuck. Brass, bronze, and other nonmagnetic materials cannot be held directly by the chuck.

The two basic types of magnetic chucks are the permanent magnet chuck and the electromagnet chuck. The permanent magnet chuck, usually made only in smaller sizes, has the advantage of not requiring a rectifier and other electrical apparatus. Permanent magnets made of special alloys are actuated by simply turning a lever 180 degrees. A permanent magnet chuck is shown in Fig. 13-5. A permanent magnet chuck can be mounted on sine plates (Fig. 13-14), which provide a convenient method of holding workpieces to be ground at an angle.

Electromagnetic chucks are made in all sizes. Since they operate on 24, 110, or 220 volt direct current, they require a rectifier to change the alternating current to direct current. A simple turn of a switch turns the chuck on or off. Large workpieces could be difficult to remove from the magnetic chuck because of residual magnetism. Some electromagnetic chucks are made to vary in holding power, which is useful in holding thin, warped workpieces.

The face of the magnetic chuck forms a true reference and locating plane on which workpieces can be accurately aligned. The magnetic

Courtesy of the Brown & Sharpe Manufacturing Company

Fig. 13-5. A permanent magnet chuck-mounted on the table of a surface grinding machine. The two castings are held to the face of the chuck by the magnetic force while grinding.

chuck is aligned on the surface grinder in a longitudinal or lengthwise direction by two keys on the bottom which fit in the T-slots of the table. Two removable stop plates attached to two sides of the chuck (Fig. 13-10) can be adjusted vertically to suit the size of the workpiece. They can also be used to align the workpiece on the chuck face.

In order to do the precise work that the surface grinder is capable of, the surface of the chuck must be parallel with the saddle and the table ways. Any scratches and burrs must also be removed from the chuck surface. For these reasons it is occasionally necessary to regrind the top surface. In any case, the top surface of the chuck should be reground every time the chuck is mounted on the table of the machine. The surface of a magnetic chuck is shown being ground true in Fig. 13-6. The chuck should be ground in the machine on which it is to be used. The grinding is done with an aluminum oxide grinding wheel with a medium grit, a medium to soft grade, and a medium structure (A-46-H-8 or similar). The grinding wheel should be dressed fairly coarse or open. A medium table speed of 35 to 60 feet per minute should be used. Sometimes an even slower table speed will give better results. The cross-feed should be approximately .005 to .015 inch per table reversal. When the finishing cuts are taken, the table speed and the cross-feed should be increased.

Courtesy of the DoALL Company

Fig. 13-6. Grinding the top surface of a magnetic chuck true.

The depth of cut or down feed should not exceed .0004 inch per cut for the first "roughing" cuts and .0001 to .0002 inch for the finish cuts. The grinding wheel should be allowed to spark out by allowing it to traverse across the entire surface of the chuck one or more times without any down feed. If available, an ample supply of coolant should be used. *It is very important to have the magnet turned on when performing this grinding operation.* The amount of stock removed from the chuck by this operation should be no more than is necessary to produce a good flat surface. Usually this can be done by grinding off no more than .001 to .002 inch. An indication that the grinding wheel has ground the entire surface of the chuck can be obtained by rubbing a thin layer of red lead, Prussian blue paste, or a proprietary compound such as Dykem Hi-Spot Blue over the surface of the chuck before the final "sparking out" cut is taken. If the entire surface has been ground the bluing or red lead will be removed. Any that remains indicates an unground surface or a low spot which must be removed by further grinding. After the surface has been ground the edges should be lightly hand-honed using a medium India oilstone or a black granite deburring stone.

The finished surface of the chuck should have a dull or nearly polished appearance. It should not be highly polished, and no burned, glazed, or dull areas should be visible. The flatness of the surface can be tested with a precision straight edge. Three narrow strips of thin paper are placed between the edge of the straight edge and the top of the chuck. One strip of paper is placed at each end of the straight edge. The strip of paper in the center must be held tight when pulled with one hand while the other hand holds the straight edge on the chuck with a light pressure. If the strip of paper in the center is loose, the surface is not flat. The chuck should be checked in several positions along two directions in this manner. Another method of checking the flatness of the chuck is to indicate the surface with a dial test indicator which is fastened to the wheel guard. The surface on the face of the magnetic chuck must be smooth and flat; otherwise it will not be possible to do very precise work on the surface grinder. Although the occurrence of scratches or dents on this surface must be avoided as much as possible by exercising care in mounting workpieces on the chuck, they do occur. Small scratches and dents that appear should immediately be removed with a fine oilstone or a black granite deburring stone.

Loading the Magnetic Chuck

The holding power of a magnetic chuck depends upon the strength of the magnetic field and upon the area of contact between the chuck and the workpiece. As previously stated, the magnetic field holds the workpiece against the face of the chuck, but the workpiece is prevented from sliding by the frictional resistance between the workpiece and the chuck face. A larger area of contact between the workpiece and the chuck face will allow the magnetic field to exert a greater pull on the workpiece, which in turn will increase the total amount of frictional resistance. If the shape of the workpiece is such that only a relatively small area is in contact with the surface of the chuck, the holding power of the chuck is reduced. In such instances the workpiece should be blocked. This is done by placing pieces of steel against the sides of the work so that in effect the workpiece is nested by the steel blocks. Such blocks must be low enough to be below the surface to be ground. It is good practice to block even larger workpieces as a safeguard against a failure of the electric current. In many cases the workpiece is placed directly onto the surface of the chuck. The frictional resistance with which the work is held can be increased by placing a sheet of good quality paper between the surface of the chuck and the workpiece. This method of mounting the workpiece on the chuck is particularly effective in holding narrow and thin workpieces.

The surface of the chuck consists of a steel plate in which brass or lead strips have been placed. On circular chucks these strips are round, forming a series of concentric rings. The steel adjacent to the brass or lead strips forms a magnetic pole when the switch is closed. The workpiece should be placed on the chuck so that it will span at least one of the brass or lead strips. Larger workpieces should be placed on the chuck to span as many of these strips as possible, and narrow and thin workpieces in particular should span as many strips as possible. They should be adequately blocked and have paper placed between the part and the surface of the chuck. If more than one workpiece is to be mounted on the chuck, they should be positioned with some regard to the brass or lead strips as explained above. Furthermore, they should be placed in a regular pattern, if possible, in order to reduce the amount of time that the grinding wheel is actually not cutting.

The magnetic chuck should be thoroughly wiped off after each load in order to dry and clean the surface. Before a new load is placed on the chuck, the surface on which the part or parts are to be placed should be checked by rubbing it with the bare hand. In this way small particles of dirt or burrs raised from small scratches or dents will be detected which would otherwise go unnoticed and which would cause inaccuracy in the location of the work on the chuck. Such inaccuracy will cause the surface to be ground not parallel with the surface placed against the chuck face. Great care should be exercised in moving workpieces, accessories, and blocks off and onto the chuck so that the surface of the chuck will not be scratched or dented. Again, it is necessary to keep the chuck face clean and free of burrs that are raised by scratches and dents if accurate work is to be done on the surface grinding machine. When the chuck is not to be used for a period of time, it should be cleaned and a thin film of oil placed over the surface of the face.

Surface Grinding Work

Surface grinding work requires good planning, cleanliness, and care. Before a job is set up on the surface grinder, the operator must know the face of the chuck is true. As the first step in setting up, the machine table and the face of the magnetic chuck are cleaned. The cleanliness of the magnetic chuck should be checked by brushing over its face with a bare hand. In a like manner, any surface on an accessory against which the workpiece is to seat, must be cleaned and checked for cleanliness. When precision work is to be done, cleanliness cannot be overemphasized.

After all of the seating surfaces have been cleaned, the workpiece is carefully placed and aligned on the magnetic chuck, so that no nicks or scratches will be made. Often the workpiece can be aligned visually, such as the two castings demonstrate in Fig. 13-5. The workpiece or the accessory may also be aligned by placing it against the side stop of the magnetic chuck, or by positioning a parallel between this stop and the piece to be worked, as shown in Fig. 13-10. In some instances the two pieces must be precisely aligned by indicating against a machined surface with a dial test indicator (see Fig. 13-7). Alignment of the workpiece may also be provided by holding it in a vise, on a V-block, against an angle plate or on any accessory that has previously been aligned by one of the methods described.

Courtesy of The L. S. Starrett Company

Fig. 13-7. Setting up a workpiece on the surface grinding machine with a dial test indicator to obtain an accurate alignment.

A typical surface grinding job is illustrated in Fig. 13-8. The task is to grind the surface of a die mounted on the lower half of a die set. This surface must have a good finish, and the edges of the die must be sharp. The die is made from hardened tool steel. It is assumed that the magnetic chuck is true and in good condition. The procedure for performing this operation is given in the following steps:

1. Select the grinding wheel. The specification of the wheel selected for this job is 32-A-46-H-8.
2. Mount the grinding wheel on the spindle.
3. True and dress the grinding wheel.
4. Clean the surface of the magnetic chuck.
5. Clean the bottom of the die set. Inspect for nicks and burrs and remove if present.
6. Carefully place the die set on the magnetic chuck to avoid scratching or denting the surface of the magnetic chuck.
7. Carefully align the die set on the chuck. In this case the workpiece can easily be aligned by positioning it against the stop plate at the rear of the chuck.

Courtesy of the Norton Company

Fig. 13-8. Sharpening a hardened die on a surface grinding machine.

8. Turn on the magnetic chuck.
9. Check the die plate to make certain that the chuck is holding by pushing against the die plate with the hand.
10. Start the grinding wheel and turn on the coolant. Allow the grinding wheel to run with the coolant on for about 1 minute.
11. Adjust the automatic cross-feed to about .015 to .020 inch per table reversal. Start the automatic table traverse.
12. Adjust the table speed to about 50 feet per minute.
13. Adjust the length of the longitudinal table travel. This is done by adjusting the position of the two dogs on the side of the table which engage the table reverse lever. The length of the table stroke should be set so that the grinding wheel runs off the end of the surface being ground a distance of about 1 inch at each end.
14. Stop the flow of the coolant and start the longitudinal table travel. Position the table so that the wheel is over the edge of the die as shown in Fig. 13-8.

15. Slowly and carefully lower the grinding wheel until it just touches the surface of the die. Look for the first sign of a slight grinding spark, and listen for the sound of contact between the wheel and the work.
16. Start the flow of the coolant.
17. Take a cut across the surface of the work by engaging the automatic cross-feed. The purpose of this cut is to establish a dimensional relationship between the wheel and the surface being ground. If the wheel starts to cut too deeply as indicated by heavy sparking, back it away from the work immediately and start over again at the high spot on the surface of the work.
18. At the end of the first pass over the work, stop the cross-feed.
19. Feed the wheel into the work .0004 inch by turning the down-feed handwheel and reading the micrometer dial.
20. Take a second cut across the workpiece by engaging the cross-feed to move in the opposite direction. At the end of the cut stop the automatic cross-feed and the automatic table traverse.
21. Inspect the workpiece. In this case the condition of the surface and the edges of the die opening are carefully checked visually. If the part is to be ground to size, it is measured at this stage and the amount of stock remaining to be ground off is determined. If the part must be removed from the machine for this measurement, the chuck and the part must be cleaned before the part is placed back on the machine. Furthermore, the part must be placed in the same position on the table as before.
22. Repeat Steps 20 and 21 until the edges of the die openings are sharp.
23. Adjust the cross-feed of the wheel to .100 inch per table reversal, and increase the table speed to 70 feet per minute.
24. Take two or three additional cuts across the workpiece, using a depth of cut or down feed of .0001 inch per cut. (Sometimes the wheel is dressed again before the finish cuts are taken.)
25. Allow the wheel to pass over the entire surface of the work two or three times without any additional down feed in order to spark out.
26. Raise the wheel and shut off the coolant. Then stop the wheel.
27. Carefully remove the die set from the chuck to avoid scratching or denting the surface of the chuck.
28. Clean and dry the surface of the table.
29. If no additional work is to be done in the machine, allow the grinding wheel to run for a few minutes with the coolant shut off. Thoroughly clean the machine. Grinding swarf should not be left on any part of the table or the chuck.

The grinding conditions (table speed, cross-feed, and down feed) recommended for the above example might have to be altered somewhat depending upon the condition of the machine, the grinding wheel actually used, and the workpiece. The suggestions listed have, however, been successfully used and should be satisfactory for the average job when performed on a machine such as shown in Fig. 13-8. It is of interest to know that the final sharpening cut on new dies as well as resharpening of used dies is often done in the manner described.

While many jobs are rather easy to set up on the surface grinder, some other job setups must be planned with much thought and care. Typical is the job of linear form grinding the slender piercing punch, shown schematically in Fig. 13-9. A profile view of the punch is shown in view A. Before this job can be started, a conforming block, shown in view E, must be made from a magnetic parallel. Frequently a profiled surface of certain tools, such as the piercing punch, must be used to locate and orient the tool to a position which enables other surfaces to be ground, but the profile does not permit this surface to be seated against a normal plane surface on a magnetic chuck, angle plate, vise, or fixture. In this event a conforming block is made against which the profiled surface can be precisely located and seated while grinding. Some conforming blocks are made of soft steel, with provisions for clamping the tool to be ground.

Referring to Fig. 13-9, view B, the hardened tool steel block is ground to a size that is equal to the width of the head of the punch. In view C it is clamped against a toolmaker's angle plate and the length dimension of the punch is ground approximately .005 to .010 inch oversize. The grinding wheel is then trued and dressed to reflect the radius in the corner joining the taper and the head of the punch. With the punch located and held on a magnetic sine plate, the first tapered side and the corner are form ground, as shown in view D, leaving approximately .005 inch stock for finish grinding. The conforming block is then located on the magnetic chuck and the punch is placed on this block as shown in view E. To begin, the second tapered side is rough ground, then both sides are ground to the finish size by turning the punch over as required. A final "spark out" grind should be made on each side using the same downfeed setting of the grinding wheel. The last operation is to form grind the radius on the nose of a punch with a grinding wheel that has been trued and dressed to the required radius as shown in view F.

Surface Grinding Accessories

The full capability of the surface grinding machine cannot be utilized without certain basic accessories such as parallels, a toolmaker's vise, angle plates, V-blocks, and radius grinding wheel truing and dressing attachment, etc. Other accessories will extend the capability of this machine and are usually selected on the basis of the type of work done in individual shops. Since all accessories are themselves precision tools they must be handled with care and stored so that they will be protected.

Precision Parallels. Heat-treated steel precision parallels find much use in surface grinding work; they are used to support and align workpieces and to align other accessories on the magnetic chuck. They are sometimes

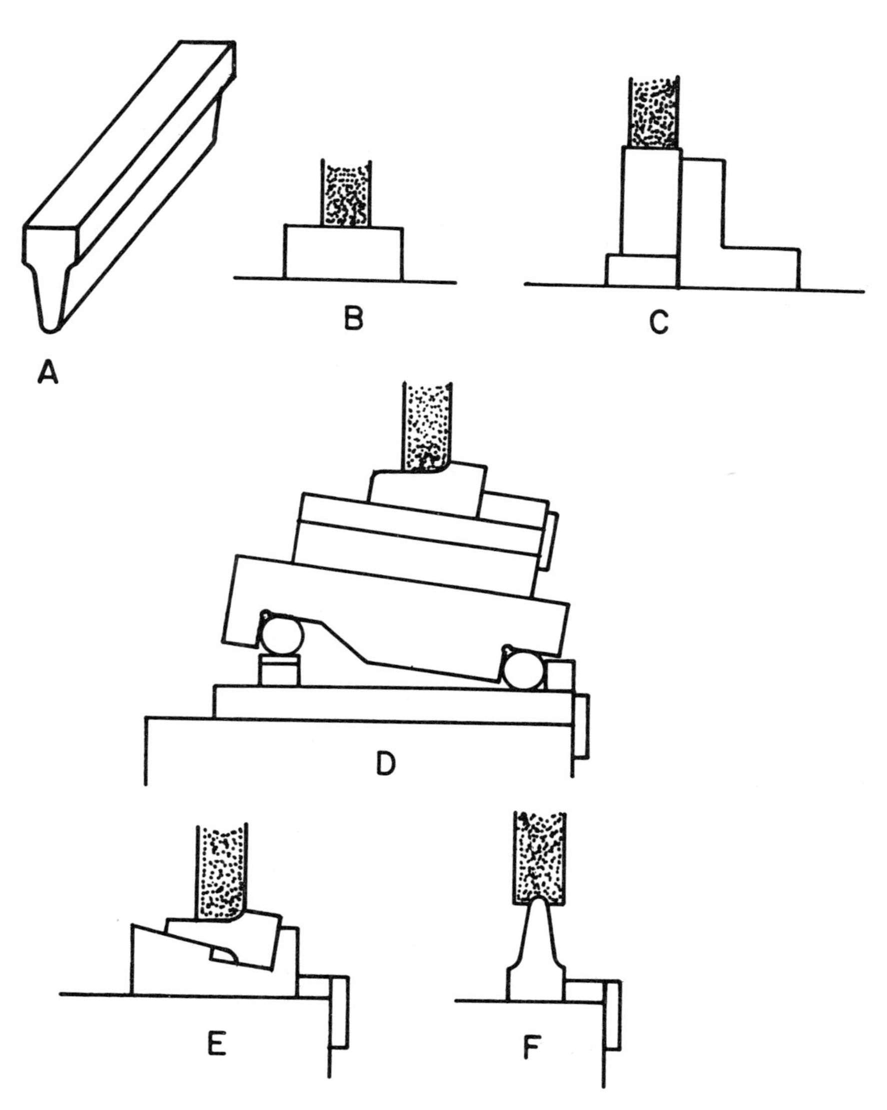

Fig. 13-9. Linear form grinding a piercing punch.

also used to decrease the magnetic pull on the workpiece by diverting this force through the parallel, which may be necessary to make certain that the workpiece will remain seated against a reference surface, such as a square-off bar on an angle plate, and not be pulled away from it.

Magnetic Parallels. These parallels are also made to precise standards of accuracy with respect to parallelism, squareness, and size. They are constructed by bonding together alternating plates of iron and brass. The brass is nonmagnetic, while the iron will conduct, but not retain, magnetism. Separated by brass plates, each iron plate will form a magnetic pole when subjected to an outside source of magnetism; when this source is removed, the parallel is essentially nonmagnetic. Magnetic parallels are used to hold and align steel and cast iron workpieces on magnetic chucks. Having more magnetic poles than a magnetic chuck, these parallels will usually hold small, thin workpieces more firmly than the chuck will. They can be machined to irregular shapes in order to make conforming blocks.

Precision Toolmaker's Vise. It is frequently more convenient to hold

Courtesy of the Engis Corporation; Diamond Tool Div.

Fig. 13-10. Precision toolmaker's vise used to hold a die section while form grinding with a grinding wheel that has been trued and dressed by a Diaform pantograph-type grinding-wheel truing and dressing attachment.

the workpiece in a vise while it is ground on the surface grinder; the job shown in Fig. 13-10 is typical. For surface grinding work the precision toolmaker's vise is most convenient to use because it is precise, yet not too heavy to handle. Often the workpiece is aligned and clamped in the vise on a surface plate, where precision tools can be used as aids in making the alignment. The workpiece in the vise is then carried over to the grinder and aligned on the magnetic chuck.

Precision Angle Plates. Precision angle plates provide a precise locating and clamping surface that is perpendicular to the face of the magnetic chuck. Some angle plates used on the surface grinder are constructed like magnetic parallels and are thus called magnetic angle plates; others are made from steel or cast iron. A most useful type of angle plate for surface grinding work is the knee type, shown in Fig. 13-11. All of the surfaces on this angle plate are machined parallel or perpendicular to each other to "tenth" (.0001 in. or 0.002 mm) tolerances. As in the case of work held in a toolmaker's vise, work held on smaller angle plates is usually set up on a surface plate and the workpiece and angle plate together are carried to the grinder to be aligned on the magnetic chuck. When the work and the angle plate are too heavy to carry, the setup must be made directly on the surface grinder. When setting up, all of the precautions with respect to cleanliness and care in aligning the workpiece must be observed.

Some examples of how a part can be set up and ground on a toolmaker's precision type angle plate are shown in Fig. 13-11. The parts are usually held with parallel clamps or with small C-clamps, which, however, are not shown in Fig. 13-11 for reasons of clarity. In view A the workpiece is aligned and raised by a precision steel parallel while clamped against the face of the angle plate. A square-off bar is attached to the angle plate shown in view B. This bar is perpendicular to the base of the angle plate and when the workpiece is seated against it, the ground surface will be perpendicular to the side of the workpiece held against the square-off bar, as well as to the face of the angle plate. A steel parallel is placed below the workpiece to reduce any tendency of the magnetism to pull the workpiece away from the square-off bar. Although the workpiece may rest on the parallel, it should not be aligned by the parallel.

The inside steps of the precision knee-type angle plate provide additional locating and clamping surfaces that are convenient for aligning and holding smaller parts. For example, in Fig. 13-11, view C, a small part is located by and held against the two surfaces of a step, one of which serves the purpose of a large parallel. The angle plate may be placed on its side, as in view D, which is a convenient way to hold the workpiece when grinding the end perpendicular to the sides. In view E the workpiece is being aligned at an angle with a sine bar to set it up for grinding an angular surface. This setup is made on a surface plate. The angle is then ground on the surface grinder, as shown in view F.

Precision V-Blocks. Precision hardened steel and magnetic V-blocks are used for a variety of purposes on the surface grinder. Their primary purpose is to locate and hold cylindrical parts, as shown in Fig. 13-12, view A.

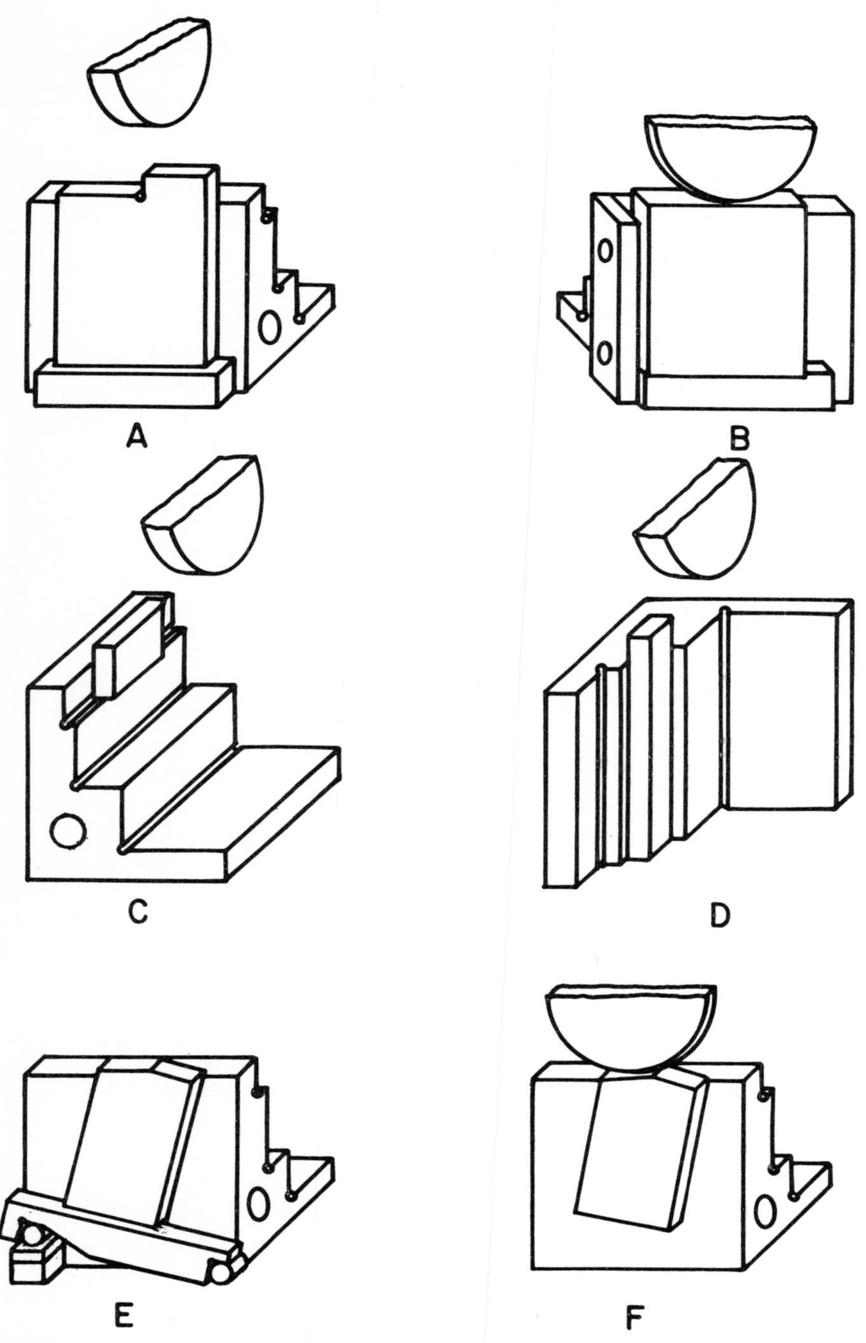

Fig. 13-11. Surface grinding setups that can be made on a toolmakers precision knee-type angle plate. The clamps holding the parts are not shown.

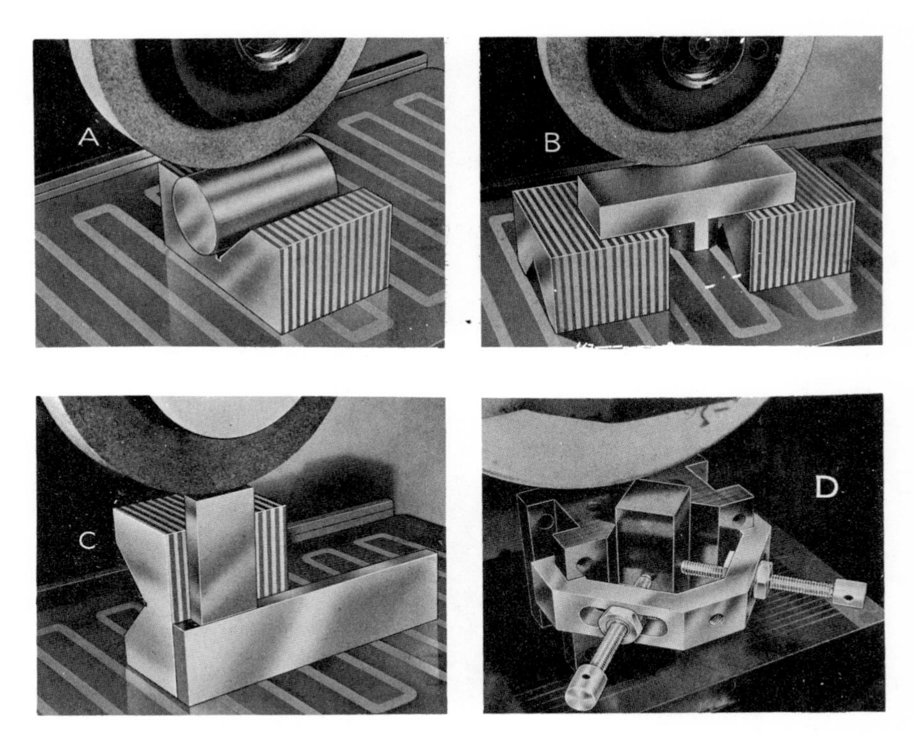

Courtesy of the Anton Machine Works

Fig. 13-12. A. Magnetic V-block holding a cylindrical workpiece. B. Magnetic V-block inverted and used as magnetic parallels. C. Magnetic V-block used as an angle plate. D. Hardened steel V-block used to hold workpiece for grinding the end face.

Longer cylindrical parts are held in two matched V-blocks. A matched pair of magnetic V-blocks is shown in view B. They are being used as magnetic parallels to raise the workpiece above the face of the chuck in order to clear a boss located on the lower surface of the workpiece. In view C the magnetic V-block is used as an angle plate, holding the workpiece by the magnetic force transferred through the V-block from the magnetic chuck. A hardened steel V-block is shown in view D holding a square workpiece; held in this manner the end of the workpiece can be ground perpendicular to the sides.

Sine Bars and Sine Plates. Sine bars and sine plates are used to hold a workpiece at a precise angle. Sine plates have a larger plane surface to which the workpiece may be clamped; they may be used directly on a machine tool to hold the workpiece, whereas sine bars are generally not used in this manner. Sine bars are primarily used to align the workpiece or to "set it up," as shown in Fig. 13-11, view E. A sine plate set up on a surface grinder is shown in Fig. 13-13; here the workpiece is held in a V-block which is clamped to the sine plate. Since only the roll attached to

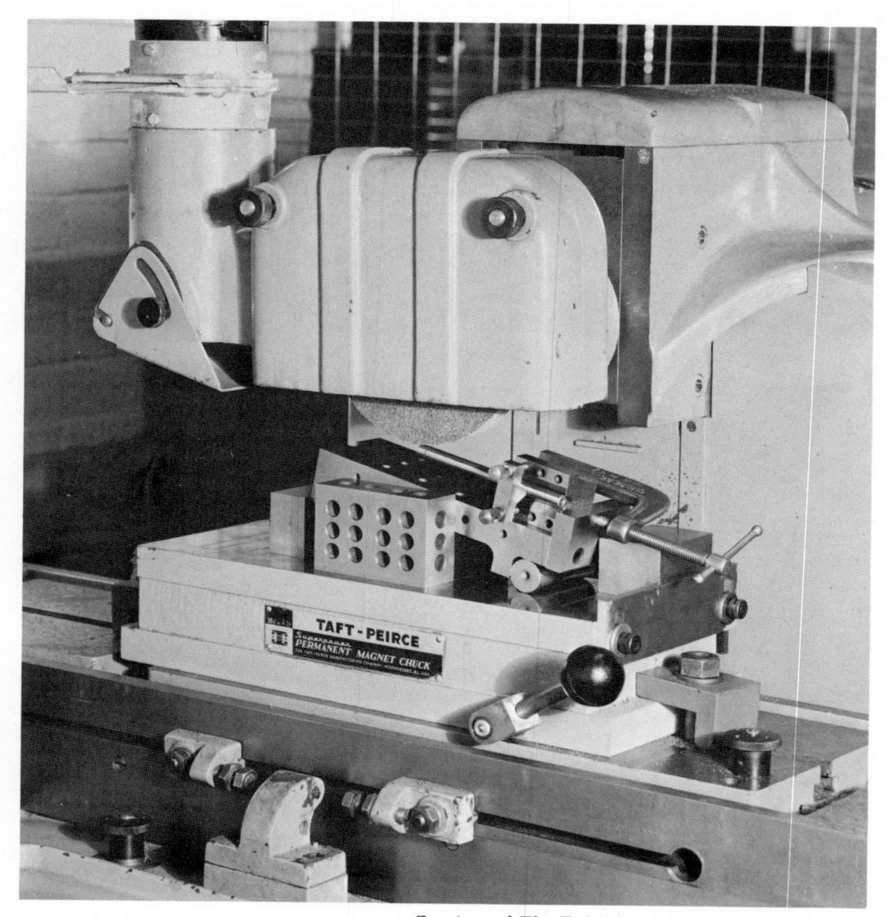

Courtesy of The Taft-Pierce Manufacturing Company

Fig. 13-13. Using a sine plate to locate the workpiece for grinding a precise angle.

the sine plate is in actual contact with the magnetic chuck and does not provide too much holding power, the sine plate must be blocked to keep the workpiece in place. Only light cuts should be taken and a relatively slow table traverse rate should be used. A compound sine plate is shown in Fig. 13-14. Compound sine plates are used to hold the workpiece at a compound angle by simultaneously tilting the work-holding plate in two perpendicular directions. On this compound sine plate the work-holding surface is a permanent magnet chuck. This sine plate also has a base plate which provides a large surface for holding it firmly on the magnetic chuck.

Sine bars and sine plates have two rolls attached to their bottom surfaces. Both rolls are exactly the same diameter and are precisely either

5 or 10 inches (125 or 250 mm) apart. The angular setting is obtained by having one of these rolls rest on a stack of precision gage blocks. The height of the gage block stack required to tilt the sine bar or sine plate at a given angle can be calculated by the following formula:

$$H = L \sin \theta \tag{13-1}$$

Where: H = The height of the precision gage block stack, in inches or millimeters
θ = The angle at which the sine bar or sine plate is to be set, in degrees and minutes
L = The distance between the rolls, in inches or millimeters

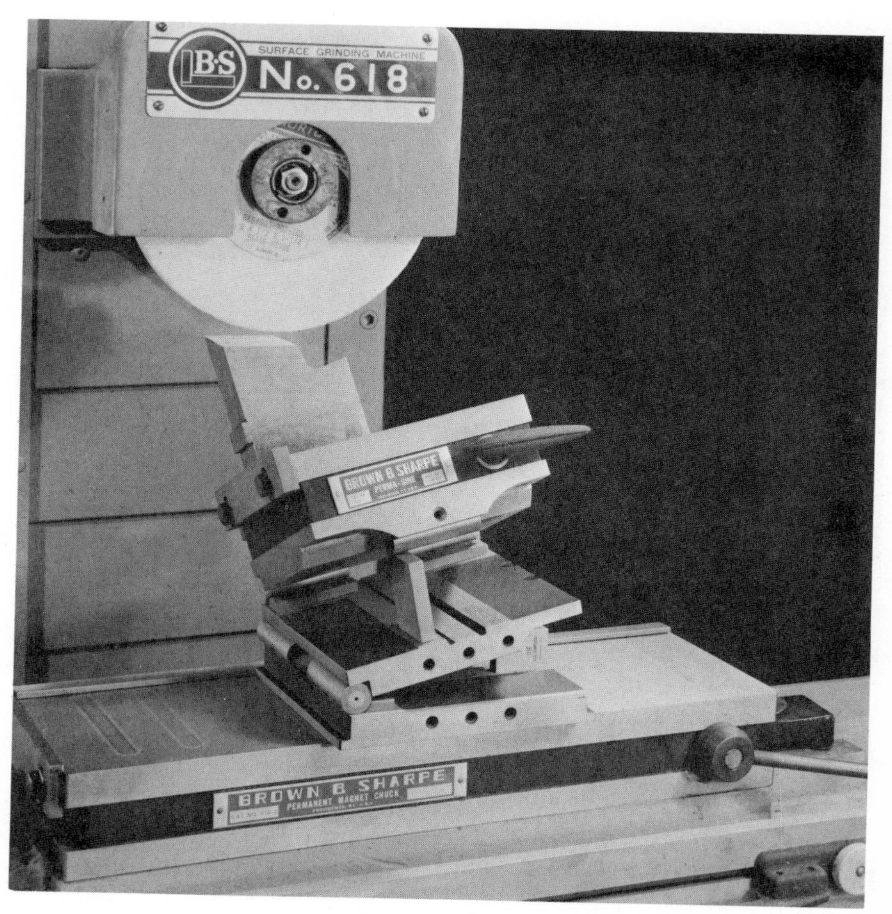

Courtesy of the Brown & Sharpe Manufacturing Company

Fig. 13-14. Compound sine plate with permanent magnet chuck on work-holding surface.

Magnetic Adjustable Work Holders. Constructed on the principle of magnetic parallels, these holders have a work-holding surface that can be tilted at an angle. An example of how these work holders are used is shown in Fig. 13-15. The required angle is obtained by placing one of the two pins located on the end of the holder against a universal precision gage or against precision gage blocks.

Magna-Lock Clamps. These clamps, as shown in Fig. 13-16, have comb-like teeth which grip the sides of the workpiece. They are made by joining two pieces of heat-treated steel with a thin piece of spring steel that is slightly bent to form a convex and a concave side to the clamps. Used in pairs, they are placed on each side of the workpiece, with the teeth against

Courtesy of the Anton Machine Works

Fig. 13-15. Magnetic adjustable work holder used to grind angles.

Courtesy of the DoALL Company

Fig. 13-16. Magna-Lock clamps used to hold thin part on a magnetic chuck.

the workpiece and the concave side facing the magnetic chuck. When the power is turned on they are pulled flat against the magnetic chuck and the teeth will grip the workpiece. Magna-Lock clamps are used to assist in holding steel and cast-iron parts; they are especially useful for holding parts made from non-magnetic materials, which cannot be held by the magnetic chuck.

Radius Truing and Dressing Attachment. This attachment is used to true and dress precise radii on grinding wheels. Some have added capability, such as the attachment shown in Fig. 13-17 which can be used to true and dress radii, tangents to a radius, and angles on the grinding wheel.

Pantograph Type Truing and Dressing Attachment. Linear form grinding requires that the grinding wheel be trued and dressed to the contoured profile that will produce the required profile on the workpiece. The pantograph type truing and dressing attachment, shown in Fig. 13-18, is designed to true and dress a precise contoured profile on the face of the grinding wheel. A sheet-metal template is made on which the profile is 10 times larger than the profile to be made on the grinding wheel. This template is mounted on the carrier slide of the pantograph attachment and the operator follows around the template with a tracer. The movement of the tracer is reduced ten times by the pantograph linkage and transmitted to two single-point diamonds which are held in separate holders. One diamond is used to rough form and the other to fine finish the profile. By making

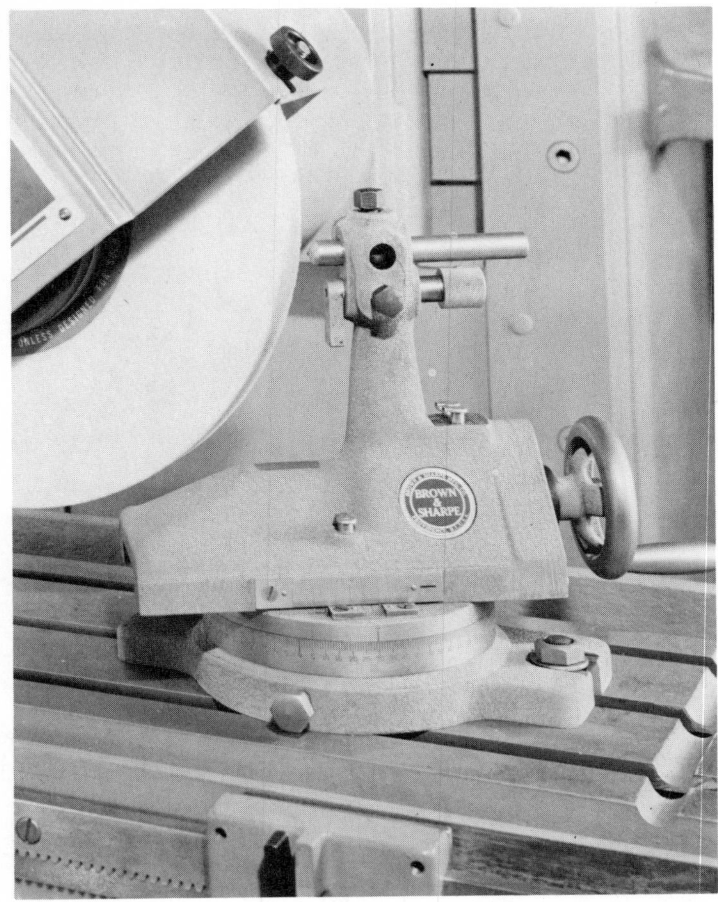

Courtesy of the Brown & Sharpe Mfg. Co.

Fig. 13-17. Radius and angle truing attachment.

successive passes of the tracer along the template, the profile is gradually produced on the grinding wheel. A form-grinding operation using a grinding wheel trued and dressed with this attachment is shown in Fig. 13-10.

Optical Projector. An optical projector system is shown in Fig. 13-19, which projects a true, magnified picture of the workpiece on the screen at all times. By using different lenses a range of magnifications is available, extending from 5:1 to 100:1. Three different procedures can be used to grind a profiled contour using the optical projection system.

One procedure is to first make an enlarged drawing of the surface to be ground. This drawing is then mounted on the screen of the projector. As the surface is being ground the operator observes the ground surface, the

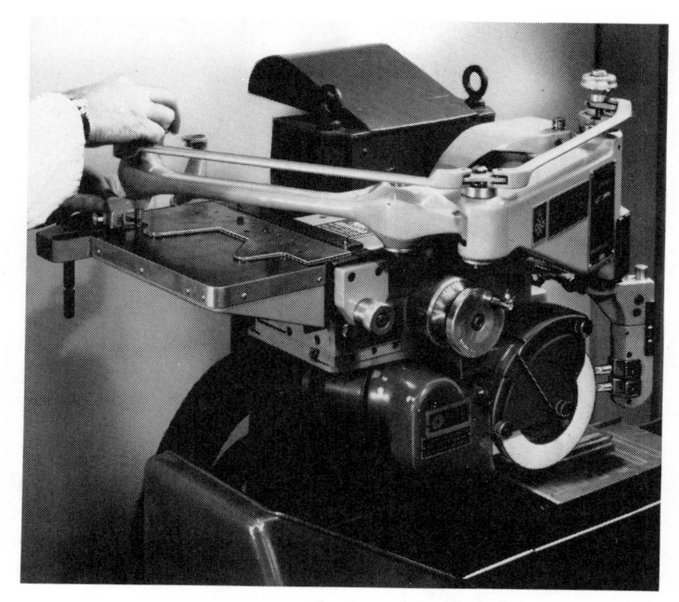

Courtesy of the Engis Corporation; Diamond Tool Div.

Fig. 13-18. Diaform pantograph type grinding wheel truing and dressing attachment used to dress profile contours on the face of grinding wheels.

drawing line, and the image of the grinding wheel on the screen. While the table is traversing back and forth, the operator manually adjusts the transverse and vertical feed until the projected image of the workpiece coincides with the line of the drawing. A second method is to make a layout of the profile on the end of the workpiece. Lines that are .001-inch (0.025-mm) wide, or less, will project on the screen in sharp detail. The workpiece is then ground to size by observing the layout line on the screen and grinding to this line. The third method is used to grind a punch to fit an existing die opening. The punch is first rough machined and heat treated. A piece of brass shim stock is epoxied onto the face of the punch. The oversize punch is pressed into the die opening about .002 to .005 inch (0.05 to 0.13 mm), which will form a shear line impression on the punch that will project on the screen. The punch can then be ground to this impression line, using the optical projector to guide the operator in manipulating the machine.

Surface-Grinding Thin Workpieces (Residual Stresses)

Residual stresses are stresses that are locked into the crystals of the metal. They are caused by rapid quenching in heat treatment, by severe cold working of the metal—such as in cold rolling or cold drawing, by the action of single- and multiple-point cutting tools, and by uneven solidification and cooling in a casting. Residual stresses are also caused by the

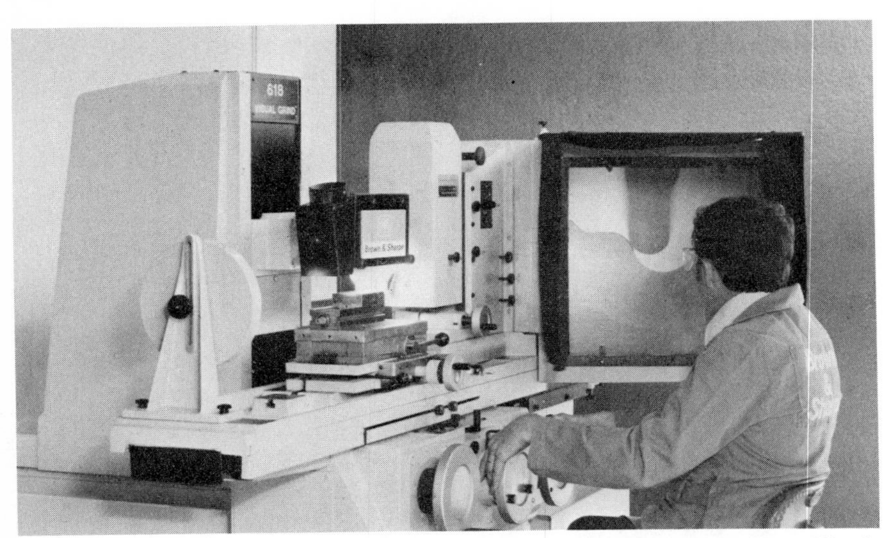

Courtesy of the Brown & Sharpe Mfg. Co.

Fig. 13-19. Visual Grind Optical System grinding a contoured profile on two identical forming die sections.

action of the grinding wheel, which heats a small section of the workpiece to a high temperature. An instant later the workpiece is rapidly cooled by the quenching action of the surrounding colder metal and by the grinding fluid. Usually residual stresses within the grains of the metal counteract each other and cause the part to be stable. In this condition they are said to be "locked in." If something occurs to unbalance the locked-in residual stresses, they will act until they are again in balance. The manner in which the unbalanced residual stresses act to balance each other is to bend or distort the workpiece until new stresses resulting from the deflection will again balance the residual stresses. The only practical method of removing the residual stresses is by a stress-relief heat treatment.

Thin pieces of cold drawn stock which have been severely cold-worked contain very large residual stresses. When a layer of this stock is ground off, the unbalanced stresses are large enough to cause it to bend when it is released from the magnetic chuck. The result is that the ground surface will not be flat. If the residual stresses in thin cold-drawn or cold-rolled stock cannot be removed by a stress-relief heat treatment prior to surface grinding, the only recourse is to grind off alternately a very small amount from the opposite sides of the workpiece.

If a thin workpiece is warped before it has been ground, it should not simply be placed on the magnetic chuck and the power turned on. The magnetic force will pull the part of the surface not touching the chuck flat against the surface of the chuck and hold it in this position while it is being ground. When the power of the chuck is released, the workpiece will spring back to approximately its original shape. The resulting face will thus be curved instead of flat. Alternately grinding the opposite sides of the workpiece will not correct this condition. This can only be corrected by placing shims below that part of the lower surface not touching the chuck when the magnetic chuck is turned off. Paper shims may be used if the original warpage is small. If it is large, metal shims will be necessary. The surface that is ground when the workpiece is properly shimmed will be flat, unless the grinding operation has caused or released some residual stresses. If the two opposite surfaces must be parallel to each other, the two sides should be alternately ground; and each time shims must be placed where required before the magnetic chuck is turned on. As already mentioned, the grinding wheel itself is the cause of residual stresses. The magnitude of these stresses is in direct relationship to the severity of the grinding action. When thin workpieces are ground, the grinding cuts should be light — otherwise the residual stresses produced by the grinding wheel will cause the part to bend.

Hardened tool steel must be carefully ground. The residual stresses caused by taking heavy cuts in hardened tool steel can be severe enough to cause the formation of a crack in the surface. Therefore, deep heavy grinding cuts should be avoided when grinding hardened tool steel.

CHAPTER **14**

Cutter and Tool Grinding

The cutter and tool grinding machine is specifically designed to sharpen a wide variety of cutting tools. It is practically a supplement to the milling machine because it is the most efficient machine for sharpening all of the many types of milling cutters. Single-point tools as well as other tools can also be sharpened on this machine. In addition, it is capable of performing a variety of other light grinding operations including cylindrical grinding, internal grinding, and surface grinding. This chapter will deal primarily with the application of the cutter and tool grinding machine in sharpening milling cutters and reamers.

Before this chapter is read, the names of the milling cutter elements given in Chapter 5 should be reviewed. Frequent reference will be made in this section to the term *milling-cutter land.* The word "land" refers to the narrow surface back of the cutting edge which is ground to provide the clearance angle. Milling cutters generally have a primary land and a secondary land. The primary land is immediately behind the cutting edge. The width of this land is kept small in order to reduce the amount of metal that must be ground. Thus less heat will be generated and the danger of overheating the cutting edge avoided. The secondary land is ground to keep the width of the primary land to a small size.

The Cutter and Tool Grinding Machine

In construction the cutter and tool grinding machine is somewhat similar to the cylindrical grinding machine, although there are some basic differences. A cutter and tool grinding machine is illustrated in Fig. 14-1. The wheel head of cutter and tool grinding machines can be raised and lowered but is not built to traverse toward the workpiece as in the case of cylindrical grinding machines. The movement of the work toward and away from the wheel is provided by a cross slide, on top of which the sliding table traverses longitudinally. On top of the sliding table rests a swivel table which can be set in angular positions. The machine in Fig. 14-1 is provided with a highly accurate device which utilizes precision gage blocks for obtaining precise angular settings of the swivel table. The wheel head of this machine can also be moved in two angular directions. There are two lower swivels: the column swivel and the eccentric swivel, both of which can be turned 360 degrees about a vertical axis. An upper swivel (Fig. 14-14) permits the wheel head to be tilted up or down about

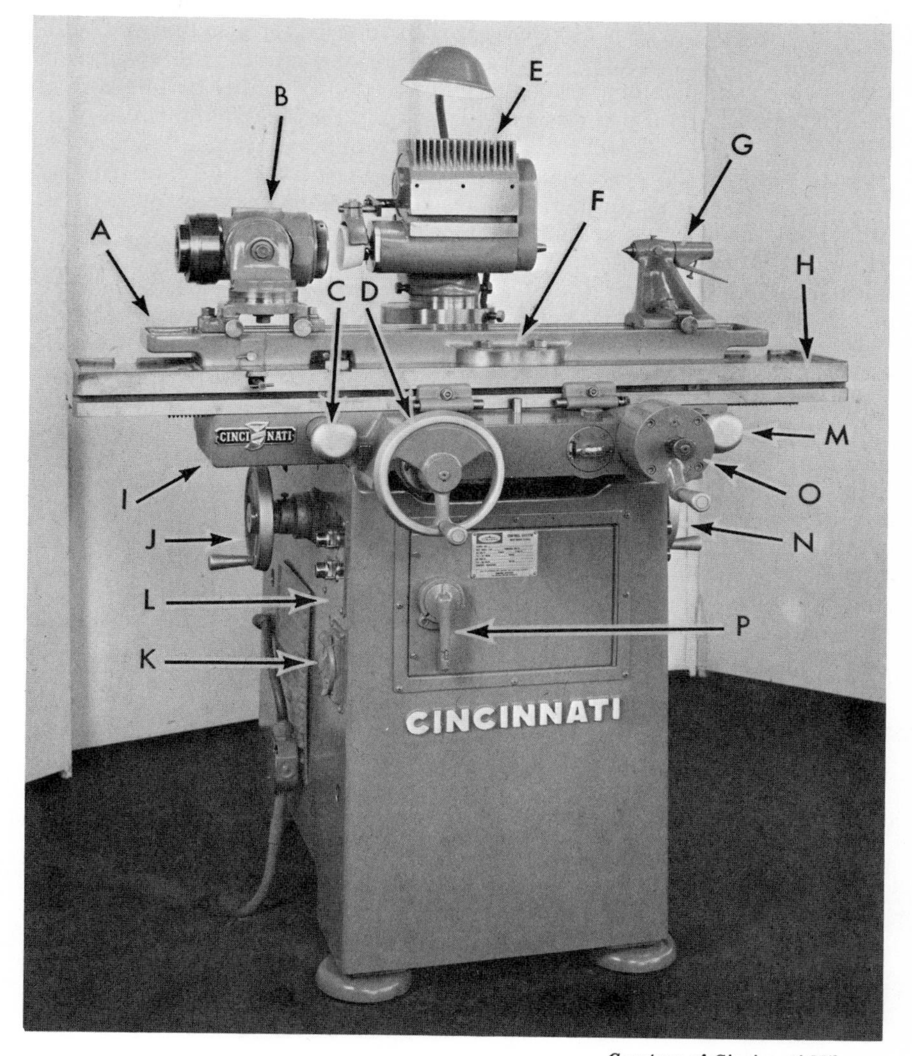

Courtesy of Cincinnati Milacron

Fig. 14-1. A cutter and tool grinding machine. Component parts are: A. swivel table; B. workhead; C. front table hand control; D. front cross slide handwheel; E. tilting wheel head; F. table swivel scale; G. right-hand tailstock; H. sliding table; I. cross slide; J. wheel-head vertical control handwheel; K. attachment power receptacle; L. main electrical control panel; M. front table hand control; N. wheel-head vertical control handwheel; O. differential table traverse control; P. main disconnect switch.

a horizontal axis, but not all cutter or tool grinding machines are provided with this arrangement (Fig. 14-9).

The workhead, which can be swiveled 360 degrees in two planes, has a graduated clearance setting dial for obtaining the desired clearance angles on milling cutters. The workhead is used primarily for grinding end milling and face milling cutters. Large face milling cutters are ground on a large face-mill grinding attachment. Most arbor-mounted milling cutters are ground with the use of two tailstocks. Only the right-hand tailstock is shown in Fig. 14-1. A great variety of additional attachments either extend the range of the work that can be done on the cutter and tool grinding machine or facilitate the performance of the more common operations. Cylindrical grinding operations can be performed using the cylindrical grinding attachment. Internal grinding (Fig. 14-2) is done with the cylindrical grinding attachment and an internal grinding spindle, which can be attached to the wheel head. A radius grinding attachment provides a means of grinding an accurate radius in the corner of milling cutters if desired. A universal vise can be used to hold the workpiece to perform surface grinding operations and to grind the angles on single-point cutting tools. These are but a few of the attachments available for use on the cutter and tool grinding machine.

In order to grind milling cutters, a group of precision arbors should be made or otherwise obtained. The diameter of these arbors should be made to provide a close sliding fit in the bores of the milling cutters to be ground. The cutters should be held firmly in place by means of nuts and collars similar to milling-machine arbors. They should have no eccentricity, and the shoulders of the arbor and the collars should be perfectly

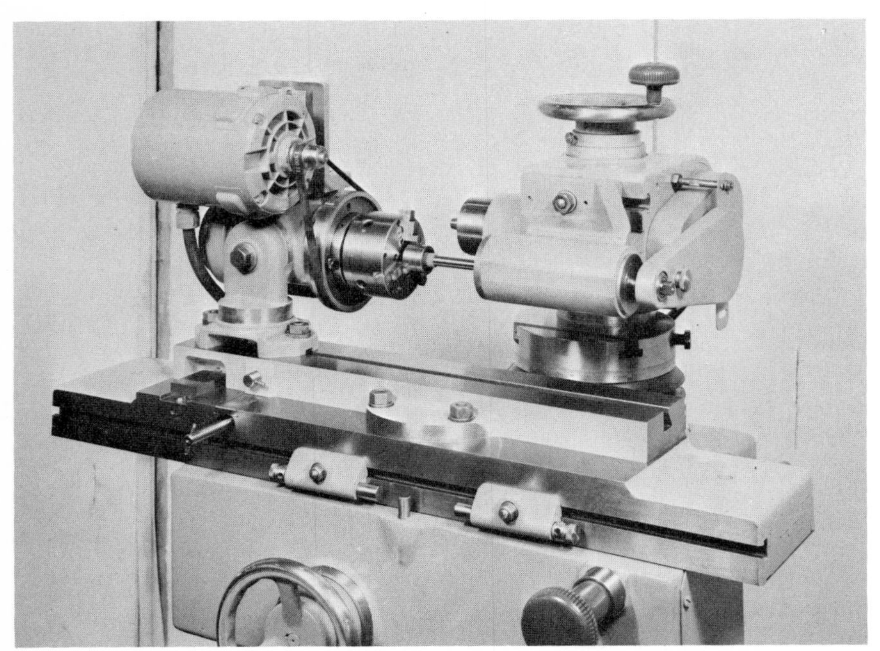

Courtesy of Cincinnati Milacron

Fig. 14-2. Performing an internal grinding operation on a cutter and tool grinder.

square. The accuracy with which the cutters are ground depends to a large extent upon the accuracy of the arbors.

Basic Methods of Grinding Milling Cutters

Four of the basic methods of grinding milling cutters are illustrated in Fig. 14-3. The milling cutter is mounted on a close-fitting arbor, and the table is traversed past the grinding wheel. Either a disc-type grinding wheel or a flaring-cup type grinding wheel can be used. The disc-type grinding wheel, A and B in Fig. 14-3, tends to produce a curvature on the ground surface that is equal to the radius of the wheel. This is not necessarily objectionable if a large-diameter wheel is used and if the width of the ground surface or land behind the cutting edge is kept small. The diameter of the disc-type grinding wheel should be at least 6 or 8 inches, and the width of the land should be between $\frac{1}{32}$ and $\frac{1}{16}$ inch. A narrow wheel should be used, or the wheel should be trued so that the grinding face of the wheel is narrow. Flaring-cup wheels should be trued so that the face is slightly concave at the grinding face. The wheel head should be

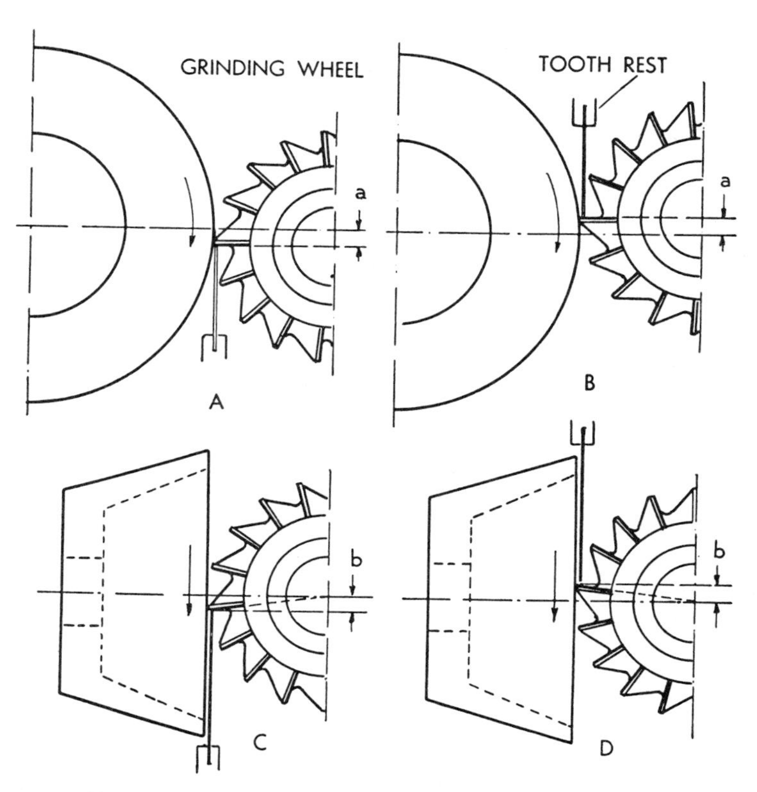

Fig. 14-3. The basic methods of grinding milling cutters.

turned 1 degree about the vertical axis so that only one face of the flaring-cup wheel will grind the cutter and the other face will clear the cutter. The milling cutter is held against the tooth-rest blade by hand. Except in certain cases, the tooth-rest blade is placed close to the wheel with approximately .003 to .004 inch clearance.

When a disc-type grinding wheel is used to grind the clearance on milling cutters as shown at A and B in Fig. 14-3, the wheel head is raised a distance, a, above or below the axis of the cutter. This distance can be calculated by the following formula:

$$a = .0087\, D_G\, C \tag{14-1}$$

where: a = The vertical distance that the grinding wheel is above or below the axis of the milling cutter, inches

D_G = Diameter of the grinding wheel, inches

C = Clearance angle to be ground on the milling cutter, degrees

The diameter of grinding wheels can be accurately measured with micrometer calipers by placing one piece of good notebook paper between the grinding wheel and the anvil of the micrometer and another between the grinding wheel and the micrometer spindle. The paper protects the precision measuring surfaces of the micrometer caliper by preventing them from contacting the abrasive particles on the grinding wheel. The thickness of the two strips of paper must be subtracted from the micrometer reading.

When grinding as shown at A, Fig. 14-3, the pressure of the grinding wheel tends to hold the milling cutter against the blade of the tooth rest. A disadvantage of this method is that since the wheel is above the axis of the cutter, the abrasive particles of the wheel are moving away from the face of the cutting edge. This tends to produce a burr on the cutting edge. From the standpoint of grinding procedure, the grinding wheel should rotate so that the abrasive particles move toward the cutting edge as shown at B. This procedure, however, has a major disadvantage in that the grinding-wheel pressure tends to pull the milling cutter away from the blade of the tooth rest and into the wheel with serious consequences. Although both procedures are used, the safer procedure shown at A is recommended.

Flaring-cup wheels as shown at C and D, Fig. 14-3, are generally recommended over disc wheels for grinding milling cutters. The milling cutter can be positioned so that the grinding pressure holds the wheel against the tooth-rest blade as at C, or it can be positioned so that the abrasive particles move toward the cutting edge as at D. The procedure shown at C is recommended. In either case the axes of the grinding wheel and the milling cutter are positioned on approximately the same horizontal plane, and the tooth rest is positioned a distance, b, above or below these axes. The distance b can be calculated by the following formula:

$$b = .0087\, D_C\, C \tag{14-2}$$

Courtesy of Cincinnati Milacron

Fig. 14-4. Clearance setting fixture used to set the milling cutter for grinding the desired clearance angle.

where: b = The height of the tooth-rest blade and the axis of the milling cutter, inches
D_C = Diameter of the milling cutter, inches
C = Clearance angle to be ground on the milling cutter, inches

When measuring the diameter of the milling cutter with micrometer calipers, place a piece of good notebook paper between each of the measuring surfaces on the micrometer and the cutter.

A clearance setting fixture, shown in Fig. 14-4, can be used to position the cutter to grind the desired clearance angle. This attachment consists of a fixed plate which has a zero reference mark on its upper surface. The fixed plate can be firmly attached to the tailstock spindle by tightening a set screw. The graduated plate is the plate ahead of the fixed plate. It has graduations in degrees marked on its top surface, and it can be secured to the fixed plate by turning a thumbscrew located on its bottom. A special dog is provided that attaches to the arbor holding the cutter. The dog has a pin which fits in a hole located in the front of the graduated plate. The procedure for using this fixture will be treated later in this chapter.

Some cutter and tool grinding machines are equipped with a wheel head that can tilt. The desired clearance angle on the milling cutter is thus obtained by tilting the wheel head as shown in Fig. 14-10. A disc wheel cannot be used when tilting the wheel head to obtain the desired clearance angle. The clearance angle can only be obtained when grinding with the face of a cup wheel. Generally a flaring-cup wheel is preferred.

These setups can be used to grind the peripheral, side, and end teeth on all milling cutters. As shown in Fig. 14-5, the cup type grinding wheel will grind a flat relief (view A), while the disc type wheel will grind a slightly concave relief (view B). Actually, the difference is extremely small since the width of the land is small in relation to the diameter of the disc type wheel.

An eccentric relief is shown in view C, Fig. 14-5. This type of relief can be ground only on the peripheral cutter teeth and only on those peripheral teeth that have been cut at a helix. Moreover, the helix angle must be at least 8 degrees and there is a minimum size relief angle that can be ground, depending on the size of the helix angle. This type of relief is used primarily on the peripheral teeth of end mills, especially on those of smaller size. A somewhat wider primary relief land is used with this type of relief and the relief angle can be at the lower range of recommended relief angles, as given later in Table 14-2.

Figure 14-6 illustrates the setup for grinding an eccentric type relief. In this setup the cutter axis, the grinding-wheel axis, and the point of contact with the cutter on the tooth-rest blade must all be in the same horizontal plane. A disc type grinding wheel must be used. It is trued so that the face of the wheel is at an angle, W; as an alternative, the grinding machine spindle head can be rotated so that the face of the grinding wheel is at this angle. The eccentric relief is ground by moving the cutter back and forth past the grinding wheel while it is resting on the stationary tooth-rest blade, which will cause it to rotate by following the path of the helix angle, H, of the flute. The action causes the wide surface of the grinding wheel to generate a convex shaped relief angle on the cutter teeth. Using the tangent function found in a table of natural trigonometric functions, the wheel angle is calculated by the following formula:

$$\tan W = \tan H \times \tan R \qquad (14\text{-}3)$$

where: W = Wheel angle
H = Helix angle of the cutter teeth
R = Relief angle to be ground on the cutter.

The wheel angles for a wide range of frequently used radial relief angles and cutter helix angles are provided in Table 14-1.

Tooth-Rest Blades and Holders

A tooth-rest blade and holder are shown in position in Fig. 14-7. In some cases the blade must be kept stationary while the milling cutter is moved back and forth across the face of the grinding wheel. In this case the tooth-rest holder is clamped to the wheel head as shown. At other times (Fig. 14-14) the blade should be moved with the milling cutter, in which case the holder is clamped to the table of the machine or onto the

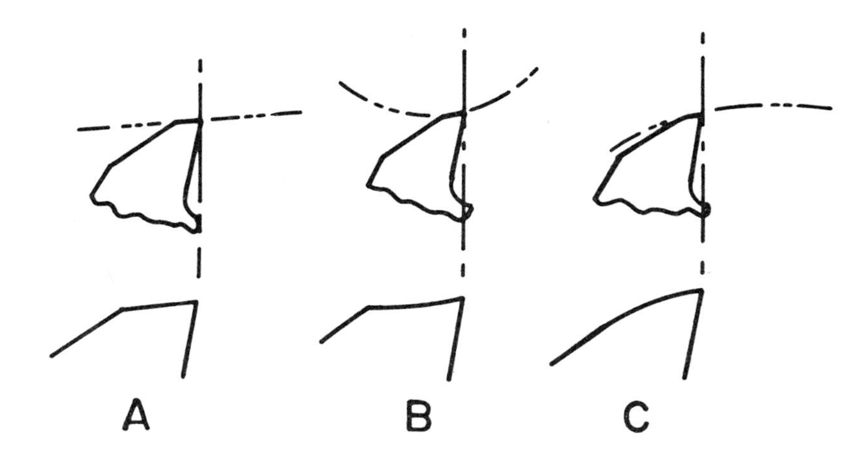

Fig. 14-5. The basic types of relief surfaces. A. Flat, as ground with a cup wheel; B. Concave, as ground with a disc wheel; C. Eccentric.

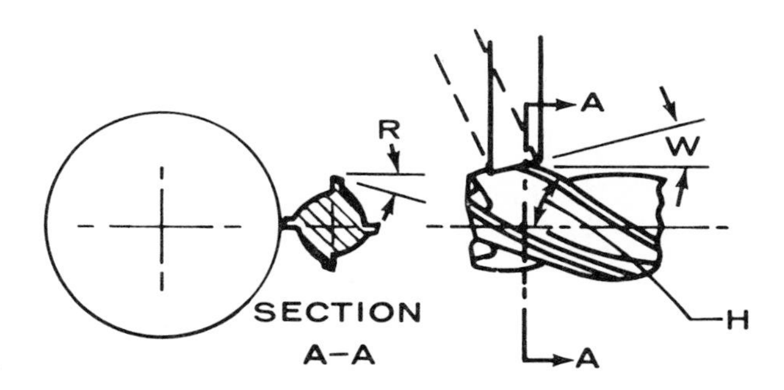

Fig. 14-6. Setup for grinding an eccentric relief.

workhead. There are two kinds of tooth-rest holders, plain and universal. A universal tooth-rest holder, shown in Fig. 14-7, is provided with a micrometer adjustment for obtaining small up and down movements in the final setting of the blade. The blade of the tooth-rest holder can, when necessary, be made to pivot aside in order to index the cutter. A spring will cause the blade to snap back in place against a fixed stop. This feature is used when the tooth rest moves along with the cutter. When the tooth rest is in a fixed position as in Fig. 14-7, the cutter is moved off the blade in order to index the cutter teeth.

The most commonly used tooth-rest blade shapes and dimensions are shown in Fig. 14-8. The plain tooth-rest blade is used when grinding milling cutters with straight teeth, and the rounded tooth-rest blade is used for cutters with helical teeth. Sometimes the tooth-rest holder will interfere by not allowing the milling cutter to be brought up to the

Table 14-1. Grinding Wheel Angles for Grinding Eccentric Type Radial Relief Angles

Radial Relief Angle, *R*, Degrees	Helix Angle of Cutter Flutes, *H*, Degrees							
	12	18	20	30	40	45	50	52
	Wheel Angle, *W*, Degrees							
1	0°13′	0°19′	0°22′	0°35′	0°50′	1°00′	1°12′	1°17′
2	0°26′	0°39′	0°44′	1°09′	1°41′	2°00′	2°23′	2°34′
3	0°38′	0°59′	1°06′	1°44′	2°31′	3°00′	3°34′	3°50′
4	0°51′	1°18′	1°27′	2°19′	3°21′	4°00′	4°46′	5°07′
5	1°04′	1°38′	1°49′	2°53′	4°12′	5°00′	5°57′	6°23′
6	1°17′	1°57′	2°11′	3°28′	5°02′	6°00′	7°08′	7°40′
7	1°30′	2°17′	2°34′	4°03′	5°53′	7°00′	8°19′	8°56′
8	1°43′	2°37′	2°56′	4°38′	6°44′	8°00′	9°30′	10°12′
9	1°56′	2°57′	3°18′	5°13′	7°34′	9°00′	10°41′	11°28′
10	2°09′	3°17′	3°40′	5°49′	8°25′	10°00′	11°52′	12°43′
11	2°22′	3°37′	4°03′	6°24′	9°16′	11°00′	13°03′	13°58′
12	2°35′	3°57′	4°25′	7°00′	10°07′	12°00′	14°13′	15°13′
13	2°49′	4°17′	4°48′	7°36′	10°58′	13°00′	15°23′	16°28′
14	3°02′	4°38′	5°11′	8°11′	11°49′	14°00′	16°33′	17°42′
15	3°16′	4°59′	5°34′	8°48′	12°40′	15°00′	17°43′	18°56′
16	3°29′	5°19′	5°57′	9°24′	13°32′	16°00′	18°52′	20°09′
17	3°43′	5°40′	6°21′	10°01′	14°23′	17°00′	20°01′	21°22′
18	3°57′	6°02′	6°45′	10°37′	15°15′	18°00′	21°10′	22°35′
19	4°11′	6°23′	7°09′	11°15′	16°07′	19°00′	22°19′	23°47′
20	4°25′	6°45′	7°33′	11°52′	16°59′	20°00′	23°27′	24°59′
21	4°40′	7°07′	7°57′	12°30′	17°51′	21°00′	24°35′	26°10′
22	4°55′	7°29′	8°22′	13°08′	18°44′	22°00′	25°43′	27°21′
23	5°09′	7°51′	8°47′	13°46′	19°36′	23°00′	26°50′	28°31′
24	5°24′	8°14′	9°12′	14°25′	20°29′	24°00′	27°57′	29°41′
25	5°40′	8°37′	9°38′	15°04′	21°22′	25°00′	29°04′	30°50′

grinding wheel. This can usually be corrected by means of an offset tooth-rest blade which is shown in use in Fig. 14-9. The hook tooth-rest blade is used for grinding cutters with a small diameter, when, because of the spacing of the teeth, the tooth-rest blade cannot be below the grinding wheel and the cutter. Blades with shapes other than those shown are sometimes required. An example of another blade shape can be seen in Fig. 14-16.

Sharpening a Plain Milling Cutter

The several methods for sharpening a plain milling cutter have already been presented in broad outline in a previous section. Two methods that are frequently used will be given in a step-by-step procedure here. One method will explain how a plain milling cutter can be sharpened on a fixed wheel-head machine, and the other will show how this can be done

Courtesy of Cincinnati Milacron

Fig. 14-7. A universal tooth-rest holder in position for grinding.

on a tilting wheel-head machine. Since the principal difference in these methods is the way in which the job is set up, the actual grinding procedure will be given only once. In starting with the fixed wheel-head machine, it is assumed that a clearance setting dial will be used. The setup is illustrated in Fig. 14-9.

1. Select the correct grinding wheel and mount it on the wheel spindle. Attach the wheel guard in position.
2. Start the machine. True and dress the grinding wheel using an abrasive stick. The face of the wheel should be made into a V-shaped profile. The sides of the V should be approximately 20 or 30 degrees with respect to the face. The apex of the V should

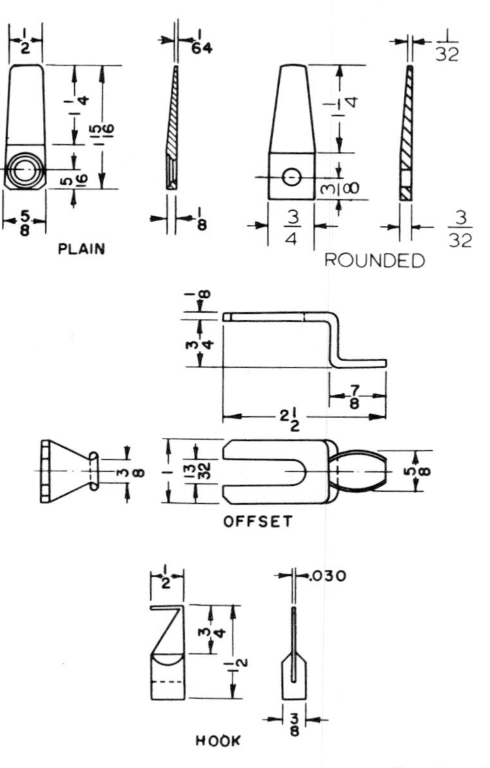

Courtesy of Cincinnati Milacron

Fig. 14-8. Types of tooth-rest blades.

be flattened to provide a grinding face on the wheel which is approximately 1/16 inch wide. It is advisable to attach a diamond to the table and to take one or two very light cuts across the face of the wheel by traversing the table. Stop the machine.

3. Attach offset blade to the tooth-rest holder and mount the holder on the wheel head as shown in Fig. 14-9. The blade should be clamped at the approximate angle of the helix of the cutter to be ground. Temporarily position the blade slightly below the center of the grinding wheel.
4. Mount the milling cutter on the arbor. A special arbor is used to hold the milling cutter in the cutter and tool grinder. It is good practice to check the arbor frequently for eccentricity with a dial test indicator accurate to .0001 inch by rotating it on centers.
5. Mount the two tailstocks on the grinding machine so that the milling cutter will be held in the approximate center of the table. The right side tailstock spindle is retractable. A spring keeps this spindle extended outward. Place the clearance setting fixture on

Courtesy of Cincinnati Milacron

Fig. 14-9. Setup for grinding the clearance angle on a fixed wheel-head machine using a disc wheel and a clearance setting fixture.

the spindle of the left side tailstock.

6. Adjust the wheel head vertically until the center of the grinding wheel is aligned with the centers of the tailstock. Most cutter and tool grinding machines are provided with a fixed height gage to do this. A fixed height gage, called a centering gage, is shown in Fig. 14-11. The base of this gage is placed on the top of the table. Two flat plates are held together on their face by a screw at a distance above the table that is equal to the height of the tailstock centers. A zero line is marked on the wheel head. The wheel head is positioned until one of the plates on the centering gage coincides with the zero line.
7. Place the milling cutter and arbor assembly on the tailstock centers. Check to see that enough spring pressure is applied by the right side spindle to hold this assembly in place.
8. Prepare the clearance setting fixture. Attach the dog to the arbor and place the pin in the hole of the graduated plate. Rotate the graduated plate until the zero graduation coincides with the index mark on the fixed plate. The graduated plate is then clamped to the fixed plate by tightening the thumbscrew. The cutter and arbor assembly is now firmly held in place by the clearance setting fixture.
9. Locate the position on the tooth of the cutter closest to the grinding wheel that is at the height of the tailstock centers. This is done

with the centering gage as shown in Fig. 14-11. The gage is moved until it touches a point on the cutter tooth. This point is carefully marked with a pencil.

10. Position the milling cutter with the mark on the tooth opposite the cutting face of the grinding wheel.
11. Clamp the tooth rest in place so that the blade will contact the tooth of the cutter just below the mark.
12. Set the wheel head so that the required clearance angle will be ground. To do this the thumbscrew holding the two plates together on the clearance setting is first loosened. Use one hand to hold the cutter in contact with the blade of the tool rest and with the other hand lower the wheel head. The tooth-rest blade, being attached to the wheel head, is also lowered when the wheel head is lowered. Thus the milling cutter is rotated and this motion transmitted to the graduated plate of the clearance setting fixture through the arbor and the dog. The angle through which the cutter and the graduated plate are turned can now be read on the graduations which are marked in degrees. Stop lowering the wheel head when the cutter has rotated through an angle equal to the required clearance angle. Tighten the thumbscrew and remove the dog.
13. Adjust the length of the table travel. Lay the cutting edge of the cutter on the tooth-rest blade, and traverse the table the full length of the cutter. Set the stop dog on the table so that the cutter will not run off the tooth-rest blade. It is not necessary to set the dog at the other end of the table since the cutter must run off the tooth-rest blade in order to be indexed. The machine is now set up to grind the required clearance angle.
14. Start the machine. Lay a tooth on the tooth-rest blade. Hold the cutter against the blade by hand and carefully move the cutter toward the grinding wheel until light sparks indicate that they have touched.
15. Keep holding the cutter against the tooth-rest blade and traverse the table until the full length of the tooth has been ground. Return the table to the end where the cutter can be disengaged from the tooth rest.
16. Check the cutter for taper, using a dial test indicator mounted on the arm of a height gage, or a surface gage. Place one of these gages on the machine table with the indicator contact point touching one end of the ground tooth. Slightly rotate the cutter back and forth and move the indicator to the position where the largest reading can be obtained. Record this reading and repeat this procedure at the other end of the ground tooth. The difference in the two indicator readings will be equal to one-half of the taper. Another method is to grind two opposite teeth, as in Step 15, and measure over these teeth with an outside micrometer caliper, using

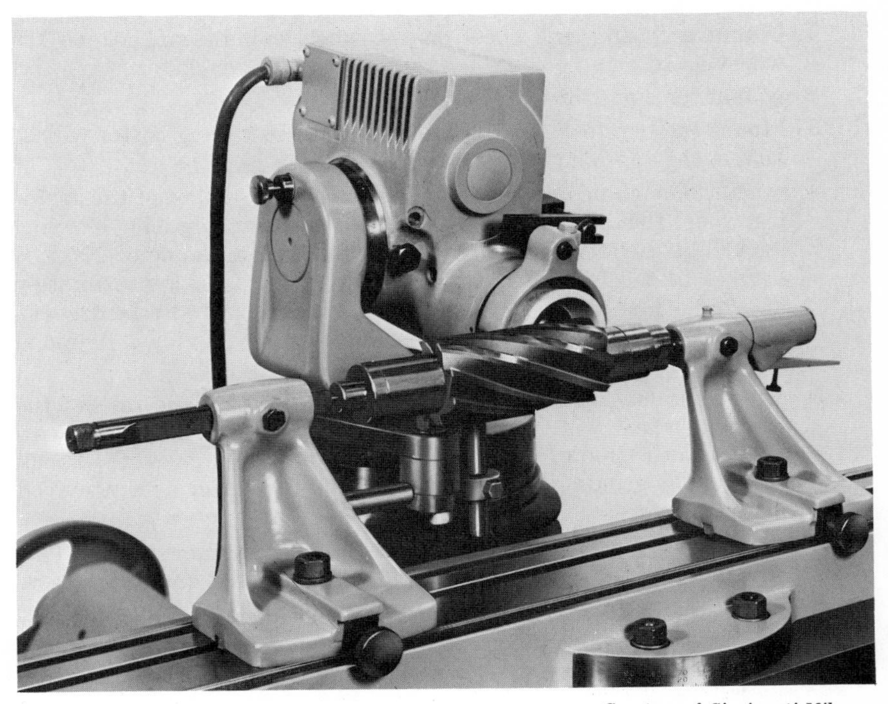

Courtesy of Cincinnati Milacron

Fig. 14-10. Setup used to sharpen a plain milling cutter on a tilting wheel-head cutter and tool grinding machine.

notebook paper to protect the measuring surfaces of the micrometer.

17. If taper is present, adjust the swivel table. Check by repeating Steps 14 through 16 until the taper is removed.
18. When the taper is removed, grind the cutter until all of the teeth are sharp. The depth of cut should not *exceed* .002 inch per pass for rough grinding and about .0005 inch per pass for finish grinding. To assure roundness start grinding on a different tooth after progressing around each full revolution of the cutter. Be sure to hold the cutter against the tooth-rest blade when traversing the wheel.
19. The secondary clearance angle which is behind the primary clearance is then ground in a similar manner. The secondary clearance should leave the primary clearance with a land width that is between $\frac{1}{32}$ and $\frac{1}{16}$ inch.

The procedure used to set up cutter and tool grinding machines with a tilting wheel head differs considerably from that described. The setup is illustrated in Fig. 14-10.

1. Mount a flaring-cup wheel on the wheel spindle and attach the wheel guard.
2. Mount the cutter on the arbor.
3. Mount the tailstocks on the table so that the milling cutter will be held in the approximate center of the table.
4. Start the machine and dress the inside conical surface of the grinding wheel that is adjacent to the face using an abrasive dressing stick. This surface should be dressed to make an angle of 20 to 30 degrees to the face of the wheel leaving the face approximately $\frac{1}{32}$ to $\frac{1}{16}$ inch wide. The face of the wheel should also be dressed. It is sometimes advisable to dress the face by moving a diamond across it with the table. Stop the machine.
5. Adjust the lower swivel of the wheel head so that the face of the grinding wheel will be positioned at an angle of 1 degree with respect to the side of the cutter. This is done to prevent the "back face" of the grinding wheel from grinding the cutter as it is moved across the wheel.
6. Set the grinding wheel to grind the desired clearance angle. This is done by swiveling the wheel head to the desired angle.
7. Mount the tooth-rest assembly on the wheel head. An offset tooth-rest blade should be used and the blade should be clamped at the approximate angle of the helix of the cutter. Temporarily position the holder so that the blade is close to and approximately on the horizontal centerline of the grinding wheel.
8. Establish the point of contact of the milling cutter tooth on the blade. To do this rub a thin coat of red lead or Prussian blue on the top of the tooth-rest blade. Lay a tooth of the cutter on the blade and traverse the entire length of the cutter over the blade. The point of contact will appear as a shiny spot on the blade.
9. Loosen the tooth-rest assembly and position the point of contact opposite the cutting face of the grinding wheel. Clamp the tooth-rest assembly in place. There should be 0.003 to 0.004 inch clearance between the face of the grinding wheel and the tooth-rest blade.
10. Position the point of contact on the tooth-rest blade on the center of the cutter. The centering gage is placed on the table as shown in Fig. 14-11. Raise or lower the wheel head until the gage plate of the centering gage is opposite the point of contact on the tooth-rest blade.
11. Set one table dog to obtain the correct length of table movement in order to prevent the cutter from dropping off the blade at one end. The cutter must be allowed to run off the tooth-rest blade at the other end so that it can be indexed.
12. Lay a tooth on the tooth-rest blade and start the machine. The machine is now set up to grind the milling center.

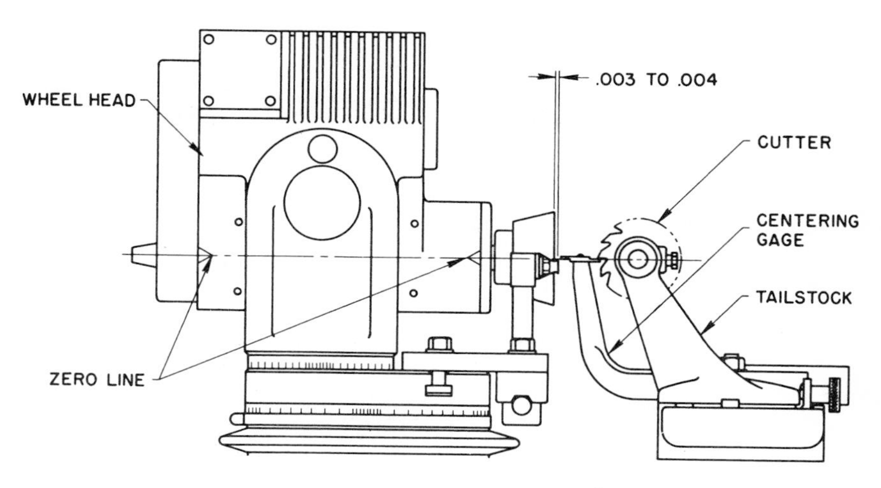

Courtesy of Cincinnati Milacron

Fig. 14-11. Setting the point of contact of the tooth-rest blade on the centerline of the cutter with a centering gage.

The procedure for grinding the cutter is the same as described in Steps 14 through 19 in the previous example. In general, the relief angles on the peripheral teeth on all milling cutters are ground to a sharp edge. Whenever a stationary tooth rest is used the blade must be wide enough and shaped so that the cutter can be run up to and off the face of the grinding wheel in such a manner that the relief at the ends will be uniform with the relief on the body of the cutter. If necessary, the blade may have to be held or formed at an angle to meet this requirement. The cutter is indexed by running it off the stationary tooth-rest blade.

Measuring the Cutter Angles

After the teeth on a milling cutter have been sharpened, the size of the relief angle should be measured. A very convenient method of measuring the relief angle on the peripheral teeth is by the *Indicator Drop Method* illustrated in Fig. 14-12. The cutter is mounted on the arbor which is held between two centers, or it may be held by its shank. One dial test indicator having a sharp contact point is mounted above the cutter, with the sharp contact point indicating a radial line passing through the center of the cutter, as shown in Fig. 14-12. The cutter is positioned so that the contact point of the indicator touches the very tip of the cutting edge. When in this position the second dial test indicator is zeroed against the face of the tooth, as close to the cutting edge as possible. The cutter is then rotated a predetermined amount called the *checking distance,* which is measured by the second indicator. After the cutter has been rotated the checking distance, the indicator drop is read on the dial test indicator positioned above

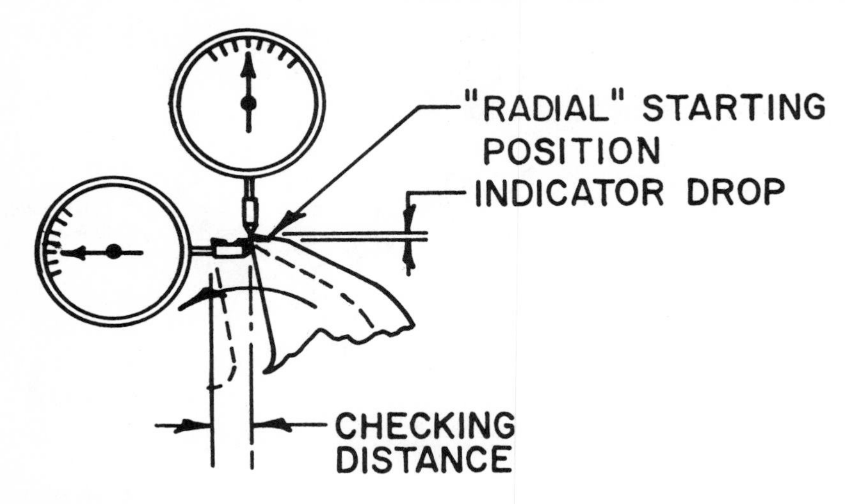

Fig. 14-12. Measuring the relief angle on the peripheral teeth on milling cutters by the Indicator Drop Method.

the workpiece to determine the relief angle.

The recommended range of relief angles on the peripheral teeth for a given cutter diameter is given in Table 14-2; the corresponding indicator drops and checking distances are also given in this table. Indicator drops on cutters ground with a flat and a concave relief (see Fig. 14-5) are essentially equal; however, for a given relief angle, the indicator drops on cutters ground with an eccentric relief are larger, as shown in the table. Cutters having an eccentric relief should be ground to a relief angle that is at the lower end of the range of recommended relief angles while the middle or higher end of the range is to be used on cutters having a flat or concave relief.

The relief angles on the side teeth and end teeth of milling cutters can also be checked by using the indicator drop method. When measuring the relief angle on these teeth the cutter is not rolled; it is moved along a straight-line path a distance equal to the checking distance, usually by moving the machine table this distance with the cutter mounted on the table. Table 14-3 provides values of the indicator drops for checking side and end teeth. The rake angle on the face of the teeth of milling cutters can also be checked by this procedure. Values of the indicator drops for checking the rake angle are given in Table 14-4.

Sharpening a Shell End Milling Cutter

The setup for sharpening the teeth on the periphery of an end milling cutter is shown in Fig. 14-13. The procedure for this operation is identical to that of sharpening a plain milling cutter, except that the cutter is mounted on the workhead rather than between centers. The cutter and

Table 14-2. Indicator Drops for Checking the Radial Relief Angle on Peripheral Teeth

Cutter Diameter, Inch	Recommended Range of Radial Relief Angles, Degrees	Checking Distance, Inch	Indicator Drop, Inches				Recommended Max. Primary Land Width, Inch
			For Flat and Concave Relief		For Eccentric Relief		
			Min.	Max.	Min.	Max.	
1/16	20-25	.005	.0014	.0019	.0020	.0026	.007
3/32	16-20	.005	.0012	.0015	.0015	.0019	.007
1/8	15-19	.010	.0018	.0026	.0028	.0037	.015
5/32	13-17	.010	.0017	.0024	.0024	.0032	.015
3/16	12-16	.010	.0016	.0023	.0022	.0030	.015
7/32	11-15	.010	.0015	.0022	.0020	.0028	.015
1/4	10-14	.015	.0017	.0028	.0027	.0039	.020
9/32	10-14	.015	.0018	.0029	.0027	.0039	.020
5/16	10-13	.015	.0019	.0027	.0027	.0035	.020
11/32	10-13	.015	.0020	.0028	.0027	.0035	.020
3/8	10-13	.015	.0020	.0029	.0027	.0035	.020
13/32	9-12	.020	.0022	.0032	.0032	.0044	.025
7/16	9-12	.020	.0022	.0033	.0032	.0043	.025
15/32	9-12	.020	.0023	.0034	.0032	.0043	.025
1/2	9-12	.020	.0024	.0034	.0032	.0043	.025
9/16	9-12	.020	.0024	.0035	.0032	.0043	.025
5/8	8.11	.020	.0022	.0032	.0028	.0039	.025
11/16	8-11	.030	.0029	.0045	.0043	.0059	.035
3/4	8-11	.030	.0030	.0046	.0043	.0059	.035
13/16	8-11	.030	.0031	.0047	.0043	.0059	.035
7/8	8-11	.030	.0032	.0048	.0043	.0059	.035
15/16	7-10	.030	.0027	.0043	.0037	.0054	.035
1	7-10	.030	.0028	.0044	.0037	.0054	.035
1 1/8	7-10	.030	.0029	.0045	.0037	.0053	.035
1 1/4	6-9	.030	.0024	.0040	.0032	.0048	.035
1 3/8	6-9	.030	.0025	.0041	.0032	.0048	.035
1 1/2	6-9	.030	.0026	.0041	.0032	.0048	.035
1 5/8	6-9	.030	.0026	.0042	.0032	.0048	.035
1 3/4	6-9	.030	.0026	.0042	.0032	.0048	.035
1 7/8	6-9	.030	.0027	.0043	.0032	.0048	.035
2	6-9	.030	.0027	.0043	.0032	.0048	.035
2 1/4	5-8	.030	.0022	.0038	.0026	.0042	.040
2 1/2	5-8	.030	.0023	.0039	.0026	.0042	.040
2 3/4	5-8	.030	.0023	.0039	.0026	.0042	.040
3	5-8	.030	.0023	.0039	.0026	.0042	.040
3 1/2	5-8	.030	.0024	.0040	.0026	.0042	.047
4	5-8	.030	.0024	.0040	.0026	.0042	.047
5	4-7	.030	.0019	.0035	.0021	.0037	.047
6	4-7	.030	.0019	.0035	.0021	.0037	.047
7	4-7	.030	.0020	.0036	.0021	.0037	.060
8	4-7	.030	.0020	.0036	.0021	.0037	.060
10	4-7	.030	.0020	.0036	.0021	.0037	.060
12	4-7	.030	.0020	.0036	.0021	.0037	.060

tool grinding machine in the illustration has a tilting wheel head which is used to obtain the required clearance angle. If the machine is equipped with a fixed wheel head, one of the methods illustrated in Fig. 14-3 must be used to grind the peripheral teeth. The tooth-rest blade is set on the center of the cutter by finding the point of contact where the milling cutter touches the blade and aligning this point with the centering gage. When using method A or B, Fig. 14-3, raise or lower the wheel head the required distance, and for method C or D, raise or lower the tooth rest the required distance.

Table 14-3. Indicator Drops for Checking Relief Angles on Side-Teeth and End Teeth

Checking Distance, Inch	Given Relief Angle								
	1°	2°	3°	4°	5°	6°	7°	8°	9°
	Indicator Drop, inch								
.005	.00009	.00017	.00026	.00035	.0004	.0005	.0006	.0007	.0008
.010	.00017	.00035	.00052	.0007	.0009	.0011	.0012	.0014	.0016
.015	.00026	.0005	.00079	.0010	.0013	.0016	.0018	.0021	.0024
.031	.00054	.0011	.0016	.0022	.0027	.0033	.0038	.0044	.0049
.047	.00082	.0016	.0025	.0033	.0041	.0049	.0058	.0066	.0074
.062	.00108	.0022	.0032	.0043	.0054	.0065	.0076	.0087	.0098

Table 14-4. Indicator Drops for Checking Rake Angles on Milling Cutter Face

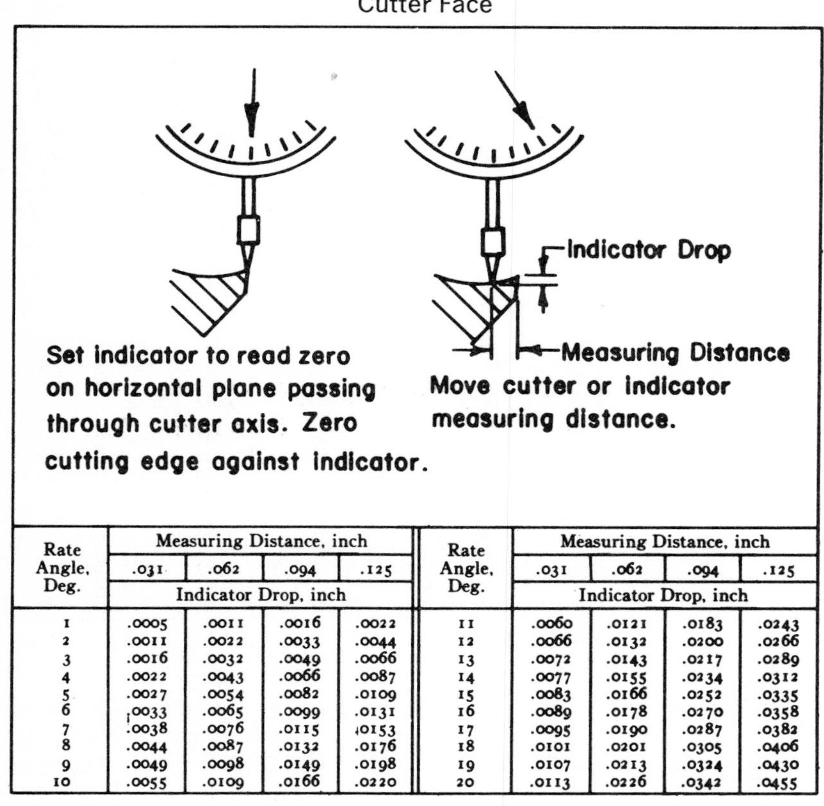

Rate Angle, Deg.	Measuring Distance, inch				Rate Angle, Deg.	Measuring Distance, inch			
	.031	.062	.094	.125		.031	.062	.094	.125
	Indicator Drop, inch					Indicator Drop, inch			
1	.0005	.0011	.0016	.0022	11	.0060	.0121	.0183	.0243
2	.0011	.0022	.0033	.0044	12	.0066	.0132	.0200	.0266
3	.0016	.0032	.0049	.0066	13	.0072	.0143	.0217	.0289
4	.0022	.0043	.0066	.0087	14	.0077	.0155	.0234	.0312
5	.0027	.0054	.0082	.0109	15	.0083	.0166	.0252	.0335
6	.0033	.0065	.0099	.0131	16	.0089	.0178	.0270	.0358
7	.0038	.0076	.0115	.0153	17	.0095	.0190	.0287	.0382
8	.0044	.0087	.0132	.0176	18	.0101	.0201	.0305	.0406
9	.0049	.0098	.0149	.0198	19	.0107	.0213	.0324	.0430
10	.0055	.0109	.0166	.0220	20	.0113	.0226	.0342	.0455

The setup for sharpening the teeth on the face of the shell end milling cutter is shown in Fig. 14-14. The procedure for grinding these teeth is given here:

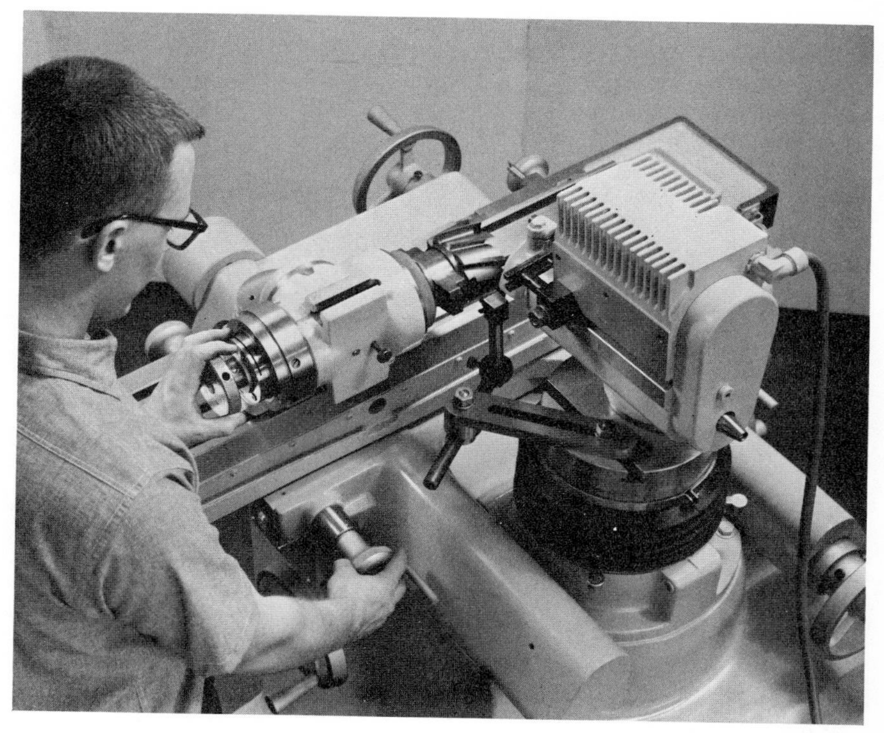

Courtesy of Cincinnati Milacron

Fig. 14-13. Sharpening the peripheral teeth of a shell end mill.

1. Mount the cutter on the workhead.
2. Position one tooth on the face of the cutter so that its cutting edge is parallel to the table and lock the spindle of the workhead. The cutting edge can be checked for parallelism by a dial test indicator placed on the wheel head.
3. Mount the universal tooth rest on the bottom of the workhead. Place a rounded tooth-rest blade against a peripheral tooth as close to the face as possible.
4. If a tilting wheel-head machine is to be used, tilt the wheel head to the desired primary clearance angle. If a fixed wheel-head machine is used, tilt the workhead to the desired primary clearance angle.
5. Grind the primary clearance on all teeth.
6. Check the face of the cutter to make sure that the teeth are ground perpendicular to the axis of the cutter. This can be done by holding a flat plate against the cutter. Rub a thin coat of Prussian blue or

Courtesy of Cincinnati Milacron

Fig. 14-14. Sharpening the teeth on the face of a shell end mill.

red lead on the plate and note how it is picked up by the cutter teeth. If a flat plate is not available, use a straight edge to check the face teeth.

7. Adjust the wheel head or the workhead to grind the secondary clearance angle.
8. Grind the secondary clearance on all teeth.
9. Set the wheel head or the workhead back to the setting for grinding the primary clearance angle.
10. Position the workhead to grind the back taper on the face of the wheel. The teeth on the face of the shell and mill should have a back taper of about 3 degrees which extends up to ⅛ inch of the corner of the cutter. When the teeth on the face are given a back taper to form a slight dishlike profile, the cutting action of the cutter will be greatly improved.
11. Regrind the primary clearance on all of the teeth up to a point that is ⅛ inch from the corner to form the back taper.

Sometimes a chamfer is ground on the corner of end milling cutters. This chamfer is ground in a manner similar to grinding the face of the teeth, except that the workhead is swiveled to the desired chamfer or corner angle. The center of the chamfer is placed at the height of the axis of the workhead. The true clearance angle on the chamfer is obtained by rolling and tilting the wheelhead. Figure 14-17 illustrates this procedure, and Table 14-5 gives the settings for the angle to tilt and the angle of roll. As was explained in Chapter 5, the chamfer width should not exceed $\frac{1}{16}$ inch.

Sharpening a Staggered-Tooth Side Milling Cutter

The setup for grinding the peripheral teeth on a staggered-tooth side milling cutter is shown in Fig. 14-15. These cutters generally have closely spaced teeth (see view D in Fig. 5-2) which will interfere with the tooth rest if it is placed below the cutter. For this reason the use of an offset tooth-rest blade, as shown in Fig. 14-16, is recommended. The blade should be ground to the form of an inverted V, and the angle on the sides should be about 5 to 8 degrees greater than the axial rake angle of the peripheral teeth. The procedure for grinding the peripheral teeth is given in the following steps:

1. Mount the flaring-cup wheel and dress the inside conical surface to an angle of 20 to 30 degrees with respect to the face of the wheel.

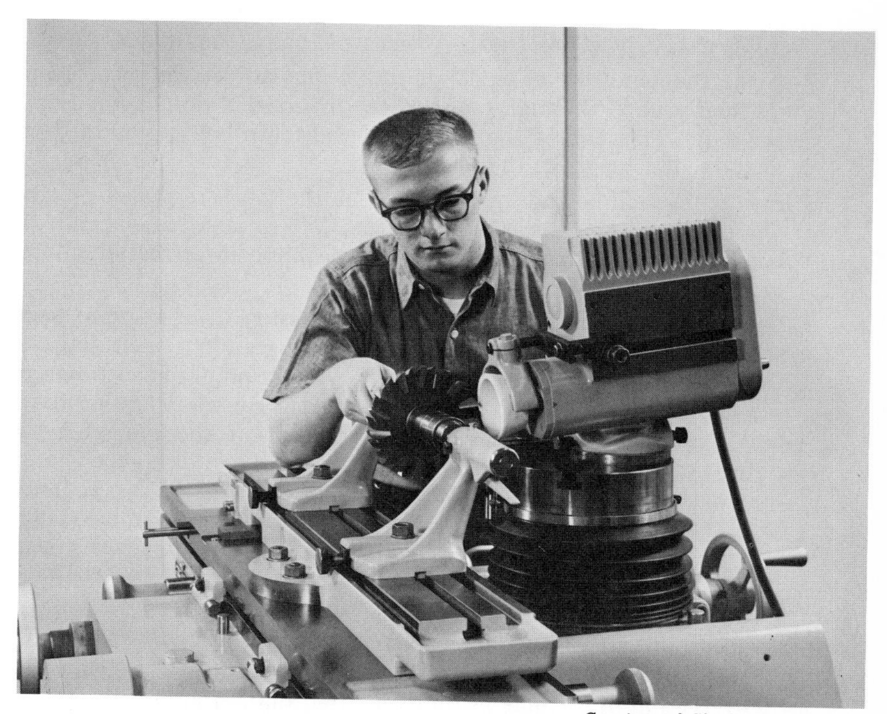

Courtesy of Cincinnati Milacron

Fig. 14-15. Sharpening the peripheral teeth on a staggered-tooth side milling cutter.

2. Mount the cutter on the arbor. Mount the tailstocks on the table. Attach the clearance setting fixture to the left side tailstock.
3. Place the cutter and arbor assembly between the tailstock centers and check the tension on the retractable tailstock.
4. Adjust the wheel head 1 degree around the vertical axis so that the cup wheel will not double cut the milling cutter.
5. Tilt the wheelhead to the desired primary clearance angle.

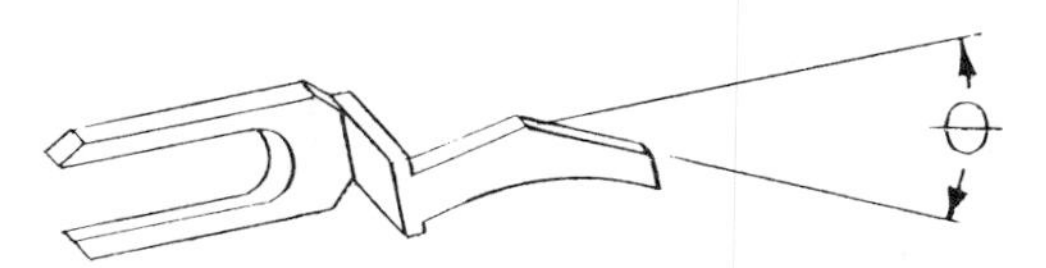

Fig. 14-16. Offset tooth-rest blade recommended for sharpening the peripheral teeth on a staggered-tooth side milling cutter.

6. Attach the tooth-rest holder to the wheelhead using the offset, inverted V, tooth-rest blade. Adjust the tooth rest until the high point of the inverted V is at the approximate horizontal center of the grinding wheel and at the center of the cutting edge.
7. Place the centering gage on the table and adjust the wheel head vertically until the high point of the inverted V is on center with the gage. This causes the portion of the tooth being ground to be at the center height of the cutter.
8. Start the machine and lay one tooth on the tooth-rest blade. Grind one tooth as the table traverses from left to right, and grind the next tooth traversing the table in the opposite direction. After grinding the two teeth stop the machine.
9. Check the comparative height or maximum radii of the two teeth that were ground. This is done by attaching a .0001-inch dial test indicator to the wheel head so that the contact point of the indicator is positioned against the edge of the teeth. By rotating one tooth and then the other over the contact point of the indicator the difference in the indicator reading is noted. This difference should not exceed .0003 inch. If it is greater than this, the tooth-rest blade must be loosened and moved slightly toward the large or "high" tooth. Steps 8 and 9 are then repeated until both teeth are the same height within .0003 inch (0.008 mm).
10. Grind the primary clearance angle on all teeth by traversing in opposite directions to grind each succeeding tooth.
11. Attach the clearance setting dog to the left end of the arbor and place the pin in the hole in the graduated plate. Remove the tooth-rest assembly. The staggered-tooth cutter has two sets of teeth, each with a different axial rake angle (Fig. 5-2). The secondary

clearance angle is obtained by grinding one set of teeth completely and then grinding the other set of teeth.

12. Select one tooth and mark the center of that tooth with a pencil. Using the centering gage, place the mark on the tooth at the center height of the cutter. Loosen the stationary plate of the clearance setting fixture and set the index mark opposite the zero graduation. Clamp it to the stationary spindle.
13. Remove the centering gage and rotate the cutter to the desired clearance angle. Lock the graduated plate to the stationary plate by tightening the thumbscrew.
14. Remount the tooth-rest assembly using a rounded tooth-rest blade. Position the blade under the side of the tooth to be ground. Remove the dog, and position the wheel head to the horizontal or zero position.
15. Swivel the table from 2 to 6 degrees, depending upon the axial rake angle of the teeth.
16. Grind the secondary clearance until the width of the primary clearance is 1/32 to 1/16 inch. Before the final width of the land is reached, check to see if it is parallel. If it is tapered, the swivel of the table is incorrect and should be adjusted.
17. Repeat Steps 11 through 16 to grind the secondary clearance on the second set of teeth.

The sides of the teeth of a staggered-tooth side milling cutter should not be ground unless it is necessary. Grinding these teeth reduces the width of the cutter, which should be avoided if possible. These teeth are ground with a very slight back taper which is usually about 1/2 degree. The back taper is generally extended to the corner of the cutter. The procedure used is similar to the procedure for grinding the teeth on the face of shell end milling cutters.

1. Mount the workhead on the table. Mount the cutter on a short stub arbor which has a No. 50 milling-machine taper at one end. Attach the cutter and arbor assembly to the workhead.
2. Level one of the teeth on the side of the cutter parallel to the table, using the centering gage to check this position. Lock the workhead spindle when the tooth is leveled.
3. Mount the universal tooth-rest assembly on the bottom of the workhead. Use a rounded tooth-rest blade, and position it on the periphery of the cutter against the tooth that has been leveled.
4. Tilt the wheel head (or the workhead) to the desired primary clearance angle.
5. Swivel the table or the workhead approximately 1/2 degree in order to grind the back taper.
6. Grind the primary clearance on all teeth.

Table 14-5. Radial Roll and Axial Tilt Required to Obtain True Clearance Angle for a Given Corner Angle *

CORNER ANGLE

	5°		10°		15°		20°		22½°		25°		30°		35°		40°		45°		
DESIRED TRUE CLEARANCE ANGLE	Axial tilt	Rad. roll	Axial tilt	Rad. roll	Axial tilt	Rad. roll	Axial tilt	Rad. roll	Axial tilt	Rad. roll	Axial tilt	Rad. roll	Axial tilt	Rad. roll	Axial tilt	Rad. roll	Axial tilt	Rad. roll	Axial tilt	Rad roll	DESIRED TRUE CLEARANCE ANGLE
½°	0	.5	.1	.5	.1	.5	.2	.5	.2	.5	.2	.5	.3	.4	.3	.4	.3	.4	.4	.4	½°
1°	.1	1.0	.2	1.0	.3	1.0	.3	.9	.4	.9	.4	.9	.5	.9	.6	.8	.6	.8	.7	.7	1°
1½°	.1	1.5	.3	1.5	.4	1.5	.5	1.4	.6	1.4	.6	1.4	.8	1.3	0.9	1.2	1.0	1.1	1.1	1.1	1½°
2°	.2	2.0	.3	2.0	.5	1.9	.7	1.9	.8	1.8	.8	1.8	1.0	1.7	1.2	1.6	1.3	1.5	1.4	1.4	2°
2½°	.2	2.5	.4	2.5	.6	2.4	.9	2.4	1.0	2.3	1.1	2.3	1.2	2.2	1.4	2.1	1.6	1.9	1.8	1.8	2½°
3°	.3	3.0	.5	3.0	.8	2.9	1.0	2.8	1.2	2.8	1.3	2.7	1.5	2.6	1.7	2.5	1.9	2.4	2.1	2.1	3°
3½°	.3	3.5	.6	3.4	.9	3.4	1.2	3.3	1.3	3.2	1.5	3.2	1.8	3.0	2.0	2.9	2.3	2.7	2.5	2.5	3½°
4°	.3	4.0	.7	3.9	1.0	3.9	1.4	3.8	1.5	3.7	1.7	3.6	2.0	3.5	2.3	3.3	2.6	3.1	2.8	2.8	4°
4½°	.4	4.5	.8	4.4	1.2	4.3	1.5	4.2	1.7	4.2	1.9	4.1	2.3	3.9	2.6	3.7	2.9	3.5	3.2	3.2	4½°
5°	.4	5.0	.9	4.9	1.3	4.8	1.7	4.7	1.9	4.6	2.1	4.5	2.5	4.3	2.9	4.1	3.2	3.8	3.5	3.5	5°
5½°	.5	5.5	1.0	5.4	1.4	5.3	1.9	5.2	2.1	5.1	2.3	5.0	2.8	4.8	3.2	4.5	3.5	4.2	3.9	3 9	5½°
6°	.5	6.0	1.0	5.9	1.6	5.8	2.1	5.6	2.3	5.6	2.5	5.4	3.0	5.2	3.5	4.9	3.9	4.6	4.3	4.3	6°
6½°	.6	6.5	1.1	6.4	1.7	6.3	2.2	6.1	2.5	6.0	2.8	5.9	3.3	5.6	3.8	5.3	4.2	5.0	4.6	4.6	6½°
7°	.6	7.0	1.2	6.9	1.8	6.8	2.4	6.6	2.7	6.5	3.0	6.4	3.5	6.1	4.0	5.8	4.5	5.4	5.0	5.0	7°
7½°	.7	7.5	1.3	7.4	2.0	7.3	2.6	7.0	2.9	6.9	3.2	6.8	3.8	6.5	4.3	6.2	4.8	5.8	5.3	5.3	7½°
8°	.7	8.0	1.4	7.9	2.1	7.7	2.8	7.5	3.1	7.4	3.4	7.3	4.0	6.9	4.6	6.6	5.2	6.2	5.6	5.6	8°
9°	.8	9.0	1.6	8.9	2.4	8.7	3.1	8.5	3.5	8.3	3.8	8.2	4.5	7.8	5.2	7.4	5.8	6.9	6.4	6.4	9°
10°	.9	10.0	1.8	9.9	2.6	9.7	3.5	9.4	3.9	9.3	4.3	9.1	5.1	8.7	5.8	8.2	6.5	7.7	7.1	7.1	10°
11°	1.0	11.0	1.9	10.8	2.9	10.6	3.8	10.4	4.3	10.2	4.7	10.0	5.6	9.6	6.4	9.1	7.1	8.5	7.8	7.8	11°
12°	1.1	12.0	2.1	11.8	3.2	11.6	4.2	11.3	4.7	11.1	5.2	10.9	6.1	10.4	7.0	9.9	7.8	9.3	8.6	8.6	12°
13°	1.2	13.0	2.3	12.8	3.4	12.6	4.5	12.3	5.1	12.1	5.6	11.8	6.6	11.3	7.6	10.7	8.5	10.0	9.3	9.3	13°
14°	1.2	14.0	2.5	13.8	3.7	13.5	4.9	13.2	5.5	13.0	6.0	12.7	7.1	12.2	8.1	11.5	9.1	10.8	10.0	10.0	14°
15°	1.3	15.0	2.7	14.8	4.0	14.5	5.3	14.2	5.9	13.9	6.5	13.7	7.6	13.1	8.7	12.4	9.8	11.6	10.7	10.7	15°
	Rad. roll	Axial tilt	Rad. roll	Axial tilt	Rad. roll	Axial tilt	Rad. roll	Axial tilt	Rad. roll	Axial tilt	Rad. roll	Axial tilt	Rad. roll	Axial tilt	Rad. roll	Axial tilt	Rad. roll	Axial tilt	Rad roll	Axial tilt	
	85°		80°		75°		70°		67½°		65°		60°		55°		50°		45°		

CORNER ANGLE

* With permission of Cincinnati Milacron

7. Remove the cutter and turn it around to grind the teeth on the second side.

Sharpening Face Milling Cutters

Small face milling cutters are ground in exactly the same manner as shell end milling cutters. The same procedure is also used for grinding large face milling cutters except that these cutters must be held on a large face mill grinding attachment in order to clear the top of the table.

Many modern face milling cutters have cemented-carbide teeth, which should be ground with diamond grinding wheels. Flaring-cup-type diamond wheels are recommended. Rough grinding can be done with a diamond abrasive grain size of 100 or 120, while 220-grain size is recommended for finish grinding.

Most face milling cutters have a chamfered corner. The corner angle is defined in Fig. 5-20. The true clearance angle on the corners of face milling cutters is the resultant of the clearance angles and is measured in the axial and the radial planes. The true clearance angle is obtained on the corners by tilting and rolling the wheel head, as shown in Fig. 14-17, in accordance with the specifications given in Table 14-5. For example, if the corner angle is 30 degrees and a 5-degree true clearance angle is desired, the axial tilt is 2.5 degrees and the radial roll is 4.3 degrees (Fig. 14-17).

As a matter of preference, sometimes a spin grinding or circle grinding operation, as in Fig. 14-18, is performed on milling cutters prior to grinding the primary clearance angle. The cutter is mounted on a cylindrical grinding attachment, and the operation is performed like a cylindrical grinding operation. While this operation is primarily performed on face milling cutters, it is also on occasion done on other types of mill-

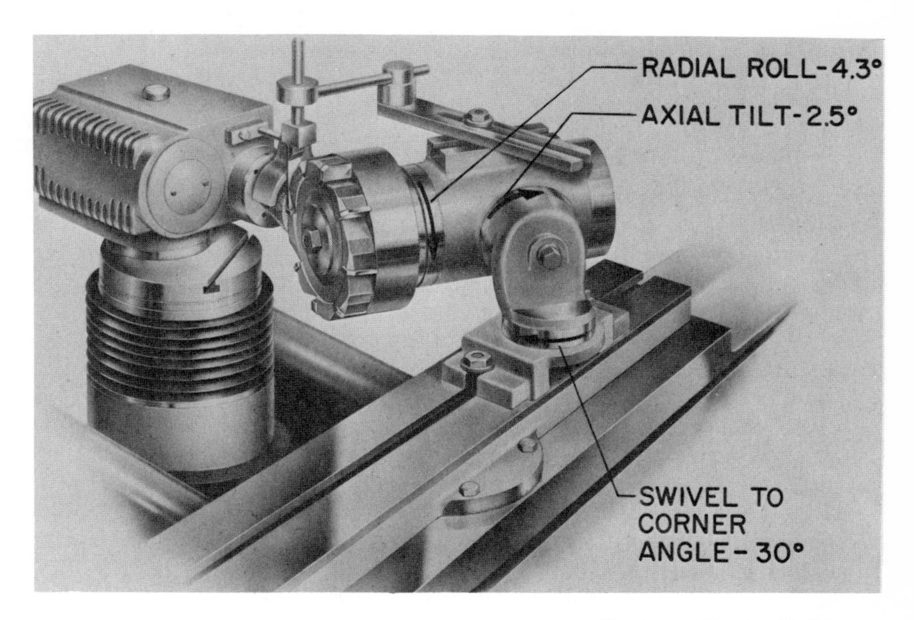

Courtesy of Cincinnati Milacron

Fig. 14-17. Setup for grinding the clearance angle on the corner chamfer of a face milling cutter.

Courtesy of Cincinnati Milacron

Fig. 14-18. Spin or circle grinding a face milling cutter.

ing cutters. The purpose of such an operation is to grind a small circular land on the tooth adjacent to the cutting edge that is concentric with the axis of the cutter; thus, the radius of each tooth is equal. When each tooth is ground, the operator can observe the gradual elimination of the circular land until a sharp cutting edge is attained. A small circular land, .001 or .002 inch wide, is sometimes left on the teeth of face milling cutters. In either case, by this procedure the operator can be sure that the radius of each finish ground tooth is equal and that there are no "high" teeth on the cutter. This procedure is particularly useful for grinding large-diameter face milling cutters which have a large number of teeth, because it eliminates the effect of wear on the grinding wheel.

After the face-mill sharpening operation has been completed, the cutter should be checked for run-out on the periphery, the corner, and the face with a .0001-inch dial test indicator. The runout should be kept to .0005 inch for cutters up to 12 inches. If the cutter is to be used for milling steel, the cutting edges should be beveled slightly with a hand hone. A diamond-honing stone will give the best results for honing carbide cutters. Do not hone the edges too much as this can reduce the life of the cutter between sharpenings. A slight bevel .002 to .003 inch in width is usually sufficient. Before a carbide cutter is removed from the grinding machine, the teeth should be inspected with a magnifying glass to make certain that there are no defects on the cutting edge.

Sharpening End Milling Cutters

Since they are essentially alike, the peripheral teeth on end milling cutters may be sharpened by the use of any of the procedures previously described for sharpening the teeth on plain milling cutters, especially the method by which an eccentric relief is ground. The primary difference is that the end mills are held by their shank instead of on an arbor. A typical setup for sharpening the peripheral teeth on an end mill is shown in Fig. 14-19. The end mill is held in a workhead spindle; it is held against a

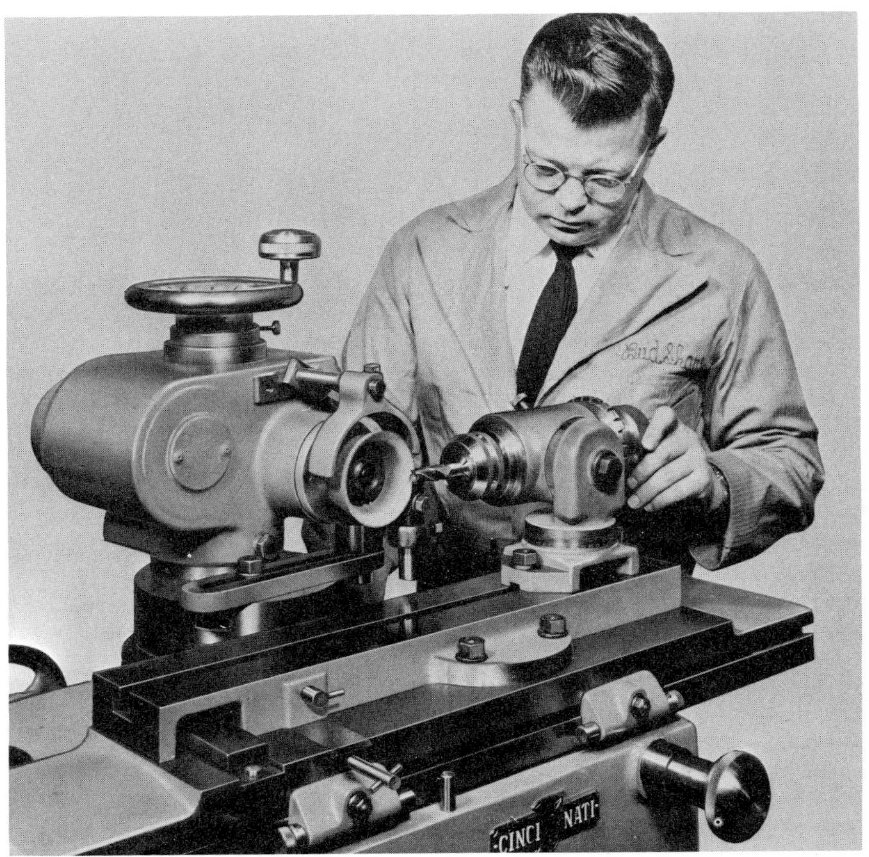

Courtesy of Cincinnati Milacron

Fig. 14-19. Setup for sharpening the peripheral teeth of an end milling cutter.

stationary tooth-rest blade by rotating the workhead spindle while the end mill is ground by moving the table of the machine back and forth. The end mill sharpening fixture shown in Fig. 14-20 has a spindle that may be mounted in an air bearing or in a plain bearing; the spindle can be rotated or moved in and out with little effort. Held in the spindle of this fixture, the peripheral teeth of the end mill are ground by moving only the spindle back and forth while at the same time rotating it to hold the cutter against the stationary tooth rest, which can be done with great sensitivity. A micrometer head used to measure the cutter size is attached to this fixture. It is especially useful for measuring the size of cutters having an odd number of teeth, such as three fluted end mills. Occasionally an end mill must be sharpened so that there is a minimum loss in diameter. In this event the peripheral teeth are sharpened by grinding the face of the tooth and a part of the flute.

The amount of stock that is removed per pass when grinding an end mill depends upon the end mill size and the material from which it is made. End mills made from general-purpose high-speed steel can have a depth of cut as much as .002 inch per pass. In sharpening end milling cutters made from highly alloyed high-speed steels—such as are used

Courtesy of The Weldon Tool Co.

Fig. 14-20. Setup for using an end mill sharpening fixture to sharpen the peripheral teeth on an end milling cutter.

to cut high-temperature alloys—a depth ranging from .0003 to .0005 inch per pass is used. The finishing cuts should also be very light in order to produce a smooth, sharp cutting edge.

The setup for sharpening the end teeth of an end milling cutter is shown in Fig. 14-21. In this setup the same fixture as shown in Fig. 14-20, is used to grind the peripheral teeth; however, to sharpen the end teeth, a collar on the spindle is adjusted so that it will not move lengthwise. The end tooth which is to be ground is held in a horizontal position by a stationary tooth rest attached to the fixture and placed in the flute as close to this

Courtesy of The Weldon Tool Co.

Fig. 14-21. Setup for sharpening the end teeth on an end mill, using an end mill sharpening fixture.

tooth as possible. In this case the fixture is tilted to the angle to be ground on the tooth; on cutter and tool grinders equipped with a tilting head, this head, instead of the fixture, may be tilted to the required angle. Each end tooth is then ground to a sharp edge by indexing the spindle against the tooth rest. The workhead shown in Fig. 14-19 can also be set up in a similar manner to grind the end teeth.

The entire job of sharpening the end teeth on end milling cutters is rather complicated and will usually require several setups similar to the setup in Fig. 14-21. It requires special knowledge and skill, as shown by the step-by-step procedures in Figs. 14-22 through 14-27. In these illustrations the shaded areas indicate the surfaces that are to be ground.

The first step in sharpening the end teeth is always to remove all of the wear on the end teeth and at the corner. Particular care must be taken to remove all of the wear on the corner. This operation is illustrated in Fig. 14-22 for a two-fluted end mill. The preliminary grinding operation on four-fluted and other end mills is made in the same manner.

The end teeth on two-fluted end mills are sharpened in three operations shown at A, B, and C in Fig. 14-23. A fourth operation, shown at D, is optional. The procedures for sharpening the end teeth of other common end milling cutters are illustrated and described in Figs. 14-24 through 14-26.

A radius grinding attachment, shown in Fig. 14-28 must be used to sharpen ball nose end mills, which are used in die sinking and other contour milling operations. The procedure for grinding these cutters is described in Fig. 14-27.

Sharpening Form-Relieved Milling Cutters

Form-relieved milling cutters are sharpened by grinding the tooth face. The relief of these cutters is made so that the profile or form of the cutter is retained if the tooth face is correctly sharpened. On many cutters of this type the tooth face is on a radial line with respect to the axis

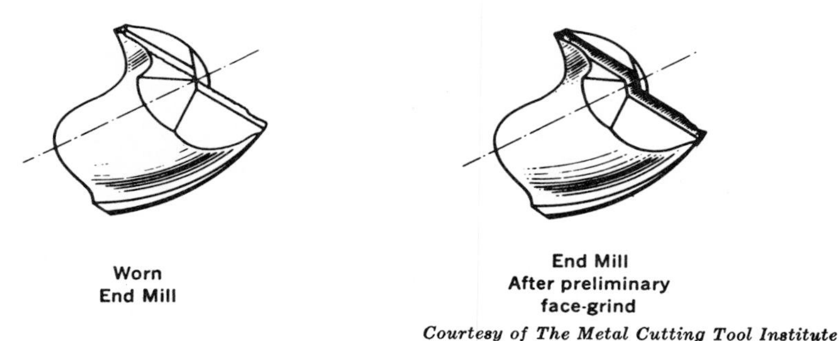

Courtesy of The Metal Cutting Tool Institute

Fig. 14-22. Two-fluted end mills before and after the preliminary grinding operation on the end teeth.

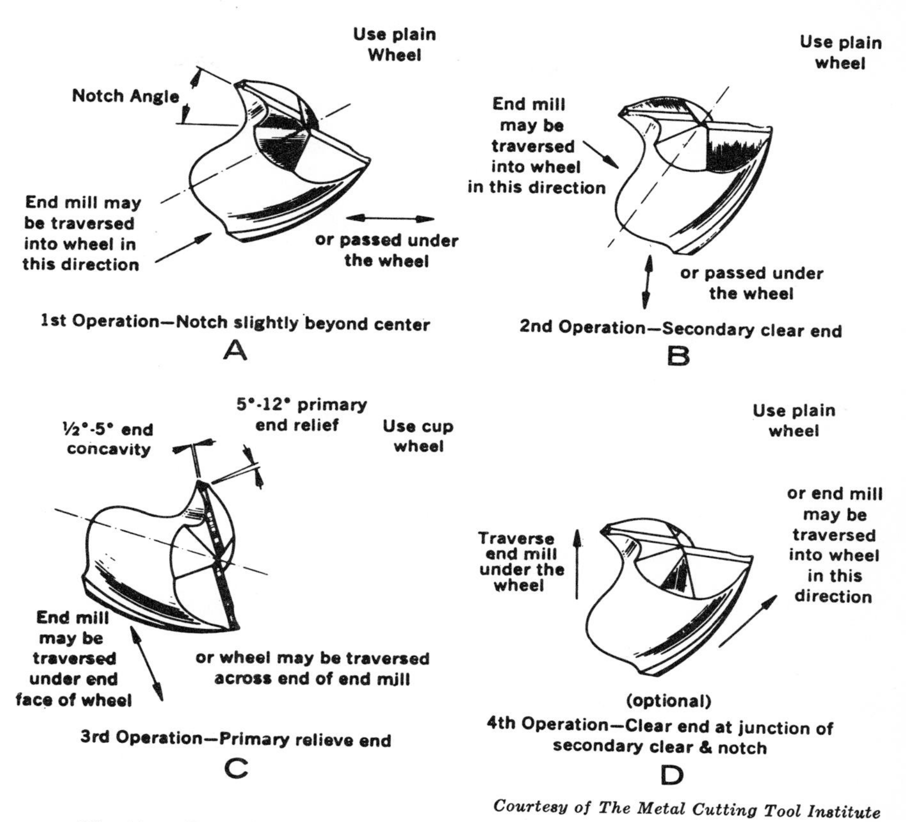

Courtesy of The Metal Cutting Tool Institute

Fig. 14-23. Procedure for sharpening the end teeth of a two-fluted end mill.

of the cutter. Because of this design, the cutter has a zero rake angle. Sometimes the tooth face is not made radial so that the cutter will have a positive rake angle which may provide an improved cutting action. It is very important to retain the original rake angle when these cutters are sharpened; otherwise the profile of the cutters will be altered.

A setup for sharpening a form-relieved milling cutter is shown in Fig. 14-29. A dish-type grinding wheel should be used. The outside diameter of this wheel should be dressed to provide a radius that will blend with the radius at the bottom of the cutter flute. The inside cone should be dressed to an angle of 5 to 10 degrees with respect to the face of the wheel. The cutting edge of the wheel should be approximately $\frac{1}{32}$ inch wide and the wheel-head spindle must be in a horizontal position. The procedure for sharpening a cutter with a radial tooth face or zero rake is given here:

1. Mount the tailstocks on the table. Do not start the grinding wheel.

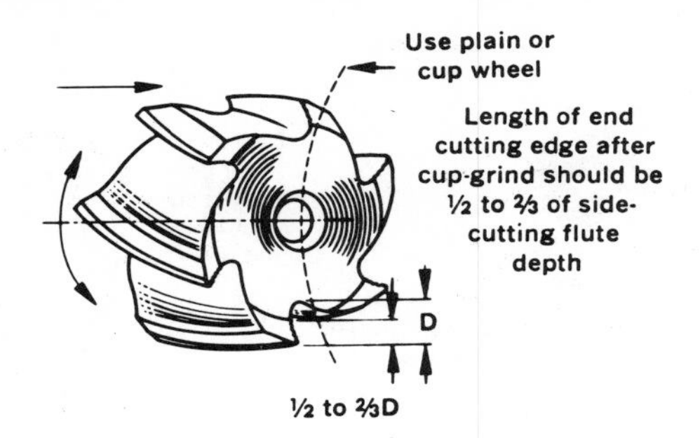

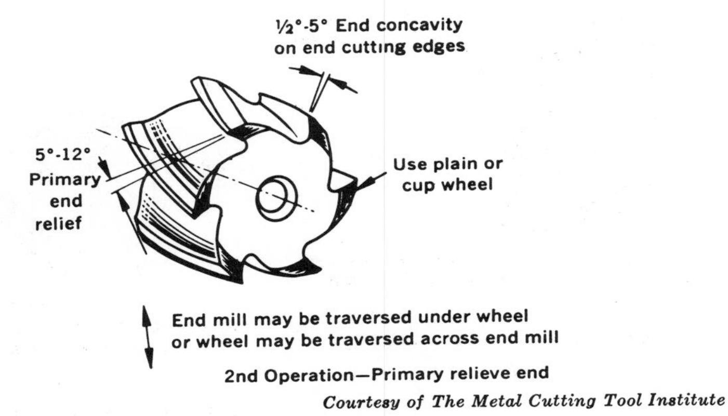

Courtesy of The Metal Cutting Tool Institute

Fig. 14-24. Procedure for sharpening the end teeth of a mutifluted cupped end mill.

2. Align the face of the grinding wheel with the point of the tailstock centers. Set the cross-slide micrometer dial to read zero.
3. Mount the cutter and arbor assembly on the tailstock centers. Raise the wheel head to prevent the grinding wheel from interfering with this operation.
4. Attach the universal tooth-rest assembly to the table. Use a plain tooth-rest blade. Do not position the blade against the cutter at this time.
5. Lower the grinding wheel into one of the tooth spaces.
6. Rotate the cutter until the tooth face touches the grinding wheel. Holding it in this position, bring the tooth-rest blade against the back of the tooth as shown in Fig. 14-23.
7. Move the cutter away from the grinding wheel a few thousandths of an inch with the cross slide and traverse to the right until the

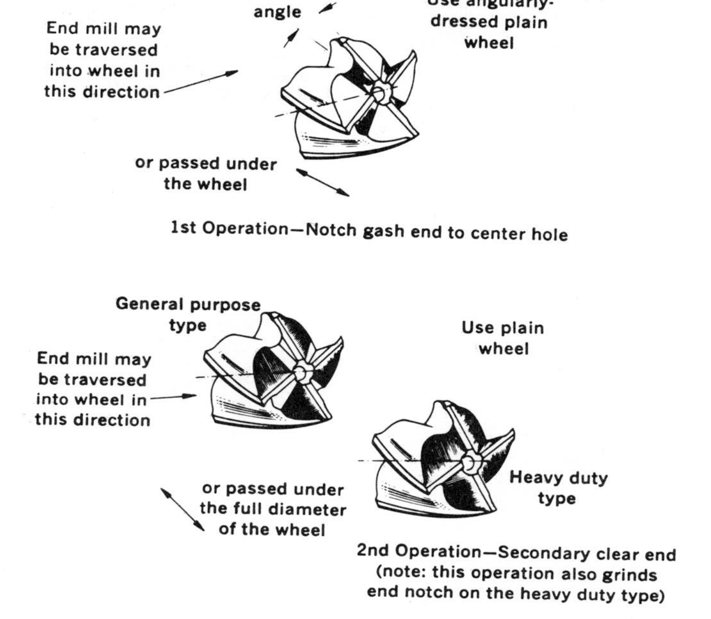

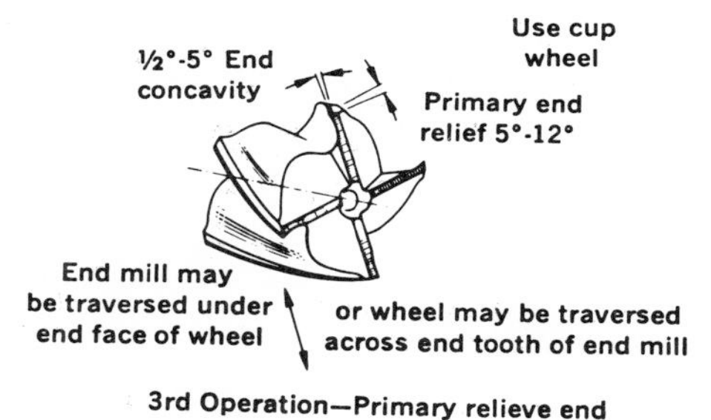

Courtesy of The Metal Cutting Tool Institute

Fig. 14-25. Procedure for sharpening the end of a four-fluted end mill which has a center hole.

grinding wheel is clear of the cutter. Reposition the cross slide to the original zero setting and start the grinding wheel.

8. From this point on, all of the grinding is done with the cross slide set at zero. The depth of each cut is obtained by advancing the tooth-rest blade with the micrometer adjustment, which will rotate the cutter about its own axis and thereby advance the tooth face toward the grinding wheel.

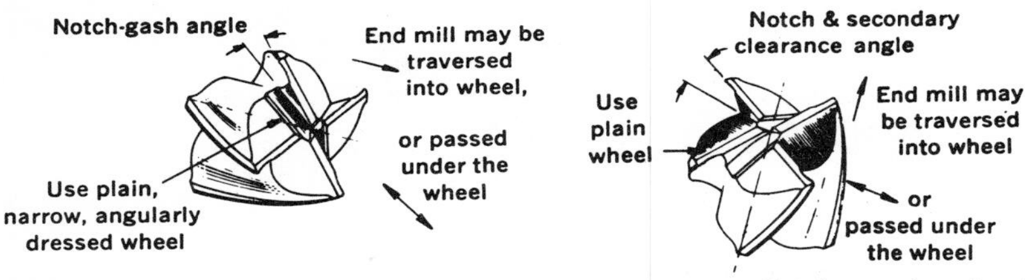

1st Operation—Gash-notch two center-cutting end teeth past center

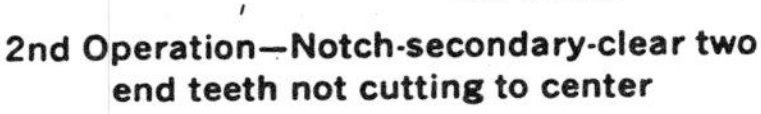

2nd Operation—Notch-secondary-clear two end teeth not cutting to center

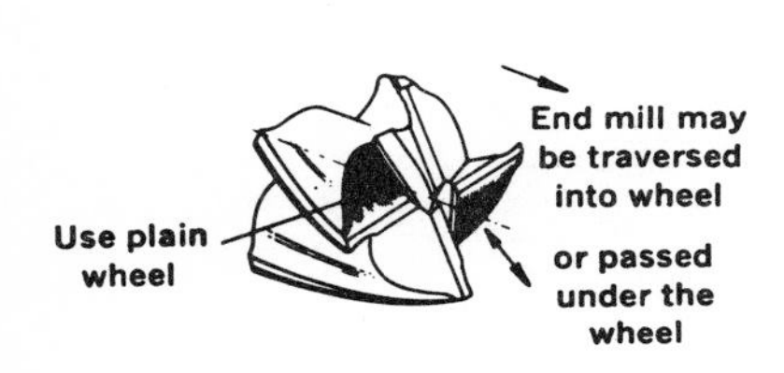

Conclusion of 2nd operation—Secondary clearing the two end teeth not cleared previously

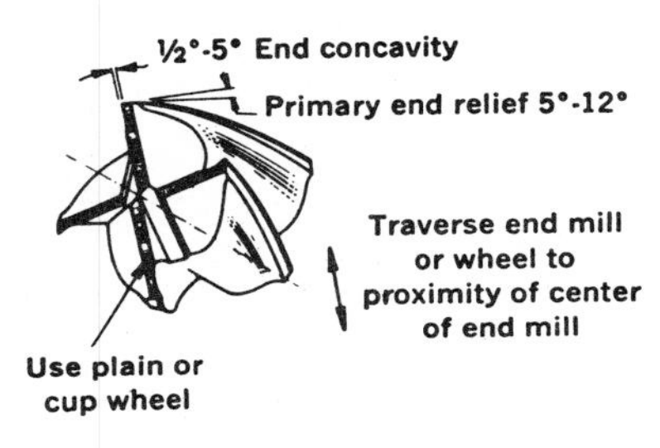

3rd Operation—Primary relieve end teeth

Courtesy of The Metal Cutting Tool Institute

Fig. 14-26. Procedure for sharpening the end of a four-fluted end mill which has center cut teeth.

9. Advance the tooth-rest blade to obtain the desired depth of cut, and grind the tooth face by one of the following methods:

   a. If a large amount must be ground off the tooth face to obtain a sharp cutting edge, the grinding wheel should approach the cutter from the top. Set the depth of cut to clean up the entire tooth face. Holding the cutter firmly against the back rest, traverse the table back and forth. At the end of each traverse, feed the grinding wheel down a few thousandths of an inch. Continue until the cutter reaches the bottom of the tooth. Rough grind each tooth in this manner.
   b. To finish grind the teeth and to remove only a small amount of stock, set the wheel to grind the entire tooth face every traverse. The depth of cut should be about .0005 inch per traverse. Traverse the table forward and back across the face of the tooth. Grind all of the teeth at the same setting before advancing the tooth rest again. Advance the tooth rest another .0005 and again grind all of the teeth. Repeat until all of the teeth are sharp. It is important for all of the teeth to be ground at the same setting on the last cut.

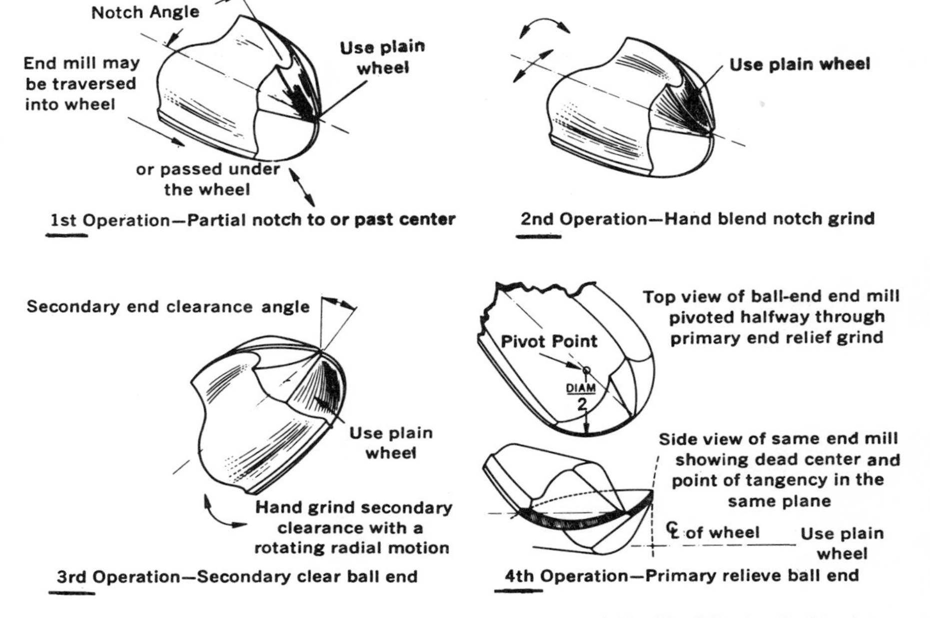

Courtesy of The Metal Cutting Tool Institute

Fig. 14-27. Procedure for sharpening the end of a two-flute, ball-nose end mill used in die sinking.

10. Check the cutting edge of each tooth with a .0001-inch dial test indicator. The concentricity of all of the teeth should be within .0005 inch when tested between the tailstock centers.

If the cutter has an axial rake angle, the grinding procedure is the same; however, a slightly different procedure is used to make the setup. This procedure is described here:

1. Align the grinding wheel with the point of the centers as before.
2. Mount the clearance setting fixture on the left side tailstock.
3. Mount the work and arbor assembly on the tailstock centers.
4. Lower the grinding wheel into one of the tooth spaces and rotate the milling cutter until it touches the cutting edge of the grinding wheel. The grinding wheel should not be running.
5. Hold the work in this position and clamp the clearance setting fixture dog on the arbor.
6. Adjust the clearance setting fixture to read zero.
7. Rotate the face of the tooth away from the grinding wheel to the correct radial rake angle as indicated by the graduations on the fixture. Clamp the graduated plate to the fixed plate.
8. Position the tooth-rest blade against the back of the tooth.

Courtesy of Cincinnati Milacron

Fig. 14-28. Radius grinding attachment set up to grind the end teeth on a ball-nose end mill. Also used to grind other radii on milling cutters.

9. Move the cross slide until the wheel just touches the face of the tooth. Set the micrometer dial of the cross-slide feed screw to zero.
10. Move the cross slide a few thousandths of an inch to back the tooth away from the grinding wheel and traverse the table so that the wheel is clear of the cutter. Reposition the cross slide to the original zero position.
11. Remove the dog and release the graduated plate. The setup is complete, and the machine is ready to grind the cutter.

A new cutter that has never been used must have the back face of each tooth ground before it is used. The back face of each tooth is the locating face for sharpening the teeth. The procedure for doing this is similar to grinding a zero rake cutter, except that the cutter is positioned in the machine with the back face facing the grinding wheel and the tooth-rest blade placed against the front tooth face. The cutter is adjusted until the face of the grinding wheel and the back face of the tooth are in the same

Courtesy of Cincinnati Milacron

Fig. 14-29. Setup for sharpening a form-relieved milling cutter.

plane. Take a light cut across each back face until it is cleaned up. The final cut on each back face must be made at the same setting.

Several attachments are available or can be built for sharpening form milling cutters. The cutters are usually held with the axis vertical in these attachments. Although the operation differs slightly from the one just described, no new principles are introduced.

Grinding Reamers

There are several different kinds of reamers, and for each of these the grinding procedure is somewhat different. The most common type of reamer is the solid machine reamer and the following instructions pertain to this reamer. Most reamers are finishing tools in which the size of the hole being finished depends upon the size of the reamer. For this reason reamers should be ground with a great deal of care and concern for accuracy.

Machine reamers are sharpened by grinding the chamfer angle on the end of the reamer. The outside diameter of the reamer teeth is not sharpened because this procedure would make the reamer undersize. Sometimes the cutting face of the reamer inside the flute is ground in order to remove a slight amount of wear or some other defect from the cutting edge. Occasionally the outside diameter of a reamer is ground undersize in order to produce an off-standard hole.

The outside diameter of a reamer is finish ground by cylindrical grinding on a cylindrical grinding machine or on a cutter and tool grinding machine. The cylindrical grinding attachment is set up, and the wheel head is centered with the centers of this attachment by using the centering gage. A dog is placed on the shank end of the reamer, and the reamer is placed between centers in the machine as shown in Fig. 14-30. Make certain that the centers are clean and in good condition before placing the reamer in the machine. The reamer should rotate in the direction opposite from which it would rotate when cutting. Only very light cuts should be taken when finish grinding the reamer to size, and care must be taken to make certain that the diameter of the chamfer end is not less than the diameter at the shank end. The diameter at the chamfer end may be ground about .0002 inch larger than the diameter at the shank end. This is called back taper, and on some reamers a definite amount of back taper is specified.

Courtesy of Cincinnati Milacron

Fig. 14-30. Cylindrical grinding of a reamer on a cutter and tool grinding machine.

The teeth on some reamers are cut so that they are not diametrically opposite each other, in which case the diameter cannot be measured with ordinary outside micrometer calipers. One method of sizing reamers with staggered teeth is to take a cylindrical piece of steel the same length as the reamer and to grind the diameter about .001 inch larger than the reamer. The reamer is then ground to this size and removed from the machine. The test bar of steel is placed in the machine, and a short length is carefully ground to the size of the reamer using very light cuts and allowing the grinding wheel to spark out. When the final cut is completed, the micrometer dial of the cross-feed hand wheel is set to zero. A .0001-inch dial test indicator is mounted on an indicator base, and the dial is set to read zero when the indicator point touches the highest point on the ground surface. The test bar is removed and the reamer placed back in the machine. The reamer is then carefully ground to size. The finished size is obtained when the micrometer dial on the cross-feed hand wheel is at zero. The size can be checked with the dial test indicator.

Courtesy of Cincinnati Milacron

Fig. 14-31. Grinding the primary relief angle on a reamer.

The cylindrical grinding operation produces a cylindrical land or margin on the peripheral surface of the reamer. Part of this circular land or margin is retained on the finished reamer. The primary relief angle is ground behind the margin, with a specified margin width left. The width of the margin will vary from .005 to as much as .040 inch, depending upon the size of the reamer and the material to be reamed. For example, the recommended margin width for reaming steel with a general purpose reamer is .005 inch for a 1/4-inch reamer and .010 for a 1-inch reamer. A slightly larger margin (.010 to .020 inch) is left on reamers specifically designed to ream cast iron, and for aluminum the margin can vary from .010 to as much as .040 inch. The primary relief angle is ground in a manner that is similar to grinding the primary relief angle on plain milling cutters. The setup for this operation is shown in Fig. 14-31. The primary relief angle depends largely on the size of the reamer. It is 15 to 20 degrees for a 1/4-inch reamer and 6 to 9 degrees for a 1-inch reamer. A secondary relief angle is usually ground behind the primary relief angle. It, too, is dependent upon the size of the reamer, varying from 40 to 12 degrees.

Since machine reamers are end cutting tools, they are sharpened by grinding the relief angle on the chamfer; if the reamer has a starting taper, this, too, is ground. The length of the end chamfer, the chamfer angle, and the chamfer relief angle must all be ground to uniform size, otherwise the reamer may cut oversize. While some machine reamers have a starting taper, most do not. On hand and machine reamers equipped

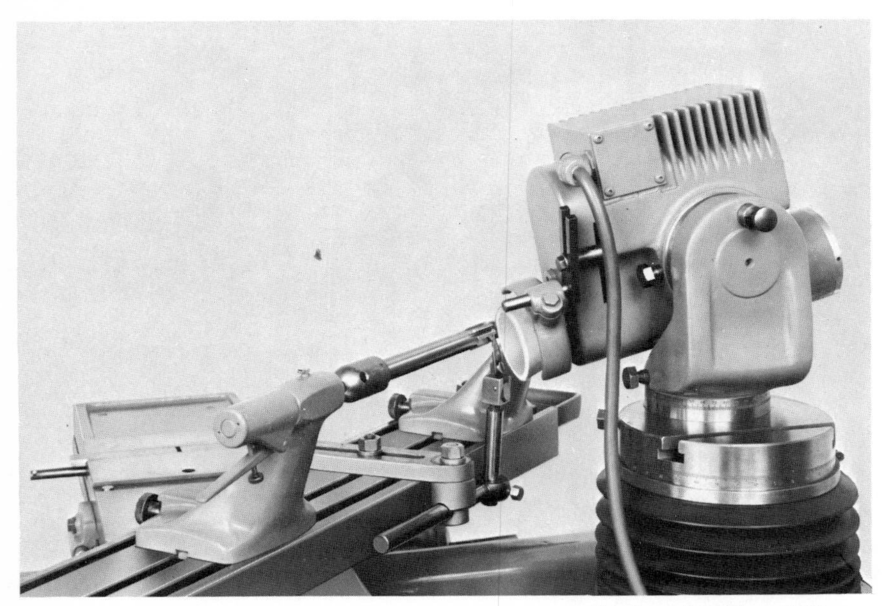

Courtesy of Cincinnati Milacron

Fig. **14-32**. Sharpening the chamfer relief angle on a reamer.

with a starting taper, the taper angle may vary from 1 to 10 degrees, measured with respect to the reamer axis. On machine reamers the starting taper is usually short, being 1/16 to 1/2 inch long, although other special modifications are sometimes used, especially on the larger size reamers. The starting taper is ground to a sharp edge, or even to a hairline edge, by using the setup shown in Fig. 14-31.

For most applications, the chamfer angle on the reamer is 45 degrees and the chamfer relief angle is from 7 to 12 degrees. On left-hand helix reamers, the chamfer angle should be about 30 degrees, or even less, if the reamer has a large helix angle, in order to have a positive effective rake angle. In general, the chamfer should be ground to a sharp edge.

Two setups for grinding the chamfer are shown in Figs. 14-32 and 14-33. In Fig. 14-32, the reamer is held between centers and the table is tilted 45 degrees for grinding the chamfer angle. The chamfer relief angle is formed by tilting the wheelhead to the required angle. In order to obtain a concentric chamfer, the operator should hold the face of the flute against the tooth-rest blade with uniform pressure when grinding each edge. Small reamers may not have a center hole in each end; they are also often difficult to grind between centers because the center interferes with the positioning of the grinding wheel to grind the chamfer angle. In this event a reamer chamfer, relief-angle grinding fixture, shown in Fig. 14-33, can be used, although this fixture is not restricted to grinding small reamers. The reamer body is held in a close fitting bushing while the shank is held on a conventional cone point center or in a cup center. The required chamfer angle and chamfer relief angle are obtained by adjusting the position of the fixture. In addition to holding the reamer against the tooth-rest blade with a uniform pressure, the operator must hold the reamer lightly against the center at the end of the fixture.

Courtesy of The Cleveland Twist Drill Co.

Fig. 14-33. Reamer chamfer relief-angle grinding fixture set up to sharpen a chamfer.

CHAPTER **15**

Numerically Controlled Machine Tools

Numerical control, or NC, is a precise electronic control system used to direct the operation of machine tools in conformance to a predetermined program; it is characterized by its ability to control the path or position of the cutting tool or workpiece along numerically defined machine tool

Courtesy of LeBlond Machine Tool

Fig. 15-1. NC lathes at work in a manufacturing department.

coordinates. Although certain modifications are required to install an NC system onto a machine tool, its basic appearance is not necessarily altered, as can be seen by the NC lathes in Fig. 15-1. There are, however, machine tools that have been specifically designed to take full advantage of NC and these machines do have a different appearance. Examples of such machines are the slant bed lathe, shown in Fig. 15-2, and the machining center, shown in Fig. 15-3. However, the basic machining operations have not changed as a result of NC. Turning, drilling, milling, and all other machining operations are performed on NC machine tools in the same manner as on their manually controlled counterparts.

NC provides many advantages. It can very rapidly position the cutting tool and the workpiece in readiness to take a cut, thereby reducing the non-cutting time of the machine. While accuracy is not necessarily increased by NC, it can often be achieved more rapidly as a result of NC by reducing the number of trial cuts required to position the cutting tool or the workpiece at the start of a cut; on NC machine tools a trial cut is usually required only on the first part to be machined and in many cases trial cuts can be entirely eliminated. The machine tool movements on NC machines are repeated precisely, resulting in a reduction in the scrap rate; for drilling operations this capability eliminates the need to use drill jigs in order to locate the positions of the holes. A single NC part program can be written to direct the machine to perform many different operations and to machine many different surfaces on a part, thereby reducing the number of setups and work-holding fixtures required. Some NC machines, such as

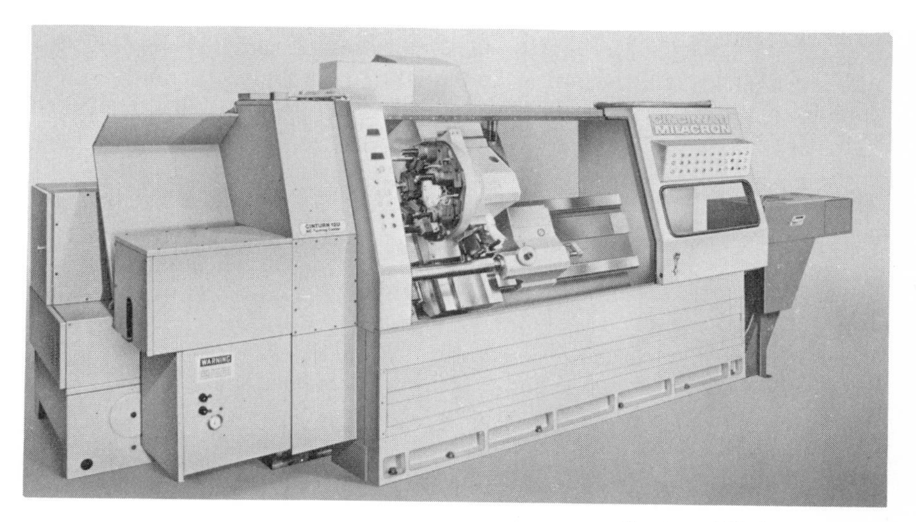

Courtesy of Cincinnati Milacron

Fig. 15-2. CNC slant-bed universal turning center for chucking and for turning on centers. In order to show tooling turrets, protective window shields have been removed. Slant bed design allows chips to fall into chip conveyor for removal.

Courtesy of Cincinnati Milacron

Fig. 15-3. Three-axis CNC machining center equipped with a rotary index table and a 30-tool random access automatic tool changer.

that shown in Fig. 15-4, are designed to generate complex machine tool motions that are far beyond ordinary human capability to control. This capability is used to machine forming dies, mold cavities, airfoil surfaces, and aircraft structural elements.

Types of Numerical Control

Numerical control systems may be classified by two basic methods of machine operation, namely, *point-to-point* and *continuous path*. The point-to-point method is also called *positioning*, and is used on machine tools that operate with the table held in a stationary position while the part is being machined, such as in the case of drilling machines. The NC system locates the table first in one and then in another position, without concern for the path followed to reach these positions. When moving from one position to the other, the cutting tool is clear of the workpiece. Many point-to-point NC machine tools are capable of a limited amount of the "picture frame" type of milling; i.e., cuts can be taken in the perpendicular directions of the table movements and at 45 degrees. As the name implies, continuous path NC or *contouring*, as it is also called, controls the path followed by the cutting tool or the workpiece at all times. This type of

NC system is used on lathes, milling machines, and other machine tools to machine contoured surfaces, as well as to take straight line cuts and to perform operations where the table is held in a stationary position.

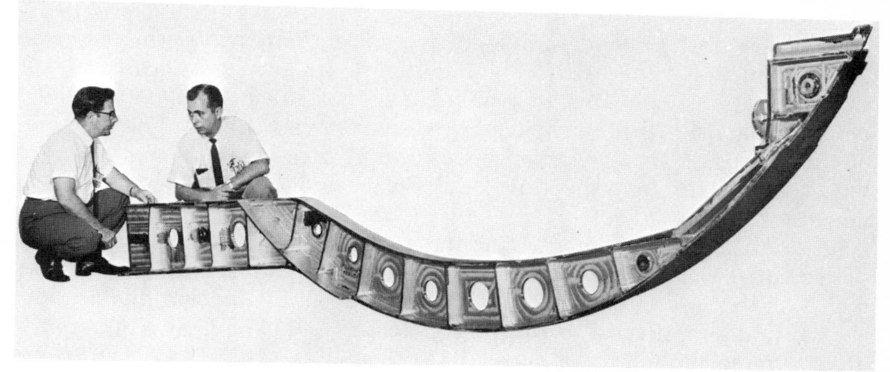

Courtesy of Cincinnati Milacron

Fig. 15-4. Aircraft fuselage ring section produced on five axis profile milling machine in two setups—one for each side.

LIST OF SOME COMMON NC SYMBOLS

BCD	Binary Coded Decimal
CAD	Computer Aided Design
CAM	Computer Aided Manufacturing
CNC	Computer Numerical Control
CPU	Central Processing Unit
CRT	Cathode Ray Tube
DNC	Direct Numerical Control
EIA	Electronic Industries Association
IC	Integrated Circuit
MCU	Machine Control Unit
NC	Numerical Control
PC	Programmable Controller
PCB	Printed Circuit Board
RAM	Random Access Memory

Another method of classifying NC is by the difference in the design of the machine control unit (MCU). This is the unit that receives the information required to machine the part and processes it into a form that can be used to control the operation of the machine tool. The part program containing the machining information is encoded on a perforated, or punched, tape, which is run through a tape reading head on the MCU in order to enter the part machining information into this unit.

There are two basic types of machine control units, hard-wired and CNC, also known as soft-wired NC. As mentioned in the List of NC Symbols, CNC stands for Computer Numerical Control. Outwardly both types of machine control units have a similar appearance; however, their inner construction is very different.

Hard-wired units have permanently wired circuits connecting the electronic elements of the MCU. These units are designed specifically to control one type of machine tool only. Many of the units have a digital readout that will display the coordinate positions of the sliding elements of the machine tool. A characteristic of hard-wired units is that the tape must be run through the tape reader each time that a part is to be machined.

CNC utilizes a dedicated computer in place of the hard-wired circuits. A dedicated computer is one that is designed to fulfill a single purpose, which, in the case of CNC, is to store and process the machining data into a form that can be used to control a machine tool. The CNC computer is often a mini-computer. As built, the computer is unable to perform any function; it must first be "set up" electronically, by having a control program entered into its memory. This is done by means of a punched tape, which is prepared by the builders of the computer or the machine tool before it reaches the user. The control program contains instructions telling the computer how to process the machining information contained in the part program; it makes the computer "think" like a lathe, milling machine, or any other designated machine tool. A computer may have any one of many different control programs entered into its memory that enable it to control any one of many different types of machine tools. The part program is entered into the computer memory by the user, also by means of a punched tape. (More than one part program can be stored in the computer memory.) This program provides the instructions required to machine the part. Since the part program is stored in the computer memory, it is only necessary to run the punched tape through the tape reader once, which is a particular feature of CNC units. Many CNC units have another prominent feature, a cathode ray tube (CRT), which can display machining information on its screen. Some CNC units have computers that have a very good computational ability, while others do not. One computer may receive information from another computer then send this data to other phases of the manufacturing system, making it possible to integrate the machine tool which it controls into a computer aided manufacturing (CAM) system.

CNC must not be confused with DNC, or direct numerical control. DNC is a numerical control system whereby the machine control units of more than one machine tool are connected to a common, centrally located computer, which is itself remote from the machines. This computer is used to store the part programs and on demand distributes the machining data to the different machine control units throughout the plant.

Another type of machine control unit is the programmable controller, or the PC. Programmable controllers are computerized soft-wired units that can control a sequence of events; applied to machine tools they control the sequence of the operation of the machine elements. They cannot control the path or position of the cutting tool or workpiece along programmed numerically defined coordinate dimensions; for this reason they are not considered to be true numerical control units. Programming a PC is simple compared to programming an NC unit. The programmer, who may be the machine operator, enters the part program into the PC unit by means of pushbuttons located on a keyboard built into the control panel of the unit. The program entry is then displayed on the screen of a CRT and stored in the computer memory for use when the machine is in operation.

NC Machining

The objective of this section is to provide an overview of the sequence of events required to machine a part on an NC machine tool. The first step in this sequence is to prepare the part program.

Preparing the Part Program. The part program is prepared by a part programmer whose qualifications will vary, depending upon the complexity of the parts that he is expected to program. In general, he must be able to read part prints, have a knowledge of machining operations and metal cutting tools, and be able to use basic mathematics—or more advanced mathematics when programming very complex parts. When using the computer assisted programming method, he must have a command of the particular processor language used.

The part programmer must visualize all of the machining operations and the cutting tools to be used in order to machine the part. Although in many cases it is the workpiece that is moved by the machine table when taking a cut, when programming, the programmer must assume that the cutting tool is moving in all cases. He must know how the part is to be held on the machine and exercise care to avoid a collision between the cutting tool and any part of the work-holding fixture. He will select the cutting tools and specify their speed and feed. Whenever a large amount of metal must be removed from the part, he should be able to estimate the power required in order to specify cuts that will not overload the machine, yet use the maximum amount of power available. With all of this in mind, he lists all of the cutting tool movements and other machine tool functions required to machine the part in logical sequence. The resulting document is called the "manuscript."

Manual Programming. There are two methods of writing the program: 1. manually, and 2. with the assistance of a computer. The steps in manual programming are shown in Fig. 15-5. Manual programming requires that all of the cutter movements and other machine functions be listed on the manuscript in sequence. Moreover, all of the numerical coordinate data must be given; the programmer must calculate all of the cutter positions and specify them on the manuscript. The manuscript is then taken directly to a tape preparation unit, which resembles an ordinary typewriter. The part program is encoded on a tape by this unit, which also prints a copy of the program on paper in common English letters and Arabic numerals; this is called a *printout.*

Computer Assisted Programming. Computer assisted programming is used to eliminate the need to make the calculations required to determine the numerical coordinate data. These calculations are often complex and time consuming, especially when programming continuous path type operations where cutter offset calculations must be made. The computer can make these calculations very rapidly and without error. With this method, the part program is written in a special processor language. There are a number of different processor languages, varying in complexity according to the complexity of the parts to be programmed by the language. Each language consists of certain symbols and English-like words that have a particular and precise meaning. The part program written in this language consists of generalized statements telling the cutter where to go, and other statements to instruct the machine on how it must operate. Processor languages greatly simplify the task of writing the program and reduce the number of entries in the manuscript.

The steps in computer assisted programming are shown in Figs. 15-6 and 15-7. When the computer is in-house, as in Fig. 15-6, the manuscript,

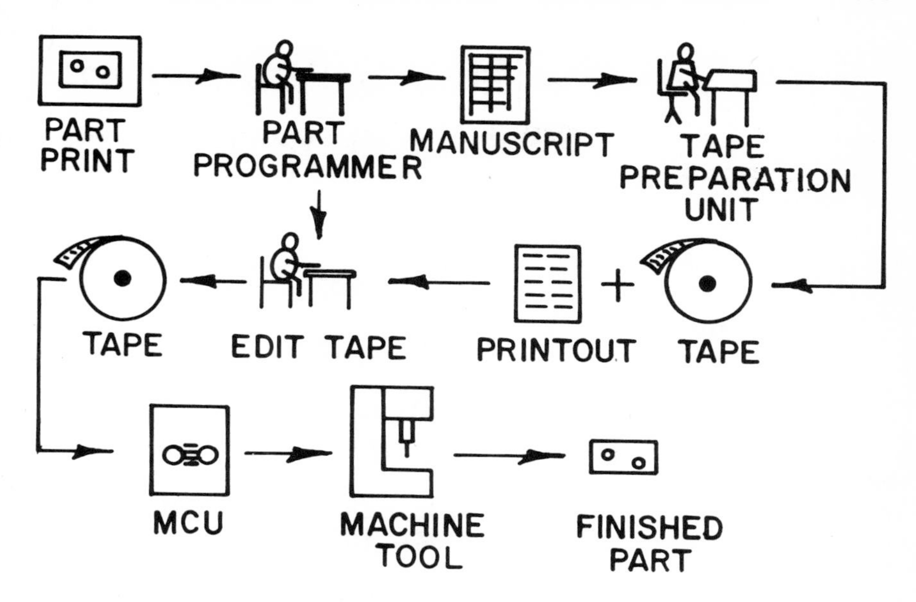

Fig. 15-5. The steps in manual part programming.

written in a processor language, is taken to a key punch, where an operator encodes the program onto tabulating cards. The tabulating cards are used to enter the part program into the computer. The information in the part program is processed in the computer from the processor language form into numerical coordinate data. A high-speed tape punch then makes a punched tape of the processed part program in its numerical form; at the same time a printout is also made.

When the computer is in a remote location, which may be as far away as another city, the procedure shown in Fig. 15-7 is used. The program manuscript, also written in a processor language, is taken to a data terminal in the plant which is connected to the computer by an ordinary commercial telephone connection. A teletype unit on the data terminal is used to transmit the part program directly into the computer, which processes the program and sends it back to the data terminal, again by telephone. As the information is received from the computer, a punched tape and a printout is made of the part program, which now contains the numerical data required to operate the machine tool.

Some CNC units have a computer that has computational capability. Part programs for these units are also written in a processor language; however, in this case a tape is made directly from the manuscript, as in manual programming. After it has been edited, this tape is read directly into the computer of the CNC unit, which processes it as required. Although this is also a form of computer assisted programming, the part

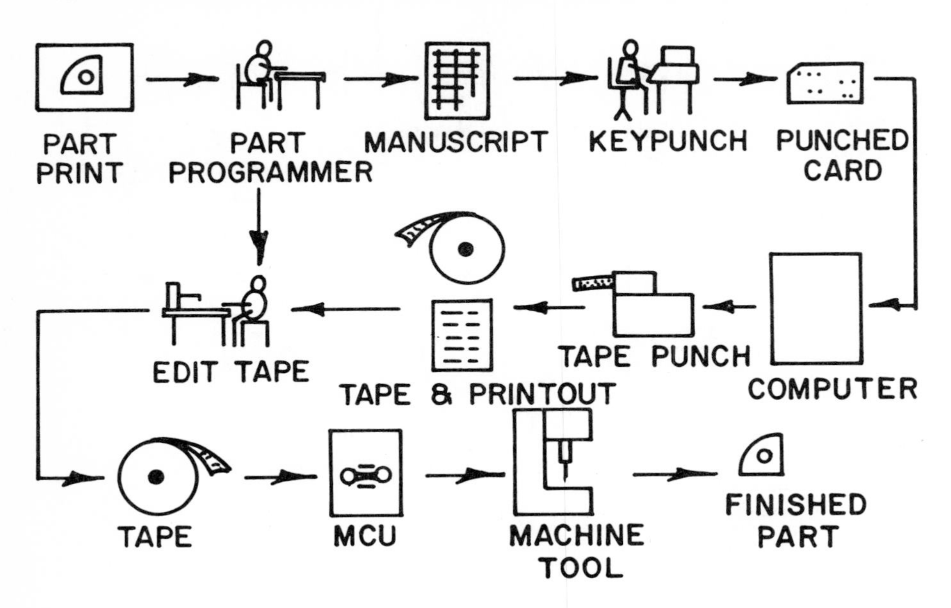

Fig. 15-6. The steps in computer assisted part programming using an in-house computer.

program is processed directly in the CNC unit and not by a separate computer. In this case, the steps are like those shown in Fig. 15-5.

Editing. Before it is used, the punched tape should be edited to find and correct any mistakes that may be on it. Sometimes, while editing, improvements in the part program become evident and can be incorporated into the program at this time. Since the printout contains the information, printed in English letters and Arabic numerals, that is coded on the tape by the pattern of the punched holes, a logical means of editing the tape is to carefully review the printout. This is a very good method of editing tapes made for point-to-point machining programs because the numerical data on the printout describes the exact cutter positions for the machining operations. Although the exact numerical positions of the cutter are also given on the printout of continuous path machining programs, it is more difficult to relate the cutter positions to the resulting dimensions of the workpiece on this type of program because of the compensations in the positions of the cutter required for cutter offsets and for the nose radius effects. If the exact cutter positions are known, then reviewing the printout is an effective method of editing continuous-path machining programs; if not, other methods are used. Tape editing units are helpful in this event. One type of tape editing unit plots the path of the cutting tool on paper and another type displays the cutter path on a CRT. These units are useful in preventing a serious collision from occurring on the machine tool; they detect decimal point errors of the type that occur when the

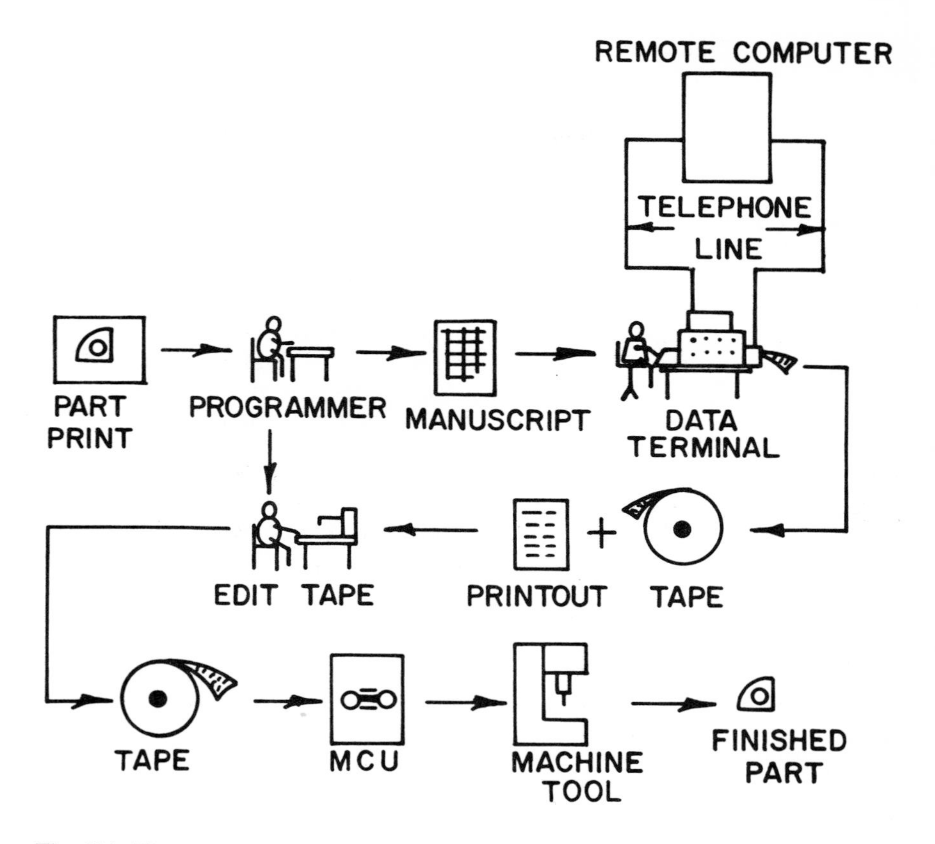

Fig. 15-7. The steps in computer assisted part programming using a computer located in a remote area which may even be in another city.

machine tool slide moves ten inches when a one-inch movement is intended. The fine editing is then done at the machine by working on a sample part. Sometimes a "dry" run is made first by running the machine tool through the programmed cycle without a part actually in place. If the machine is controlled with a CNC unit, the corrections and changes can be made in the part program while it is stored in the computer memory. Some CNC units have a high-speed tape punch, which is used to make a corrected tape from the corrected part program stored in the computer memory. This tape is used as a permanent storage medium for the corrected part program. It is used for re-entering the program into the computer memory, when required. Editing is also often done after the actual operation of machining the parts has started in order to improve the efficiency of the operation or to correct any small errors. This may include changes in the cutting speed, in the feed rate, or small changes in the cutter position.

The NC Machine Operator. The machine operator fulfills an important function in NC machining which is beyond merely loading and unloading the part on the machine. He must know how to operate and take care of the machine, read part prints, know how to use and take care of precision tools, and have an understanding of cutting tools. The operator should be provided with a complete set of instructions for setting up and machining each part.

As always, the first step in doing the job at the machine is to set up the work-holding fixture, the workpiece, the cutting tools, and the machine tool. On the first part to be machined it is often necessary to establish the position of the cutting tool in relation to the workpiece by taking a trial cut, especially for finishing cuts on close-tolerance operations. A short oversize cut is taken, or when boring a hole the first cut is made undersize. The surface that has been cut is then measured and a compensation is made in the position of the cutting tool or the workpiece so that the workpiece can now be machined to size. On close tolerance work similar compensations may have to be made at various times to compensate for tool wear. The operator should watch the cutting tools for signs of wear and replace any that are dull. He should watch the machine tool and the machining operations for possible malfunctions. In some shops, the operator inspects a finished part during the time that another part is being machined. This mode of operation allows the operator to detect out-of-tolerance operations immediately and to make the necessary corrections in the machine settings, thereby reducing or eliminating out-of-tolerance parts from coming off the machine. An intelligent and conscientious operator may sometimes be able to make valuable suggestions for improving the part program.

Rectangular Coordinates

All NC measurements are based on rectangular coordinates. Rectangular coordinates are uniformly calibrated distances along two or three mutually perpendicular axes. A system of two-dimensional rectangular coordinates is shown in Fig. 15-8. By convention, the two axes are called the x axis and the y axis, and their intersection is called the *origin* or *zero point*. Distances along the x axis which are to the right of the origin are considered to be positive; those to the left of the origin are negative. Likewise, along the y axis, distances above the origin are positive and those below the origin are negative. The position of any point located in the plane of the coordinate axis can be specified by the coordinate dimensions along the x and y axes. For example, in Fig. 15-8, point A is located by moving $+3$ units in the x direction and $+2$ units in the y direction; its position is specified by $x+3$ and $y+2$. In the same manner, point B is specified by $x-2$ and $y+3$, while point C is specified by $x-3$ and $y-2$.

A three-dimensional space system of rectangular coordinates is shown in Fig. 15-9. Here a third axis, called the z axis, is made perpendicular to both the x and the y axes. By specifying the coordinate dimensions along

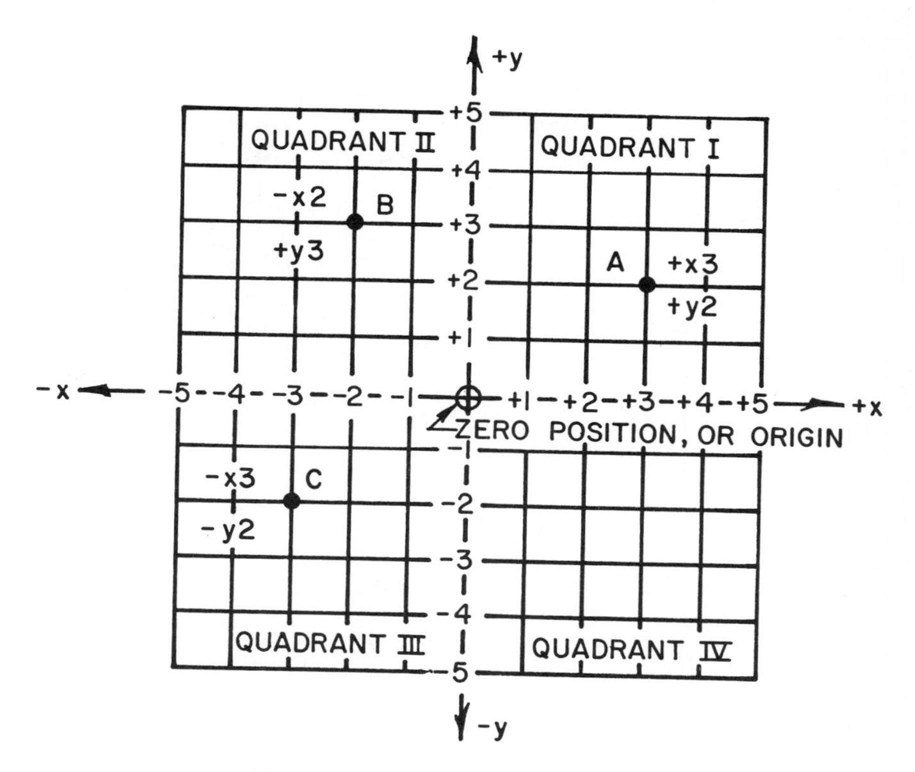

Fig. 15-8. Rectangular coordinate system.

the x, y, and z axes, the position of any point in space can be determined.

Referring again to Fig. 15-8, the two coordinate axes are considered to divide the plane into four quadrants. Thus, point A is located in the first quadrant, point B in the second quadrant, and point C is located in the third quadrant; but there is no point in the fourth quadrant. Another system of coordinate dimensioning, called polar coordinates, is sometimes used to define the position of a rotating member, as in the case of indexing. Polar coordinates specify the position of a point by its distance from the center or axis of rotation and by the angle through which it is rotated.

Since machine tool slides are usually made to move in mutually perpendicular directions, the rectangular coordinate system can readily be adapted to define the positions of these slides. For example, the table of a drilling machine will move in the x and y coordinate directions and its spindle will move in the z direction, as shown in Fig. 15-11, view B. In order to define the axes of motion of machine tools more precisely, some modification in the definition of the coordinate axes are made in practice, as will be shown in a following section.

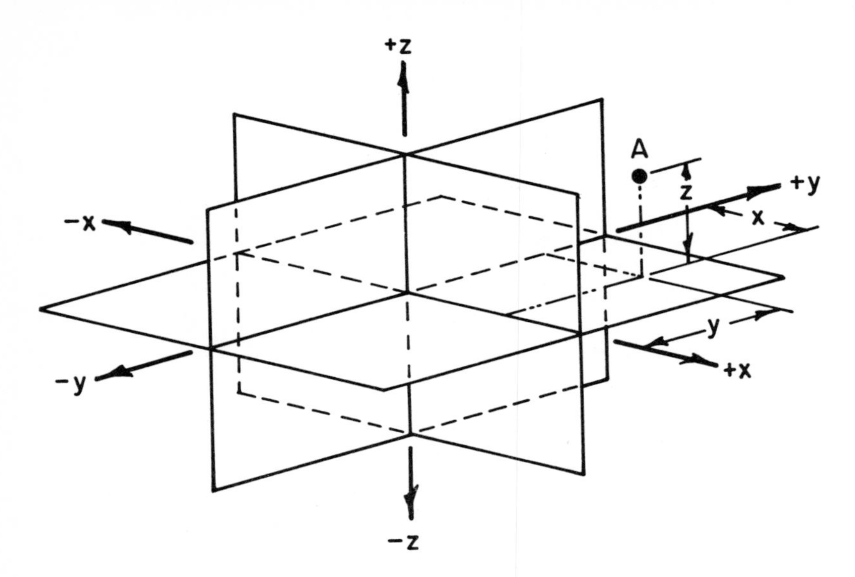

Fig. 15-9. Three-dimensional space system of rectangular coordinates. The point in space is specified by the x, y, z coordinate distances.

Axis and Motion Designations

The axis and motion designations of NC machine tools have been standardized by the Electronic Industries Association (EIA RS-267-A). They are used to identify the motions when writing the program. Some typical examples of these designations are shown in Figs. 15-10, 15-11, and 15-12. All of the motion directions in these illustrations are shown in their positive directions; motion in the opposite direction is negative. Some of the motion directions are unprimed while others are designated by a primed letter, such as x and x'. The programmer, setup man, and the machine operator should think exclusively in terms of the unprimed directions. When writing the part program, the programmer must always direct the cutting tool to move while the workpiece is considered to be stationary. On some machine tools, such as lathes, the cutting tool is the actual moving element. If, however, the machine element moves the workpiece instead of the cutting tool, the control system must respond to the part program commands in the opposite direction to that defined by a moving tool. An arrow with a primed letter, such as $+x'$, is the direction of motion of a *moving workpiece* in response to a command calling for positive motion of the tool; an arrow with an unprimed letter, such as $+x$, is the direction of motion of the *cutting tool* in response to the same command.

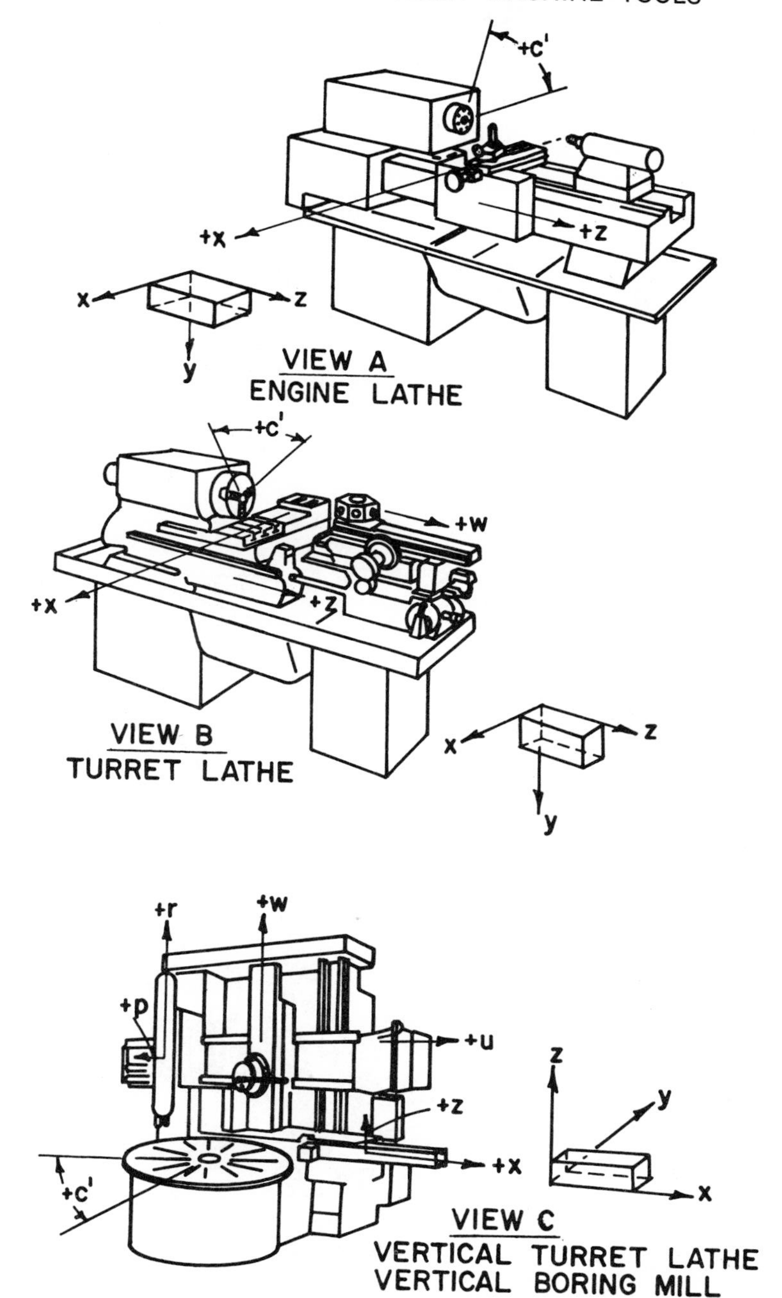

Fig. 15-10. Axis designations for turning type machine tools.

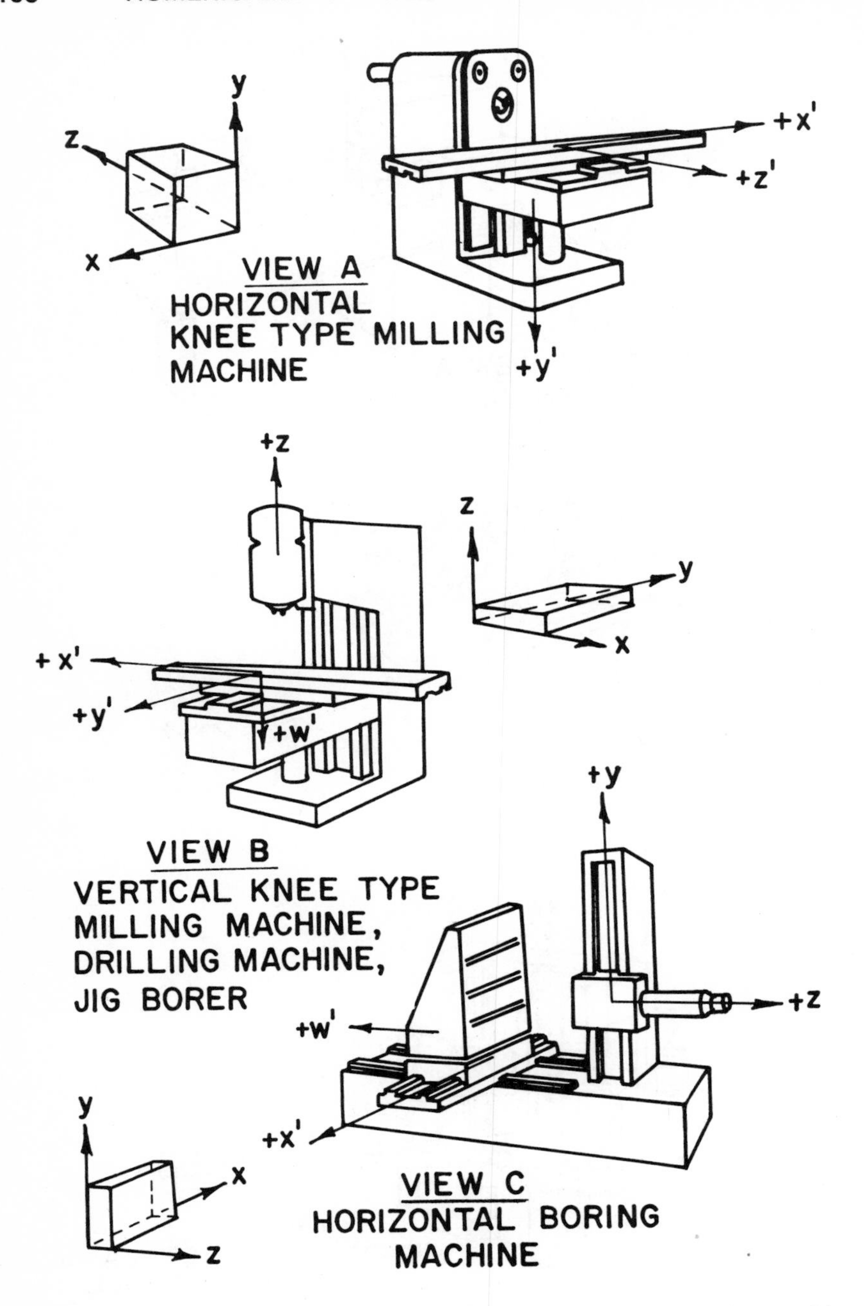

Fig. 15-11. Axis designations for milling, drilling, and horizontal boring machines.

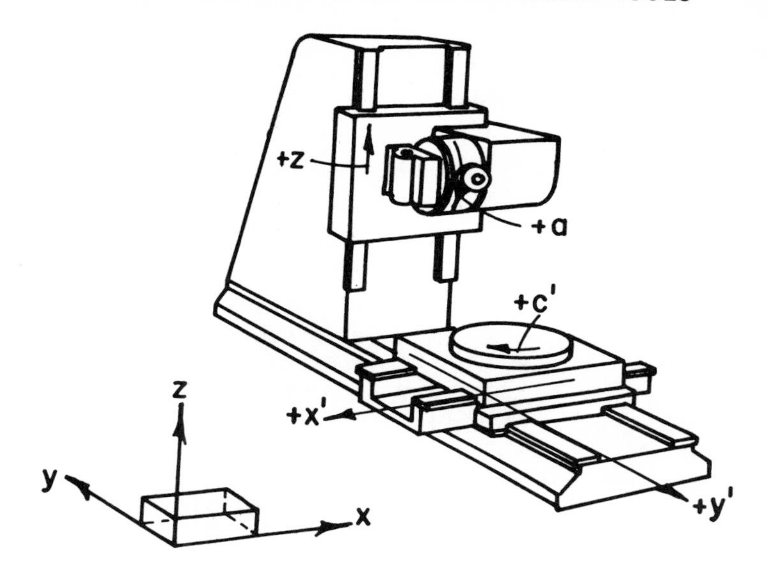

Fig. 15-12. Axis designations for 5-axis profile and contour mill.

The three principal axes of motion are the x, y, and z axes. The z axis is always parallel to the machine tool spindle; it is always positive from the work-holding means toward the tool-holding means. Motion that is horizontal and parallel to the work-holding surface is always designated as the x axis. The y axis is always perpendicular to both the x and z axes. Secondary motions parallel to the x, y, and z axes are designated as u, v, and w; tertiary motions in these directions are designated p, q, and r, respectively. The secondary and tertiary motion designations are also used to define motions that are not parallel to the x, y, and z axes, if such motion exists. Rotary motion around axes parallel to the x, y, and z axes are designated by the letters a, b, and c, respectively.

Binary Numbers

Binary numbers and binary notation are the basis by which numerical data are entered into the computer. A knowledge of binary numbers is not essential to programming or to operating NC machine tools; however, it will provide a better insight into how computers and NC control units operate.

The binary numbering system is based on the powers of the number two. For example, the number 2 is expressed by 2^1 in the binary system, the number 4 by 2^2, etc. Obeying the laws of exponents, the number 1 is expressed in binary by 2^0. Any number can be expressed by the sum of appropriate binary numbers. As an example, the numbers one through ten are expressed by binary numbers in the following manner:

$$1 = 2^0 \qquad 6 = 2^2 + 2^1$$
$$2 = 2^1 \qquad 7 = 2^2 + 2^1 + 2^0$$
$$3 = 2^1 + 2^0 \qquad 8 = 2^3$$
$$4 = 2^2 \qquad 9 = 2^3 + 2^0$$
$$5 = 2^2 + 2^0 \qquad 10 = 2^3 + 2^1$$

Binary notation is an organized manner of expressing numerical values by means of characters which represent binary numbers. The binary numbers are assumed to be placed from right toward the left in order of their increasing power, starting with 2^0. Instead of writing the actual binary number, two characters, called bits, are used, which are usually 1 and 0. The character 1 indicates that the binary number is present and should be counted, while the 0 indicates that it is not present and should not be counted. The sum of the binary numbers that are present in a row is the numerical value of the binary notation. A typical example is given below:

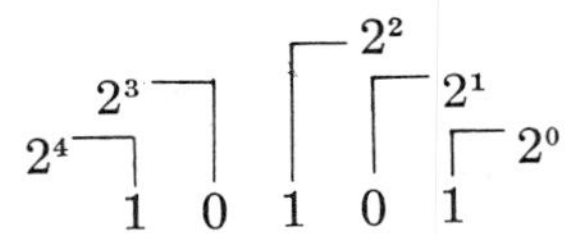

The numerical value of this binary notation is $2^4 + 0 + 2^2 + 0 + 2^0 = 21$. A few other examples of binary notation are shown below:

Binary Notation	Numerical Value
0 =	0
10 =	$2^1 + 0 = 2$
11 =	$2^1 + 2^0 = 3$
101 =	$2^2 + 0 + 2^0 = 5$
1110 =	$2^3 + 2^2 + 2^1 + 0 = 14$

The advantage of binary notation is that any number can be expressed by using only two characters or bits. Electronically these characters can be represented by a switch that is either open or closed. In a computer the bits may be a plus or minus charge of a ferrite core, magnetic film, or by the charge of another material. Binary notation is used on the punched tape; in this case the two characters, or bits, are indicated by the presence or absence of a punched hole.

The NC Tape

The NC tape may be made of paper, a laminate of paper and plastic, or aluminum and plastic. The width, thickness, and the size and spacing of the holes have been standardized. The standard width of the tape is one inch and it has eight tracks or channels, not including the small sprocket holes used to feed the tape.

Tape Format. The tape format is the physical arrangement of the data on the program tape and the overall pattern in which it is organized and presented. Several different tape formats have been used, including the EIA RS-244-A (see Fig. 15-13A), EIA RS-273-A, EIA RS-274-B, EIA RS-358 (see Fig. 15-13B), and ASCII. ASCII is also called ANSCII, meaning American National Standard Code for Information Interchange. The EIA RS-358 format (Fig. 15-13B) is a sub-set of the ASCII format, and is now considered to be the standard format. Older machine control units require a specific tape format to be used; many newer machine control units can handle either EIA or ASCII coding.

Binary Coded Decimal. All tape formats use the binary coded decimal, or BCD method of individual digit coding. Starting from the right (at the top in Figs. 15-13A and 15-13B), the first channel is assigned the value 1, the second channel, 2; the third channel, 4; and the fourth channel, 8. It should be noted that these values correspond to 2^0, 2^1, 2^2, and 2^3. The sixth channel is assigned the numerical value of 0. Thus, these channels can be used to designate any number between 0 and 9 in the manner described in the previous section and shown in Fig. 15-13. The numerical quantities are expressed in binary notation running the length of the tape. Each number is expressed in terms of a given number of digits, which is usually six. Decimal points are not shown but are understood; in the case of a six-digit number, the decimal point is understood to be between the second and third digit. Numbers, letters, signs, and other symbols are encoded on the tape by the pattern of the horizontal rows (shown vertically) in Fig. 15-13.

Parity Check. Parity check is a method of automatically checking the tape for errors caused by the malfunction of the tape punch. Each horizontal row must have either an odd or an even number of punched holes, depending on the tape format used. The EIA RS-244-A tape format has an odd number of punched holes for the parity check, while the EIA RS-358 format has an even number of holes for parity. Failure to have the required even or odd number of holes in a row will cause the control system to stop.

The Word. The characters or bits in NC tape language are used to make up an NC word. Each NC word has enough characters in a logical sequence to cause a specific machine tool action. For example, an NC word might be $x+05000$, meaning that the x coordinate of a point is 5.000 inches.

The Block. The block, like a sentence, contains a sufficient number of words to provide the instructions for a single operation at a given location. For example, the block might contain the x, y, and z coordinates plus additional instructions for machining the part at this location.

Auxiliary or Miscellaneous Functions. These functions are words used to transmit machine operating instructions other than coordinate dimensions. They are identified by a lower case letter followed by a code number. Some of the more frequently used functions will be described in the

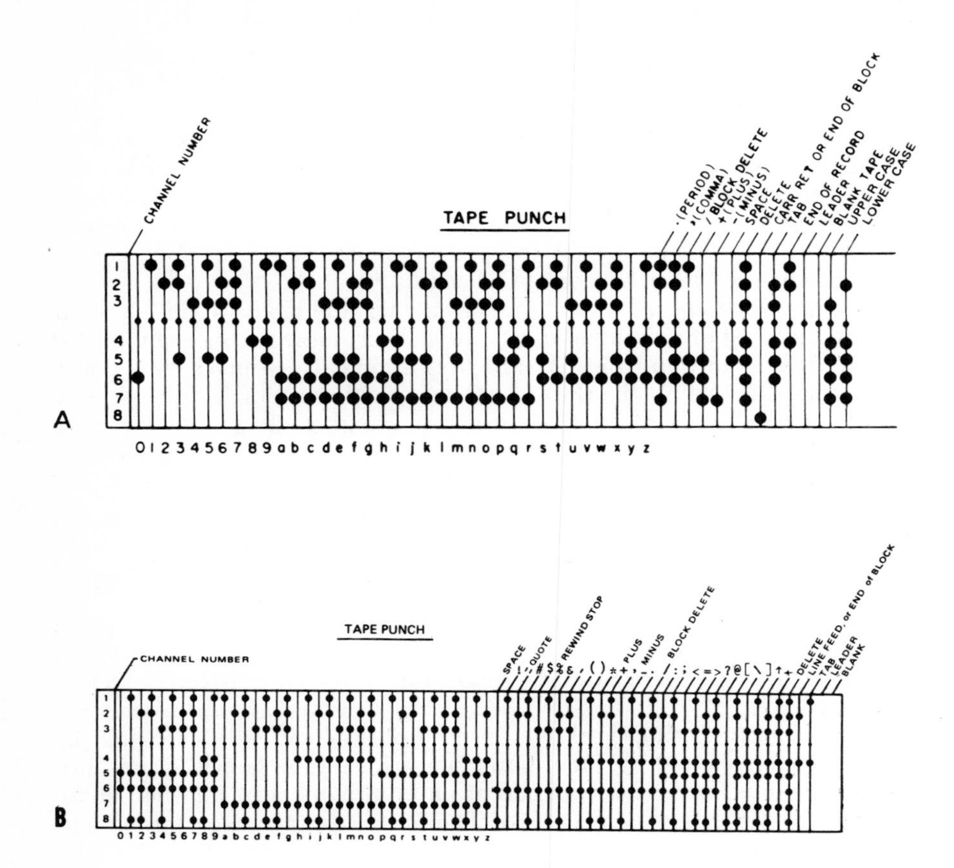

Fig. 15-13. Tape formats: A. EIA RS-244-A; B. EIA RS-358.

following lines. In all cases the machine tool control unit must have the ability to perform the designated functions.

The f Word. The feed rate is specified by the *f* word.

The g Word. This is a preparatory word; it is used to prepare the control system for instructions that are to follow. For example, *g*90 indicates that all of the following coordinate dimensions are expressed as absolute dimensions; *g*91 indicates that they will be expressed as incremental dimensions.

The m Function. The *m* function is used to control miscellaneous machine tool functions, such as turning on the coolant (*m*07), or to engage the rapid traverse or feed motion in the + (*m*15) or the − (*m*16) direction.

The n Word. The *n* word is used to identify the sequence numbers of the blocks. Although not mandatory, it is common practice to identify every block on the tape with an n word that is usually made of three digits; e.g., n001, n002, n003, etc.

The s Word. The s word is used to identify the code that specifies the spindle speed of the machine.

The t Word. This word is used with machines having automatic tool changers or turrets which can be indexed in order to position a particular cutting tool in preparation for taking a cut. The t word calls out the particular tool that is to be used.

In addition to those listed, other address characters are used on machine tools that have the capability of implementing the commands. For example, angular dimensions about the x, y, and z machine tool axes are addressed by a, b, and c words, respectively.

The Machine Control Unit

The machine control unit, or MCU, converts the instructions contained in the part program into a form that can be used to control the machine tool. There are two types of machine control units, hard-wired and CNC, or computer numerical control, which is also called soft-wired NC. Hard-wired units use integrated circuit (IC) digital logic packages which are mounted in a fixed and permanent arrangement on plug-in printed circuit boards (PCB). The PCB connectors that receive these boards are also wired together permanently to connect the electronic components in a fixed and permanent manner. Each hard-wired unit is designed and built to control only one type of machine tool. Input signals that activate the control functions must be derived from the tape; i.e., the tape must be run through the tape reader each time a part is to be machined.

The distinguishing feature of CNC or soft-wired units is the built-in computer, which is often a mini-computer. Unlike hard-wired units, each CNC unit is inherently able to control more than one type of machine tool. It is adapted to control a particular type of machine tool by entering an appropriate control program in the computer memory, which is done by the builder of the CNC unit or the machine tool before the unit is delivered to the user. The control program is also called the "executive" program or the "software" program; it consists of an ordered set of instructions which provides the computer based control system with the capability of executing the commands of the part program. It is stored in the computer memory for an indefinite period. It may be electronically erased and replaced with another control program to enable the CNC unit to control another type of machine tool, although this feature is seldom used. Modifications to the control program can be made; this is usually done by the builder of the computer or the machine tool and very seldom by the user.

The part program is first entered into the computer and stored in the computer memory by running a tape through the tape reader. After this has been accomplished the tape is no longer required, except to serve as a permanent storage medium for the part program. More than one part program can be held in the computer memory at one time; the number of different part programs that can be stored in the computer memory at one time depends on their length and upon the storage capacity of the computer. A feature called random access memory (RAM) allows any stored part program to be called upon for use whenever it is needed. In this way it is possible to store all of the part programs that are to be used to operate the machine for a day, a week, or a month. A permanent record of each part program must be kept because when it is stored in the computer memory it may have to be electronically erased to make room for another part program. The punched tape used to enter the program into the computer is a good permanent storage medium. Some CNC units are equipped to handle diskettes, commonly called "floppy discs." A diskette is a seven-inch disc on which the same amount of information can be permanently stored as on 2,000 or 3,000 feet of punched tape. When required, the part program can be re-entered into the computer memory from either the punched tape or the diskette.

Many CNC units have a cathode ray tube (CRT) on which information can be displayed. There are many ways in which the CRT is of help to the machine operator and to the part programmer. While the machine is operating, the CRT can display on its screen the machine slide positions for all axes, the feed rate, and other information. A message to the operator can be programmed in the part program, which is displayed on the CRT screen at the appropriate time. The part numbers of all programs stored in the computer memory can be displayed. Some units can display the number of characters in each program and the number of characters available to store additional programs. Other information that may be displayed includes tool offsets, tool length compensations, cutter radius compensations, and fixture offsets. Some CNC units have a computerized self-diagnostic program used to isolate and identify malfunctions in the NC system; the CRT is used to display the diagnostic information.

An important application of the CRT is to display part program information when editing. Information may be added, deleted, or changed in the part program while it is stored in the computer memory without involving the tape. Some CNC units have a high-speed tape punch which is used to make a tape of the edited and revised program for permanent storage.

The computers used in CNC units may or may not have the ability to perform computations from information given in the part program in the form of generalized processor language statements. Those that do not, must have the input data on the tape include the numerically defined coordinate positions for each movement of the machine tool slides. The degree to which this information is required or not depends on the computational ability of the computer. Some computers have only a limited amount of computational ability; they may be able to calculate bolt hole circles, pocket milling sequences, or make similar calculations from a single statement on the program. Others can calculate the numerical coordinate data from generalized processor language statements for entire programs of varying complexity.

Computers

Computers are used for two distinct purposes in NC. In computer assisted programming, computers having computational ability are used to generate numerical part program data in a form that can be used by the MCU of a machine tool. These computers are not a part of the MCU; they may be situated in a remote location. Another application of computers is in CNC; these computers are a part of the MCU. Generally they are smaller computers, often mini-computers. In CNC applications the computers are primarily used to store part programs and to process these programs in order to generate output signals, which when amplified, control the operation of a particular machine tool. In some designs the computer performs the entire control function, while in others it performs a storage and "read" function, with the control function performed by a hard-wired circuit. Although many computers in CNC units have only a very limited or no computational ability, others do have some computational ability and are able to calculate specific coordinate positions from generalized processor statements.

Computers utilize binary notation. The Arabic numerals are converted to binary notation by the computer. Likewise, the output data intended to be read by humans is converted back into Arabic numerals by the computer. It is not necessary for the programmer to understand or to use binary notation. Binary notation is used because it is made up of only two digits. The two binary digits correspond to two conditions at which the electronic components of the computer can exist, namely, on or off, charged or discharged, a positive or negative charge, or conducting and non-conducting. For example, a binary zero (0) may be represented by a positive charge and a binary one (1) by a negative charge. The positive or negative binary notation is called a "bit."

The basic memory and logic section of the computer is called the central processing unit, or CPU, which includes all of the circuits that control the processing and execution of the instructions entered into the computer. The computer logic elements are composed of many small electrical cores or films that can change from one condition to the other almost instantaneously and with no apparent movement. The computer operates by simply adding numbers. Only nanoseconds, or billionths of a second, are required to make simple additions. Subtraction is accomplished by adding a negative to a positive number; multiplication is accomplished by making a series of additions; and division is accomplished by a series of subtractions. A unique feature of the computer is its ability to utilize its own developed output data as additional input data. It can add a series of numbers, store the sum, and retrieve this sum at a later date for additional processing.

Several methods can be used to enter information into the computer and for receiving the processed output. Information may be entered and received by means of a punched tape, tabulating cards, a teletype machine, diskettes or "floppy" discs, magnetic tape, and by signals from or to another computer. Other common output forms are printout sheets and electrical signals that can be used to control the operation of machine tools. Some computers have provisions for buffer storage; i.e., information to be used in a program can be stored separately and in advance so that it is immediately available for action after appropriate instructions have been given. With buffer storage it is not necessary to wait for information to be read either from a tape or from tabulating cards.

Frequently encountered in dealing with computers are the two terms *hardware* and *software*. Hardware is the term used to describe the physical component parts of the computer. Software is the term used to describe all of the program manuscripts, tapes, decks of tabulating cards, and all other documentation associated with computers and NC.

Software items essential to the operation of a computer are the control program and the general processor. These two programs are entered into the computer by the builder, and without them the computer would not be able to function. The control program, also called "the executive program," is used in CNC; it consists of an ordered set of instructions in computer language and format which provides the computer based control system with the capability of properly executing the system functions and the commands of the part program. The general processor is also called a *processor program, master program,* or just simply, *processor*. It is a program used by computers for the purpose of calculating coordinate data when programming by the computer assisted method. The general processor instructs the computer how to perform the computations specified by the part program, which are written in generalized processor language; it prepares the cutter location data for a particular part without reference to the machine tool on which the part is to be machined. Before the output data developed by the generalized processor can be used it must be post-processed; i.e., it must be processed in the computer by another program which is called the *post processor*. The post processor is a computer program which adapts the output of the data of the generalized processor to the requirements of a particular combination of machine tool and MCU. Thus, the data resulting from the post processor program is in a form that can be used by a particular NC machine tool to machine the part. There are many different general processor programs; some are for general use while others have been developed by the builders of machine tools and computers for use on their own equipment, along with the required post processor.

The basic unit of computer logic is the *word,* which is composed of a number of *bits*. The size of the computer is rated by the number of words of memory or capacity that it has. A mini-computer will have 4K to 8K capacity, where K denotes kilo, or a thousand; i.e., the mini-computer will have 4,000 to 8,000 word capacity. Medium sized computers will have

16K, 32K, or 64K capacity and large computers will have up to 1,000K capacity, or more.

Interpolation

Interpolation means the approximation of curved sections by a series of straight lines; i.e., curved sections are cut by means of a series of straight-line cuts. As shown in Fig. 15-14, four straight line cuts do not produce an acceptable circular section. Six straight-line cuts are only slightly better, and twelve straight-line cuts produce a rough approximation of a circle. A very close approximation of a circle can be obtained if a sufficient number of short, straight-line cuts can be taken. The number of cuts required is dependent on the largest acceptable error between the chord and the theoretical arc of the circle, as expressed by the dimension T in Fig. 15-14. This dimension is determined by the requirements of the part. Obviously, the smaller the allowable T dimension is, the more straight-line cuts will be required to generate the curve.

Linear interpolation is a method by which the coordinate positions of each straight-line cut must be programmed and made to appear on the program tape or in the computer storage. While this method is used, it can require a vast amount of numerical data to be generated and stored, which limits the amount of sculpturing work that can be done by this method.

Circular interpolation greatly reduces the amount of numerical data required. Although the curved surfaces are generated by a series of short straight-line cuts as before, a computer component in the MCU computes a sufficient number of coordinate positions to describe the circular path and it then generates the controlling signals required to develop the cut. The computer element will break up the circular span into small straight-line cuts which are often only .0002 in. (0.005 mm) in length. The programmer specifies circular interpolation by means of a preparatory function ($g17$, $g18$, or $g19$). As shown in Fig. 15-15, he then is required to

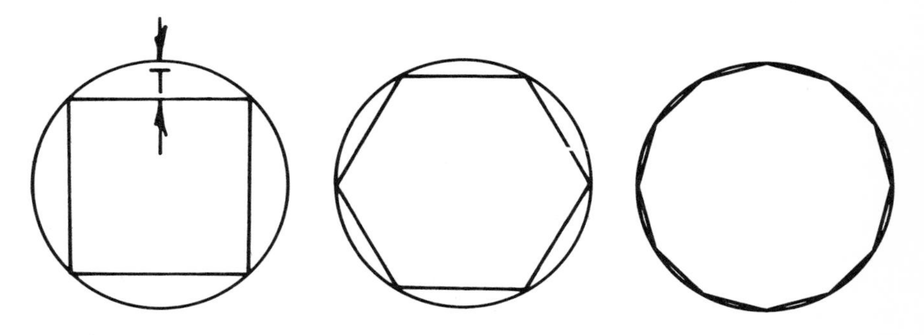

Fig. 15-14. Linear interpolation. Increasing the number of straight-line chords results in a closer approximation to a circle or an arc.

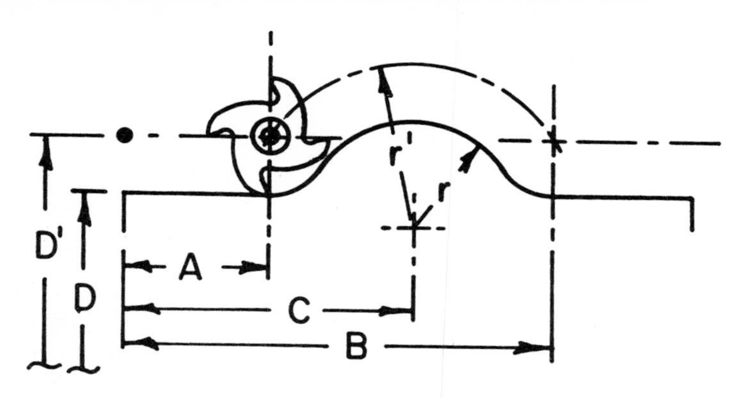

Fig. 15-15. Circular interpolation. Dimensions *A, B, C, D,* and *r* are programmed to define the arc or circle to be cut. If cutter compensation is not available, *D'* and *r'* are specified instead of *D* and *r*.

specify only five dimensions, which are the end points of the arc and the direction of the cutter travel; i.e., whether clockwise or counterclockwise. The programmer must allow for the cutter offset, unless this can be done automatically by the computer. Often the five blocks of information required with circular interpolation can replace a thousand or more blocks of information required with linear interpolation. Circular interpolation can be used to approximate second- and third-degree curves and many free-form shapes. It is especially useful for machining parts that are limited to straight-line cuts, circles, and arcs.

Parabolic and cubic interpolation are highly specialized methods of generating curved surfaces that are limited to a few industries that require the machining of exotic shapes. Cubic interpolation is used to machine such parts as automotive sheet metal forming dies and certain aircraft structural elements. This method will not only machine curved surfaces, but will also blend one curved surface into the next without a visible demarcation point.

Part Programming

There are two basic types of part programs, these are called point-to-point or "positioning," and continuous path or "contouring." Operations such as drilling a series of holes, where the workpiece is held stationary as each hole is drilled, are typically programmed by the point-to-point method. Certain NC machine tools, such as NC drilling machines, although intended to operate in this mode, are also able to take simple straight-line cuts. Other NC machine tools are meant to be able to machine contours as well as to take straight-line cuts and perform point-to-point type operations; these machines are programmed by the continuous path method, by which the path of the cutting tool is controlled at all times. Point-to-point programming may be done entirely manually or with the assistance of a computer. Although continuous path operations may be programmed manually, computer assisted programming is much to be preferred in order to save time, to simplify the programming procedure, and to reduce the risk of error.

The first step in programming a part is to study the part print and to become thoroughly familiar with all of the part features. Next, the programmer must determine how the part is to be machined, the cutting tools to be used, the cutting speed and feed, and how the part is to be held in the machine. Sometimes this is done by other persons, such as a process planner and a tool designer, in which case these plans must be obtained. The programmer then documents each operation in its logical sequence on the program manuscript. When the part is programmed manually, each coordinate machine tool position must be determined and documented on the program manuscript. If the computer assisted method is used, the programmer documents the coordinate data and the operating instructions, using a processor language to enter this information in the manuscript. In either case, the programmer always assumes that all of the movements are made by the cutting tool, even though it may actually be stationary and the movements are made by the workpiece. When heavy cuts are to be taken, he should check to see if the machine tool has sufficient power to take the cut and to specify cuts that will utilize the maximum power available.

Each part program is unique, and for computer assisted programming there are many different processor languages. It is not possible within the scope of this book to treat all of these languages and all of the techniques that are used; however, a few simple examples will follow to serve as an illustration.

Point-to-Point Programming. In the part shown in Fig. 15-16, two ½-inch holes are to be drilled on an NC drilling machine. This machine has a full floating zero NC system, meaning that the zero point from which all dimensions are referenced can be located anywhere on the machine tool table, within a specific range. Some machines have a fixed zero system, which requires that the coordinate dimensions to the holes be programmed from the fixed zero point on the machine. Also, it is necessary to locate the part on the machine table at exact coordinate distances from the fixed zero point. This is not necessary in this case; the part will be oriented on the machine table along the axes shown in the top view of the part in Fig. 15-16. The part program manuscript is shown in view A, of Fig. 15-17; and the first block of information as it appears on the tape is shown in view B.

The program in Fig. 15-17 is written using the absolute coordinate system. In this system all of the coordinate locations are programmed from a single zero point. An alternate method, called the incremental system, is to program each location from the previous position and not from a single zero point. Using this method the coordinates for sequence number 001 would be the same as shown in Fig. 15-17; for sequence number 002 the x coordinate would be 2.000 and the y coordinate would be

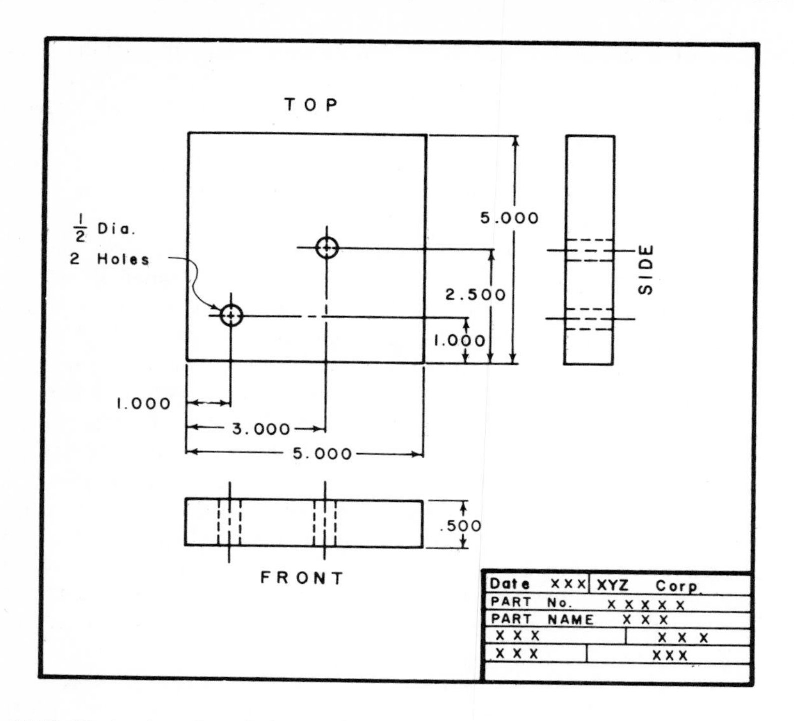

Fig. 15-16. Part print of workpiece to be programmed. The two holes are to be drilled through the part.

1.500. Neither system is inherently correct and each system has its logical area of application. Sometimes both systems are used in a single part program. Most NC systems are capable of using either system.

Continuous Path Programming. Computer assisted part programming is the method used in most instances for continuous path programming. The part programmer must be thoroughly conversant with the processor language that is being used. There are many different processor languages and it is not possible here to treat any one in detail. Each processor language consists of a number of English-like words that have a specific and precise meaning. One of the first and perhaps, most complete, processor languages is called APT, meaning Automatic Programmed Tool system. The following example will be written in the APT language.

Four types of statements are required to write a complete APT program, these are:

1. Motion statements, which describe the position of the cutting tool, such as GØRGT/BSURF, or GØTØ/P1
2. Geometry statements, which describe the configuration of the part, such as P1=PØINT/5.625−3.5

Part No. 12345 MANUSCRIPT Date x/x/xx

Part Name FLAT PLATE XXXX MACHINE Prepared By AB

Checked By EF

Sequence No.	TAB or EOB	x Coordinate	TAB or EOB	y Coordinate	TAB or EOB	m Word	TAB or EOB	Comments
RWS								
001	TAB	+1.000	TAB	+1.000	TAB	13	EOB	USE ½" HSS DRILL
								SET FEED DEPTH FOR
								DRILL TO CLEAR
								BOTTOM OF WORKPIECE
								SET SPEED AT
								1040 rpm.
002	TAB	+3.000	TAB	+2.500	EOB			
003	TAB	0.000	TAB	0.000	TAB	30	EOB	MACHINE STOPS AND TAPE AUTOMATICALLY
								REWINDS. REMOVE COMPLETED PART AND
								PUT IN NEW PART.

A

RWS 0 0 1 tab + 1 0 0 0 tab + 1 0 0 0 tab 1 3 EOB 0 0 2 tab

B

Fig. 15-17. A. Manuscript of the point-to-point part program for drilling the holes in part shown in Fig. 15-16. B. The first block of information of the part program, as it would appear on the tape.

3. Post Processor statements, which apply to the machine tool and control system, such as FEDRAT/.005 IPR
4. Auxiliary statements, which convey information not provided by the other statements, such as PARTNØ, or FINI.

The part shown in Fig. 15-18 is to be turned on an NC lathe. Figure 15-19 illustrates the path of the cutting tool and identifies the surfaces to be cut. The first step is to make a programming layout of the part as shown in Fig. 15-20. Each surface to be machined and each end point is identified on the layout, as well as the starting point (SP). After the layout has been made, the necessary post processor and auxiliary statements are listed in the manuscript, as follows:

```
PARTNØ RØUND SHAFT NØ1
MACHIN/DEF
INTØL/.001
ØUTØL/.002
CUTTER/.062
CØØLNT/ON
CLPRNT
FEDRAT/4, IPM
SPINDL/300, SFM
```

In the APT program, the nose radius (.031 in.) of the cutting tool is designated by the diameter of a theoretical circle; hence, CUTTER/.062.

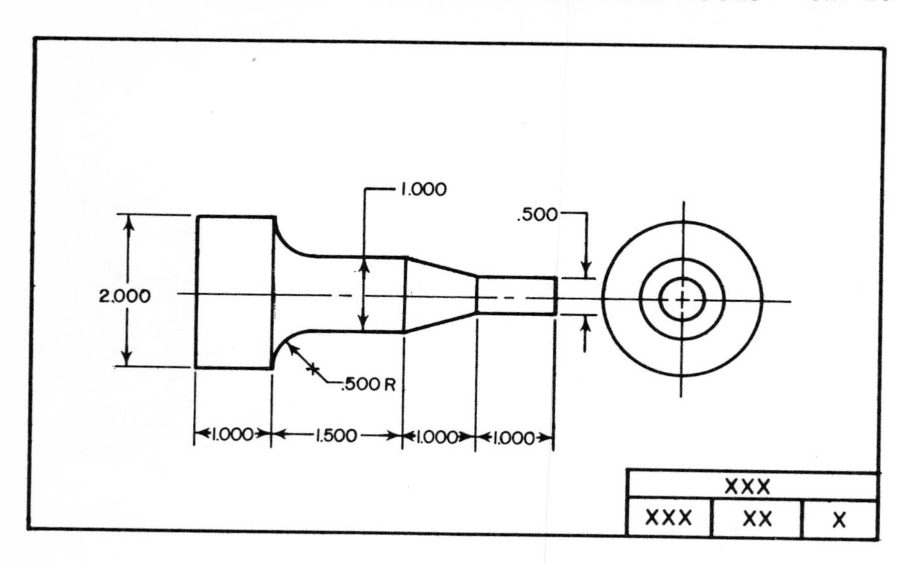

Fig. 15-18. Part print of workpiece to be turned on a lathe.

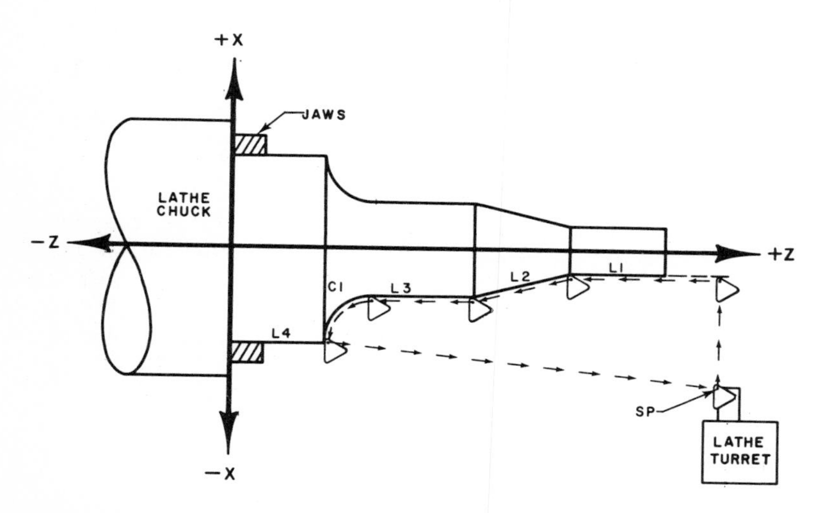

Fig. 15-19. The path of the cutting tool (center of the nose radius) for turning the workpiece in Fig. 15-18.

INTØL and ØUTØL refer to the inner and outer tolerances respectively. CLPRNT instructs the computer to print out the coordinate dimensions of all end points and straight line moves. Referring to Figs. 15-19 and 15-20, the next step is to list the geometry statements.

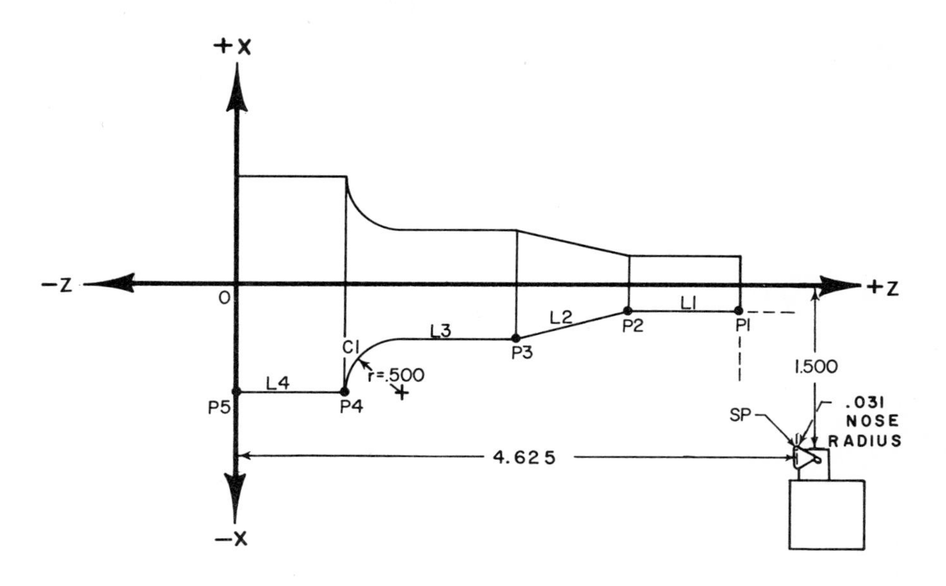

Fig. 15-20. Programming layout of workpiece to be turned, showing each surface to be turned, each end point, and the starting point (SP).

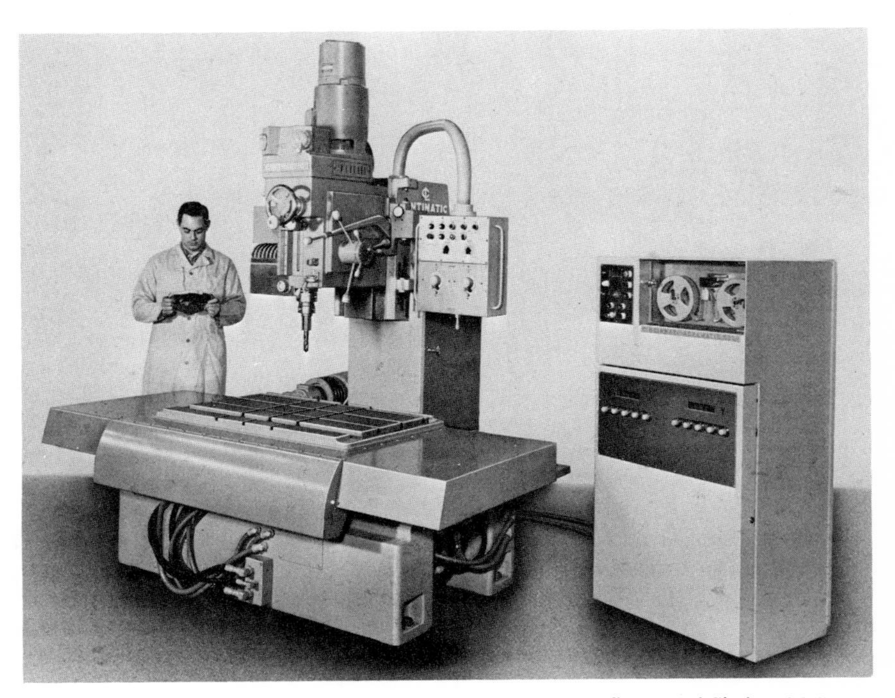

Courtesy of Cincinnati Milacron

Fig. 15-21. Drilling machine equipped with a point-to-point numerical control system.

Courtesy of the G. A. Gray Company

Fig. 15-22. Vertical boring machine equipped with a continuous path numerical control system.

```
SP= PØINT/4.625, −1.5
P1= PØINT/4.5, −.25
P2= PØINT/3.5, −.25
P3= PØINT/2.5, −.5
P4= PØINT/1, −1
P5= PØINT/0, −1
L1= LINE/P1, P2
L2= LINE/P2, P3
C1= CIRCLE/1.5, −1, .5
L3= LINE/P3, RIGHT, TANTØ, C1
L4= LINE/P4, P5
```

Line 3(L3) is defined by making a "right turn," as in an automobile, from line 2, at point 3, and extending to the tangent of the circle, C1. In the

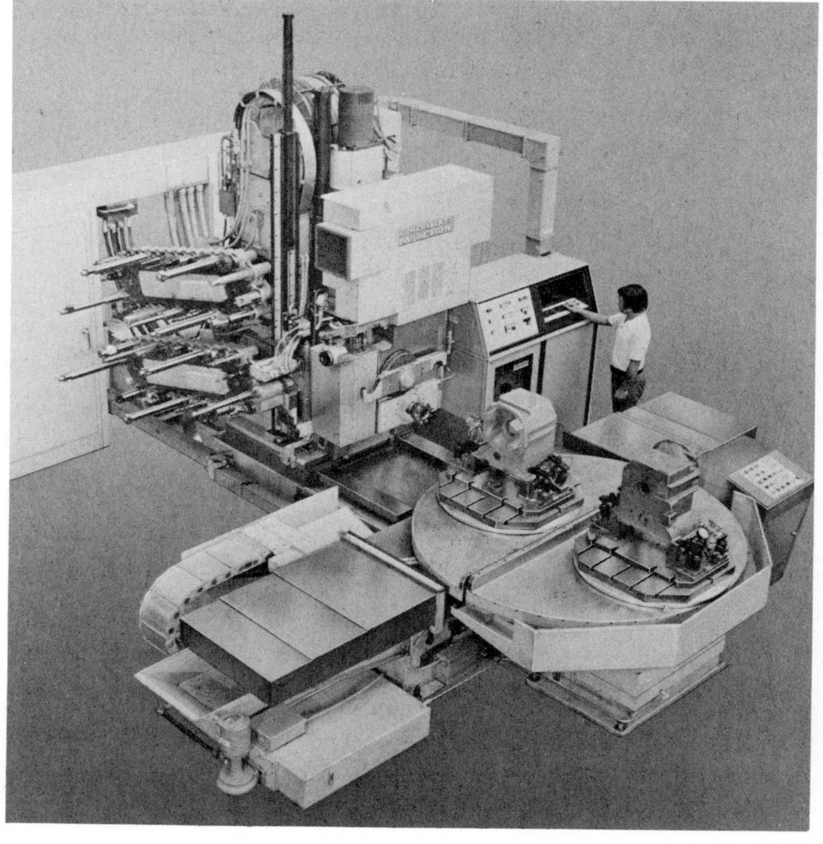

Courtesy of Cincinnati Milacron

Fig. 15-23. Three-axis CNC machining center equipped with a shuttle table and an automatic tool changer.

motion statements shown below, it will be necessary to specify a "right turn to line 3."

```
FRØM/SP
GØ/TØ, L1
GØLFT/L1, TØ, L2
GØLFT/L2, PAST, L3
GØRGT/L3, TANTØ, C1
GØFWD/C1, PAST, L4
GØTØ/SP
```

Courtesy of Cincinnati Milacron

Fig. 15-24. Large, three-axis, traveling column CNC machining center equipped with two indexing rotary tables.

Finally, the closing auxiliary statements are added:

```
COOLNT/OFF
FINI
```

NC Machine Tools

Many conventional machine tools have been altered to enable them to be adapted to NC. Except for the provision of control units on the slides and a cabinet to house the MCU, they are not different in appearance than conventional machine tools. Some machine tools have retained their basic elements, but have been extensively redesigned for NC. Typical examples of such machines are the NC drilling machine in Fig. 15-21 and the NC vertical boring machine in Fig. 15-22.

The flexibility and control obtainable by means of NC has resulted in the design of machine tools specifically for NC. Whereas manually controlled machine tools must have all of the control levers, hand wheels, and handles readily accessible to the operator, this is not required on NC

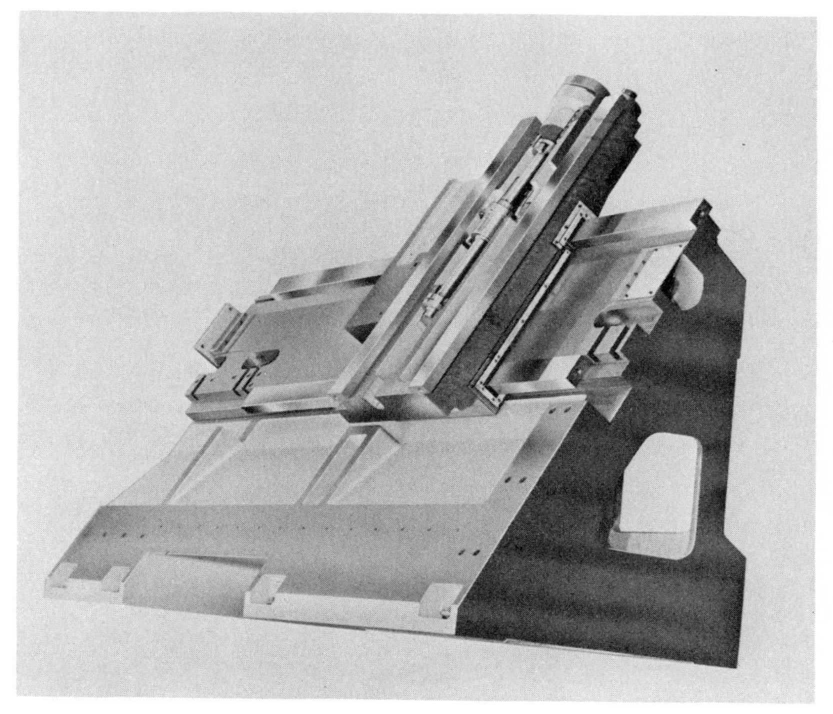

Courtesy of Cincinnati Milacron

Fig. 15-25. Slant bed and cross slide of slant-bed lathe.

machine tools. The NC machine tool can be designed to provide the greatest possible access by the cutting tools to the workpiece surfaces; it is only necessary to allow for the easy loading and unloading of the workpiece. An example of such machine tools is the machining center, which has been developed as a result of NC. Typical machining centers are used to machine as many surfaces as possible on the workpiece in a single setup, and to perform many different operations, which include milling, drilling, boring, counterboring, tapping, and facing. Usually, but not always, these machines have automatic tool changers and provisions for storing a large number of cutting tools. The machining center in Fig. 15-23 has a shuttle table, which allows the operator to set up a part in one workholding fixture while a second part is being machined in the other. In some cases two different fixtures are used, each holding the workpiece in a different orientation thereby enabling all of the surfaces on the part to be machined. Large machining centers, shown in Fig. 15-24, may have a

Courtesy of Cincinnati Milacron

Fig. 15-26. Chucking operation performed on universal type CNC slant-bed lathe.

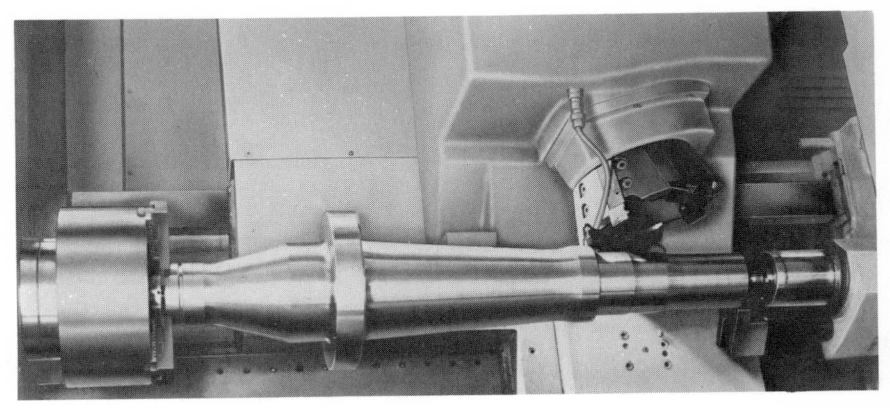

Courtesy of Cincinnati Milacron

Fig. 15-27. Shaft type slant-bed NC turning center turning a shaft that is held on centers. A universal floating jaw chuck drives the workpiece without exerting a force on the headstock center.

traveling column. The column slides in two perpendicular directions and the spindle housing moves up and down on the column; the spindle can also move in and out of the spindle housing. Two indexing rotary tables are provided; a part can be machined on one, while a part can be set up on the other. In the setup shown in Fig. 15-24, two parts are set up on each table, one on each face of a box angle plate which is mounted on the rotary table.

Another machine tool designed specifically for NC is the slant bed lathe shown in Fig. 15-2. The slant bed and the cross slide of this lathe are shown in Fig. 15-25. This design provides an easy access to the cutting tools and to the workpiece, for loading and unloading. It also allows the chips to fall unobstructedly away from the cutting area. Slant bed lathes may be designed specifically for turning on centers, for chucking work, or as universal models that can be set up to perform both types of operations. Universal turning centers, Fig. 15-2, designed for turning on centers and for chucking have two indexing turrets mounted on the cross slide which provide tooling stations on which cutting tools are mounted. A universal turning center is shown in Fig. 15-26 performing a chucking operation. In Fig. 15-27 a slant bed lathe designed for turning on centers is shown turning a shaft. The workpiece is mounted on two centers and is driven by a compensating, or floating-jaw type chuck, which allows the workpiece to remain accurately seated on the centers while at the same time it firmly grips the workpiece, thus providing a smooth drive. While special toolholders are sometimes used on these lathes, most jobs can be done with standard tool holders. The standard tool holders are qualified; i.e., they have toleranced dimensions on their locating surfaces to enable them to be positioned accurately in the indexing turrets.

CHAPTER **16**

Surface Plate Inspection Methods

The surface plate is an indispensable tool in machine shops and in tool and die shops. It establishes a reference plane from which precision measurements can be made. Formerly precision surface plates were made from seasoned gray cast iron. It was rough and finish machined on the planer and finally hand scraped to the required flatness. Although cast-iron surface plates are still used, in recent years there has been a trend toward using granite as a surface plate material. Black granite has become the material from which a majority of the surface plates are made. Granite surface plates can be made with a flatness within millionths of an inch, and this flatness can be preserved because granite does not corrode or rust. Also, they are harder and resist wear better than metal plates. They do not burr, gall, or crater. Granite plates are nonmagnetic and have exceptional thermal stability.

Surface plates are made in a large variety of sizes ranging from 8 by 12 inches to 72 by 144 inches. Larger surfaces are obtained by linking several plates together in proper alignment with each other. Some surface plates are provided with threaded inserts, T-slot shaped inserts, dovetail grooves, or slots which can be used to hold a variety of standard clamps, gaging accessories or the workpiece. A large variety of accessories such as parallels, V-blocks, and angle plates as well as a variety of precision measuring tools are used in doing surface plate work. The ability to use these tools accurately and intelligently is the mark of a craftsman.

Surface Plate Accuracy

Contractors doing work for the armed forces or the aerospace program must have surface plates meeting Federal specifications of accuracy to assure that the supplies and services meet the quality standards established by the contract. Surface plates are made to three grades of accuracy. The most accurate plates, Grade AA, are made to a tolerance of ±.000025 inch per 2 square foot area. Grade A has an accuracy of ±.000050 inch per 2 square foot area and Grade B has a flatness tolerance of ±.0001 inch per 2 square foot area. A simple shop test for the accuracy of a surface plate is to measure the same part in the same manner on different areas of the surface plate. This test for the repeatability of the measurement is an indication of the precision built into the surface plate.

Granite surface plates are finished by lapping with fine abrasives. They are measured for flatness and calibrated. Figure 16-1 shows an autocollimator, an instrument frequently used to calibrate a surface plate. This instrument has a built-in light source and a lens system which causes the rays of light to leave the autocollimator in parallel paths. The rays of light are directed at a target mirror and are reflected back into the autocollimator and are viewed in the eyepiece. If the mirror target is not exactly at right angles to the optical line of sight, the reflected image will

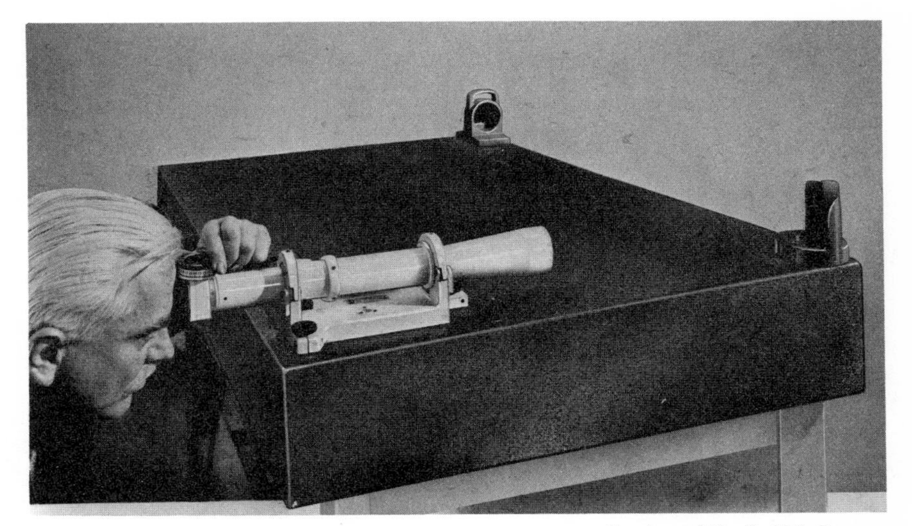

Courtesy of the DoALL Company

Fig. 16-1. Calibrating a black granite surface plate with an autocollimator.

appear displaced from the cross hairs in the eyepiece. A micrometer on the autocollimator is used to read this deviation to .2 second of arc. A corner target mirror reflects the light rays from the autocollimator to the reflecting target mirror shown in the opposite corner of Fig. 16-1. The rays are reflected from the target mirror to the corner target mirror and back to the autocollimator.

A total of eight lines of reading is taken. These readings are around the four edges of the surface plates, along the two diagonals, and across the two center lines. A number of readings are taken along each line of reading. These readings are plotted on graph paper to form a profile of the surface plate along the eight lines of sighting.

Another method of calibrating granite surface plates is by means of a laser-powered interferometric surface contour projector. This method is very fast. A 36- by 48-inch surface plate, which would take from 4 to 8 hours to autocollimate, record, and graph, can be calibrated in 10 to 15 minutes.

Another accepted and very fast method of calibrating surface plates is by means of the Rahn Planekator ™ shown in Fig. 16-2. This instrument consists of a very precise straightedge made of a hard ceramic material, two supports, and a height gage, which has a very sensitive, .000020 inch (0.0005 mm), Mahr indicator. Pencil lines are first drawn on the surface plate to indicate where the readings are to be made. The straightedge is then placed on top of the two supports along a diagonal line, as shown in Fig. 16-2. One of the supports is adjustable vertically and this support is moved until the indicator on the height gage reads the same at both ends of the straightedge. Indicator readings are then taken in various positions along the straightedge and the values are recorded. This procedure is followed along the other diagonal and along six, inch-spaced lengthwise and crosswise axes of the surface plate.

For very precise inspection measurements, it is necessary to know the variations that can be expected in the measurements when the workpiece is measured in different locations on the surface plate; this is called the repeat measurement accuracy and can also be determined by the Repeat-O-Meter, shown in Fig. 16-3. This meter has two fixed and one floating contact, which are placed on the surface plate and the movement of the floating contact is measured by a very sensitive dial test indicator as the meter scans the surface of the surface plate. The difference between the largest plus and minus indicator readings is the maximum possible error that can occur when measuring a part on the surface plate.

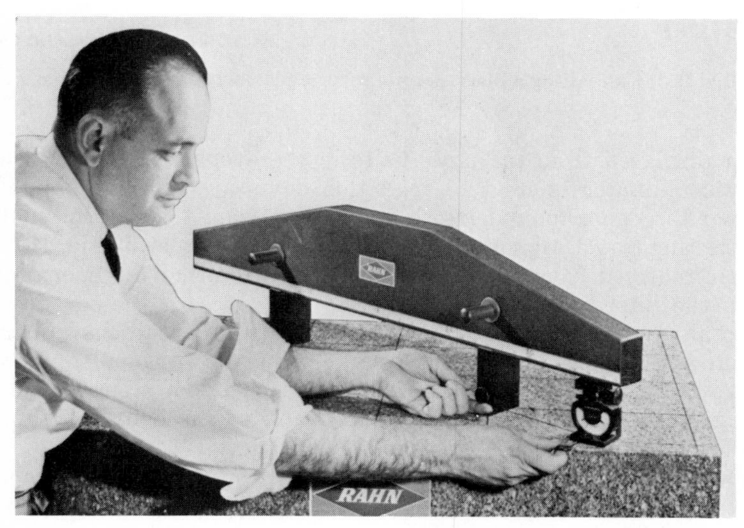

Courtesy of the Rahn Granite Surface Plate Co.

Fig. 16-2. Calibrating a surface plate with a Planekator ™ and a sensitive .000020 in. indicator.

Courtesy of the Rahn Granite Surface Plate Co.

Fig. 16-3. Checking the repeat measurement accuracy of a surface plate with a Repeat-O-Meter.

Care of Surface Plates

The surface plate is a very precise tool which is an integral part of any measurement made on its surface. The condition of this surface is a very important factor in the accuracy of these measurements. Good housekeeping and cleanliness are essential in making any precision measurement. This applies in particular to surface plates. Dust, dirt, and oil should not be allowed to collect on the surface. It should frequently be wiped clean. The use of a special proprietary surface plate cleaner is recommended. A thin coat of oil should be placed on gray cast-iron surface plates when they are not being used for longer periods of time or when left standing overnight.

The table on the base supporting a larger surface plate should support the surface plate in the same manner as when it was made. To assure this, the stands developed by the surface plate manufacturer should be used whenever possible. The surface plate should be carefully leveled and kept in this condition.

Chips, scratches, and burrs should be prevented from occurring on the surface plate, since a slight burr or a speck of dust can cause an error in a precise measurement. Any burr should be removed from the workpiece before it is placed on the surface plate. Rough castings should not be placed directly on the surface plate, but should be elevated above the surface by parallels or other accessories. Extreme care should be used in lifting workpieces and gages on or off the surface plate to avoid chipping or scratching the surface. Particular care should be taken to avoid heavy contact with the edges of the plate. The surface plate should not be overloaded. Only the workpiece, surface plate accessories, and measuring tools required for the job should be placed on the surface. The full surface of the plate should be utilized. This will distribute the wear and avoid concentrations in any one area.

Surface Plate Work—Checking for Parallelism

The setup for checking a master-box parallel for parallelism on a surface plate is shown in Fig. 16-4. An electronic height gage is used in this case because the degree of sensitivity and accuracy obtained is greater than can be obtained with an ordinary dial test indicator. The height gage is coupled to an electronic amplifier. The amplifier shown in the illustration is powered by a self-contained rechargeable nickel-cadmium battery so that it can be used in locations that are not close to an electrical outlet. If desired, it can also be operated on 115V AC current. It has four scales, ranging in sensitivity from .0005 to .000010 inch.

To check the parallel for parallelism, the contact point of the gage head is placed on the top surface of the box parallel and the amplifier is set on the .0001-inch scale, and the indicating needle on the face of the amplifier is zeroed. The contact point is then carefully moved back and forth and across over the top surface of the box parallel. The reading on the .0001-inch scale on the amplifier will show any deviations from parallelism. When matched sets of box parallels are to be checked, the second parallel is checked by means of the same zero setting established during the check of the first parallel.

Surface Gage Work—Checking for Squareness

The squareness of two surfaces with respect to each other is frequently expressed as the *deviation per six inches of length.* While the squareness can be checked with a precision machinist's square, the exact amount of deviation can not be determined. When it is necessary to determine the exact deviation, the method illustrated in Fig. 16-5 can be used. A master cylindrical square and a transfer gage is used. The transfer gage has a locating button on the front face of the base which is placed against the cylindrical square. With the gage in this position, the dial test indicator is adjusted to read zero. The transfer gage is then placed against the right-angle block as shown, and any deviation from squareness can be read directly on the indicator. This procedure is repeated at the midpoint height of the right-angle block to determine if this face is either bowed or dished.

Surface Gage Measurement—Checking for Roundness

Roundness may be described as the measurement of a cylindrical part, or part feature—such as a ring, shaft, or hole, for its uniformity to that

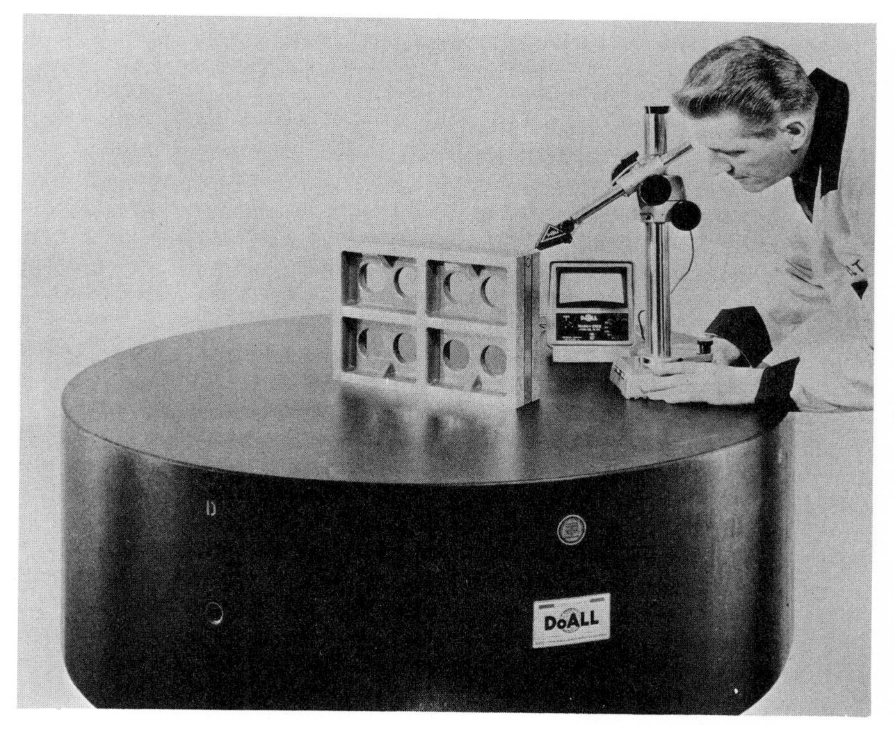

Courtesy of the DoALL Company

Fig. 16-4. Checking a master-grade box parallel for parallelism.

of a true circle. It can also be considered a measurement of its true diameter. Roundness should be checked with the surface in question resting in a precision V-block as shown in Fig. 16-6. It could also be checked while held between precision bench centers, but the accuracy is influenced by the condition of the center holes in the part and by any eccentricity between the cylindrical surface being checked and the center holes.

To check for roundness, the part is placed in the precision V-block and the contact point of a dial test indicator is positioned on the centerline of the part. The indicator is positioned on the centerline of the part by moving the contact point over the outside diameter to the position where the maximum reading occurs. Zero the dial indicator by rotating the face. Then carefully rotate the part by hand, and note any deviation from zero on the dial indicator. Make sure that the V-block is not moved and that the part does not move lengthwise in the V-block when it is rotated. If the part is moved lengthwise, any taper may give an erroneous reading.

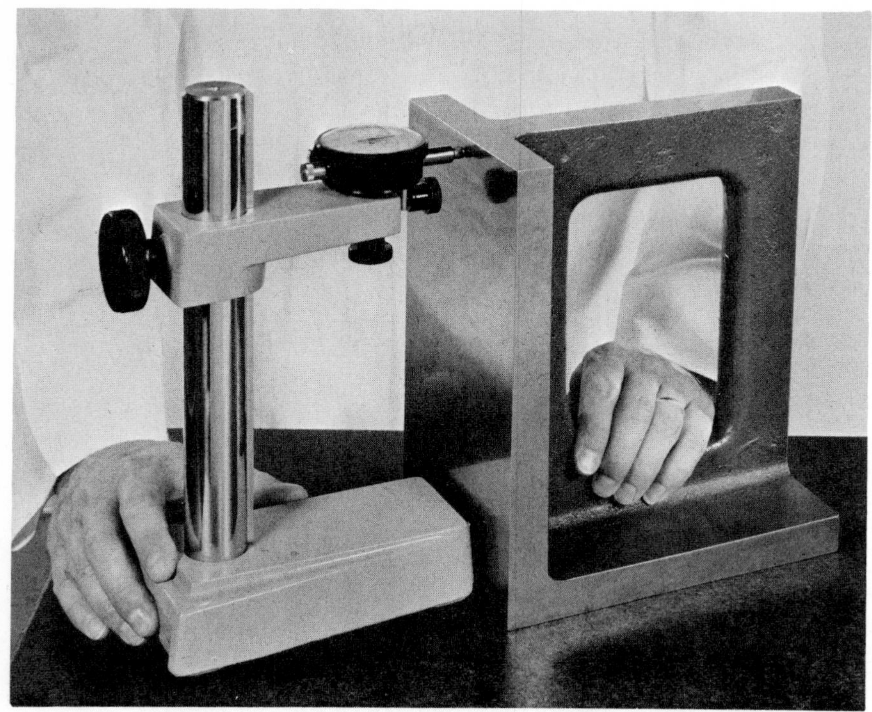

Courtesy of the DoALL Company

Fig. 16-5. Checking a precision right-angle block for squareness.

Courtesy of the DoALL Company

Fig. 16-6. Checking a precision ground bearing for roundness in a precision black granite V-block.

Surface Gage Work—Length and Height Measurement

Among the most common precision measurements made on a surface plate are the determination of height and length. These measurements may be made from one surface to another on the same workpiece as in Fig. 16-7; or they may be made directly from the top of the surface plate to the surface to be measured as in Fig. 16-8. In either case three basic procedures can be used to make these measurements.

One procedure, shown in Fig. 16-7, is to use a vernier height gage with a small dial test indicator attached to the movable arm. The indicator

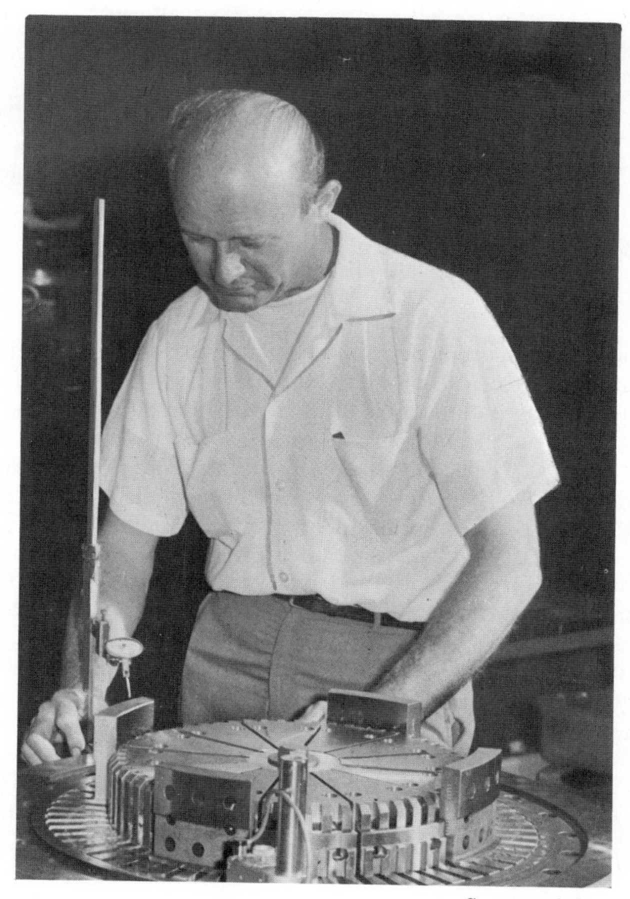

Courtesy of the DoALL Company

Fig. 16-7. Measuring the height of a pad attached to the disc using a vernier height gage.

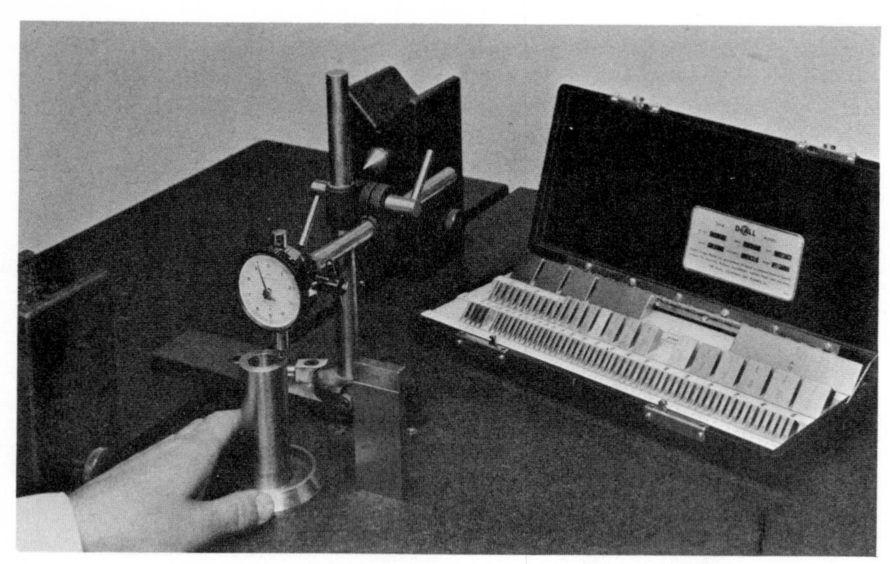

Courtesy of the DoALL Company

Fig. 16-8. Measuring the length of a part by measuring the height above the surface plate.

does not make a measurement; it serves only to zero the reading of the vernier height gage. In Fig. 16-7, the distance from the face of the large disc to the top of the pad is to be measured. Here the vernier height gage is placed directly on the face of the disc in order to reach the pad. On smaller workpieces, where the surface to be measured can be reached, it would be preferable to place the vernier height gage directly on the surface plate. In either case, the measurement is made by reading the scale on the vernier height gage when the indicator is zeroed on the surface on which the height gage is resting and when the indicator is zeroed on the top of the pad. The difference in these two readings is the distance to be measured.

The second procedure is similar except that the first step is to lower the arm until the reading of the vernier height gage scale is zero. The indicator is then adjusted to read zero while it is in contact with the surface on which the vernier height gage is resting. Next, the arm is raised and lowered again to zero the dial test indicator on the surface to be measured. The vernier height gage will then read the distance to be measured directly.

The third method of measuring, Fig. 16-8, is a comparative method. Very precise results can be obtained through the use of precision gage blocks and a dial test indicator that reads in "tenths" (.0001 inch). Often the required accuracy of the measurement is in terms of one-thousandth or one-half-thousandth of an inch, in which case a thousandth (.001 inch) or a one-half-thousandth (.0005 inch) indicator can be used. The precision gage blocks are assembled together to form a length which is equal to the length or height to be measured. The contact point of the indicator is positioned over the gage blocks, and the indicator is adjusted to read zero. The contact point is then placed on the surface to be measured, and the indicator reading will show the deviation, plus or minus, from the specified size.

Test types of dial indicators such as shown in Fig. 16-9 A and B are sometimes used for this sort of work. They have an advantage of being

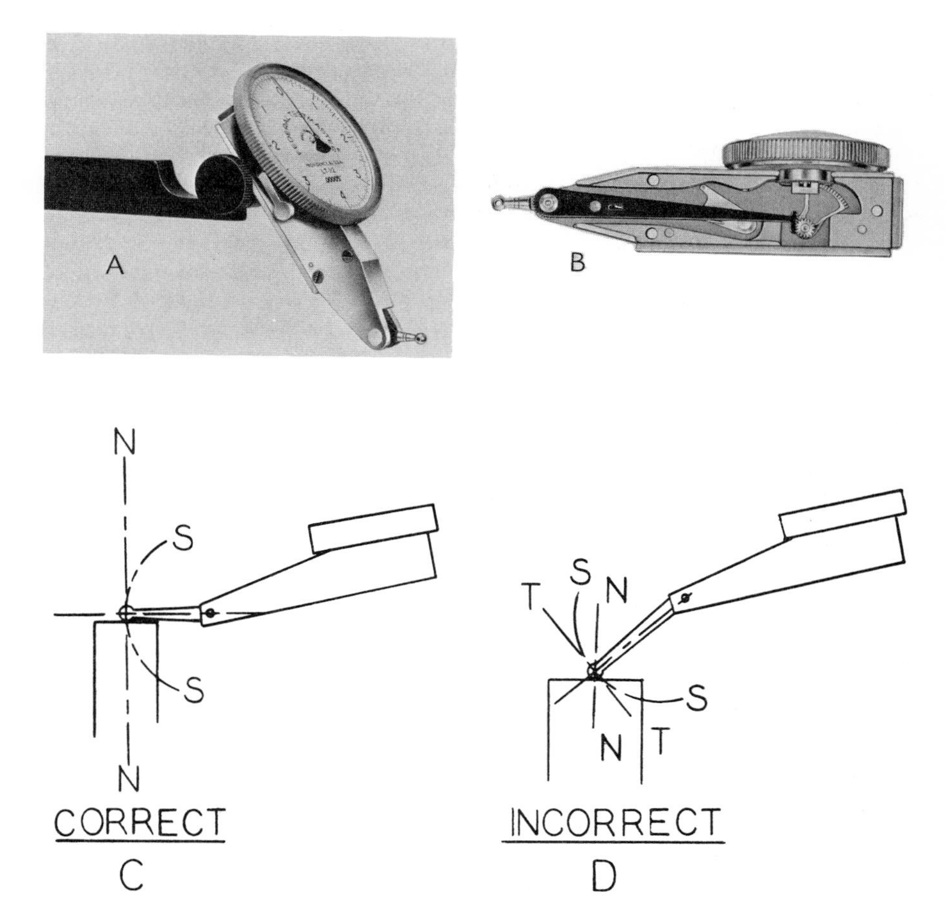

Views A and B courtesy of Federal Products Corp.

Fig. 16-9. A. Test-type dial indicator. B. Test-type dial-indicator mechanism. C. Correct use of test-type indicator. D. Incorrect use of test-type indicator.

light and easy to attach to surface gages and height gages. While they are very accurate, caution must be exercised when they are used to determine exact distances. The contact point on this type of indicator is attached to a lever which pivots about a point inside the body of the indicator instead of to a stem which moves in and out. When this kind of indicator is used to obtain direct readings, they must be used as shown at C, Fig. 16-9. The lever of the indicator causes the contact point to swing along the arc S–S. The measurement must be made along the line N–N which is perpendicular to the surface being measured. When this surface is measured, the lever of the indicator should be positioned so that a small movement will cause it to move in a direction that is as nearly as possible along N–N as shown at C, Fig. 16-9. If the indicator is positioned as shown, the contact point D will move in the direction of T–T when it is displaced and the indicator reading will not represent the correct distance along N–N. This error is sometimes called the cosine effect.

Another measuring instrument used to make height measurements on the surface plate is the micrometer height gage shown in Fig. 16-10. Heights can be read directly by means of this gage. The height gage and indicator are only required to transfer the reading to the work. The gage is housed in a rigid frame which is supported on three ground and lapped pads. Inside the frame is a column of 1-inch precision gage blocks which are permanently wrung together and which form a series of 1-inch steps. Because the lower gage block in the column is only .090-inch thick, parts .100 inch in height can be checked. The contact point of a dial test indicator or of an electronic height transfer gage can be placed on the steps in order to zero these instruments.

The gage block assembly is moved up and down over a 1-inch range by the micrometer head located on top of the frame. Below the micrometer head is a digital readout which displays the micrometer reading in thousandths of an inch. The large-diameter micrometer head has numbered divisions which are also in thousandths of an inch as well as smaller divisions which are equal to ten-thousandths of an inch. The micrometer is read by reading the inches on the scales at the side of the frame, the thousandths of an inch on the digital readout, and the final ten-thousandths of an inch on the head. With this instrument the exact measurement is obtained, as in the case of the vernier height gage. Readings over and under the required height do not have to be obtained with the dial test indicator. The indicator or electronic height gage is used only to transfer the zero reading from the workpiece to the gage block on the micrometer height gage.

Surface Plate Work—Measuring the Location of Holes

It is frequently necessary to measure the precise location of a hole when doing precision machine work or tool and die work. This type of measurement can readily be made on a surface plate. The procedures used to measure the location of a hole are very similar to those used to measure lengths and heights. The principal difference is that the measurement of

Courtesy of The L. S. Starrett Company

Fig. 16-10. Checking the height of a boss on a jet engine part with an electronic height gage and a micrometer height gage.

the hole location is not direct, since the distance actually measured is to the bottom of the hole or to the top of a pin inserted in the hole instead of directly to the center of the hole.

As an example, the location of the two large holes in Fig. 16-11 must be measured. This workpiece has two perpendicular edges that act as reference edges from which the dimensions to the holes are given. Before the part is placed on the surface plate, the diameter of each hole is measured with a precision measuring tool. The workpiece is then clamped to a precision angle plate with one of the reference edges resting against the surface plate, as shown in Fig. 16-11. A measurement is made from the top of the surface plate to the bottom of the hole, and the distance to

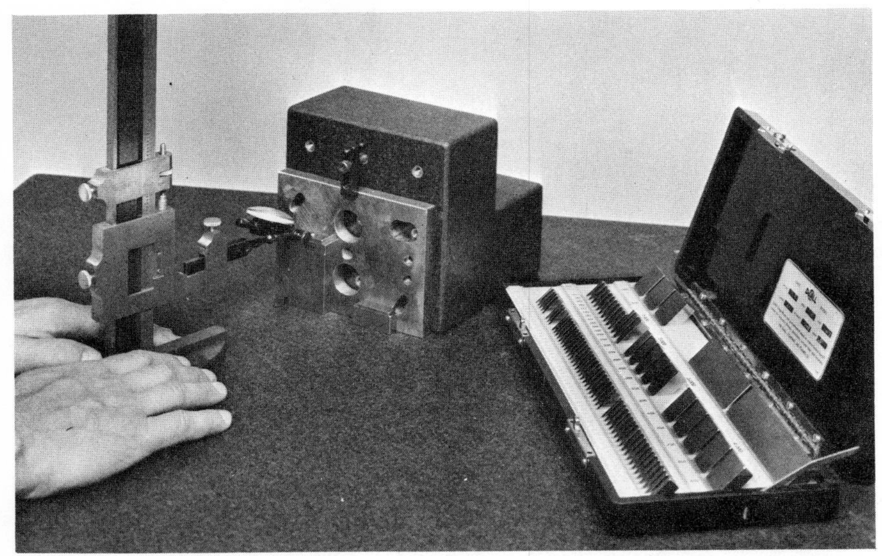

Courtesy of the DoALL Company

Fig. 16-11. Measuring the location of a hole.

the center of the hole is found by adding one-half of the diameter of the hole to this measurement. This procedure is repeated for the second hole and again for both holes, with the second reference edge resting on the surface plate.

As shown in Fig. 16-11, the measurements are made with a vernier height gage to which a test-type indicator has been attached. This type of dial test indicator is used because the arm can reach inside of the hole. The position of the contact point should be adjusted to avoid the cosine effect. The actual measurement may be made by reading the vernier height gage scale when the indicator is zeroed against the top of the surface plate and again when it is zeroed against the lower surface of the hole. The distance to the lower surface, or "bottom," of the hole from the surface plate and the edge of the workpiece is equal to the difference in these readings. Another method is to use precision gage blocks, which are stacked together so that their length is equal to the difference between the specified distance to the center of the hole and one-half of the measured hole diameter. The gage blocks are placed adjacent to the hole, as shown in Fig. 16-11, and the height of the bottom of the hole is compared to the height of the gage blocks, using the height gage and the indicator. The indicator readings will give the plus and minus derivations from the specified height of the hole above the surface plate. As a final check it is advisable to re-assemble the gage block stack so that its length is equal to the actual height found by the previous measurement; then the hole is checked again, as before, to find out if the indicator will read zero both in the hole and over the gage block stack. When indicating the bottom of the hole the height gage should be moved to the position where the smallest reading of the indicator occurs.

Sometimes it is difficult or impossible to position the contact point of the indicator inside a small hole. In this event, as a temporary measure, a pin can be pressed into the hole with a light press or push fit. Then instead of measuring to the bottom of the hole, the measurement is made from the top of the surface plate to the highest point over the pin, using either of the methods previously described. The distance to the center of the hole will then be the difference between the distance from the surface plate to the top of the hole and one-half of the diameter of the pin.

Frequently the center distance between two holes must be found. The same measuring procedure as described is used to measure from the top of the surface plate to the bottom of the two holes, or to the top of pins inserted in the holes. Also, diameters of the holes or the pins must be measured and the formulas given in Fig. 16-12 can then be used to calculate the center distance. The appropriate formula should be used, depending on the four different situations shown in Fig. 16-12. These formulas will provide the true center distance if the centerline connecting the axes of the holes is in a vertical position when the measurements are made. If this centerline is not in a vertical position, the formulas will give the center distance in the direction perpendicular to the reference edge resting on the surface plate. Two sets of measurements are required in this case, each set is made from one of the two perpendicular reference edges, as shown in Fig. 16-13. The true center distance is found by calculation, using the Pythagorean theorem or simple trigonometry.

Example 16-1:

Calculate the true center distance of the holes shown in Fig. 16-13. The diameter of the large hole is 1.252 in. and the smaller hole diameter is .874 in. The measured distances to the bottom of the holes are shown in Fig. 16-13 when the workpiece is resting on each of two perpendicular reference edges.

For Position A:

$$L_A = M_1 - M_2 + \frac{D_1}{2} - \frac{D_2}{2} = 2.807 - .502 + \frac{1.252}{2} - \frac{.874}{2}$$
$$= 2.494 \text{ in.}$$

For Position B:

$$L_B = M_1 - M_2 + \frac{D_1}{2} - \frac{D_2}{2} = 4.103 - .987 + \frac{1.252}{2} - \frac{.874}{2}$$
$$= 3.305 \text{ in.}$$

Using the Pythagorean theorem to find the true center distance:

$$\text{True Center Distance} = T = \sqrt{L_A{}^2 + L_B{}^2} = \sqrt{2.494^2 + 3.305^2}$$
$$= 4.140 \text{ in.}$$

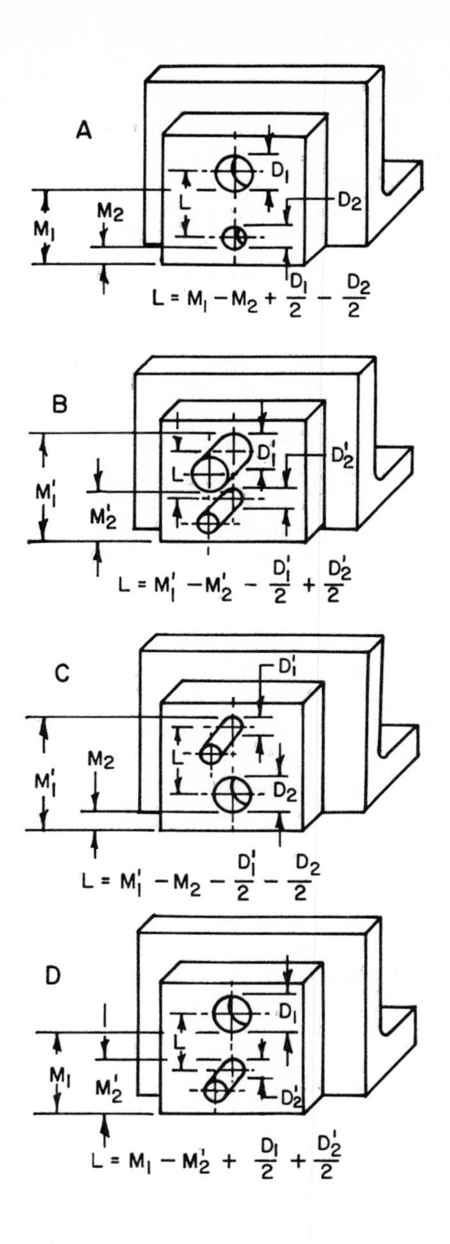

Fig. 16-12. Formulas for calculating the center distance between holes: A. with measurements to bottom of both holes; B. with measurements over pins in both holes; C. with measurements over upper hole pin and bottom of lower hole; and D. with measurements to bottom of upper hole and over lower hole pin. L = Center distance; D = Hole diameter; M = Measurement. M and D are measurements in hole; M' and D' are measurements over pins.

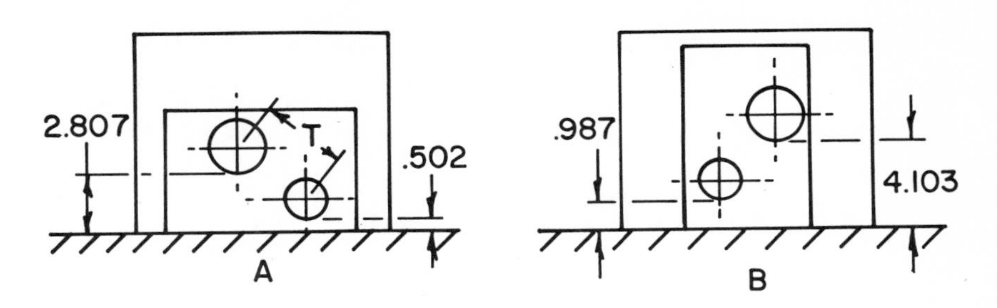

Fig. 16-13. Example 16-1. Measurements to the bottom of two holes with workpiece in position A and position B, in order to find the true center distance, *T*.

Surface Plate Work—Concentricity, T.I.R., and Coaxiality

T.I.R. means Total Indicated Runout, which in shop terminology is also sometimes called total indicator runout. This measurement, made with a dial test indicator or similar instrument, refers to total eccentricity of a surface about an axis of rotation and is given in thousandths or ten-thousandths of an inch. T.I.R., then, expresses the accuracy of the concentricity of a cylindrical surface about an axis of rotation. Coaxiality means a common axis in a part with two or more cylindrical surfaces.

A common method of testing the concentricity of a cylindrical surface which has been turned on centers is to mount it on a bench center as shown in Fig. 16-14. The bench centers have two dead centers about which the workpiece can rotate. In the illustration, the concentricity of the cylindrical surfaces of a small crankshaft is being measured. Either a dial test indicator or an electronic height gage can be used to measure the T.I.R. In either case the contact point of the instrument used must be placed on the exact center of the shaft. When in this position the instrument will read the total indicated runout. The contact point is placed in this position by moving it across the top of the shaft until it reaches the position where the maximum reading is obtained. The shaft is then rotated by hand, and the T.I.R. can be read directly on the dial of the indicator or on the electronic amplifier unit. This should be done in several positions on each of the diameters of the shaft.

The axial runout of shoulders and flanges on shafts is also sometimes very important. The runout can be measured by placing the contact point of the indicator or the electronic height gage on the shoulder, as shown in Fig. 16-15, and rotating the shaft. When done as illustrated, the shaft should be in a true vertical position.

Although the concentricity of two or more surfaces on a shaft may be some indication of their coaxiality, it is not a direct measurement of this characteristic. A direct method of measuring this characteristic is to place one of the cylindrical surfaces in a precision V-block and to rotate it while indicating the other surface. If the two surfaces are internal cylin-

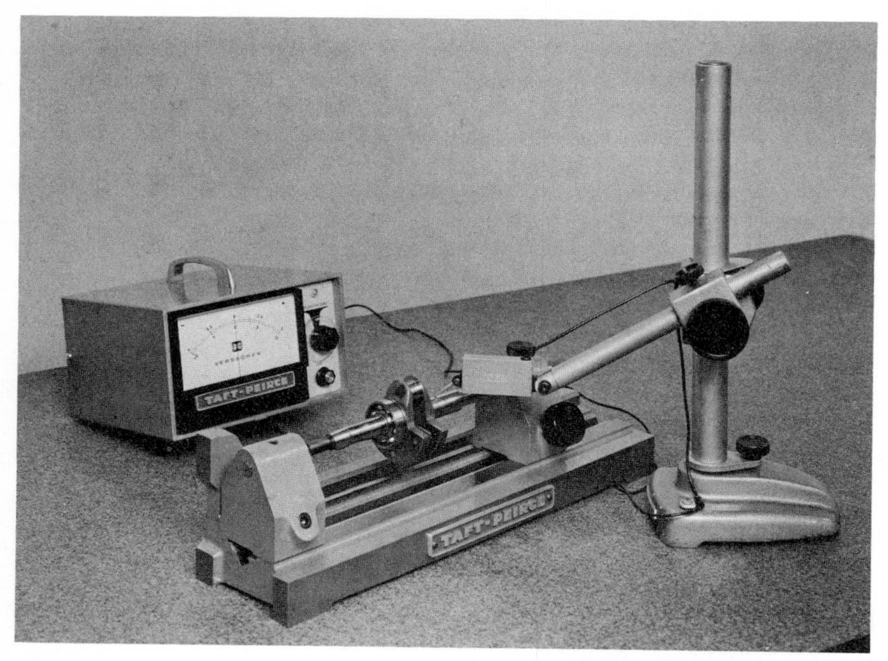

Courtesy of The Taft-Pierce Manufacturing Company

Fig. 16-14. Measuring the concentricity of a small crankshaft on precision bench center.

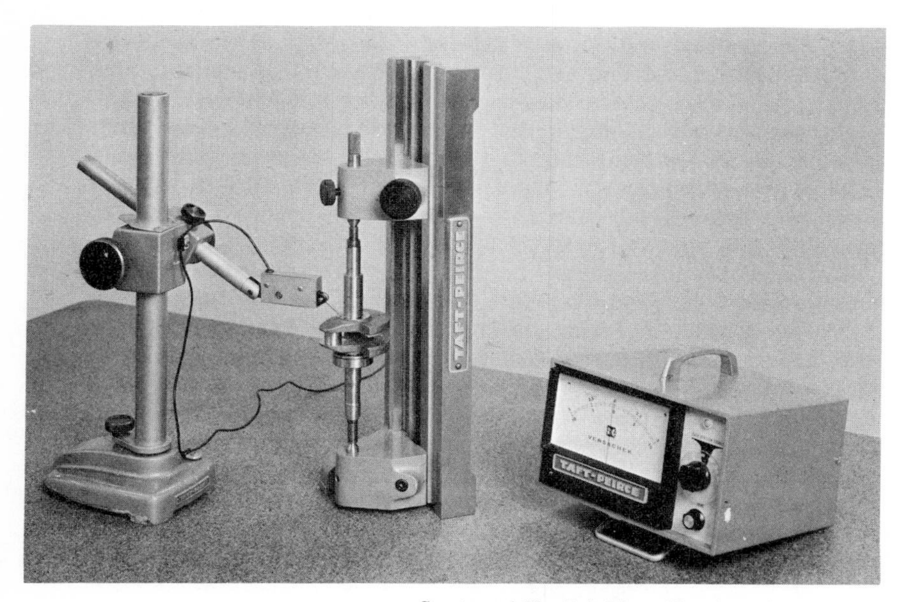

Courtesy of The Taft-Pierce Manufacturing Company

Fig. 16-15. Measuring the shoulder runout of a small crankshaft with the bench center placed in an upright position.

drical surfaces, this is done indirectly. In Fig. 16-16, the coaxiality of two bores in a pump housing is being measured. A plug is placed in the smaller bore with a light press or push fit so that one end extends beyond the housing. The V-block is clamped in place on the surface plate, and the housing and plug assembly is lightly clamped in the V-block. The contact point of the electronic height gage is positioned at the lowest point of the bore, which places it in the center of the bore. The coaxiality is then measured by slowly and carefully rotating the housing.

Surface Plate Work—Measuring Angles and Tapers

Several methods are available for measuring angles, perhaps the most common of which is to use a protractor. There are many different kinds of protractors; however, only precision protractors with machine-cut graduations should be considered for machine shop work. The most accurate machinist's protractor is the vernier bevel protractor. It is graduated to read an angle as small as 5 minutes. Precision machinist's protractors, such as the vernier bevel protractor in Fig. 16-17, are frequently used to measure angles on the surface plate. Another method of measuring angles on certain types of workpieces is the dividing head. It can be placed on

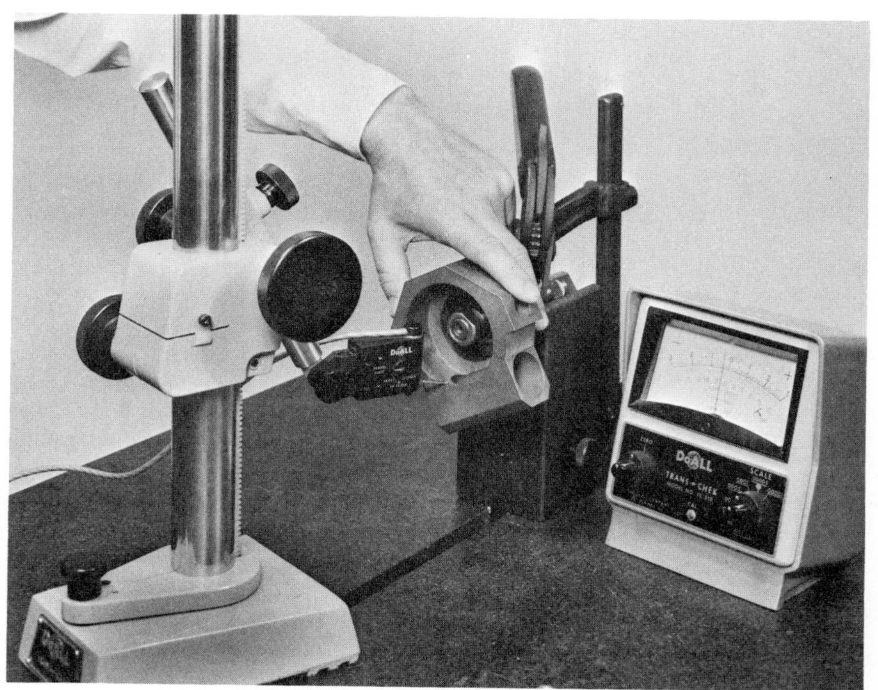

Courtesy of the DoALL Company

Fig. 16-16. Measuring the coaxiality of two bores in a pump housing.

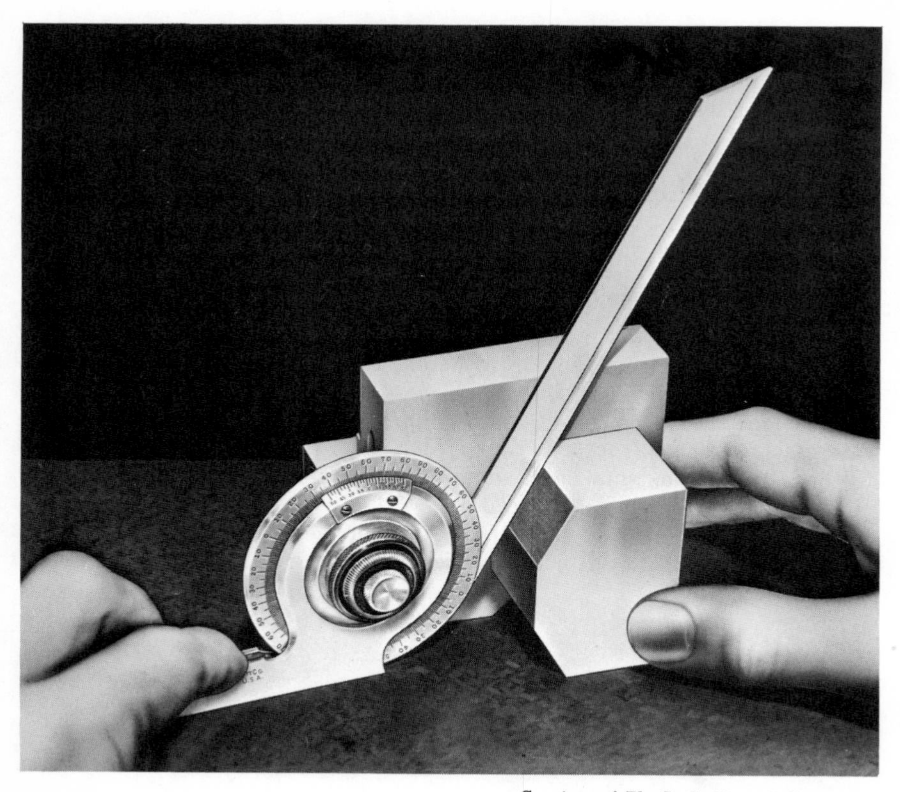

Courtesy of The L. S. Starrett Company

Fig. 16-17. Measuring an angle on a surface plate with a vernier bevel protractor.

the surface plate. The part is held on the dividing head, and the surface plate provides a precise reference plane from which the measurements can be made, as shown in Fig. 16-18.

Still another method for measuring angles on the surface plate is the sine bar or sine plate. In Fig. 16-19 a small sine bar is shown measuring the angle of a precision gage block scriber attachment. The sine bar is tipped to the required angle by the gage blocks placed underneath one of the rolls. For any sine bar or sine plate the height of the stack of gage blocks is determined by the following formula:

$$H = L \sin \theta \tag{16-1}$$

where: H = Height of gage block stack, inches
θ = Angle to be measured, degrees and minutes
L = Length or distance between the rolls on the sine bar or sine plate, inches

Courtesy of the Anton Machine Works

Fig. 16-18. Using a dividing head on a surface plate to measure an angle. An Air Glide transfer gage, which slides on a film of air, is used to transfer the measurements from the gage blocks to the workpiece.

The angle on the workpiece can be checked by placing the contact point of the dial test indicator at one end as shown in Fig. 16-19 and by moving it along this surface to the other end. When the angle on the scriber is exactly equal to the angle at which the sine bar is set, this surface should be parallel to the surface plate. To measure this angle, two check points should be established on the angular surface of the scriber point. The distance between these check points must be known exactly. The indicator or electronics amplifier is zeroed at one of the check points, and the amount of deviation from the zero reading at the other check point is determined. The true angle on the part being measured is then calculated by the following formula:

$$\sin \theta' = \sin \theta \pm \frac{DL}{S} \tag{16-2}$$

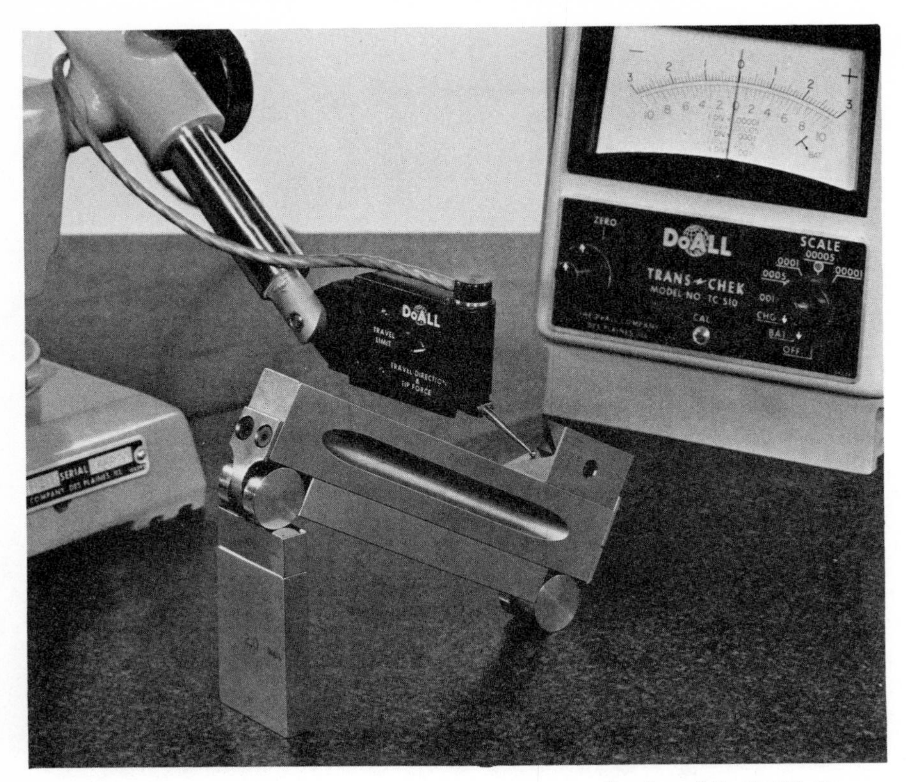

Courtesy of the DoALL Company

Fig. 16-19. Measuring an angle on a surface plate using a sine bar and an electronic height gage.

where: θ' = True angle of the part being measured, degrees and minutes

θ = Angle to which the sine bar or sine plate has been set, degrees and minutes (same as θ in Formula 16-1)

L = Length or distance between the rolls on the sine bar or sine plate, inches

D = Deviation in dial test indicator or electronic amplifier reading, inches

S = Distance between check points, inches

The term DL/S in Formula 16-2 is added to sin θ if the deviation is in the plus direction and subtracted from sin θ if the deviation is in the minus direction, when the end adjacent to the highest part of the sine bar or sine plate is checked first.

Example 16-2:

The angle of the scriber point shown in Fig. 16-19 is to be 35 degrees.

1. Calculate the height of the gage block stack required to obtain this angle when using a 5-inch sine bar.

$$\begin{aligned} H &= L \sin \theta \\ &= 5 \times .57358 \\ &= 2.8679 \text{ inches} \end{aligned}$$

2. Calculate the true angle of the scriber point if the height gage reads zero at one check point and +.0001 inch at the other check point. The distance between the check points is ½ inch.

$$\sin \theta' = \sin \theta \pm \frac{DL}{S}$$

In this case the term DL/S is added to sin θ. If the deviation was −.0001 inch, this term would be subtracted. Thus:

$$\sin \theta' = \sin \theta + \frac{DL}{S}$$

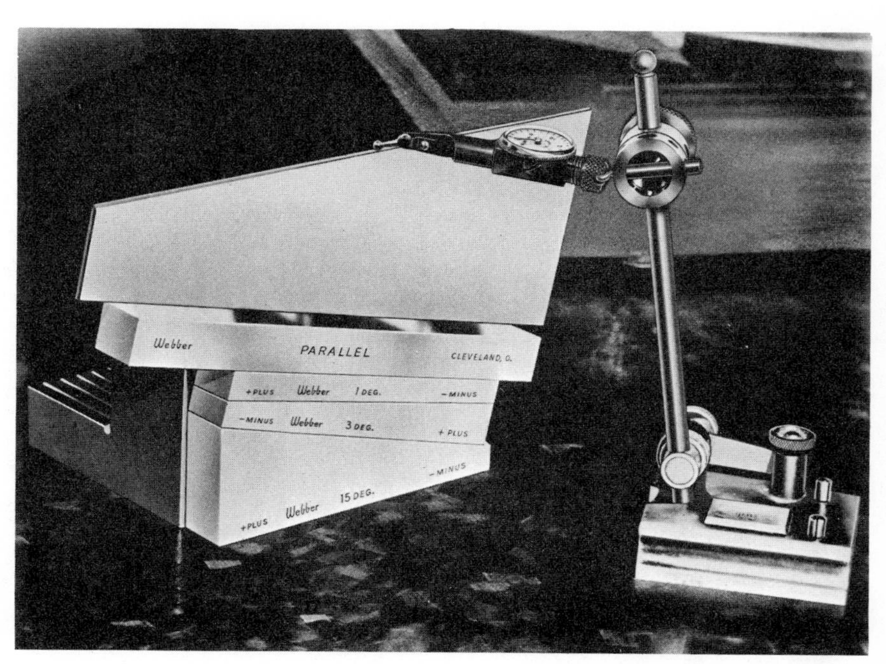

Courtesy of The L. S. Starrett Company

Fig. 16-20. Measuring an angle using precision angle gage blocks.

$$= .57358 + \frac{.0001 \times 5}{.500}$$

$$\sin \theta' = .57458$$

$$\theta' = 35°04'$$

The answer 35°04′ is correct to the nearest minute (sin 35°04′ = .57453; sin 35°03′ = .57429). If greater accuracy is required, it is necessary to use interpolative methods or a more accurate table of trigonometric functions.

Precision angle gage blocks, Fig. 16-20, are the most precise method of measuring angles. Only 16 blocks are required to construct any angle from zero to 99 degrees in 1-second increments, or 356,400 different angles. This is possible because any angle can be added or subtracted from any other angle. For example, in Fig. 16-20 the gage blocks are constructed to an angle of 13 degrees. This is done by adding and subtracting as follows:

$$13 = 15 - 3 + 1$$

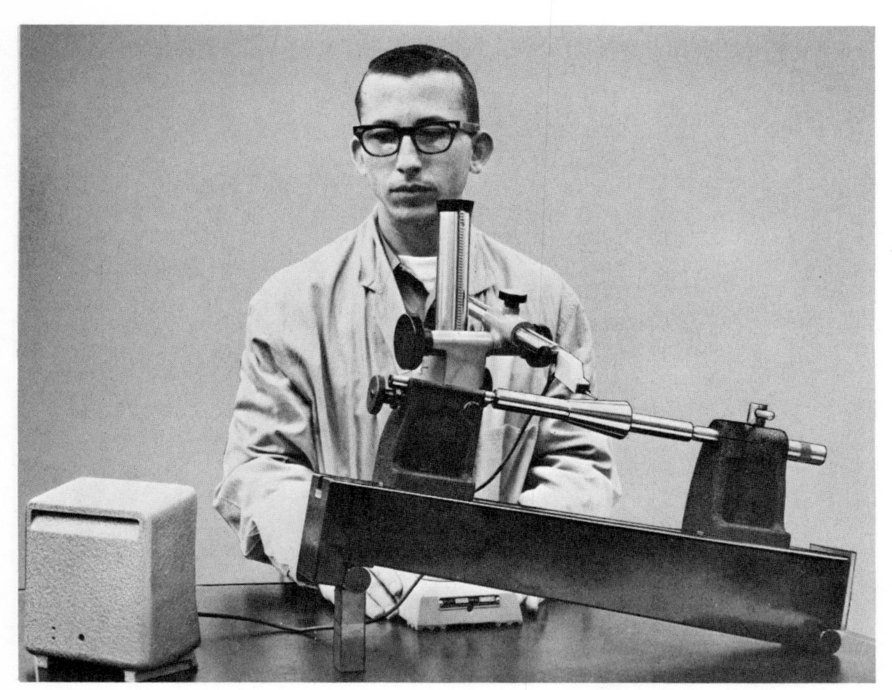

Courtesy of the DoALL Company

Fig. 16-21. Measuring the angle of a taper using a master grade granite 20-inch sine bar with metal T-slot containing a set of matched bench centers.

The 15-degree block is resting on the surface plate. A 3-degree block is placed on the 15-degree block so that the minus end of the 3-degree block is in contact with the plus end of the 15-degree block. A 1-degree block is placed on the 3-degree block so that the plus end of the 1-degree block contacts the minus end of the 3-degree block. Each end of these blocks is marked plus or minus. In the illustration a precision gage-block grade parallel is placed on the 1-degree block which provides a seat for the part being measured.

In addition to the 6-inch parallel block and a 6-inch knife edge block, the full set contains the following blocks from which all of the angles can be constructed:

1 degree	1 minute	1 second
3 degrees	3 minutes	3 seconds
5 degrees	5 minutes	5 seconds
15 degrees	20 minutes	20 seconds
30 degrees	30 minutes	30 seconds
45 degrees		

Angle gage blocks are available in the following three grades of accuracy: laboratory master grade, ±¼ second; inspection grade, ±½ second; tool-

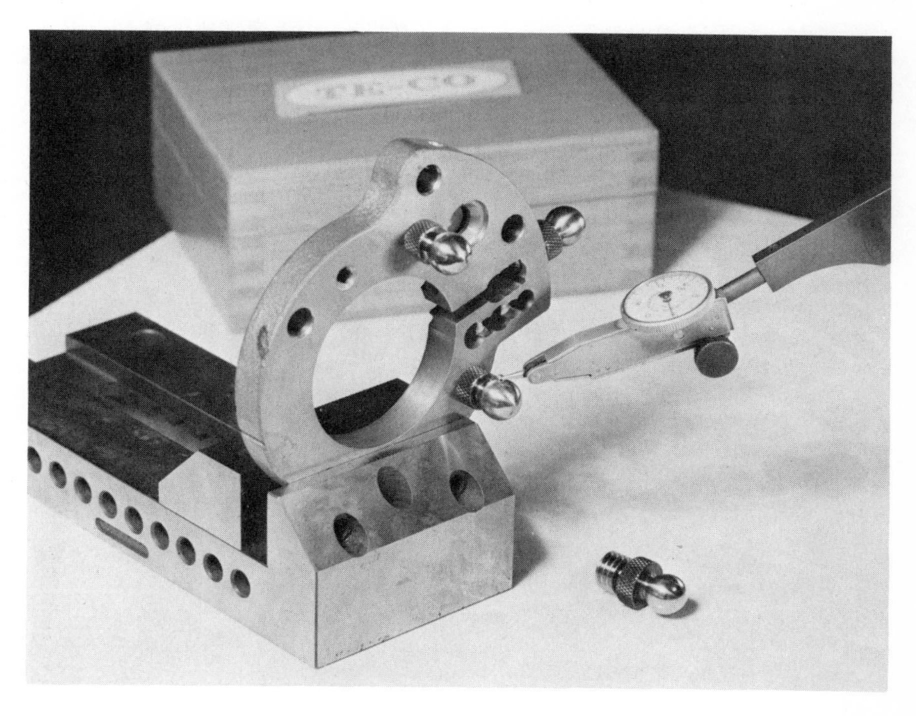

Courtesy of TE-CO.

Fig. 16-22. Precision tooling balls used to measure angular hole locations.

room grade, ±1 second. In addition to measuring angles these blocks can be used to check the settings of sine bars and sine plates. They can also be used to set adjustable angle plates, to position tool slides accurately, to measure the angle of tapers, etc.

The angle of taper is usually measured with a sine bar or a sine plate in combination with a bench center, as shown in Fig. 16-21. The procedure is similar to measuring angles, except in one small detail. First, the part is mounted snugly on the bench centers, and the sine bar is tipped to the required angle by placing a stack of precision gage blocks of specified length below one of the rolls. In measuring the deviation at either check point, the contact point of the dial test indicator or the electronic height gage must be moved across the workpiece in order to obtain the highest reading. The difference in the highest reading at each end of the workpiece is the deviation. Formula 16-2 can then be used to calculate the true angle. The usual procedure to find out if there is any deviation is to check random points at both ends of the taper. If the deviation is zero,

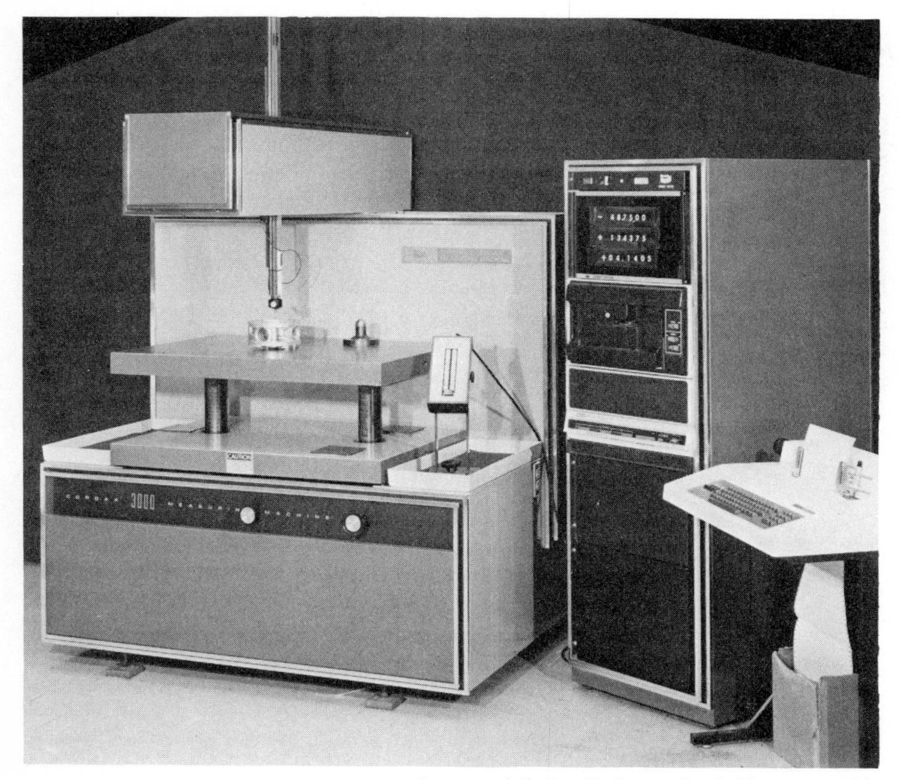

Courtesy of the Bendix Automation & Measurement Div.

Fig. 16-23. Computer controlled coordinate measuring machine.

no further inspection is warranted. If there is a deviation, measurements are made at accurately located check points and the true angle is calculated.

Tooling Balls

Tooling balls, Fig. 16-22, are very precisely ground and lapped tools used to measure the distance between holes that have axes which are not parallel to each other. While primarily an inspection tool, they are also used to align workpieces on a machine tool when setting up. The relationship of surfaces that are at an angle to each other can also be measured by tooling balls located in construction holes machined in these surfaces for this purpose. Tooling balls are made in one piece, consisting of a spherical head and a stem. The stem may be cylindrical or threaded; cylindrical stems are pressed in holes with a light fit and threaded stems are screwed into tapped holes. Measurements are made over the largest diameter of the spherical head with a micrometer caliper, or more often by means of a dial test indicator attached to a height gage.

Coordinate Measuring Machines

Coordinate measuring machines, Fig. 16-23, are used to inspect machined parts of any configuration. The workpiece is mounted securely on the table, which serves as a surface plate, and a probe tip is placed in the probe shaft. The probe is then moved to a reference position located on the table, or against the first point on the surface to be checked. From this position the probe is moved progressively to the various positions on the workpiece that are to be measured. A close-up view of the part on the machine is shown in Fig. 16-24. The center of the "tooling ball" in this setup is used as the reference point from which all measurements are made.

Measurements are made in the x, y, and z directions. In each direction there is a steel grating that has 1000 lines per inch. A corresponding grating segment is attached to the movable measuring head. As this grating segment passes over the steel grating, a beam of collimated light produces a pattern of dark and light bands, called a Moiré fringe pattern. The fringe pattern is converted into electrical signals by photocells and the output signals are used to indicate the direction of the motion and to measure the distance moved. In this manner a continuous readout of the position of the probe is obtained to a resolution of .0001 or .00025 inch (0.002 or 0.006 mm).

Coordinate measuring machines may be equipped with a simple digital readout for manual operation, or with a computer. In manual inspection the operator moves the probe manually from point to point, taking readings on the digital readout to obtain the part dimensions. The probe is also moved manually when measuring by the computer assisted method; with this method the part dimensions are verified in relation to a program that has been entered into the computer. Direct computer control is a

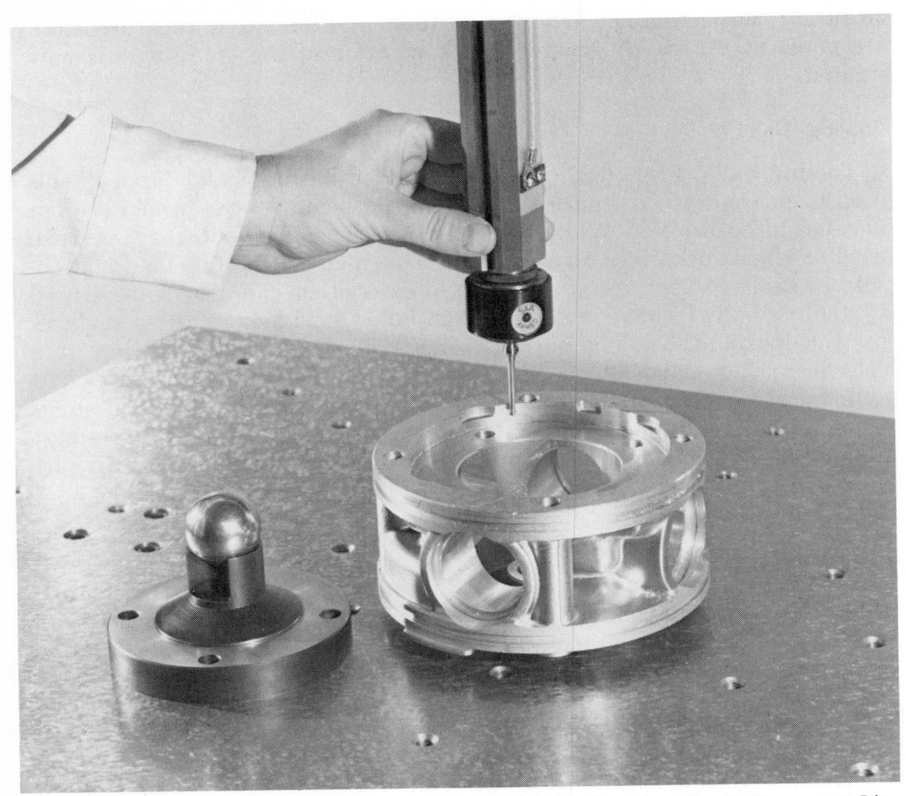

Courtesy of the Bendix Actomation & Measurement Div.

Fig. 16-24. Setup for measuring part on a coordinate measuring machine. Tooling ball is the reference point for measurements to the part surfaces.

method that does not require an operator to move the probe. The probe is moved automatically by the computer in the x, y, and z directions, which performs the necessary calculations to obtain the part dimensions and produces a printout to document the final results. Many accessories are available with these machines, including standard and special probe tips, a right-angle attachment for attaching probe tips, a microscope, an optical viewing screen, and even a small hole-drilling attachment for drilling holes up to ⅛ inch in diameter.

Appendix 1

Calculating Angle of Table Swivel for Helical Milling

The angle a_c, illustrated in Fig. 9-10, may be used to obtain greater accuracy when calculating the swivel angle for helical milling. (See Formula 9-11.) It *must* be used in calculating the transverse and vertical cutter offsets when the method of making this offset from predetermined calculation is used. (See also Appendix 2.) Formula 9-11 is repeated here:

$$\tan b = \tan c \cos (r + a_c)$$

$$\text{if } a_c = 0$$

$$\tan b = \tan c \cos r$$

The formula for calculating angle a_c and an example of how the method of successive approximations is used to solve this formula is given below:

$$\tan a = \frac{\tan a_c}{\sqrt{1 + \tan^2 c \cos^2 (r + a_c)}} \qquad \text{(A-1)}$$

where:

a = Side angle on the side of the fluting cutter that will cut the face of the cutting edge of the cutter blank, degrees (see Fig. 8-15)
a_c = Angle a of the fluting cutter projected to the axis of the cutter blank, degrees (see Fig. 9-10)
b = Angle of table swivel
c = Helix angle to be cut, degrees
r = Radial rake angle to be cut on the cutter blank, degrees

Since a_c appears in two different terms in this formula which cannot be combined into one term, a direct solution for a_c is impossible. Since it is known that the value of angle a_c will be slightly greater than the angle a, a value for a_c is assumed and the right side of the equation is solved. This answer is then compared to the value of the tangent of angle a which is on the left side of the equation. This process is repeated until the answer is equal to or very closely equal to the tangent of angle a.

An example of how Formula A-1 is solved will be given below. This example will correlate with Fig. 9-7, page 246.

Here $a = 15°$, $c = 25°$, and $r = 10°$. To start, a value for angle a_c is assumed which should be close to the value of angle a, or in this case, 15°. At the beginning, then, assume that angle $a_c = 16°$.

$$\tan a = \frac{\tan a_c}{\sqrt{1 + \tan^2 c \cos^2 (r + a_c)}}$$

$$\tan 15° = \frac{\tan 16°}{\sqrt{1 + \tan^2 25° \cos^2 (10° + 16°)}}$$

$$.26795 = \frac{.28674}{\sqrt{1 + (.46331)^2 (.89879)^2}}$$

$$.26795 = \frac{.28674}{\sqrt{1 + (.217445)(.807823)}} = \frac{.28674}{1.08428}$$

$$.26795 \neq .26445$$

Since the two sides of the equation are not equal (the equation does not balance), the assumed value for a_c is incorrect. Another value is assumed for a_c, and this approximation should be better than the first because it is obvious that angle a_c must be larger. Try $a_c = 16°15'$:

$$\tan 15° = \frac{\tan 16°15'}{\sqrt{1 + \tan^2 25° \cos^2 (10° + 16°15')}}$$

$$.26795 = \frac{.29147}{\sqrt{1 + (.46631)^2 (.89687)^2}} = \frac{.29147}{1.08393}$$

$$.26795 \neq .26891$$

The assumed value of a_c is too large. Try $a_c = 16°10'$.

$$\tan 15° = \frac{\tan 16°10'}{\sqrt{1 + \tan 25° \cos (10° + 16°15')}}$$

$$.26795 = \frac{.28990}{\sqrt{1 + (.46631)^2 (.89751)^2}} = \frac{.28990}{1.08405}$$

$$.26795 \approx .26742 \text{ (very nearly)}$$

Thus, the value of the angle a_c is very nearly equal to 16°10′, which is close enough for all practical purposes. It will now be possible to calculate the angle of swivel of the table using Formula 9-11.

$$\begin{aligned} \tan b &= \tan c \cos (r + a_c) \\ &= \tan 25° \cos (10° + 16°10') \\ &= (.46631)(.89751) \end{aligned}$$

$$\tan b = .41852$$

$$b = 22°43'$$

Thus, the angle, b, to which the table must be swiveled, is 22° 43′ when the more nearly correct value of angle a_c or 16° 10′ is used. When a_c is assumed to be 15°, the calculated angle of swivel, b, is equal to 22° 55′ as was shown in Chapter 9. Since this difference is small, it is not necessary to calculate a more correct value for angle a_c unless this must also be done in order to calculate the table offsets, as shown in Appendix 2.

Appendix 2

Calculating the Transverse and Vertical Cutter Offsets for Helical Milling

A convenient method of offsetting fluting cutters prior to milling helical flutes in milling cutter blanks is to calculate these offsets in advance and then to position the table in accordance with the calculated values. Formulas 8-4 and 8-5 (Chapter 8) with only a slight modification are used as the basis for these calculations. This modification involves the substitution of the fluting cutter angle a_c for the angle a, which is necessary because the table is swiveled for cutting the helix.

$$n = \frac{D}{2} \sin (a_c + r) - d \sin a_c - R (\cos a_c - \sin a_c) \qquad \text{(8-4B)}$$

$$m = \frac{D}{2} [1 - \cos (a_c + r)] + d \cos a_c - R (\cos a_c + \sin a_c - 1) \qquad \text{(8-5B)}$$

where:
- n = Transverse offset, inches
- m = Vertical offset, inches
- D = Finished diameter of the milling cutter, inches
- d = Depth of the flute, inches
- r = Radial rake angle, degrees
- a_c = Side angle of fluting cutter, a, projected to the axis of the cutter blank, degrees
- R = Radius of the fluting cutter teeth, inches

The following example will correlate with Fig. 9-7, found on page 246. In this example $d = .500$ inch; $D = 4.000$ inches; $R = .125$ inch; $a_c = 16°10'$; $r = 10°$. The diameter of the cutter blank is 4.030 inches. However, the diameter of the cutter when it is completely finished must be used in these calculations—otherwise there will be a slight error in the resulting rake angle and in the depth of flute on the finished milling cutter. There should be a compensating table adjustment before the offset adjustments are made in order to correct for the difference in the diameter of the cutter blank and the finished cutter. This compensating adjustment should also correct for the thickness of the paper feeler used to "touch up" the fluting cutter on the top of the cutter blank.

$$n = \frac{D}{2} \sin (a_c + r) - d \sin a_c - R (\cos a_c - \sin a_c)$$

$$= \frac{4}{2} \sin (16°10' + 10°) - .500 \sin 16°10'$$

$$- .125 (\cos 16°10' - \sin 16°10')$$

$$= 2 (.44098) - .500 (.27843) - .125 (.96045 - .27843)$$

$$= .88196 - .13922 - .08525$$

$$n = .6575 \text{ inch}$$

$$\mathrm{m} = \frac{\mathrm{D}}{2} [1 - \cos (\mathrm{a_c} + \mathrm{r})] + \mathrm{d} \cos \mathrm{a_c} - \mathrm{R} (\cos \mathrm{a_c} + \sin \mathrm{a_c} - 1)$$

$$= \frac{4}{2} [1 - \cos (16° 10' + 10°)] + .500 \cos 16° 10'$$

$$- .125 (\cos 16° 10' + \sin 16° 10' - 1)$$

$$= 2 (1 - .89751) + .500 (.96045) - .125 (.96045 + .27843 - 1)$$

$$= .20498 + .48023 - .02896$$

$$= .6554 \text{ inch}$$

A considerable difference occurs in the answers for m and n if 15 degrees instead of 16°10′ is used as the value for angle a_c. The answers would be: $n = .6274$ inch and $m = .6432$ inch. Thus, when this method is used to offset the cutter blank relative to the fluting cutter, the angle a_c should first be calculated. Using angle a in place of angle a_c will result in an error which may be significant.

Index